INTEREST RATES AND COUPON BONDS
IN QUANTUM FINANCE

The economic crisis of 2008 has shown that the capital markets need new theoretical and mathematical concepts to describe and price financial instruments.

Focusing on interest rates and coupon bonds, this book does not employ stochastic calculus – the bedrock of the present day mathematical finance – for any of the derivations. Instead, it analyzes interest rates and coupon bonds using quantum finance. The Heath–Jarrow–Morton model and the Libor Market Model are generalized by realizing the forward and Libor interest rates as an imperfectly correlated quantum field. Theoretical models have been calibrated and tested using bond and interest rates market data.

Building on the principles formulated in the author's previous book (*Quantum Finance*, Cambridge University Press, 2004), this ground-breaking book brings together a diverse collection of theoretical and mathematical interest rate models. It will interest physicists and mathematicians researching in finance, and professionals working in the finance industry.

BELAL E. BAAQUIE is Professor of Physics in the Department of Physics at the National University of Singapore. He obtained his B.S. from Caltech and his Ph.D. from Cornell University. His specialization is in quantum field theory, and he has spent the last ten years applying quantum mathematics, and quantum field theory in particular, to quantitative finance. Professor Baaquie is an affiliated researcher with the Risk Management Institute, Singapore, and is a founding Editor of the *International Journal of Theoretical and Applied Finance*. His pioneering book *Quantum Finance* has created a new branch of research in theoretical and applied finance.

Cover illustrations: Shanghai skyline and the Bund.

INTEREST RATES AND COUPON
BONDS IN QUANTUM FINANCE

BELAL E. BAAQUIE

National University of Singapore

CAMBRIDGE
UNIVERSITY PRESS

CAMBRIDGE
UNIVERSITY PRESS

Shaftesbury Road, Cambridge CB2 8EA, United Kingdom

One Liberty Plaza, 20th Floor, New York, NY 10006, USA

477 Williamstown Road, Port Melbourne, VIC 3207, Australia

314–321, 3rd Floor, Plot 3, Splendor Forum, Jasola District Centre, New Delhi – 110025, India

103 Penang Road, #05–06/07, Visioncrest Commercial, Singapore 238467

Cambridge University Press is part of Cambridge University Press & Assessment,
a department of the University of Cambridge.

We share the University's mission to contribute to society through the pursuit of
education, learning and research at the highest international levels of excellence.

www.cambridge.org
Information on this title: www.cambridge.org/9780521889285

First published 2010

A catalogue record for this publication is available from the British Library

Library of Congress Cataloging-in-Publication data
Baaquie, B.E.
Interest rates and coupon bonds in quantum finance / Belal E. Baaquie.
p.   cm.
ISBN 978-0-521-88928-5
1. Interest rates. 2. Zero coupon securities. 3. Finance.   I. Title.
HG1621.I586 2009
332.8–dc22

2009024540

ISBN   978-0-521-88928-5   Hardback

This book is dedicated to my wife Najma Sultana Baaquie,
my son Arzish Falaqul Baaquie, and my daughter Tazkiah Faizaan Baaquie.
Their precious love, affection, support, and optimism
have made this book possible.

Contents

Prologue

The 2008 economic crisis has shown that the capital markets need new and fresh theoretical and mathematical concepts for designing and pricing financial instruments. Focusing on interest rates and coupon bonds, this book does not employ stochastic calculus – the bedrock of the present-day mathematical finance – for any of the derivations. Interest rates and coupon bonds are studied in the self-contained framework of quantum finance that is independent of stochastic calculus. Quantum finance provides solutions and results that go beyond the formalism of stochastic calculus.

It is five years since *Quantum Finance* [12] was published in 2004 and it is indeed gratifying to see how well it has been received. No attempt has been made to re-work the principles of finance. Rather, the main thrust of this book is to employ the methods of theoretical physics in addressing the subject of finance. Theoretical physics has accumulated a vast and rich repertoire of mathematical concepts and techniques; it is only natural that this treasure house of quantitative tools be employed to analyze the field of finance, and the debt market in particular.

The term 'quantum' in *Quantum Finance* refers to the use of *quantum mathematics*, namely the mathematics and theoretical concepts of quantum mechanics and quantum field theory, in analyzing and studying finance. Finance is an entirely classical subject and there is no $\hbar$ – Planck's constant, the *sine qua non* of quantum phenomena – in quantum finance: the term 'quantum' is a *metaphor*. Consider the case of classical phase transformations that result from the random fluctuations of classical fields; critical exponents, which characterize phase transitions, are computed using the mathematics of nonlinear quantum field theories [95]. Similar to the case of phase transitions, quantum mathematics provides powerful theoretical and mathematical tools for studying the underlying random processes that drive modern finance.

The principles of quantum finance provide a comprehensive and self-contained theoretical platform for modeling all forms of financial instruments. This book, in particular, is focused on studying interest rates and coupon bonds. A detailed analytical, computational, and empirical study of debt instruments constitutes the main content of this book.

The Libor Market Model and the Heath–Jarrow–Morton model, which are the industry standards for modeling interest rates and coupon bonds, are both based on exactly correlated Libor and forward interest rates. The book makes a quantum finance generalization of these models to *imperfectly correlated* interest rates by modeling the forward interest rates as a quantum field. Empirical studies provide strong evidence supporting the imperfect correlation of interest rates. Many groundbreaking results are obtained for debt instruments. In particular, it is shown that quantum field theory provides a generalization of Ito calculus that is required for studying imperfectly correlated interest rates.

In the capital markets, interest rates determine the returns on cash deposits. Coupon bonds, on the other hand, are loans that are disbursed – with the objective of earning interest – against promissory notes. In principle, the interest paid on cash deposits and the interest earned on loans are equivalent. However, all interest rates are only defined for a *finite* time interval – of which the minimum is overnight (24 hours). In particular, all interest rate *derivatives* are based on benchmark interest rates for cash deposits of a duration of 90 days. The bond (derivatives) markets, in contrast, have no such minimum duration. The existence of a finite duration for the (benchmark) interest rates creates *two distinct sectors* of the debt derivatives market, namely derivatives of interest rates and derivatives of coupon bonds – with a nonlinear transformation connecting the two sectors.

Numerous and exhaustive calculations are carried out for diverse forms of interest rate and coupon bond options. Complicated concepts and calculations that are typical for debt instruments are introduced and motivated, in some cases by first discussing analogous and simpler equity instruments. It is my view that only by actually working out the various steps required in a calculation can a reader grasp the principles and techniques of what is still a subject in its infancy. Almost all the intermediate steps in the various calculations are included so as to clear the way for the interested reader. A few key ideas are repeated in the various chapters so that each chapter can be read more or less independently.

The material covered in the book is primarily meant for physicists and mathematicians engaged with research in the field of finance, as well as professional theorists working in the finance industry. Specialists working in the field of debt instruments will hopefully find that the theoretical tools and mathematical ideas developed in this book broaden their repertoire of quantitative approaches to finance. The material could also be of interest to physicists, probabilists, applied mathematicians, and statisticians – as well as graduates students in science and engineering – who are thinking of pursuing research in the field of finance.

One of the aims of this book is to be self-contained and comprehensive. All derivations and concepts are introduced from first principles, and all important results are derived *ab initio*. Given the diverse nature of the potential audience, fundamental concepts of finance have been reviewed for readers who are new to this field. Appendix A reviews the essential mathematical background required for following the various derivations and is meant to introduce specialists working in finance to the concepts of quantum mathematics.

Acknowledgements

I thank Bloomberg, Singapore and Capital Fund Management, France for providing data used in the empirical studies.

I thank Cui Liang for many discussions; it has been a great pleasure to have had a fruitful and enjoyable collaboration. I thank Chua Wee Kang, Cao Yang, and Tang Pan for their useful input.

I would like to thank Jiten Bhanap and Sanjiv Das for generously sharing their insights on finance and financial instruments. I thank Jean-Philippe Bouchaud, Ebrahim M. Shahid, Bertrand Roehner, Carl Chiarella, and Lim Kian Guan for useful discussions, Frederick H. Willeboordse for many fruitful interactions and Mitch Warachka and Arzish F. Baaquie for providing valuable feedback.

It is a pleasure to acknowledge the encouragement and friendship of Zahur Ahmed, Yamin Chowdhury, and Bal Menon.

I am indebted to Shih Choon Fong for his invaluable support and visionary leadership. And lastly, I want to record my heartfelt esteem, respect, and gratitude to my father, Mohammad Abdul Baaquie, for being a lifelong source of inspiration, guidance and lofty ideals.

1

Synopsis

The book consists of *three major themes*. Any one of the three components can be read without many gaps in the analysis.

1 The introductory chapters are primarily intended for readers who are unfamiliar with the *fundamental concepts of finance*. The principles and mathematical expressions for debt instruments, which are analyzed in later chapters, are reviewed in Chapters 2, 3, and 4. Options are briefly discussed and the Black–Scholes option theory is given a path integral formulation.

2 A major subject matter of the book is the theory of coupon bonds. A quantum field theory of the *bond forward interest rates* $f(t, x)$ is developed in Chapter 5 and forms a core chapter. It provides a model for the study of coupon and zero coupon bonds. Many of the derivations in later chapters are based on the quantum finance model of bond forward interest rates.

3 The quantum finance formulation of Libor interest rates is another major topic. The Libor Market Model is formulated in Chapter 6; the nonlinear Libor forward interest rates $f_L(t, x)$ that it is based upon are transformed into *logarithmic Libor interest rates* $\phi(t, x)$. In Chapter 7 some empirical properties of the Libor Market Model are studied and in Chapter 8 the prices of Libor options are obtained by using techniques of quantum field theory. A derivation of the Libor Market Model's nonlinear drift term is given in Chapter 15, based on the Libor Hamiltonian and state space of $\phi(t, x)$.

The inter-connection of the various chapters is shown in the flowchart given overleaf.

Chapter dependency flowchart

2

Interest rates and coupon bonds

Interest rates, coupon bonds, and their derivatives are the main instruments of the debt markets, which constitute well over 60% of the entire capital markets. A brief discussion locates the debt markets in the general framework of finance and points to the growing importance of the debt markets in the global economy. Interest rates are a measure of the returns on cash deposits, whereas coupon bonds are a measure of the present value of future cash flows. From this intuitive and apparently simple idea flow all the various definitions of interest rates and coupon bonds. The fundamental concept of forward interest rates that describe the bond market is introduced. The interest rate markets are driven by Libor and Euribor; these two instruments are defined and a few of their important features are discussed.

2.1 Introduction

Finance is the discipline that studies the borrowing, lending, and investing of money capital. The main form of money capital is *paper* issued by various governments and private organizations, which includes corporations and individuals. The three pillars of finance are equity, debt, and foreign exchange and are the basis of all financial instruments. Financial markets, collectively known as the capital markets, trade in these instruments [31].

Capital in economics represents the collection of productive assets required for carrying out economic activities. Financial 'paper' is not merely ordinary paper, but, rather, the preferred form of money capital that is used for *representing* value: a value based not on how it has been generated but, rather, on its present day and future value in the capital markets – and in economic activity in general. Money capital carries an intrinsic risk since expectations of what can be realized in the present and future are always subject to uncertainties inherent in any form of forecasting. Unlike traditional economies – where finance is a passive force and auxiliary to the real economy – the capital markets today are one of the most powerful and dynamic components of the modern global economy and a potent force for economic growth and expansion. The capital markets are expected to become increasingly important with the increasing inter-connections of the global economy. However, there is a downside to the increasing importance of finance. Due to the inherently uncertain and random nature of money capital, the capital markets have an uncontrollable and unpredictable component that can wreak havoc on the real economy. Advanced theories of money capital are required for creating financial instruments that can be used for managing risk and reducing instabilities of the capital markets – and thus help to *tame* the destabilizing spikes, bubbles, meltdowns, and crashes of the financial markets.

Money capital comes in many forms with the main three forms being stocks and shares of companies, debt instruments, and cash of various currencies. Money, or more precisely money capital that is seeking returns from the economy, is a dynamic quantity – with opportunities for money to yield profit constantly changing with time. Interest rates reflect the relation of the value of money with time and quantify the time-dependent and dynamic aspect of money.

Debtors pay a return – the amount depending on the interest rate – to the providers of credit. Debt and surplus capital are two sides of the same coin, since debt for one party is the complement of the credit that the other party has provided. The world's debt market is an expression of the net savings that the world economy has generated.

One needs sophisticated and effective models of interest rates to manage and expand the net global savings so as to maximize its returns. It is from this perspective – of optimizing the management of the international debt markets – that quantum finance models of interest rates and coupon bonds have been developed and form the main content of this book.

Optimizing the management of international liquidity will result in better allocation and returns on investments as well as create conditions for the prosperity of society at large. In particular, managing flows of international capital to developing and other higher risk economies, using customized financial instruments, would result in a larger fraction of mankind having access to investment capital – leading to the betterment of people's lives and wealth.

2.2 Expanding global money capital

The nature of finance has undergone a radical change in the last 30 years, with the financial sector of the economy becoming increasingly more important. There are many indicators that point to this fundamental change in the financial superstructure of economically advanced countries.

In 2006, the world economy generated about US$65 trillion worth of goods and services, of which raw materials (taken directly from nature) constituted about two-thirds (US$43 trillion) of the total value. The remaining one-third (US$22 trillion) was the value added by human labor. For example, in 2007 – based on daily production of 85 million barrels (about 31 billion barrels a year) – the sale of petroleum at around US$100 per barrel generated a revenue worth about US$3.1 trillion, with a large part of this revenue being invested in the capital markets.

In general, a substantial fraction of the net profit generated by the world economy as well as the savings and net accumulated surplus capital of many individuals, organizations, and countries is held in the form of money capital. In particular, cash rich oil and gas producers as well as East Asian economies (with substantial national reserves) have created 'sovereign funds' for investing their surplus in the capital markets. Money capital is bound to be increasingly important; due to the enormous scale of the global economy and the net savings it generates, there is not enough gold or other precious commodities that can hold this value. Paper seems the only way to represent and store the generated global surplus value.

Risk management, based on models that quantify the degree of risk, allows many institutional investors to convert net savings into money capital. Better risk management instruments have drawn risk-averse investors, such as insurance companies and pension and sovereign funds amongst others, to place their assets in the capital markets, contributing to the current explosion of the money capital.

IMF estimates that in 2005 the total value of the stocks, bonds, and bank loans worldwide was about US$165 trillion. The global bond (debt) market's share was close to US$104 trillion – by far the largest component of the global capital markets – accounting for over 63% of the total; banking credit in 2008 amounted to about US$23 trillion. In 2005, cross-border money flows (stocks, bonds, real estate, and so on) amounted to about US$6 trillion. The foreign exchange markets have also undergone a phenomenal increase, with about US$3 trillion being traded daily in 2007.

In 2007, global stocks were worth about US$56 trillion – about 35% of the capital markets – with the US and Eurozone each accounting for US$18 trillion and the rest of world accounting for US$20 trillion. The US capital markets had a total worth of US$42 trillion of which US$24 trillion was in the bond (debt) market and US$18 trillion in stocks (equity).

In 2006 global debt issuance rose to a record US$6.9 trillion with the global syndicated loan volume exceeding US$3.2 trillion. During the period of 2000–2005 nonfinancial companies worldwide issued $19.3 trillion worth of debt, in the form of corporate medium-term notes (MTNs), with the biggest issuers being the automotive industry, issuing 70 MTNs worth US$4.54 trillion followed by insurance companies issuing 26 MTNs worth US$4.49 trillion.

Market liquidity and risk management – two of the current lynch pins of the financial system – require the participation of speculators. A speculator, who can be an individual, a corporation, or a financial institution, makes an estimate of the future and if right profits and if wrong loses. Speculating on the capital markets usually means taking high risks since the future is always uncertain. Speculative positions create market liquidity as well as provide a mechanism for sharing risk, which, for example, a (manufacturing) business, not having in-house expertise in risk management, may want to dispense with.

Although the term 'speculator', to some, carries a negative connotation, the market needs both informed and uninformed, traders. Speculators are not inside traders but, instead, should be called uninformed traders, in contrast to informed traders who buy or sell a specific instrument. If only informed traders were market players, any move to buy or sell would lead to slippage in the offered prices, leading to the informed traders being held to ransom by the market. Uninformed traders provide the 'veil', a background of 'noise', that allows informed traders to enter the market without causing major slippages in prices. One needs both the informed and uninformed traders for the market to function efficiently.

2.2.1 Securitization

Another reason for the expansion of the capital markets is that financial engineering has created instruments that allow diverse forms of future cash flows to be used for issuing vast amounts of securitized debt. Securitization is the consolidation and structuring of cash-flow producing financial instruments, called asset-backed securities, that can then be traded in the capital markets. For example, the securitization of cash flows, such as mortgage payments and rentals, has allowed these to be traded in the capital markets – adding to the depth and liquidity of the capital markets.

Securitization is a relatively new concept in finance, having gained acceptance only over the last 20 years. Securitized debt has grown in the issuance of new loans and covers such diverse sectors as residential mortgages, commercial real estate, corporate loans, auto loans, student loans, and so on. In 1990, just 10% of mortgages in the United States were securitized, compared to 70% in 2007. It is estimated that by the middle of 2008 there were asset-backed securities worth US$10.2 trillion in the US and US$2.3 trillion in Europe. In 2007, new issues of asset-backed securities amounted to US$3.5 trillion in the US and US$650 billion in Europe. Securitization has had a major setback due to the 2008 US economic crisis, with the issuance of new mortgage-backed securities dropping by almost 85% in the first half of 2008 compared to the same period in 2007. The 2008 subprime crises in US home mortgages is claimed by some critics to be a negative example of securitization; this is not entirely correct and is discussed in Appendix B.2.

The lack of securitization can be a formidable barrier to economic development. It has been argued by Soto [91] that the securitization of third world developing countries' real estate, and of property in general, into tradable financial instruments could release vast amounts of capital. It was estimated that, in 1997, capital worth about US$9.3 trillion was locked up due to lack of securitization, an amount twice of the then total US public debt [91]. This 'dead capital', if securitized, could play a major role in the economic growth of the developing countries. Mortgages are fungible (a commodity that is freely interchangeable with another in satisfying an obligation) only in countries where the rule of law is well established and the legal system guarantees ownership. To securitize real estate assets in third world countries, hence, requires a stable political system that is accountable and relatively free from corruption. For these reasons, third world countries will have to overcome many major hurdles before they are in a position to create mortgage and other asset-backed securities, which would in turn release presently inert capital.

2.2.2 *Profitability of the financial sector*

At present, the rate of return of the financial sector and services in general is about 20% for the advanced economies of the US, Europe, and Japan – much higher than the 8–10% returns from manufacturing.[1] For example, from 2002 to 2006 five leading US investment banks – Goldman Sachs, Merrill Lynch, Morgan Stanley, Lehman Brothers, and Bears Stern – had an average return on equity of about 22%, amounting to US$30 billion – rivaling returns for such profitable industrial sectors as pharmaceuticals and energy.

The increasing volume of financial money capital reflects the overall expansion of the world economy, with vast amounts of surplus finding its way to the capital markets. The high rates of return from finance capital is one of the reasons for the immense infusion of savings and other assets into the global capital markets. The higher rate of return is thought to be due to the finance industry not being as mature as manufacturing and is taken to indicate a shift of the global economy to a new regime. There is, however, a contrarian view that the high returns from finance are primarily the result of the formation of an asset bubble – and hence intrinsically unstable and not sustainable.

The September–October 2008 global financial meltdown seems to provide strong evidence in support of the contrarian view. By the end of the September–October

---

[1] The rate of return on manufacturing is thought to be low due to the increasingly large capital investment required for setting up and upgrading modern industries.

2008 US financial meltdown, all the five US investment banks had ceased to exist – with Goldman Sachs and Morgan Stanley having converted themselves into bank holding companies. The consistently high returns of 22% from 2002 to 2006 shown by the five investment banks, with hind sight, is seen to completely coincide with the formation and expansion of the US subprime mortgage loans' financial bubble and may have simply been a result of this bubble.

Finance may still give a return higher than manufacturing due to the creation of new financial instruments, but in the current climate of financial turmoil and contraction it will be a while before such innovations find acceptance in the capital markets.

2.3 New centers of global finance

The United States (US) capital markets, since 1945, have been the most important component of the global capital markets, playing a central role in shaping and developing the international financial system. In Appendix B, the structure of the US debt markets is briefly reviewed.

The US is losing its pre-eminent position in the global capital markets due to the following reasons: (a) massive financial losses caused by the 2008 economic meltdown – in the US stock market, for example stocks on the Dow Jones lost 34% of their value in 2008 (the largest drop since 1931), and in the bankruptcy of major US financial institutions; (b) the rise of other capital markets and centers of wealth. The year 2007 saw a sea change in the distribution of global wealth. Largely due to the rise of China and India and investments by oil and gas producing countries, for the first time since the Second World War (1945) London displaced New York to become the center of the global capital markets. Over 40% of the world's foreign equities were traded in London, more than New York. Over 30% of the world's foreign currency trading took place in London, being larger than New York and Tokyo combined.

The US capital markets, in 2007, were worth US$42 trillion of which US$7.3 trillion was owned by foreigners, namely 17%, who also held 44% of the US national debt. In contrast to both New York and Tokyo, which depend largely on their domestic and East Asian markets, 80% of London's business comes from international sources, spread widely over many regions and countries.

The shift away from a US-centered global financial system can also be seen in the emergence of the Euro as an international reserve currency, as can be seen from Table 2.1. The Euro was introduced in 1999 and by 2008 had appreciated over 50% against the US Dollar. International reserves are now held in both the US Dollar and the Euro, with estimates that by 2010 about 34% will be held in Euro and 54% in the US Dollar, in contrast to 2000 when 71% of world reserves were held in

Table 2.1 *International reserve in Euros and US Dollars, and the projected currency distribution of these reserves by the year 2010.*

	Currency of international reserve		
	2000	2007	2010 (projected)
US Dollar $	71%	63%	54%
Euro €	18%	26%	34%

US Dollars. Some economists have predicted that, by as early as 2015, the Euro may overtake the US Dollar as the main international reserve currency provided two conditions hold: (a) more countries, including the UK, join the Eurozone countries and (b) the 2008 US economic crisis causes a deterioration in the value of the US Dollar.

With the increasing pace of globalization, one can expect the emergence of new international centers of finance in Shanghai, Hong Kong, Singapore, Mumbai, Dubai, Sao Paolo, and so on.

2.4 Interest rates

Interest rates, in essence, represent the interplay of time with economic activity, money capital, and real (tangible) assets.

The money form of capital represents real productive assets of society that can erode over time; furthermore, other factors like inflation, currency devaluations, new technologies, and so on make the value represented by financial assets a variable quantity that responds to changing circumstances. Financial assets represent the ability to initiate or facilitate economic activities, opportunities for which are tied to many social factors. For these and many other reasons, the effective value of money is strongly dependent on time.

How does one estimate the time value of money? From economic theory, the sum total of all the endogenous and exogenous factors that affect the time value of money are contained in the interest rates that one earns on cash deposits or on Treasury Bonds. Money invested in other financial instruments is more complicated to value as risk premiums are involved, perceptions of which differ between investors. Ultimately, the time value of money involves discounting expected future cash flows from bonds to obtain its present-day value; or, inversely, compounding present-day cash deposits for obtaining its expected future value.

Interest rates fix the cost of borrowing capital, the 'cost' of money, and are determined by both, the supply and demand for money – which depend on the prevailing interest rates – and by the macroeconomic policies of central banks.

Central banks would, ideally, like to hold down inflation while at the same time engendering economic growth; central banks balance inflation against the rate of economic growth by regulating the supply of money. One of the major tools for influencing the supply and demand for money is by setting interest rates.

Market forces of supply and demand and central banks' setting of interest rates are in a state of constant tension. Market forces sometimes force the central banks to change the interest rates so as to bring them in line with the market; at other times, central banks intervene by changing the interest rates and thus affecting the market's demand for money.

The concepts of *discounting* and *compounding* are fundamental to finance. However, contrary to what one intuitively expects, the relation turns out to be far more complex than discounting and compounding simply being the inverse of each other. The different forms of compounding (discounting) present (future) cash flows provide different ways by which interest rates are defined.

Consider the future value of a fixed deposit that is rolled over continuously; a constant interest rate leads to an exponential compounding of the value of the initial fixed deposit. Discounting, on the other hand, is the procedure that yields the present-day value of a pre-fixed future cash flow and is exponentially smaller for constant interest rates. In essence, all measures of interest rates arise by either discounting expected future cash flows to obtain their present-day value or by compounding the present-day value of fixed deposits to obtain the value of future cash flows.

2.5 Three definitions of interest rates

The following procedures for defining interest rates are widely used in the financial markets, with an interest rate 'yield curve' for each case.

- Returns on cash deposits using simple interest rates. This is the basis of defining Libor and Euribor, the two fundamental market determined interest rates.
- Discrete compounding of cash deposits and discrete discounting of bonds. This procedure is the basis for the definition of the zero coupon yield curve (ZCYC), which is fundamental to the interest rates and bond markets.
- Instantaneous compounding and discounting future cash flows. This definition leads to the concept of instantaneous forward interest rates, the main theoretical construct of the bond market.

To simplify the discussion of the central concepts, all interest rates for now are taken to be constant. The more complex generalizations of these concepts are discussed in the later sections.

2.5.1 Simple interest rates

Consider a principal sum of amount M, kept in a bank fixed deposit at time t and earning a simple interest at the rate of L per year. After a period of say T years, the initial amount M increases to $M[1 + (T - t)L]$. Conversely, if one is to receive a pre-fixed amount B at time T in the future, the value of that amount at time t is given by $B/[1 + (T - t)L]$. In summary

$$M \text{ at time } t = M[1 + (T - t)L] \text{ at time } T$$

$$\frac{B}{[1 + (T - t)L]} \text{ at time } t = B \text{ at time } T \qquad (2.1)$$

2.5.2 Discrete compounding and discounting: yield to maturity

Consider a fixed deposit made at time t; the principal earns a *yield to maturity* z, a dimensionless quantity that is a measure of simple interest for a period, usually taken to be one year. At the end of one year, the interest earned is compounded – namely, the interest earned is *added* to the principal sum. At the end of the first year $M(1 + z)$ is the amount in the fixed deposit; at the end of the second year the amount in the fixed deposit is $M(1 + z)^2$, and so on. For a deposit of duration $T - t$ years, there are $[T - t] = (T - t)/1$ number of compounding.[2]

Hence, at time T, the discretely compounded amount for a fixed deposit made at time t is given by

$$M \text{ at time } t = M(1 + z)^{[T-t]} \text{ at time } T$$

$$\frac{B}{(1 + z)^{[T-t]}} \text{ at time } t = B \text{ at time } T \qquad (2.2)$$

where the last equation gives the discretely discounted value at time t of a pre-fixed payment B at time T.

2.5.3 Continuous compounding and discounting

Consider the case of discrete compounding, but now let ϵ be an infinitesimal period of discrete compounding. Consider the limit of $\epsilon \to 0$; simple interest payments are now given by $z = \epsilon r$; r is the instantaneous spot interest rate and has the dimension of 1/time. The interest generated in the time interval t to $t + \epsilon$, is $M\epsilon r$ and the fixed deposit is compounded to yield $M(1 + \epsilon r)$. For the time interval $T - t$, the

---

[2] Note $[T - t]$ is always an *integer*.

number of times the principal is compounded is $(T - t)/\epsilon$; hence the value of the continuously compounded fixed deposit at time T is given by

$$\lim_{\epsilon \to 0} M(1 + \epsilon r)^{(T-t)/\epsilon} = M e^{r(T-t)}$$

In summary, for continuously compounded interest rates

$$M \text{ at time } t \;\; = \;\; M e^{r(T-t)} \text{ at time } T$$
$$\frac{B}{e^{r(T-t)}} \text{ at time } t \;\; = \;\; B \text{ at time } T \tag{2.3}$$

All the different ways of defining interest rates are of course consistent. Any inconsistency or incompatibility in the different definitions of interest rates leads to arbitrage opportunities in the prices of debt instruments.[3] This in turn would lead to trades that remove any pricing inconsistency.

2.6 Coupon and zero coupon bonds

Cash represents present-day value, whereas bonds represent future cash flows.

Bonds are fundamental instruments of debt; the seller of a bond issues a promissory note to the buyer that states the seller's (legal) obligation to make a future payment of a certain pre-determined amount. The amount includes a component that is the return on the bond and reflects the interest rate paid by the issuer of the bond.

One of the primary financial instruments of the national and international debt markets are government and corporate bonds. Interest rates can be derived from the market prices of bonds. Given the vast diversity of the bond market, only those aspects of bonds are discussed that are of direct relevance to the material covered in this book. The readers are referred to the extensive literature on bonds [73].

A *zero coupon bond* is a financial instrument that gives a single pre-determined payoff, of say €1, called the principal amount, when it matures at some fixed future time T; its price at earlier time $t < T$ is given by $B(t, T)$. Note that for a zero coupon bond there are no coupon payments and hence the name.

At time t there are, in principle, infinitely many zero coupon bonds with varying maturities; that is, bonds $B(t, T)$, in principle, exist for all $T \in [t, t + \infty]$ years. In practice, in the capital markets, the zero coupon bonds are usually issued with maturity from one day to about 30 years in the future and hence $T \in [t, t+30]$ years. The collection of the prices of all zero coupon bonds $B(t, T)$, with maturity from present time t to a maximum time T is called the zero coupon bond term structure.

---

[3] Arbitrage opportunities means that one can make risk-free profit that is higher than the (risk-free) rate of return on fixed deposits. See Section 3.5.

Consider a *coupon bond*, denoted by $\mathcal{B}(t)$, that pays a principal of L when it matures at time T, and pays fixed dividends (coupons) a_i at times $T_i, i = 1, 2, \ldots, N$. The value of the coupon bond at time $t < T_i$ can be shown [63, 65] to be equivalent to a portfolio of zero coupon bonds with maturities coinciding with the payment dates of the coupons. Quantitatively

$$\mathcal{B}(t) = \sum_{i=1}^{N} a_i B(t, T_i) + L B(t, T) = \sum_{i=1}^{N} c_i B(t, T_i) \qquad (2.4)$$

For simplicity of notation, the time of maturity of the coupon bond is taken to be the date of the last coupon payment, that is $T = T_N$. The final payment is included in the sum by setting $c_i = a_i$; $c_N = a_N + L$.

Intuitively, the reason that a portfolio of zero coupon bonds is equal to a coupon bond is because the two instruments have the same cash flow. Every coupon payment for the coupon bond is equivalent to a zero coupon bond maturing at the time of the payment. A fundamental theorem of finance states that any two financial instruments that have the same cash flow are identical [63]. The proof follows from the fact that, otherwise, arbitrage opportunities would exist for the prices – which is ruled out in an efficient market.

2.6.1 *Coupon bond yield-to-maturity y*

Given the wide variety of coupon bonds, with different face values L, different amounts and number of coupon payments a_i and N respectively, it is difficult to compare the *rates of return* of two different coupon bonds. For this reason, a generalization of the zero coupon bond yield-to-maturity z, given in Eq. (2.23), is defined for coupon bonds and denoted by y.

Coupon bond yield-to-maturity y is the annual yield such that, at time t, the present values of the future cash flows, discretely discounted yearly by y, equal the face value of the coupon bond. For coupon bonds with N number of (annual) payments, the yield-to-maturity is defined as follows

$$\mathcal{B}(t) = \sum_{i=1}^{N} \frac{a_i}{(1+y)^i} + \frac{L}{(1+y)^N}$$

Given the values of $\mathcal{B}(t), a_i$, and N, it is in general a nonlinear problem to evaluate y, and is usually done numerically. Once the y value of a coupon bond is determined, one can accurately compare it with other coupon bonds with very different cash flows. One can readily generalize the definition of the coupon bond yield-to-maturity y for coupons that are paid out c times a year and so on.

From Eq. (2.4) one can conclude that the zero coupon bonds are the fundamental instruments of the bond market. If one can model the behavior of the zero coupon bonds, one automatically has, in principle, a model for the coupon bonds. However, as is to be expected, the coupon bond is a much more complex instrument than the zero coupon bond.

All bonds have a credit risk, which is the likelihood of default, due by the possible inability of the issuer to pay either the coupons or the principal amount. Credit risk arises from various sources and the financial consequences of default are taken into account in the pricing models of such defaultable, or risky, bonds; in particular, the higher the possibility of default, the higher the interest rate that has to be paid out by the issuer of the bond.

An important class of both coupon and zero coupon bonds are those that carry no risk of default; such bonds are called *Treasury Bonds*. In practice, bonds issued by the US federal government are taken to be risk-free Treasury Bonds and consequently have the lowest interest rates in the debt market. Almost all the discussions on bonds, in the later chapters, are confined to the study of risk-free Treasury Bonds.

Since bonds generate pre-fixed (series of) cash flows, they belong to the larger class of financial instruments called *fixed-income securities*. The ownership of a fixed-income security is often, erroneously, considered to be less risky than the ownership of equity since – short of the issuer going bankrupt – the owner of a fixed-income security is guaranteed a return. However, due to interest rate risk, credit risk, and currency risk (for the bonds that are issued in a foreign currency), a bond portfolio before maturity can lose as much value, or even more, than a portfolio of equities.

2.7 Continuous compounding: forward interest rates

The present-day value of a bond is obtained by discounting future cash flow(s) using various methods, with each method providing a definition of interest rates.

Consider the simplest case of an economy that has a constant interest rate r. As discussed in Eq. (2.3) a continuously compounded fixed cash deposit of €1 made at time t will yield, at time T in the future, a cash of amount $\exp\{(T-t)r\}$. Hence a zero coupon bond yielding €1 at time T has a present value of

$$B(t,T) = e^{-(T-t)r}$$

In general, a real economy never has an interest rate that is constant over future time. Instead, for each future time T, there is a separate effective interest rate, denoted by $r(t, T)$, called the *term structure of interest rates* and also known as the *interest yield curve*. The zero coupon bond is given by

$$B(t, T) = e^{-(T-t)r(t,T)}$$

$$\Rightarrow r(t, T) = -\frac{1}{T-t} \ln B(t, T) \tag{2.5}$$

The interest yield curve can also be used for determining the future value of a fixed deposit that is continuously compounded; for €1 deposited at time t and continuously compounded, its future value at time T is locked in at time t to be equal to $\exp\{(T - t)r(t, T)\}$.

Forward interest rates, denoted by $f(t; T_1, T_2)$, are continuous rates that are available in the debt market such that one can lock-in, at time t, the interest rate for a deposit from future time T_1 to T_2, with $T_2 > T_1$.

To understand the relation of $f(t; T_1, T_2)$ to zero coupon bonds, consider two zero coupon bonds $B(t, T_1)$ and $B(t, T_2)$, with $T_2 > T_1$. The definition of bonds in terms of the interest yield curve given in Eq. (2.5) yields

$$B(t, T_1) = e^{-(T_2-T_1)f(t;T_1,T_2)} B(t, T_2)$$

$$\Rightarrow f(t; T_1, T_2) = -\frac{1}{T_2 - T_1} \ln \left[\frac{B(t, T_1)}{B(t, T_2)}\right] \tag{2.6}$$

Discounting of bonds, from future to present time, is shown in Figure 2.1.

For a deposit made at time t, the future value at times T_1 and T_2 are $\exp\{(T_1 - t)r(t, T_1)\}$ and $\exp\{(T_2 - t)r(t, T_2)\}$, respectively. However, the value of the two deposits are related, as shown in Figure 6.6, since one can take the cash obtained at time T_1 and lock-in the interest at time t, for the duration from T_1 to T_2 using $f(t; T_1, T_2)$. The principle of no-arbitrage yields

$$e^{(T_2-t)r(t,T_2)} = e^{(T_1-t)r(t,T_1)} e^{(T_2-T_1)f(t;T_1,T_2)} \tag{2.7}$$

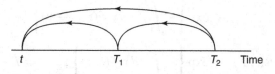

Figure 2.1 The discounting of bond payoff directly from time T_2 to time t and via an intermediate time T_1.

2.8 Instantaneous forward interest rates

Forward interest rates play a central role in the study of interest rates and coupon bonds. The forward interest rates provide a representation of zero coupon term structure that is analytically and conceptually very useful in the study of the bond market.

To derive the instantaneous forward interest rates from the term structure of the zero coupon bonds, consider two bonds that are mature at infinitesimally separated future times. More precisely, in Eq. (2.6) let $T_2 = T_1 + \epsilon$; hence one obtains the following[4]

$$B(t, T + \epsilon) = e^{-\epsilon f(t,T,T+\epsilon)} B(t, T) \tag{2.8}$$

The limit of forward interest rates

$$f(t, T) \equiv \lim_{\epsilon \to 0} f(t, T, T + \epsilon) \tag{2.9}$$

defines the instantaneous *forward interest rates*, namely $f(t, T)$. Instantaneous forward interest rates $f(t, T)$ are the rate, fixed at time t, for instantaneous loans at future time $T > t$; as expected $f(t, T)$ has the dimensions of 1/time. All forward interest rates are always positive and hence

$$f(t, T) > 0 \quad \text{for all } t, \ T \tag{2.10}$$

The spot interest rate $r(t)$ is the instantaneous interest rate at time t; the definition of the instantaneous forward rates yields

$$r(t) = f(t, t)$$

Eq. (2.8) provides a recursion equation. Let maturity time be discretized into a lattice with $T - k\epsilon$ points; then, since $B(t, t) = 1$, Eq. (2.8) yields the following

$$B(t, T) = \exp\{-\epsilon f(t, T - \epsilon)\} B(t, T - \epsilon)$$

$$= \exp \left\{ -\epsilon \sum_{k=1}^{(T-t)/\epsilon} f(t, T - \epsilon k) \right\} B(t, t)$$

$$\rightarrow \exp \left\{ -\int_0^{T-t} dy\, f(t, T - y) \right\}$$

$$\Rightarrow B(t, T) = \exp \left\{ -\int_t^T dx\, f(t, x) \right\} ; \quad x = T - y \tag{2.11}$$

---

[4] In practice, one takes $\epsilon = 1$ day $= 1/360$ year.

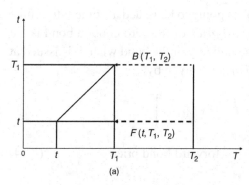
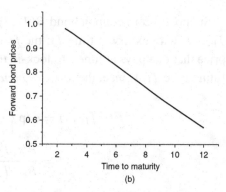

(a) (b)

Figure 2.2 (a) The forward interest rates, indicated by the dashed lines, that define a zero coupon bond $B(T_1, T_2)$ and its forward price $F(t, T_1, T_2)$. (b) The forward bond price $F = F(t, T_1, T_2)$ for zero coupon bonds maturing at different times T_2, with $T_1 - t = 2$ years in the future. The forward interest rates $f(t, T)$ were obtained from the US$ zero coupon yield curve for $t = 29$ January 2003.

Figure 2.2(a) graphically represents the forward interest rates that define a zero coupon bond $B(T_1, T_2)$.

It is worth noting that one can directly obtain the current value of the bond $B(t, T)$ by discounting the €1 payoff taking infinitesimal backward time steps ϵ from maturity T to present time t, which yields[5]

$$B(t, T) = e^{-\epsilon f(t, t+\epsilon)} e^{-\epsilon f(t, t+2\epsilon)} \ldots e^{-\epsilon f(t, x)} \ldots e^{-\epsilon f(t, T)}$$

$$\Rightarrow B(t, T) = \exp\left\{ -\int_t^T dx f(t, x) \right\} \tag{2.12}$$

$$\frac{\partial B(t, T)}{\partial T} = -f(t, T) B(t, T) \tag{2.13}$$

In fact, the result given above, using the concept of discounting, is obtained more formally in Eq. (2.11), using the recursion equation.

Eq. (2.12) shows that $f(t, x)$ is a set of variables equivalent to the zero coupon bonds. From the definition of the instantaneous forward interest rates given in Eq. (2.12), the forward interest rate and the interest yield curve are given by the following

$$f(t : T_1, T_2) = \frac{1}{T_2 - T_1} \int_{T_1}^{T_2} dx f(t, x)$$

$$r(t, T) = \frac{1}{T - t} \int_t^T dx f(t, x)$$

---

[5] The fixed payoff €1 is assumed and is not written out explicitly.

Suppose a zero coupon bond $B(T_1, T_2)$ is going to be issued at some future time $T_1 > t$, with expiry at time T_2; the *forward price* of the zero coupon bond is the price that one pays at time t to lock-in the delivery of the bond when it is issued at future time T_1. Hence, the *forward bond price* is given by

$$F(t, T_1, T_2) = \exp\left\{-\int_{T_1}^{T_2} dx f(t, x)\right\}$$

$$= \frac{B(t, T_2)}{B(t, T_1)} : \quad \text{forward bond price} \qquad (2.14)$$

In terms of the forward interest rates the forward bond price is given by

$$F(t, T_1, T_2) = \exp\{-(T_2 - T_1) f(t; T_1, T_2)\}$$

Figure 2.2(b) shows the forward bond price $F = F(t, T_1, T_2)$ of the bond $B(T_1, T_2)$. The values of the forward bond price are plotted in Figure 2.2(b), as a function of maturity time. It can be seen that the forward price falls rapidly, as is expected, given the exponential discounting of the bond prices.

At any instant t, the capital markets (implicitly) have instantaneous forward interest rates from present t out to a time T_{FR} in the future; for example, if t refers to present time t_0, then one has forward rates from t_0 till time $t_0 + T_{FR}$ in the future. In the market, T_{FR} is at least about 30 years, and hence we have $T_{FR} > 30$ years. In general, at any time t, all the forward interest rates $f(t, x)$ exist till time $t + T_{FR}$ and, hence, have future time x with $t < x < t + T_{FR}$.

2.9 Libor and Euribor

The two main international currencies are the US Dollar and the Euro, which is the currency of the European Union. As can be seen from Table 2.1, almost 90% of international cash reserves are in the form of US Dollars or Euros. Cash fixed deposits in these currencies account for almost 90% of simple interest rates that are traded in the capital markets. Cash deposits in US Dollars as well as British Pounds earn simple interest at the rate fixed by Libor and deposits in Euros earn interest rates fixed by Euribor.

2.9.1 Libor

The interest rates offered for time deposits are often based on Libor, the *London Interbank Offered Rate* [12]. Libor is one of the main instruments for interest rates in the debt market, and is widely used for multifarious purposes.

Libor was launched on 1 January 1986 by the British Bankers' Association. Libor is a daily quoted rate based on the interest rates at which *commercial banks* are willing to lend funds to other banks in the London interbank money market. The minimum deposit for a Libor has a par value of $1,000,000. Libor is a simple interest rate for fixed bank deposits and the British Bankers' Association has daily quotes of Libor for loans in the money market of the following duration: overnight; one and two weeks; one, three, four, five, six, nine, and 12 months. Libors of longer duration are obtained from the interest rate swap market and are quoted for future loans of duration from two years to 30 years. A Libor zero coupon yield curve is constructed from the swap market and is quoted by vendors of financial data. The Libor market is active in maturities ranging from a few days to 30 years, with the greatest depth in the 90- and 180-day time deposits.

The three-month Libor is the benchmark rate that forms the basis of the Libor derivatives market. All Libor swaps, futures, caps, floors, swaptions, and so on are based on the three-month deposit. The main focus of this book is Libor derivatives and the term Libor will be taken to be synonymous with the three-month Libor.

In 1999 the open positions on Eurodollar futures had a par value of about US$750 billion, and has grown tremendously since then. The Chicago Mercantile Exchange (CME) Libor futures represent one-month Libor rates on a $3 million deposit. In 2008, CME had Eurodollar futures and options on Libor with open interest of over 40 million Libor contracts and an average daily volume of 3.0 million. Libor is amongst the world's most liquid short-term interest rate futures contracts. Interest rate swaps, with Libor taken as the floating rate, currently trade on the interbank market for maturities of up to 50 years.

Market data on Libor futures are given for daily time t in the form of $L(t, T_i - t)$, with fixed dates of maturity T_i (March, June, September, and December) and shown in Figure 2.3(a). The shortest maturity time is $\theta_{min} = 3$ months, and the spot rate is taken to be $r(t) = f(t, \theta_{min})$.

2.9.2 Euribor

Euribor (Euro Interbank Offered Rate) is the benchmark rate of the Euro money market, which has emerged since 1999. Euribor is simple interest on fixed deposits in the Euro currency; the duration of the deposits can vary from overnight, weekly, monthly, three monthly out to long duration deposits of ten years and longer. Euribor is sponsored by the Financial Markets Association (ACI) and by the European Banking Federation (FBE), which represents 4,500 banks in the 24 member states of the European Union and in Iceland, Norway, and Switzerland. Euribor is the rate at which Euro interbank term deposits are offered by one prime bank to another.

The choice of banks quoting for Euribor is based on market criteria. These banks are of first-class credit standing. They are selected to ensure that the diversity of the Euro money market is adequately reflected, thereby making Euribor an efficient and

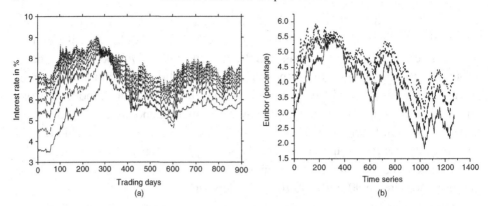

Figure 2.3 (a) Daily Eurodollar futures for Libor rates $L(t, t+7\text{ years}), \ldots L(t, t+ 6\text{ years}), \ldots L(t, t + 1\text{ year})$, and $L(t, t + 0.25\text{ years})$ with $t \in [1996, 1999]$. (b) Euribor maturing one, two, and three years in the future, from 26 May 1999 to 17 May 2004.

representative benchmark. All the features discussed for Libor can also be applied to Euribor.

Euribor was first announced on 30 December 1998 for deposits starting on 4 January 1999. Figure 2.3(b) shows daily values for three Euribor forward interest rates for 90-day deposits one, two, and three years in the future. Since its launch, Euribor has been actively trading on the options markets and is the underlying rate of many derivatives transactions, both over-the-counter and exchange-traded. Euribor is one of the most liquid global interest rate instruments, second only to Libor. The Euribor zero coupon yield curve, based on the rates being contracted in the Euribor swaps market, extends out to 50 years in the future.

2.10 Simple interest rate

Cash deposits can earn simple interest rates for a given period of time. For example, one can lock-in at time t, a simple interest rate, denoted by $L(t; T_1, T_2)$, for a fixed deposit from future time T_1 to T_2. The period of the deposit, namely $T_2 - T_1$, is called the *tenor* of the simple interest rate.

A deposit of €1, made from time T_1 to T_2, will increase, as in Eq. (2.1), to an amount $1 + (T_2 - T_1)L(t; T_1, T_2)$. Similarly, the present-day value of a zero coupon bond $B(t, T)$ is given by

$$B(t, T) = \frac{1}{1 + (T - t)L(t; t, T)}$$

and more generally

$$B(t, T_2) = B(t, T_1) \frac{1}{1 + (T_2 - T_1)L(t; T_1, T_2)}$$

From the definition of zero coupon bonds given in Eq. (2.12) the simple interest rates are given in terms of the instantaneous forward interest rates by the following

$$\frac{1}{1 + (T_2 - T_1)L(t; T_1, T_2)} = \exp\left\{-\int_{T_1}^{T_2} dx f(t, x)\right\}$$

$$\Rightarrow L(t; T_1, T_2) = \frac{1}{T_2 - T_1}\left[\exp\left\{\int_{T_1}^{T_2} dx f(t, x)\right\} - 1\right] \tag{2.15}$$

From Eq. (2.10) one has $f(t, x) > 0$ and this leads to

$$L(t; T_1, T_2) > 0 \tag{2.16}$$

The forward interest rates for returns on fixed cash deposits are the same as $f(t, x)$; these rates are, in principle, identical to the forward interest rates discussed in Section 2.8.

Consider a future time falling within the fixed maturity times, say $\theta = x - t$, with $T_i - t \leq \theta \leq T_{i+1} - t$; to obtain $L(t, \theta)$ with fixed θ, a spline interpolation for the values of the Libor yields the values of $L(t, T)$ for continuous future time T. The spline interpolation is necessary since the Libor data are provided only for discrete maturity times $T_i - t$, whereas for empirically studying interest rates, data are required for constant θ. The daily interpolated data, from 1996 to 1999 for the Libor rates, is plotted in Figure 2.3(a).

A futures contract is an undertaking by participating parties, entered into at time t, to loan or borrow a fixed amount of principal at an interest rate fixed by Libor $L(t, T_1, T_2)$; the contract is executed at a specified future date $T_1 > t$. Consider a futures contract entered into at time t for a 90-day deposit of the principal P, from future time T to $T + \ell$ ($\ell = 90/360$ year). On maturity, an investor who is long on the contract receives P plus simple interest I; hence

$$P + I = P[1 + \ell L(t; T, T + \ell)] \tag{2.17}$$

where $L(t; T, T + \ell)$ is the (annualized) three-month Libor interest rate. For simplicity of notation, a 90-day tenor is written as ℓ.

One can express the principal plus interest based on compounding by instantaneous forward interest rates and obtain

$$P + I = Pe^{\int_T^{T+\ell} dx f(t, x)} \tag{2.18}$$

Define the *benchmark* three-month Libor by

$$L(t;T) \equiv L(t;T,T+\ell) \tag{2.19}$$

The relationship between Libor and forward interest rates, from Eqs. (2.17) and (2.18), is given by

$$1 + \ell L(t,T) = e^{\int_T^{T+\ell} dx f(t,x)}; \quad \Rightarrow L(t,T) = \frac{e^{\int_T^{T+\ell} dx f(t,x)} - 1}{\ell} \tag{2.20}$$

Note that the above equation is a special case of the relation between $L(t;T_1,T_2)$ and $f(t,x)$ given in Eq. (2.15) with $T_1 = T$ and $T_2 = T + \ell$.

Forward interest rates $f(t,x)$ can be extracted from Libor futures data. Since Libor is determined on a daily basis, the data for the forward interest rates are given only for discrete calendar time. Future time is also discrete, with the benchmark Libor given at 90-day intervals.

In terms of zero coupon bonds $B(t,T)$, from Eqs. (2.12) and (2.20), Libor has the following representation

$$L(t,T) = \frac{1}{\ell} \frac{B(t,T) - B(t,T+\ell)}{B(t,T+\ell)} \tag{2.21}$$

It is sometimes assumed that the Libor futures prices are approximately equal to the forward interest rates. More precisely, from Eq. (2.20)

$$L(t,T) = \frac{e^{\int_T^{T+\ell} dx f(t,x)} - 1}{\ell} \simeq f(t,T) + O(\ell) \tag{2.22}$$

The errors in setting Libor equal to the forward interest rates are usually negligible, given the other errors that arise in the empirical study; the justification for this assumption is discussed in [12]. In summary, Libor can be identified with the forward interest rates, but sometimes it is more appropriate to use the full expression for $L(t,T)$.

2.11 Discrete discounting: zero coupon yield curve

Recall that, from Section 2.5, the *yield-to-maturity* z of a zero coupon bond is the annual simple interest that is discretely compounded every year. Let T,t be the maturity and issue date of the bond; as before, let $[T - t] = (T - t)/\text{year}$ be an integer equal to the number of years. On maturing, the bond value of €1 will compound to $(1 + z)^{[T-t]}$. Since, on maturity, the payoff of the bond is €1, the relation of z to the price of the zero coupon bond at t is given by

$$B(t,T) = \frac{1}{(1 + z)^{[T-t]}} \tag{2.23}$$

Note the yield-to-maturity varies for the different bonds; hence, a more precise notation is to have a *term structure* for the yield-to-maturity, called the zero coupon yield curve (ZCYC) and denoted by $Z(t, T)$; similar to z, $Z(t, T)$ is dimensionless. Eq. (2.23), for a ZCYC that is annually compounded, has the following generalization

$$B(t, T) = \frac{1}{(1 + Z(t, T))^{[T-t]}} \qquad (2.24)$$

Equation (2.24) states that $Z(t, T)$ is the dimensionless yield-to-maturity, compounded annually, that is earned by the zero coupon bond $B(t, T)$. If the interest is paid out c times a year, then the number of payments is $c[T - t]$ with each payment of interest being $Z(t, T)/c$; hence, for a ZCYC for interest that is compounded c times a year, the bond is given by

$$B(t, T) = \frac{1}{\left(1 + \frac{1}{c} Z(t, T)\right)^{c[T-t]}} \qquad (2.25)$$

In the bond market, for semi-annual (six monthly) payments, $c = 2$ and hence

$$B(t, T) = \frac{1}{\left(1 + \frac{1}{2} Z(t, T)\right)^{2[T-t]}} \qquad (2.26)$$

Market data for $Z(t, T)$ from the bond market are given in Figure 2.4(a) for fixed future remaining time, that is for $Z(t, t + \theta)$, with future remaining time θ ranging

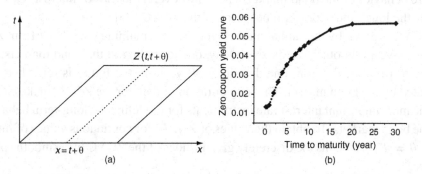

Figure 2.4 (a) The zero coupon yield curve (ZCYC) $Z(t, T)$ is given along the lines of constant $\theta = x - t$; the diagram shows $Z(t, t + \theta)$ for $\theta = $ constant. (b) The spline curve fit for the US Treasury Bonds semi-annually compounded zero coupon yield curve (ZCYC) $Z(t, T)$, with market data given by the filled squares. The curve is given for calendar time $t = 29$ January 2003 out to 30 years into the future. The market values of the ZCYC are given for discrete future remaining time θ equal to 3m, 6m, 1y, 2y, 3y, 4y, 5y, 6y, 7y, 8y, 9y, 10y, 15y, 20y, 30y.

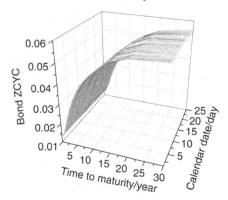

Figure 2.5 Zero coupon yield curve obtained from US Treasury Bonds. The calendar date is from 29 January 2003 to 4 March 2003, with a total of 25 trading days.

from three months out to 30 years. Figure 2.4(b) shows the structure of the ZCYC $Z(t, T)$ for US Treasury Bonds as a function of remaining future time $\theta = T - t$ with calendar time t fixed at 29 January 2003; market data are the discrete points and the interpolation curve is the result of a spline fit.[6]

Figure 2.5 shows the ZCYC obtained from US Treasury Bonds. The ZCYC rises and flattens out, as expected, since forward interest rates for the future are in general higher than near-term loans and spot rates, given that risks accumulate further out into the future. However, there are cases in which the ZCYC may have an inversion in the future reflecting some regulation or other exogenous factors that affect the future behavior of the bond market. There are several theories of interest rates that study the long- and short-term behavior of the ZCYC [73].

ZCYC is given by the capital markets for future remaining time $T - t$ from zero up to 30 years; in other words, every day, due to trading in the bond markets, the value of the $Z(t, T)$, from one day up to 30 years in the future is refreshed and updated by the bond market. The long duration of data for the ZCYC makes it one of the most important interest rate instruments for modeling the long-term behavior of the bond markets. To obtain the values of $Z(t, T)$ for continuous values of future time $\theta = T - t$ one fits the discretely given values of the ZCYC by a smooth spline curve.[7]

---

[6] For the empirical studies in later chapters, daily Treasury Bond ZCYC data for calendar times from 29 January 2003 to 28 January 2005 were used.

[7] It is assumed that the ZCYC rates are smooth; the assumption is a reasonable one to make as one would intuitively expect that the ZCYC, say three years into the future would not be too different from that of three years and one month into the future. The loss in accuracy due to the spline interpolation is unimportant since the future times at which $Z(t, T)$ is specified, namely values of T, are separated by at least by three months. The market data that are being studied have random errors larger than the errors introduced by the spline interpolation. See Section 2.14.

Both the bond markets as well as the Libor markets provide a ZCYC. For the case of Libor, the British Bankers' Association quotes daily interest rates for overnight (24 hours) deposits up to rates for deposits made one year in the future. Libors for deposits made at future time from one to 30 years are obtained from the interest rate swaps market. All the Libors are combined to produce a single ZCYC that is semi-annually compounded to produce the effective Libor zero coupon bonds. More precisely

$$B_L(t, T) = \frac{1}{\left(1 + \frac{1}{2}Z_L(t, T)\right)^{2[T-t]}} \tag{2.27}$$

where the subscript L indicates Libor. In Chapter 6 the Libor forward interest rates that are derived from $B_L(t, T)$ are discussed. In principle, $B(t, T) = B_L(t, T)$, but there are differences related to the risk of default in the Libor market being greater than in the Treasury Bond market. Figure 2.6(a) shows the Libor ZCYC for two days five years apart and Figure 2.6(b) shows the Libor ZCYC $Z_L(t, x)$ for 25 consecutive days until 8 August 2008.

The term structure of the zero coupon bonds $B(t, T)$, for some fixed time t, consists of the prices for all $T \in [t, t + 30$ years$]$. The market usually gives the term structure of the zero coupon bonds $B(t, T)$ in terms of the ZCYC. Figure 2.7(a) shows the term structure of zero coupon bonds as reconstructed from the US Treasury Bonds' ZCYC. Figure 2.7(b) shows the term structure for Libor zero coupon bonds $B_L(t, T)$ for two days five years apart; the shape of the Libor is different from the Treasuries result.

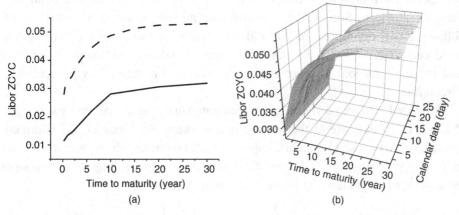

Figure 2.6 (a) A graph of Libor $Z_L(t, T)$. The solid line is for 8 August 2008 and the dashed line is for 28 October 2003. (b) A graph of Libor $Z_L(t, T)$, with future time $T - t$ shown along the x-axis out to 30 years; the daily values for $t -$ for 25 subsequent days until 8 August 2008 – are shown along the y-axis.

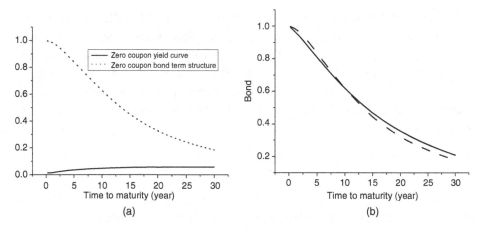

(a)   (b)

Figure 2.7 The ZCYC data are for $t = 29$ January 2003 and with $T - t$ up to
30 years into the future. (a) The term structure zero coupon bonds $B(t, T)$ (dotted
line) obtained from ZCYC $Z(t, T)$ data (unbroken line). Note $B(t, T)$ falls off
exponentially due to exponential discounting. (b) The term structure, up to 30 years
for Libor zero coupon bonds $B_L(t, T)$ – obtained from the Libor ZCYC. The solid
line is for 8 August 2008 and the dashed line is for 28 October 2003.

2.12 Zero coupon yield curve and interest rates

Both the ZCYC and the forward interest rates are descriptions of the same financial
instrument, namely the zero coupon bonds; in the case of Libor, the ZCYC does not
correspond to any actual traded zero coupon bonds, but, rather, is a compact way
of expressing market data on the term structure for all the Libors taken together.

The two different descriptions, namely the ZCYC and the forward interest rates,
are useful for representing different aspects of the interest rate and bond markets. Recall, from Section 2.5, that discounting future cash flows provides the
following two definitions of the underlying interest rates: (a) the zero coupon
yield curve (ZCYC) $Z(t, T)$, defined by the annual or semi-annual discounting
and (b) instantaneous forward interest rates $f(t, T)$, defined by instantaneous
discounting.

The traded zero coupon bond prices are quoted in the bond markets by specifying
the ZCYC. In the case of the interest rate markets, the Libor ZCYC is directly
quoted, based on a semi-annual compounding for obtaining the hypothetical Libor
zero coupon bonds. Eqs. (2.12) and (2.25) are the key for relating the zero coupon
bond price to the underlying interest rates and yield

$$B(t, T) = \frac{1}{\left(1 + \frac{1}{c}Z(t, T)\right)^{c[T-t]}} = \exp\left\{-\int_t^T dx f(t, x)\right\}$$

$$\Rightarrow \int_t^T dx f(t, x) = c[T - t] \ln\left(1 + \frac{1}{c}Z(t, T)\right) \qquad (2.28)$$

Eq. (2.28) is dimensionally consistent. The left-hand side is dimensionless; as required, the right-hand side is also dimensionless; c is the dimensionless number of payments per year, $Z(t, T)$ is dimensionless, and, furthermore, the integer $[T - t]$ is also dimensionless.

From Eq. (2.28) one has, by differentiating on future time T, the following

$$f(t, T) = \frac{c}{\epsilon} \ln \left(1 + \frac{1}{c} Z(t, T) \right) + \frac{[T - t]}{1 + \frac{1}{c} Z(t, T)} \frac{\partial Z(t, T)}{\partial T} \qquad (2.29)$$

One can numerically differentiate the ZCYC to extract $f(t, T)$; this procedure does yield an estimate of $f(t, x)$ from Eq. (2.29), but with such large errors that it makes the estimate quite useless for any empirical purpose.

The zero coupon bonds $B(t, T)$ are reconstructed directly from the ZCYC using Eq. (2.25) in Figure 2.8(a) (continuous line) and from forward interest rates $f(t, T)$ (dotted line), which have been extracted from the ZCYC using Eq. (2.29). One can see from Figure 2.8(a) that one gets large and systematic errors by using $f(t, T)$: the longer the time in the future the larger the systematic errors.

Both the interest rate and bond markets directly provide the ZCYC that is the *integral* of the forward interest rates over an interval of future time $[t, T]$. Hence, to minimize errors, all the numerical procedures that employ the ZCYC data should, as far as possible, directly employ the ZCYC data. One needs to avoid numerically differentiating $Z(t, T)$.

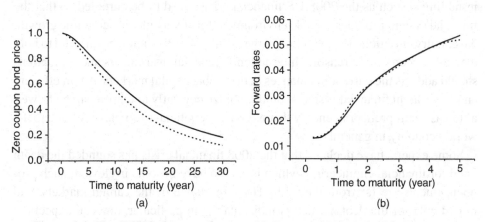

Figure 2.8 The ZCYC data are for $t = 29$ January 2003 and with $T - t$ up to 30 years into the future. (a) Zero coupon term structure $B(t, T)$ constructed from ZCYC (unbroken line) and from the forward interest rates (dotted line). (b) Forward interest rates with maturity time up to five years in the future, constructed from Libor $L(t, T)$ (dotted line) and from ZCYC $Z(t, T)$ (unbroken line).

Figure 2.8(b) shows the zero coupon term structure $B(t, T)$, obtained empirically from the ZCYC and from Libor. It is seen, as expected, that for relatively short remaining future time of $T - t \leq 5$ years the two curves give almost identical results. The ZCYC produce zero coupon term structure for maturities of up to 30 years in the future and one needs to generate the long duration zero coupon bond directly from the Libor ZCYC and not from market Libors, which are usually of a duration of up to ten years.

There is an empirical difference between Libor and Treasury Bonds. Libor has a finite probability of default, whereas Treasuries are risk free; the difference between these two rates is expressed by the TED (Treasury Eurodollar) spread. The difference is the spread between the Libor and forward interest rates derived from the ZCYC and is a measure of the risk of default of financial institutions that lend and borrow at Libor; in most cases, the spread is negligible and is ignored in all of the later discussions.

2.13 Summary

A brief review of finance shows the key role that the debt markets play in the capital markets and in the global economy. The changing nature of global capital markets and, in particular, the shift of the international capital markets to new centers were briefly discussed.

Given the growing importance of financial markets for the global economy, instabilities, such as the 2008 US financial crisis, need to be curtailed so that the financial system provides a stable environment for steady global economic growth. No country or region, no matter how large or 'important', should be allowed to hold the global economy to ransom. International financial instruments and regulations should address the current imbalances in the global capital markets. A fair, efficient and transparent financial system would mobilize currently untapped capital as well as release entrepreneurial energy that would be beneficial to all players – and to the world economy in general.

Some experts have declared that the 2008 financial crisis has sounded the death knell for financial engineering, which is said to have become irrelevant; such pronouncements are far from the truth. The importance of the capital markets, and in particular of the debt market, is indisputable; in particular, one can expect the global debt markets to play an increasingly important role in the international capital markets and in the world economy in general. Far from financial engineering being irrelevant, powerful quantitative financial models will continue to be indispensable in managing risk and maximizing returns on capital. Interest rate models of increasing sophistication will be required for designing and pricing ever-more complex debt instruments as well as for efficiently deploying a vast and expanding mountain of debt capital.

The chief component of the global capital markets is the debt market, which in turn consists mainly of the bond and interest rate markets. Bonds and interest rates are fundamental financial instruments of the debt market and reflect the time value of money. The different ways of defining interest rates and yields of bonds are the result of different ways of either discounting future cash flows or compounding present-day cash deposits or other tradable assets. The two ways of defining the future value of time lead to forward interest rates.

Coupon bond and forward interest rates and their derivatives will be discussed at length in the following chapters. Libor and Euribor were briefly discussed as these are the most important interest rate instruments, having the greatest liquidity and being the most widely traded. The three-month Libor and Euribor are taken to be the benchmark interest rates earned on cash deposits as these are the most relevant for the interest rate derivatives markets.

2.14 Appendix: De-noising financial data

All the values of financial instruments are influenced by background random noise. Consider for example the market value of a 90-day Libor $L(t, T_0)$ that matured at a fixed date of $T_0 = 16$ December 2003. The original data series on Libor is for the period from 14 June 2000 to 16 December 2003. The daily Libor is plotted from 14 June 2000 to 10 June 2002 in Figure 2.9. One can see the value of Libor is jagged (nondifferentiable) on a small time scale and regular on a long time scale.

It is assumed that Libor, and in general the price of all financial instruments, is composed of its true value, denoted by $s(t)$ and superimposed on it is noise, denoted by $w(t)$. In other words, one has [43]

$$L(t, T_0) = s(t) + w(t)$$

It is assumed that $w(t)$ is *white noise*, specified by the normal random variable given by $N(\mu, \sigma)$; at every instant, the smooth component of the market price, namely $s(t)$, has added to it a noise that is drawn from a normal (Gaussian) random variable. The random noise is assumed to be centered around the market price $s(t)$ and hence it is expected that $\mu = 0$. In other words, the observed random market price for Libor, based on the assumptions discussed, is given by

$$L(t, T_0) = s(t) \pm \sigma \text{ with } 66\% \text{ likelihood}$$

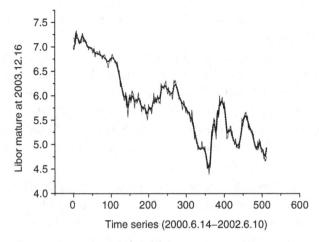

Figure 2.9 Original and de-noised Libor $L(t, T_0)$ maturing at *fixed time* in the future given by $T_0 = 16$ December 2003 and for the time period $t \in [14$ June 2000, 10 June 2002].

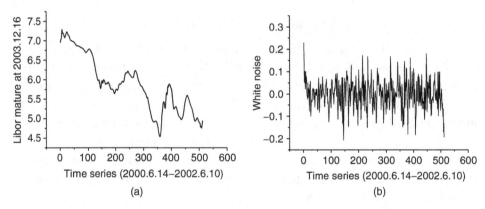

Figure 2.10 (a) The smooth component $s(t)$ of Libor $L(t, T_0)$. (b) Gaussian white noise $w(t) = N(\mu, \sigma)$ inherent in the market value of Libor $L(t, T_0)$, with $\mu = -1.4 \times 10^{-6}\%$ per year and $\sigma = 0.0629\%$ per year.

The behavior of the smooth portion $s(t)$ of Libor can have complicated dynamics and, in particular, is expected to be mean reverting.

De-noising consists of *subtracting*, at *each* instant, the white noise component from the market value of $L(t, T_0)$ and thus obtaining its smooth component, namely $s(t)$. For many purposes, it is the smooth component of the market value of a financial instrument that is required. One of the most efficient procedures is to use wavelet analysis to filter out white noise. There are many different 'basis' states that one can use to transform the market price to its smooth component and the Debauche wavelets D8 were used.

Figure 2.9 shows the original (jagged) swaption market price and the de-noised smooth curve as well. Figure 2.10(a) shows the de-noised swaption price $s(t)$ and Figure 2.10(b) shows the noise component $w(t)$. The distribution of white noise $w(t)$ is given by $N(-1.4 \times 10^{-6}, 0.0629)$, where the units for the parameters of the normal distribution are % per year.

Note that the typical value of Libor, as given in Figure 2.9, is of the order of 5%; the noise component is given by $\sigma = 0.06\%$, which is small – about 1% of the market price. This is what one expects since noise is supposed to be a small background component of the market price. Furthermore, $\mu = -1.4 \times 10^{-6}\%$ per year, which is completely neglible compared to the price of the daily value of Libor, hence confirming the assumption that the random noise is symmetrically distributed about the smooth curve $s(t)$.

3

Options and option theory

Financial derivatives – and options in particular – form an important component of financial instruments. Considerable negative criticism has been directed at options and derivatives in light of the 2008 economic crisis, some of it being justified and some being off the mark. As long as there are assets and liabilities, there will be a need to protect the future value of assets, as well as of finding ways for maximizing returns on assets. Derivatives play a central role in achieving these twin objectives.

Given the uncertainties of the financial markets, there is a strong demand from banks, financial organizations, and investors for predicting the future behavior of securities. Derivative instruments, and options in particular, are a response to this need of the market and are widely traded in the financial markets.

Options and other derivatives of underlying financial securities have contributed significantly to the explosion of the capital markets and their general principles are discussed. There are three broad categories of derivatives, namely forwards, futures, and options. Option theory is developed for equities using a path integral formulation of white noise.[1]

A series expansion of an option's price is defined for a generic case, in powers of the underlying security's volatility. The volatility expansion is of great generality and will turn out to be crucial in developing approximation schemes for a variety of interest rates and coupon bond options.

3.1 Introduction

Financial derivatives, or derivatives for short, form the bedrock of modern finance and have played a key role in providing the tools for managing returns on

---

[1] Interest rate and coupon bond options are discussed in some detail in Chapter 4.

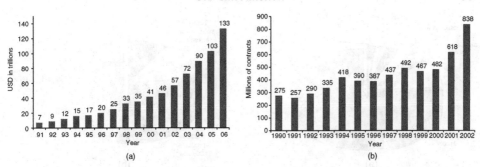

Figure 3.1 (a) US total derivatives' annual outstanding notional value, for 1991–2006. (b) US total futures' annual volume for 1990–2002.

capital, quantifying and managing risk, and for speculating on the capital markets. Derivatives are *derived* from other underlying financial instruments: the cash flows of a derivative depend on the price of underlying instruments [59, 65].

A brief review is given of the US derivatives' market as it is globally the leading one. Figure 3.1(a) shows the phenomenal increase in the notional value of US derivatives from 1991 to 2002 and Figure 3.1(b) gives the total annual volume, from 1990 to 2002, of US futures' contracts.

On 11 April 2007, the *Wall Street Journal* estimated that the global capitalization of the derivatives' markets (futures, options, swaps, etc.) exceeded US$450 trillion dollars. The over-the-counter derivatives' market in 2007, in the midst of market turmoil, increased by 15% and by December 2007, reached a truly astronomical notional value of US$596 trillon [25]. In contrast, in 2005 the total value of the stocks, bonds, and bank loans worldwide was about US$165 trillion. By 2008 it is estimated that the total notional value of the derivatives' market has crossed the US$600 trillion benchmark.

Some leading indicators of the US derivatives' and options' markets are the following. Figure 3.2(a) gives a breakdown of US derivatives used in different commodities and Figure 3.2(b) gives a breakdown of the types of credit derivatives used in the capital markets. Figure 3.3(a) gives a breakdown of the US derivatives' volume by largest contracts for 2002 and Figure 3.3(b) gives a breakdown of derivatives by the US exchanges on which they are traded.

The total global financial derivatives' market consists of three sectors, namely derivatives written on interest rates, currency exchange rates, and on equities, commodities, and so on. Figure 3.4(a) shows the notional value of the global derivatives' market that, in 2004, was about US$197 trillion; interest rate derivatives accounted for 72% of the market. The total interest rate derivatives' market had a notional value of US$142 trillion, as shown in Figure 3.4(b). Interest rate swaps accounted

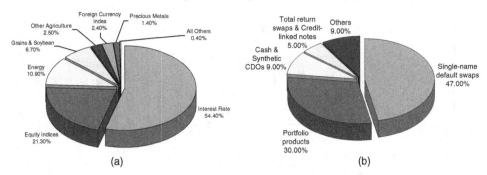

Figure 3.2 (a) US derivatives' volume breakdown, by commodity group, for 2002.
(b) US credit derivatives' notional breakdown by product category, for 2003. CDOs
stands for credit default options.

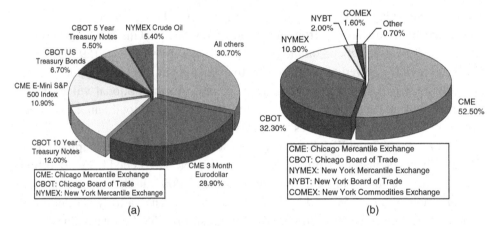

Figure 3.3 (a) US derivatives' volume breakdown by largest contracts, for 2002.
(b) US derivatives volume breakdown by exchange, for 2002.

for 78%, the lion's share of the market, with interest rate options accounting for
14% with a notional value of about US$14 trillion.

3.2 Options

Options have many uses, from being an instrument that is used for hedging a
portfolio (in order to reduce risk) to the use of derivatives as a tool for speculation.
The two main forms of derivatives are futures' and options' contracts.

In case of the forward and futures' derivatives, the seller is obliged by the contract
to take delivery of the asset in question. An investor may be more interested in the

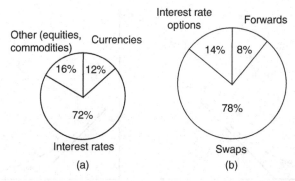

Figure 3.4 (a) A breakdown of the 2004 global derivatives' market, with a notional value of US\$197 trillion, into three primary sectors, namely interest rates, foreign exchange, and equities, commodities, and others. (b) A breakdown of the interest rate derivatives' sector of the global derivatives' market, having a notional value of US\$142 trillion.

profit that can be made by entering into a contract, rather than actually possessing the asset. For such an investor, options provide the appropriate instrument.

Options are derivatives that can written on any security, including other derivative instruments. For the sake of simplicity, the discussion in this section focuses mainly on options on stocks (equity). Interest rate and coupon bond derivatives and options are discussed in the next chapter.

An option C is a contract to buy or sell a security, called a call or a put option respectively, that is entered into by a buyer and seller. For a *call option* the seller of the option is obliged to provide the stock of a company S at some pre-determined price K and at some fixed time in the future; the buyer of the option, on the other hand, has the right to either *exercise* or *not exercise* the option. If the price of the stock on maturity is less than K, then clearly the buyer of a call option should not exercise the option. If, however, the price of the stock is greater than K, then the buyer makes a profit by exercising the option. Conversely, the holder of a *put option* has the right to sell or not sell the security at a pre-determined price to the seller of the put option.

In summary, an *option* is a contract with a fixed maturity, and in which the buyer has the option to either buy or sell a security to the seller of the option at some pre-determined (but not necessarily fixed) strike price [59]. The precise form of the strike price is called the option's *payoff function*. There are a great variety of options, and these can be broadly classed into *path independent* and *path dependent* options.

Options are either traded in the highly diversified and rapidly growing derivatives' market or are negotiated OTC (over the counter) between two parties.

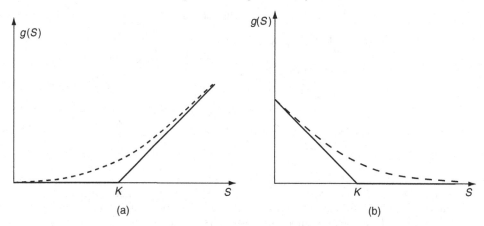

Figure 3.5 (a) Payoff for call option $\mathcal{P}(S) = g(S)$. Dashed line is possible values of the option before maturity. (b) Payoff for put option $\mathcal{P}(S) = g(S)$. Dashed line is possible values of the option before maturity.

3.3 Vanilla options

Vanilla options usually refer to the simplest of options and are defined by a pay-off function that is *path independent*, namely depends on only the value of the underlying security at the time of maturity. In other words, the payoff function is *independent* of how the security arrived at its final price.

The most widely used vanilla option is the *European option*, which comes in two varieties, namely the call and the put options.

Consider the price of an underlying security $S(t)$. The price of a European *call option* at time t depends on $S(t)$ and is denoted by $C(t) = C(t, S(t))$. The call option gives the owner of the instrument the option to buy the security at some future time $t_* > t$ for the strike price of K.

At time $t = t_*$ the option matures and the value of the call option $C(t, S(t))$ is given by

$$C(t_*, S(t_*)) = \begin{cases} S(t_*) - K, & S(t_*) > K \\ 0, & S(t_*) < K \end{cases} = [S(t_*) - K]_+$$

$$= \mathcal{P}(S)$$

where $\mathcal{P}(S) \equiv \mathcal{P}$ is the payoff function and is shown in Figure 3.5(a). Clearly, the European call (and put) options are path independent since the payoff function depends only on the final price of the security.

The positive valued function is defined as follows

$$[a - b]_+ = (a - b)\Theta(a - b) = \begin{cases} a - b, & a > b \\ 0, & a < b \end{cases}$$

where the Heavyside step function $\Theta(x)$ is defined in Eq. (A.3).

A European put option, denoted by $P(t)$, has a payoff function $[K - S(t_*)]_+$ is similar to the call option payoff and shown in Figure 3.5(b); the holder has the option to sell a security S at a price of K.

3.3.1 Put–call parity

The call and put option obey a relation called put–call parity. Put–call parity can be derived using arguments based on the requirement that the option price is free from arbitrage opportunities. Suppose the spot interest rate is given by r, and is a constant. A simple no-arbitrage argument [59] shows that

$$C(t) + Ke^{-r(t_*-t)} = P(t) + S(t); \quad t \le t_* \tag{3.1}$$

and is put–call parity for the call and put options.

Put–call parity puts nontrivial constraints on any perturbation expansion for the price of an option: the expansion needs to satisfy, order by order, put–call parity.

3.4 Exotic options

Exotic options [59] are defined by payoff functions that, in general, have a complicated dependence of the underlying security. In particular, exotic options can be path dependent, depending on the path that the security takes from the time it is issued to the time at which the option expires. There is an almost endless variety of exotic options such as the look-back options, quanto options, basket options, hybrid options, dual-strike options, and so on. There are also OTC options that are customized for the specific needs of investors [59, 37], or which are primarily designed to serve the specific needs of a niche market.

Three primary exotic options are the following:

- American option
- Asian option
- Barrier option

The *American option* is widely used in the financial markets; it has the same payoff as the European option *except* that it can be exercised at any time before the expiry of the contract. A variant of the American option is the Bermudan option, which has the same payoff as a European option, except that it can be exercised at a series of pre-fixed times before it expires.

The American option is clearly path dependent, since the choice of an early exercise depends on the value of the security for the entire duration of the option until its expiry. No-arbitrage arguments [59] show if a security does not pay a dividend,

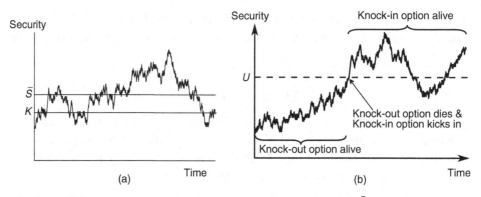

Figure 3.6 (a) The trajectory of a stock price, its average value $\bar{S}$ and strike price K. (b) Parity for the knock-out and knock-in barrier option, namely $C_{KO} + C_{KI} = C_{European}$, is shown in the diagram. Barrier is at the value of U of the security. Since at any moment one of the options has a nonzero value, the sum, in effect, is equal to the European call option.

the American call option has the same value as the corresponding European call option. However, it can be shown that the American put option pays more if it is exercised before maturity. The European put option is a special case of the American option, with the holder of the American put option having greater choice. Hence, an American *put option* has a price *higher* than the corresponding European put option.

Another widely traded exotic option is the *Asian option*. It has a payoff function that depends on the average value of the security taken over the whole period of its duration, namely from the time it is written till the time it expires. Figure 3.6(a) shows, for an Asian option, a typical path taken by the stock till maturity, its mean value $\bar{S}$, and its strike price K.

Asian options smooth out the payoff function so that, unlike the European option, the final value of the underlying security does not play a crucial role. The Asian option prevents option traders from trying to manipulate the price of the underlying security on the exercise date, for example by buying or selling large quantities of the underlying security.

Due to the fact that the payoff is a function of the security's average value over the Asian option's duration, the effective volatility of the underlying security is reduced; for this reason, an Asian option is *cheaper* than a similar European or American option. The price of exotic put options obey the following inequalities

$$P_{Asian} \leq P_{European} \leq P_{American}$$

Barrier option is a distinct class of exotic options, which come in two varieties, namely 'knock-out' or 'knock-in'. For knock-out options C_{KO}, a value is pre-set for the value of the underlying security, called the barrier. The knock-out barrier option has the same payoff as a European option, except that the barrier option is terminated with zero value the moment the price of the underlying security exceeds the barrier. The barrier option is, in general, *cheaper* than a European option since it allows less variation in the paths that the underlying security can take in reaching the option's maturity. Hence

$$C_{KO} \leq C_{European}$$

The knock-in barrier option C_{KI} has the same payoff as an European option, except that it has zero value until the security crosses a pre-set barrier, when it takes on a nonzero value and continues to be 'alive' until the option matures. It can be shown that the following parity, illustrated in Figure 3.6(b), exists between the knock-in, knock-out barrier option, and an European option having the same payoff function [59]

$$C_{KO} + C_{KI} = C_{European} \tag{3.2}$$

3.5 Option pricing: arbitrage

The payoff function $\mathcal{P}$ of, say, a call option $C(t)$ specifies the value of the option at some future time t_*. The pricing problem is to find the value of the option $C(t)$ at present time $t < t_*$. The essential idea in the pricing of all options, including those on debt instruments, is that of discounting a pre-specified future price of the option, namely the payoff function, to its present value. In this sense, all of option theory is a final value problem.

How should we discount $\mathcal{P} = C(t_*, S(t_*))$ – the option's payoff function specified at future time t_* – to obtain $C(t, S(t))$ – the value of the option at earlier time $t < t_*$? Discounting a payoff function to its present value requires the following:

- The numeraire for the discounting.
- The processes (evolution equations) that propagate the payoff function from the future calendar time to the present.

One could naively expect that evolving the underlying security $S(t)$ back, from future time to the present, determines the option price, which is the current value of the payoff function. In analogy with a fixed bank deposit, the future value of the stock $S(t_*)$ should be discounted by its expected rate of return μ to obtain its present value, namely that

$$S(t) = \exp\{-\mu(t_* - t)\}S(t_*)$$

Hence, one might expect that the present value of the option price is given by $\mathcal{P}(t, S(t))$. There are, however, two flaws in this line of reasoning.

Firstly, a fixed deposit in the bank earns a risk-free rate of return equal to the spot rate r, whereas owning a stock $S(t)$ is full of risk with no guarantee that its returns will be μ. This uncertainty in the returns on a stock is precisely why investors buy stocks in the first place since the investor expects a return higher than the risk-free r.

The second flaw is that a stock in an efficient market evolves not deterministically but randomly and $S(t)$ is modeled to follow a stochastic process. Hence its present value has to be evaluated by taking an average over its random evolution.

Options are bought by investors who are risk takers as well as those who are risk averse. Since there is only one price for every financial instrument, both these investors need to agree on the price of an option. The price of an option, consequently, has to be risk neutral or risk free.

A fundamental principle of finance is the *principle of no arbitrage* which states that no risk-free financial instrument can yield a rate of return above that of the risk-free rate r. The theorem of no-arbitrage is a formal result that follows from the obvious fact that all players in the financial market would like to have more money than less money. In other words, there is no free lunch – if one wants to earn higher returns, one has to take commensurate higher risks.

Similar to the mathematical models in scientific theories, which are tested by experiment, mathematical models in finance are tested by the capital markets. Any shortcoming of a mathematical model, for say the price of a financial instrument, leads to arbitrage opportunities – thus showing the incorrectness of the mathematical model in question.

3.6 Martingales and option pricing

Martingales form the theoretical cornerstone of option pricing and are discussed briefly in Appendix A.2.

An important result of theoretical finance is the following: in an efficient market the price of a financial instrument is free from any possibility of arbitrage if and only if the evolution of the discounted value of the financial instrument follows a *martingale process* [59, 88]. A martingale evolution is a process that is risk free. The price of an option is determined by assuming a martingale evolution of the underlying security.

It should be noted that the *actual* market evolution of a security, for example a stock, *does not* follow a martingale process since there would then be no risk premium for owning such a security. Instead, the martingale risk-free evolution of a security is a *theoretical construct* – necessary for pricing a derivative instrument such that its price is free from arbitrage opportunities. The theory of option pricing hinges on the property of martingales as it provides a price that is free from arbitrage opportunities.

A stochastic process is said to have a martingale evolution if its present value is equal to the expectation value of its future (discounted) value. Let the numeraire for discounting $S(t)$ be $\exp\{rt\}$; a discounted stock $e^{-rt}S(t)$ has a martingale evolution if it satisfies the following

$$e^{-rt}S(t) = E[e^{-rT}S(T)] \Rightarrow S(t) = E[e^{-r(T-t)}S(T)] \qquad (3.3)$$

where the symbol $E[\ldots]$ represents the expectation value over the stock's evolution. It can be shown that a stock price $S(t)$, discounted using a risk-free rate of return r, undergoes a martingale evolution.

The martingale condition for zero coupon bonds $B(t, T)$, similar to the martingale evolution for a stock as in Eq. (3.3), is given by

$$B(t, T) = E\left[e^{-\int_t^{t_*} dt' r(t')} B(t_*, T)\right] \qquad (3.4)$$

$$= E\left[e^{-\int_t^{T} dt' r(t')}\right]: \qquad (3.5)$$

$$\text{discounted value of } B(T, T) = 1$$

For the general case of the interest rates being stochastic, the expectation value has to be performed over random spot interest rate $r(t)$.

For evaluating the *price of an option* $C(t, S(t))$, the underlying stock price $S(t)$ is evolved using a martingale process. Since one is comparing the value of the payoff at two different times, the time value of money needs to be accounted for. The present value of the option can be obtained by discounting the value of the payoff by the (money market) numeraire.

Consider a payoff function of an option $C(t, T; K; S(t))$ that matures at calendar time T and is a general function of the security, which could be path dependent or independent, namely[2]

$$C(T, T; K) = \mathcal{P}[K; S(\cdot)]$$

The price of the option is given by the martingale condition; the money market numeraire $\exp(rt)$, where r is the spot interest rate taken to be a constant, yields the following

$$\frac{C(t, T; K; S(t))}{\exp(rt)} = E\left[\frac{C(t, T; K; S(\cdot))}{\exp(rT)}\right] = E\left[\frac{\mathcal{P}[K; S(\cdot)]}{\exp(rT)}\right] \qquad (3.6)$$

---

[2] The notation $S(\cdot)$ means that the payoff depends on all the values that the stock price takes from t to T, namely on $S(\tilde{t})$; $\tilde{t} \in [t, T]$.

Hence, the martingale evolution of the underlying security $S(t)$ yields the following option price

$$C(t,T;K;S(t)) = E[\exp\{-r(T-t)\}\mathcal{P}[K;S(\cdot)]] = e^{-r(T-t)}E[\mathcal{P}] \quad (3.7)$$

All options are priced in a similar manner; the payoff function is discounted using an appropriate numeraire and the underlying security is evolved using a martingale process.

3.7 Choice of numeraire

The concept of *discounting* requires a discounting factor, or equivalently a numeraire, which so far has been taken to be $\exp\{\int_t^T dt'r(t')\}$. One Euro cash deposited in a bank at time t will compound to $\exp\{\int_t^T dt'r(t')\}$ at time T; hence, discounting by the spot interest rate is said to be using the money market numeraire.

Discounting by the numeraire $\exp\{r(T-t)\}$ is quite arbitrary and any numeraire satisfying some general requirement is also adequate. The generality of choosing a numeraire is addressed in Chapter 9.

The property of a numeraire generalizes the concept of martingales. Suppose that instead of the money market numeraire one chooses a zero coupon bond $B(t,t_*)$ as the numeraire, called the forward bond numeraire. For an appropriate choice of the evolution of the underlying stock, the combination $C(t)/B(t,t_*)$ can be made into a martingale. A salient property of martingales is that the expectation value of the future (random) value of the martingale is equal to its present value; hence, since $B(t_*,t_*) = 1$, one has the following

$$\frac{C(t)}{B(t,t_*)} = E\left[\frac{C(t_*)}{B(t_*,t_*)}\right] = E[\mathcal{P}] \quad (3.8)$$

$$\Rightarrow C(t) = B(t,t_*)E[\mathcal{P}] \quad (3.9)$$

where $\mathcal{P} = C(t_*)$ is the payoff function.

Eq. (3.8) shows that, due to the martingale property of the options, its price can be obtained by discounting the payoff function using a numeraire from a large class.

3.8 Hedging

All financial instruments are subject to the random evolution of underlying instruments such as stock prices, interest rates, exchange rates, etc. There are many ways of defining risk as discussed in Bouchaud and Potters [28]. Hedging is the general procedure for reducing, if not completely eliminating, an investor's exposure to risk. Derivative instruments are essential for hedging. For example, to hedge a portfolio that contains a security one also needs to include (in the portfolio) another financial asset that moves in the opposite direction. When the stock moves either up or down, the financial asset moves in the opposite direction. Such a negatively correlated financial asset is a derivative of the stock in question.

For interest rate derivatives, there are many underlying sources of risk, such as interest rate risk, liquidity risk, risk of default, currency risk, and so on. Interest rate risk and hedging interest rate options are discussed in Chapter 14.

The 1973 seminal paper of Black and Scholes [26] was the first to recognize that perfectly hedging a derivative enables one to price the derivative using the concept of no-arbitrage. Specifically, in the absence of market friction such as short-selling constraints, the ability to hedge a derivative security coincides with one's ability to replicate its payoff. The seller of an option assumes the risk of a potential liability at its maturity. In particular, the buyer of a call option is entitled to receive a nonnegative payoff from the seller if the stock price is above a certain threshold. Thus, an increase (decrease) in the stock price increases (decreases) the value of a call option.

The terminal value of a call option can be replicated by buying stock and borrowing from the money market account (temporary cash loan). In particular, there exists a trading strategy involving the stock and money market account for creating a replicating portfolio that mimics the call option's value across time.

Intuitively, including a specific amount of the underlying stock in the portfolio leads to fluctuations in the value of the portfolio that are identical to those of the call option. Therefore, if one sells a call option, one can hedge this possible liability by replicating the option's payoffs to ensure one has the required funds available to pay the buyer. Hence, selling a call option while replicating its payoffs creates a riskless portfolio containing the call option, a certain amount of stock, and the money market (cash) account. The critical amount of stock that needs to be purchased and included in the replicating portfolio is referred to as the option's *delta-hedge* parameter. Similarly, a portfolio consisting of bond and interest rate instruments can be replicated and hence hedge interest rate options.

Ascertaining the trading strategy and the delta-hedge parameter, which replicates a derivative, enables one to price this security using the principle of no-arbitrage. Risk preferences become irrelevant once a risk-less portfolio is created by hedging. Moreover, the initial cost involved in forming a replicating portfolio, which provides identical payoffs as the derivative, must equal the price of the derivative by the principle of no-arbitrage (law of one price).

Consider a security, say a stock, and an option of the stock; at time t let the price of the stock and the option be denoted by $S(t)$ and $C = C(t, S(t))$, respectively. The stock evolves randomly, specified by the following stochastic differential equation

$$\frac{dS(t)}{dt} = \alpha S(t) + \sigma S(t) R(t) \tag{3.10}$$

where α is the drift and σ the volatility of the stock. $R(t)$ is Gaussian white noise that is completely specified by its first two moments[3]

$$E[R(t)] = 0; \quad E[R(t)R(t')] = \delta(t - t') \tag{3.11}$$

White noise, for equal time given by $t = t'$, is singular. From Eq. (A.34), to leading order in ϵ, $R^2(t)$ is *deterministic*; in other words

$$R_n^2 = \frac{1}{\epsilon} + \text{random terms of } O(1) \tag{3.12}$$

White noise is given a path integral formulation in Appendix A.4.

3.9 Delta-hedging

One would like to create a portfolio $\Pi(S, t)$ that is independent of the random fluctuations (changes) in the stock value $S = S(t)$. Since the price of the option C is correlated with the price of the stock, it is natural to try and form a portfolio in which the random changes in the price of the stock are precisely *canceled* out by the random changes in the price of the option. Consider the portfolio

$$\Pi(S(t), t) = C(S(t), t) + \Delta S(t) \tag{3.13}$$

The total instantaneous change in the value of the portfolio, for $\dot{S} = dS/dt$, is given by

$$\frac{d\Pi(S(t), t)}{dt} = \frac{dC(S(t), t)}{dt} + \Delta \dot{S} \tag{3.14}$$

$$\Rightarrow \frac{dC(S(t), t)}{dt} = \frac{1}{\epsilon}[C(S(t) + \epsilon \dot{S}, t + \epsilon) - C(S(t), t)]$$

$$= \frac{\partial C(S(t), t)}{\partial t} + \dot{S}\frac{\partial C(S(t), t)}{\partial S} + \frac{\epsilon}{2}\dot{S}^2\frac{\partial^2 C(S(t), t)}{\partial S^2} + O(\epsilon) \tag{3.15}$$

From Eqs. (3.10), (3.12), and (A.34)

$$\dot{S}^2 = \sigma^2 S^2 R^2(t) + \text{non-singular terms}$$

$$= \frac{\sigma^2}{\epsilon}S^2 + \text{non-singular terms} \tag{3.16}$$

---

[3] The Dirac-delta function $\delta(t - t')$ is reviewed in Appendix A.1 and is essential for understanding the derivations in this book. This appendix is strongly recommended for readers unfamiliar with the Dirac-delta function.

Hence, from Eqs. (3.14), (3.15), and (3.16), one obtains[4]

$$\frac{d\Pi}{dt} = \frac{\partial C}{\partial t} + \left[\frac{\partial C}{\partial S} + \Delta\right]\dot{S} + \frac{\sigma^2}{2}S^2\frac{\partial^2 C}{\partial S^2} \tag{3.17}$$

The only dependence of the portfolio Π on the random changes in the price of the stock S comes from the $\dot{S}$ term in Eq. (3.17) above. Delta-hedging is achieved by choosing Δ such that the coefficient of the $\dot{S}$ term is zero. Hence, one has

$$\Delta = -\frac{\partial C}{\partial S} \quad : \text{ delta-hedging} \tag{3.18}$$

$$\Rightarrow \frac{d\Pi_H}{dt} = \frac{\partial C}{\partial t} + \frac{\sigma^2}{2}S^2\frac{\partial^2 C}{\partial S^2} \tag{3.19}$$

and hence, from Eq. (3.13), the delta-hedged portfolio is given by

$$\Pi_H = C - \frac{\partial C}{\partial S}S \tag{3.20}$$

The hedged portfolio Π_H consists of owning an option C and short-selling $\partial C/\partial S$ amount of stock S.

The change in the portfolio's value $\delta\Pi$ over a small interval ϵ, given in Eq. (3.17), can be written more succinctly as follows

$$\Pi = C + \Delta S; \quad \delta\Pi = \frac{\partial\Pi}{\partial t}\delta t + \frac{\partial\Pi}{\partial S}\delta S$$

$$\text{delta-hedging: } \frac{\partial\Pi}{\partial S} = 0 = \frac{\partial C}{\partial S} + \Delta \tag{3.21}$$

$$\Rightarrow \Delta = -\frac{\partial C}{\partial S}$$

Gamma-hedging of a portfolio is required if variations in S result in big changes in Δ; that is, $|\partial\Delta/\partial S| >> 1$. To create a portfolio Π that is gamma-hedged, one can take an option together with two other instruments and impose the following two conditions

$$\text{delta-hedging: } \frac{\partial\Pi}{\partial S} = 0$$

$$\text{gamma-hedging: } \frac{\partial^2\Pi}{\partial S^2} = 0 \tag{3.22}$$

This creates a portfolio that is delta- and gamma-hedged.

---

[4] The notation is simplified by letting the dependence of Π and C on S and t be implicit.

3.10 Black–Scholes equation

The Black–Scholes equation, which forms one of the pillars of option theory, can be derived using the concept of delta-hedging.

Note that delta-hedging portfolio Π has resulted in removing all the random terms from $d\Pi/dt$. The right-hand side of Eq. (3.19) is completely deterministic – its value being fixed by the price of the stock $S(t)$ and option $C(t, S)$, both of which are known at time t. Delta-hedging has resulted in a *risk-free* portfolio. The principle of no-arbitrage demands that, if the market is to be free from arbitrage opportunities, all risk-free portfolios must give a rate of return that is equal to r, the spot interest rate.[5] Hence, from Eqs. (3.19) and (3.20)

$$\frac{d\Pi_H}{dt} = r\Pi_H \tag{3.23}$$

$$\Rightarrow \frac{\partial C}{\partial t} + \frac{\sigma^2}{2}S^2\frac{\partial^2 C}{\partial S^2} = r\left[C - \frac{\partial C}{\partial S}S\right]$$

$$\Rightarrow \frac{\partial C}{\partial t} + rS\frac{\partial C}{\partial S} + \frac{\sigma^2}{2}S^2\frac{\partial^2 C}{\partial S^2} = rC \quad : \text{ Black–Scholes eq.} \tag{3.24}$$

The specific nature of the option, namely whether it is a call or a put option or a barrier option and so on, is defined by imposing appropriate boundary and final conditions on the option C.

The Black–Scholes equation is independent of the drift term α in Eq. (3.10), which determines the rate of growth of the stock $S(t)$. The value of α depends on the risk propensities of the investor, with one investor for example expecting $S(t)$ to grow and hence taking α to be positive, whereas another investor may take the opposite view and model α as being negative. The equation for the option on $S(t)$, in contrast, is independent of α, which has been replaced by the risk-neutral rate of return r.[6] The reason being that the price of the option must necessarily be risk neutral, since both the buyer and seller of the option must agree on its price, regardless of whether they are risk averse or risk prone.

The discounted portfolio price $\tilde{\Pi}(t) = e^{-rt}\Pi_H(t)$ follows a martingale evolution. To prove this, one has, from Eqs. (3.23) and (A.17), that

$$\frac{d}{dt}\tilde{\Pi}(t) \equiv \frac{d}{dt}\left[e^{-rt}\Pi_H(t)\right] = 0; \quad \Rightarrow E\left[\frac{d\tilde{\Pi}}{dt}\right] = 0$$

$$\Rightarrow E[\tilde{\Pi}(t)] = \tilde{\Pi}(t_0) : \quad \text{martingale}$$

---

[5] The spot rate r is taken to be a constant; the general case of r being a stochastic quantity gives a similar result but needs a more complicated derivation.

[6] Hedging a portfolio consists of removing the $dS(t)/dt$ term, which contains α, as in Eq. (3.18). The singular piece in $S^2(t)$, which contributes to the second derivative term, depends only on σ, the coefficient of white noise $R(t)$.

From Eq. (3.3), the risk-neutral evolution of the discounted stock price $e^{-rt}S(t)$ is a martingale. Since $\Pi = C + \Delta S$, it follows that option price $C(t)$ must also follow a martingale process; the expected future value of $C(t)$, discounted by the money market numeraire $\exp\{rt\}$, is equal to its present value $C(t_0)$. Eq. (3.7), hence, yields for the option price

$$\frac{C(t_0)}{e^{rt_0}} = E\left[\frac{C(t)}{e^{rt}}\right] \quad : \text{martingale}$$

and which can be shown to be equivalent to the Black–Scholes equation given in Eq. (3.24).

3.10.1 Black–Scholes equation for N securities

The Black–Scholes equation can readily be generalized to the case of a derivative that depends on N underlying correlated securities S_i, $i = 1, 2, \ldots, N$, such as the stocks of companies, stock market index, bonds, and so forth.

Eqs. (3.10) and (A.32) have the following generalizations

$$\frac{dS_i(t)}{dt} = \alpha_i S_i(t) + \sigma_i S_i(t) R_i(t); \quad i = 1, 2, \ldots, N \tag{3.25}$$

$R_i(t)$ are N correlated Gaussian white noises, given in Eq. (A.40), and specified by

$$E[R_i(t)] = 0; \quad E[R_i(t)R_j(t')] = \rho_{ij}\delta(t - t')$$

The correlation matrix ρ_{ij} is real and symmetric, with $\rho_{ij}^2 < 1$.

Similar to Eq. (A.34), the equal time product of correlated white noise, to leading order in ϵ becomes *deterministic* and yields, for $t = n\epsilon$

$$R_{ni} R_{nj} = \frac{\rho_{ij}}{\epsilon} + \text{random terms of } O(1) \tag{3.26}$$

Form the hedged portfolio $\Pi(t) \equiv \Pi(S_1, S_2, \ldots, S_N; t)$ from the underlying securities and an option given by $C(t) \equiv C(S_1, S_2, \ldots, S_N; t)$ as follows

$$\Pi = C + \sum_{i=1}^{N} \Delta_i S_i$$

In general, to delta-hedge a portfolio of N-equities $\Pi(S_1, S_2, \ldots, S_N; t)$ requires, similar to Eq. (3.21)

$$\text{delta-hedging}: \quad \frac{\partial \Pi}{\partial S_i} = 0; \quad i = 1, 2, \ldots, N \tag{3.27}$$

$$\Rightarrow \Delta_i = -\frac{\partial C}{\partial S_i}$$

Analogous to Eqs. (3.23) and (3.24), option $C(t)$ satisfies the following equation

$$\frac{d\Pi_H}{dt} = r\Pi_H \tag{3.28}$$

$$\Rightarrow \frac{\partial C}{\partial t} + \frac{1}{2}\sum_{ij=1}^{N}\sigma_i\sigma_j\rho_{ij}S_iS_j\frac{\partial^2 C}{\partial S_i\partial S_j} = r\left[C - \sum_{i=1}^{N}\frac{\partial C}{\partial S_i}S_i\right]$$

$$\Rightarrow \frac{\partial C}{\partial t} + r\sum_{i=1}^{N}S_i\frac{\partial C}{\partial S_i} + \frac{1}{2}\sum_{ij=1}^{N}\sigma_i\sigma_j\rho_{ij}S_iS_j\frac{\partial^2 C}{\partial S_i\partial S_j} = rC \tag{3.29}$$

Similar to the single equity case, for the N-equity case the specific nature of the option C is fixed by its (final value) payoff function and by boundary conditions.

A simple choice for the correlation matrix ρ_{ij} is given by

$$\rho_{ij} = \begin{cases} 1, & i = j \\ \rho, & i \neq j \end{cases} = \delta_{ij}(1-\rho) + \rho; \quad \rho^2 < 1 \tag{3.30}$$

Eq. (3.29) is the Black–Scholes equation for N correlated securities. In Chapter 15 the Black–Scholes equation will be re-cast as a special of Euclidean time Schrodinger equation, driven by the Black–Scholes Hamiltonian H_{BS}.

3.11 Black–Scholes path integral

The Black–Scholes option price C, in principle, can be obtained by solving the partial differential equation given in Eq. (3.29). The option price can also be given an integral closed form expression using the Feynman path integral, which has been discussed in great detail in [12, 15]. In this section, the N-security case is briefly discussed as a preparation for the more complex discussion in Chapter 5 on the quantum field theory of forward interest rates.

The path integral formulation of white noise is discussed in Appendix A.4. From Eq. (A.40), N correlated Gaussian white noises $R_i(t)$ are specified by

$$E[R_i(t)] = 0; \quad E[R_i(t)R_j(t')] = \rho_{ij}\delta(t-t'); \quad 0 \leq t,t' \leq T$$

The path integral, which yields the white noise correlators, from Eq. (A.40) is given by the following

$$E[R_i(t)R_j(t')] = \frac{1}{Z_R}\int DR\, e^S\, R_i(t)R_j(t') = \rho_{ij}\delta(t-t')$$

$$S = \int_0^T dt \mathcal{L}; \quad \mathcal{L} = -\frac{1}{2} \sum_{ij=1}^{N} \rho_{ij}^{-1} R_i(t) R_j(t) \; : \; \text{Lagrangian}$$

$$\int DR = \prod_{t=0}^{T} \prod_{i=1}^{N} \int_{-\infty}^{+\infty} dR_i(t); \quad Z_R = \int DR e^S$$

For defining the path integral in terms of equities S_is, it convenient to change the time variable t to *remaining time* variable given by $\tau = T - t$; the option matures at real time $t = T$ that yields $\tau = 0$; the option at present time $t < T$ is given by a value of $\tau > 0$. For option pricing, all the equities must have a martingale evolution; this in turn fixes the drift term as follows

$$\alpha_i = r \tag{3.31}$$

where r is the risk-free spot interest rate.

In terms of remaining time τ the evolution equation for S_i, from Eq. (3.25), is the following

$$-\frac{dS_i(\tau)}{d\tau} = r S_i(\tau) + \sigma_i S_i(\tau) R_i(\tau); \quad i = 1, 2, \ldots, N \tag{3.32}$$

Since S_is are strictly positive random variables, a change of variables to z_i is defined by $S_i = e^{z_i}$; hence[7]

$$-\frac{dz_i(\tau)}{d\tau} = -\frac{d \ln S_i(\tau)}{d\tau} \equiv -\frac{1}{\epsilon} [\ln S(\tau) - \ln S(\tau - \epsilon)] \tag{3.33}$$

$$= \frac{1}{\epsilon} \left[\ln \left(1 - \epsilon \frac{dS/d\tau}{S} \right) \right] = -\frac{dS/d\tau}{S} - \frac{\epsilon}{2} \left[\frac{dS/d\tau}{S} \right]^2 + O(\epsilon)$$

From Eqs. (3.32) and (3.26)

$$\left[\frac{dS/d\tau}{S} \right]^2 = \sigma_i^2 R_i^2(\tau) = \sigma_i^2 \frac{\rho_{ii}}{\epsilon}$$

Hence, from Eq. (3.32)

$$-\frac{dz_i(\tau)}{d\tau} = r - \frac{1}{2} \rho_{ii} \sigma_i^2 + \sigma_i R_i(\tau); \quad i = 1, 2, \ldots, N$$

In the scheme of stochastic quantization [95] the evolution equation for the N-equities is viewed as a change of variables from R_i to S_i. Imposing the change

---

[7] Consistent with τ being remaining time, the backward finite difference is used for defining $dz(\tau)/d\tau$.

of variables by introducing Dirac-delta functions into the white noise path integral given in Eq. (A.40) yields[8]

$$\int DZ \int DR \prod_{t=0}^{\tau} \prod_{i=1}^{N} \delta\left(\frac{dz_i(t)}{dt} + r - \frac{1}{2}\rho_{ii}\sigma_i^2 + \sigma_i R_i(t)\right) e^S \qquad (3.34)$$

This Dirac-delta function procedure gives the *backward* Fokker–Planck Lagrangian [95]; it has been shown in [12] to be the one appropriate for option pricing since the final value, namely the payoff, has to be propagated backwards in time to obtain the present-day price of the option.

The functional integral over white noise $\int DR$ can be performed exactly and, in effect, in the integrand of the path integral, one has the following identity

$$R_i(\tau) = -\left[\frac{dz_i(\tau)/d\tau + r - \frac{1}{2}\rho_{ii}\sigma_i^2}{\sigma_i}\right] \qquad (3.35)$$

Hence, the Black–Scholes *action* S_{BS}, Lagrangian $\mathcal{L}_{BS}$ and *partition function* Z_{BS} are given by the following[9]

$$Z_{BS} = \int DZ e^{S_{BS}}; \quad S_{BS} = \int_0^{\tau} dt \mathcal{L}_{BS} \qquad (3.36)$$

$$\mathcal{L}_{BS} = -\frac{1}{2} \sum_{ij=1}^{N} \rho_{ij}^{-1} \left[\frac{dz_i(t)/dt + r - \frac{1}{2}\rho_{ii}\sigma_i^2}{\sigma_i}\right]\left[\frac{dz_j(t)/dt + r - \frac{1}{2}\rho_{jj}\sigma_j^2}{\sigma_j}\right]$$

$$\int DZ = \prod_{t=0}^{\tau} \prod_{i=1}^{N} \int_{-\infty}^{+\infty} dz_i(t)$$

The *boundary conditions* on the path integral measure $\int DZ$ are that the initial values $S_i(t = 0)$ are fixed. Since $t = 0$ corresponds to τ, the boundary values $S_i(\tau) = e^{z_i(\tau)}$ are fixed.

In the path integral approach the probability for the security $S(t)$ to take a particular path from t to T is given by $\exp\{S_{BS}\}/Z_{BS}$ and the expectation value of any function of the security $S(\cdot)$ is given by summing over all possible paths, which is

---

[8] All arguments of $z(\tau)$ are always for remaining time. In Eq. (3.34) and in the Lagrangian, the symbol t in term $z(t)$ refers to remaining time and *not* calendar time. This abuse of the use of t is adopted for simplifying the notation for integrations in the action S.

[9] It can be shown that the change of variables has a constant Jacobian, so that $\int DR = \int DZ$ for the case of constant volatility σ_is. The *forward* Fokker–Planck Lagrangian is appropriate for initial value problems and differs from the derivation of the backward Fokker–Planck Lagrangian in that there is a nontrivial Jacobian is going from $\int DR$ to $\int DZ$ [95].

achieved by performing the functional integration $\int DZ$. Hence, the average value of any function of the underlying security, denoted by $\mathcal{O}[z]$ is given by

$$E[\mathcal{O}[z]] = \frac{1}{Z_{BS}} \int DZ e^{S_{BS}} \mathcal{O}[z] \tag{3.37}$$

As realized in Eq. (3.31), option pricing is driven by a theoretical martingale evolution of all underlying securities, which states that the discounted expectation value of the future value of the security is equal to its present. Hence, the expectation in Eq. (3.3) can be written out explicitly using Eq. (3.37) and yields the following

$$e^{-rt} S_i(t) = E[e^{-rT} S_i(T)]$$

$$\Rightarrow e^{z_i(t)} = \frac{e^{-r(T-t)}}{Z_{BS}} \int DZ e^{S_{BS}} e^{z_i(T)} \tag{3.38}$$

In fact, it can be shown that Eq. (3.38) yields $\alpha_i = r$ if one starts without assuming Eq. (3.31).

3.11.1 Equity Lagrangian: stochastic volatility

The special case of one equity is given by taking $\rho_{ij} = \rho_{11} = 1$; the Black–Scholes action and Lagrangian are given by [12]

$$S_{BS} = \int_0^\tau \mathcal{L}_{BS}; \quad \mathcal{L}_{BS} = -\frac{1}{2\sigma^2} \left[\frac{dz(t)}{dt} + r - \frac{1}{2}\sigma^2 \right]^2 \tag{3.39}$$

Let $S = \exp\{z_1\}$ be equity and $\sigma_1^2 = \exp\{z_2\}$ be its stochastic volatility – the two being correlated by ρ. The correlation matrix is given by

$$[\rho]_{ij} = \begin{pmatrix} 1 & \rho \\ \rho & 1 \end{pmatrix}; \quad [\rho^{-1}]_{ij} = \frac{1}{1-\rho^2} \begin{pmatrix} 1 & -\rho \\ -\rho & 1 \end{pmatrix} \tag{3.40}$$

Suppose only equity is traded and stochastic volatility is not. Then, only the drift of the security $\alpha_1 = r$, whereas $\alpha_2 = \mu$ needs to be fixed from the market. Let the volatility of volatility $\sigma_2 = \xi$ be a constant. Hence, from Eq. (3.36), the Merton–Garman Lagrangian for stochastic volatility is given by

$$\mathcal{L}_{MG} = -\frac{1}{2(1-\rho^2)} \left\{ e^{-z_2} \left[\frac{dz_1(t)}{dt} + r - \frac{1}{2}e^{z_2} \right]^2 + \frac{1}{\xi^2} \left[\frac{dz_2(t)}{dt} + \mu - \frac{1}{2}\xi^2 \right]^2 \right\}$$

$$+ \frac{\rho}{\xi(1-\rho^2)} e^{-z_2/2} \left[\frac{dz_1(t)}{dt} + r - \frac{1}{2}e^{z_2} \right] \left[\frac{dz_2(t)}{dt} + \mu - \frac{1}{2}\xi^2 \right]$$

The Jacobian of the transformation from DR to DZ is nontrivial, due to σ_1^2 being stochastic and is given, for an interval $[t_0, t_*]$, as follows

$$DZ = \left[\prod_{t=t_0}^{t_*} e^{z_2(t)/2} \right] DR \equiv e^{z_2/2} DR$$

and yields, for $\tau = t_* - t_0$, the Merton–Garman path integral given by

$$Z_{MG} = \int DZ e^{-z_2/2} \exp\left\{ \int_0^\tau dt \mathcal{L}_{MG} \right\} \tag{3.41}$$

The Merton–Garman path integral given in Eq. (3.41) was obtained in [12, 15] using techniques based on the Hamiltonian.

3.12 Path integration and option price

The path integral formulation of option price is a powerful analytical and computational tool that has been discussed in detail in [12]. The path integral representation of the option price is derived from the Black–Scholes path integral.

Consider a payoff function of an option that matures at calendar time T and is a general function of the security, which could be path dependent or independent, namely

$$\mathcal{P}[K; z(\cdot)] = \mathcal{P}[K_i; z_i(t); \ 0 \le t \le \tau]$$

The price of the call option, $C(t, T; K; S(t)) = C(\tau; K; z(\tau))$, is given in Eq. (3.7) by the martingale condition. The money market numeraire $\exp(rt)$, where r is the spot interest rate taken to be a constant, yields the following

$$\frac{C(t, T; K; S(t))}{\exp(rt)} = E\left[\frac{\mathcal{P}[K; z(\cdot)]}{\exp(rT)} \right] \tag{3.42}$$

The expectation value $E[\ldots]$ is obtained by the path integral as in Eq. (3.37) and gives the following path integral realization of the option price (recall $\tau = T - t$)

$$C(\tau; K; z(\tau)) = \frac{e^{-r\tau}}{Z_{BS}} \int DZ e^{S_{BS}} \mathcal{P}[K; z(\cdot)] \tag{3.43}$$

$$\text{Boundary condition} : z_i(\tau) = z_i \ : \ \text{fixed} \tag{3.44}$$

The path integral measure $\int DZ$ sums the integrand $e^{S_{BS}} \mathcal{P}[K; z(\cdot)]$ over all possible functions (paths) $z_i(t)$; $0 \le t \le \tau$ with the condition that $z_i(\tau)$ has the fixed value of z_i; the value of the $z_i(0)$ is arbitrary (random). The path integral

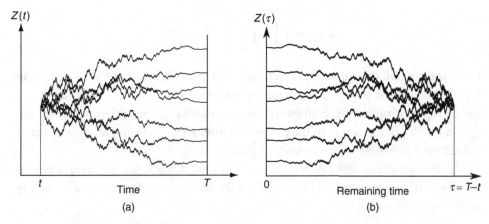

Figure 3.7 Random paths of the security $S = e^z$ evolving forward in calendar time t in Figure (a) and *backward* in remaining time $\tau = T - t$ in Figure (b). The velocity dz/dt is zero for $t = T$, or, equivalently, $dz/d\tau$ is zero for $\tau = 0$; the random paths have been magnified near $t = T$ and $\tau = 0$ to make the boundary condition more transparent.

measure $\int DZ$ has a term equal to $\int_{-\infty}^{+\infty} dz(0)$. A remarkable fact is that letting $z_i(0)$ be an integration (random) variable is equivalent to imposing the boundary condition that $dz_i(0)/d\tau = 0$ is obeyed by all the paths, as shown in Figure 3.7.

All the paths over which the path integration $\int DZ$ is performed obey the following boundary conditions

$$\frac{dz_i(0)}{d\tau} = 0; \quad z_i(\tau) = z_i \tag{3.45}$$

All possible functions $z_i(t); \; 0 < t < \tau$ that satisfy the two boundary conditions as in Eq. (3.45), are given by the following discrete Fourier expansion

$$z_i(t) = z_i + \sum_{n=0}^{\infty} z_{ni} \sin\left(\frac{2n+1}{2}\frac{\pi(t-\tau)}{\tau}\right); \quad 0 \le t \le \tau \tag{3.46}$$

$$-\infty \le z_{ni} \le +\infty; \quad i = 1, 2, \ldots, N; \quad n = 0, 1, 2, 3, \ldots \infty$$

The path integral integration measure factorizes, up to an irrelevant constant that cancels out, into infinitely many integrations, namely $\int DZ = \prod_{n=0}^{\infty} \prod_{i=1}^{N} \int_{-\infty}^{+\infty} dz_{ni}$. This in turn yields the following

$$C(\tau; K; z) = \frac{e^{-r\tau}}{Z_{BS}} \prod_{n=0}^{\infty} \prod_{i=1}^{N} \int_{-\infty}^{+\infty} dz_{ni} e^{S_{BS}} \mathcal{P}[K; z(\cdot)] \tag{3.47}$$

$$Z_{BS} = \prod_{n=0}^{\infty} \prod_{i=1}^{N} \int_{-\infty}^{+\infty} dz_{ni} e^{S_{BS}}$$

Note the remarkable and important fact that the Fourier expansion given in Eq. (3.46) and the expression for the option price given in Eq. (3.47) are valid for all types of options, regardless of the complexity of the action S driving the option pricing or the nature of the payoff function $P[K; z(\cdot)]$. In particular, even if the action S is nonlinear, the Fourier expansion is valid.

The Fourier expansion given in Eq. (3.46) provides a powerful computational tool for developing perturbation expansions for nonlinear systems, such as an equity with stochastic volatility.[10]

3.13 Path integration: European call option

The Black–Scholes path integral for $C(\tau; K; z(\tau))$ is explicitly evaluated to illustrate certain key features of the path integral formulation for the option price. Consider, for the sake of simplicity, a European call option for a single equity with payoff function given by

$$P = \left(S(T) - K\right)_{+} = \left(e^{z(0)} - K\right)_{+} \tag{3.48}$$

The action for the single equity, for $\alpha \equiv r - \sigma^2/2$ and from Eq. (3.39), is given by

$$S_{BS} = -\frac{1}{2\sigma^2} \int_{0}^{\tau} dt \left(\frac{dz}{dt} + \alpha\right)^2 \tag{3.49}$$

The path integral expression for the option price, from Eqs. (3.43) and (3.48), is given by

$$C(\tau; K; z) = \frac{e^{-r\tau}}{Z_{BS}} \int DZ e^{S_{BS}} \left(e^{z(0)} - K\right)_{+}$$

Boundary condition : $z(\tau) = z$: fixed $\tag{3.50}$

Consider the orthogonality relation

$$I = \int_{0}^{\tau} dt \cos\left[\frac{(2n+1)\pi t}{2\tau}\right] \cos\left[\frac{(2m+1)\pi t}{2\tau}\right] \tag{3.51}$$

$$= \frac{\tau}{2}\delta_{n-m}; \quad m, n = 0, 1, 2, \ldots \infty$$

---

[10] *All* option pricing path integrals, including nonlinear cases like stochastic volatility, have a Fourier expansion. In contrast, for nonlinear path integrals in physics, the quadratic part of the action has to be analyzed for its eigenfunctions and only then can an eigenfunction expansion – of which the Fourier expansion is a special case – be defined for the path integral [95].

Eqs. (3.46) and (3.51) yield the following for the Black–Scholes action

$$S_{BS} = -\frac{1}{2\sigma^2} \int_0^\tau dt \left(\left[\frac{dz}{dt}\right]^2 + 2\alpha \frac{dz}{dt} + \alpha^2 \right)$$

$$= -\frac{1}{2\sigma^2} \sum_{n=0}^\infty \left(\frac{\pi}{\tau}\right)^2 \left(\frac{2n+1}{2}\right)^2 \cdot \frac{\tau}{2} z_n^2 - \frac{\alpha}{\sigma^2}[z - z(0)] - \frac{\tau\alpha^2}{2\sigma^2} \quad (3.52)$$

The payoff function is re-written using the Dirac-delta function's integral representation given in Eq. (A.7) and yields

$$C(\tau; K; z) = \frac{e^{-r\tau}}{Z} \int_{-\infty}^{+\infty} \frac{d\eta}{2\pi} \int_{-\infty}^{+\infty} dw \int DZ e^{S_{BS}+i\eta[z(0)-w]}(e^w - K)_+$$

$$= e^{-r\tau} \int_{-\infty}^{+\infty} \frac{d\eta}{2\pi} \int_{-\infty}^{+\infty} dw e^{F(\eta,w)}(e^w - K)_+ \quad (3.53)$$

where $e^{F(\eta,w)} = \frac{1}{Z} \int DZ \, e^{S_{eff}}$ \quad (3.54)

Since $\sin[(2n+1)\pi/2] = (-1)^n$, the value of $z(0)$ is given by Eq. (3.46) as

$$z(0) = z + \sum_{n=0}^\infty (-1)^n z_n$$

The effective action, from Eqs. (3.52), (3.53), and (3.54), is given by

$$S_{eff} = S_{BS} + i\eta[z(0) - w]$$

$$= \sum_{n=0}^\infty S_n - \frac{\alpha^2\tau}{2\sigma^2} + i\eta(z - w)$$

$$S_n = -\frac{1}{2\sigma^2\tau} \cdot \frac{\pi^2}{8}(2n+1)^2 z_n^2 + (-1)^n z_n \left(\frac{\alpha}{\sigma^2} + i\eta\right)$$

The effective action S_{eff} has completely factorized into terms that depend on only one n. The path integral is, hence, given by

$$\frac{1}{Z_{BS}} \prod_{n=0}^\infty \left[\int_{-\infty}^{+\infty} dz_n e^{S_n} \right] = \prod_{n=0}^\infty e^{F_n}$$

The path integral is a product of infinitely many independent Gaussian integrations and, from Eqs. (A.21) and (3.54), yields the result

$$F(\eta, w) = \sum_{n=0}^{\infty} F_n - \frac{\alpha^2 \tau}{2\sigma^2} + i\eta(z - w)$$

$$= \frac{\sigma^2 \tau}{2} \frac{8}{\pi^2} \left(\frac{\alpha}{\sigma^2} + i\eta \right)^2 \sum_{n=0}^{\infty} \frac{1}{(2n + 1)^2} - \frac{\alpha^2 \tau}{2\sigma^2} + i\eta(z - w) \quad (3.55)$$

Since

$$\sum_{n=0}^{\infty} \frac{1}{(2n + 1)^2} = \frac{\pi^2}{8} \quad (3.56)$$

one has the following simplification

$$F(\eta, w) = -\frac{\sigma^2 \tau}{2} \eta^2 + i\eta(z + \tau\alpha - w) \quad (3.57)$$

Performing the remaining η integration in Eq. (3.53) using Eq. (3.57) gives the famous Black–Scholes result for the European call option

$$C = \frac{e^{-r\tau}}{\sqrt{2\pi\sigma^2\tau}} \int_{-\infty}^{+\infty} dw \, e^{-\frac{1}{2\sigma^2\tau}(z+\tau\alpha-w)^2} (e^w - K)_+ \quad (3.58)$$

$$= SN(d_+) - e^{-r(T-t)} K N(d_-) \quad (3.59)$$

Recall $\alpha = r - \sigma^2/2$, $\tau = T - t$ and

$$d_\pm = \frac{\ln\left(\frac{S}{K}\right) + \left(r \pm \frac{\sigma^2}{2}\right)(T - t)}{\sigma\sqrt{T - t}}; \quad S = e^z \quad (3.60)$$

The cumulative distribution for the normal random variable $N(x)$ is defined by

$$N(x) = \frac{1}{\sqrt{2\pi}} \int_{-\infty}^{x} e^{-\frac{1}{2}z^2} dz \quad (3.61)$$

3.14 Option price: volatility expansion

For all financial instruments, intrinsic volatility is a small parameter. This is especially true for interest rate and bond options, with the market forward interest rate volatility being typically of the order of 10^{-2}/year. One can, hence, define a generic and rapidly convergent expansion of the option price in powers of the volatility.

Consider an equity, interest rate, or coupon bond option that has a payoff function $\mathcal{P}$, which matures at time t_* with strike price K and is given by

$$\mathcal{P} = [S(t_*) - K]_+$$

$S(t_*)$ is the financial instrument on which the option is written. The price of the option at earlier time t_0, using the forward bond numeraire, is given by Eq. (3.8) as

$$C(t_0, t_*, K) = B(t_0, t_*) E[S(t_*) - K]_+$$

Let $S(t_0)$ be the price of the instrument at time t_0: one expects that $S(t_*) - S(t_0)$, up to factors depending on drift, is of the order of volatility since all fluctuations away from the initial value are due to nonzero volatility. One has the following

$$C(t_0, t_*, K) = B(t_0, t_*) E[S(t_*) - K]_+$$

$$= B(t_0, t_*) E[V - \tilde{K}]_+$$

$$V = S(t_*) - S(t_0); \quad \tilde{K} = K - S(t_0) \qquad (3.62)$$

The (random) quantity $V = S(t_*) - S(t_0)$, up to factors of drift that will be accounted for, has an order of magnitude value equal to $O(\sigma)$, the volatility of $S(t)$. The option price will be obtained in powers of V, which in turn, after the expectation value is taken, will lead to the option price as a power series in σ.

Using the representation of the Dirac-delta function given in Eq. (A.7)

$$\delta(Q) = \frac{1}{2\pi} \int_{-\infty}^{+\infty} d\eta e^{i\eta Q} \qquad (3.63)$$

yields the following expression for the payoff function

$$\mathcal{P} = \left(S(t_*) - K\right)_+$$

$$= \int_{-\infty}^{+\infty} dQ \delta(V - Q)(Q - \tilde{K})_+$$

$$= \int_{-\infty}^{+\infty} dQ \frac{d\eta}{2\pi} e^{i\eta(V-Q)} (Q - \tilde{K})_+ \equiv \int_{Q,\eta} e^{i\eta(V-Q)} (Q - \tilde{K})_+$$

$$\simeq \int_{Q,\eta} e^{-i\eta Q} (Q - \tilde{K})_+ \left[1 + i\eta V + \frac{i^2}{2} \eta^2 V^2 \cdots + \frac{i^n}{n!} \eta^n V^n + O(\sigma^{n+1}) \right]$$

$$(3.64)$$

Note V is the only random quantity in the payoff function and is an effective 'potential' for option pricing. One has the following expansion for the option price

$$\frac{C(t_0, t_*, T, K)}{B(t_0, t_*)} = E[\mathcal{P}] \tag{3.65}$$

$$\simeq \int_{Q,\eta} e^{-i\eta Q} (Q - \tilde{K})_+ \left(E[1] + i\eta E[V] + \frac{i^2}{2} \eta^2 E[V^2] \cdots + \frac{i^n}{n!} \eta^n E[V^{(n)}] \right)$$

$$\simeq \int_{Q,\eta} e^{-i\eta Q} (Q - \tilde{K})_+ \left(C_0 + i\eta C_1 - \frac{1}{2} \eta^2 C_2 + \cdots + \frac{i^n}{n!} \eta^n C_n + O(\sigma^{n+1}) \right)$$

To evaluate the option price to an accuracy of $O(\sigma^n)$, one needs to evaluate the coefficients

$$C_0 = E[1] \tag{3.66}$$

$$C_1 = E[V] : D_1 = C_1/C_0 \tag{3.67}$$

$$C_2 = E[V^2] : D_2 = C_2/C_0 \tag{3.68}$$

$$\ldots C_n = E[V^n] : D_n = C_n/C_0$$

For most cases, the option price is obtained by evaluating the coefficients C_0, C_1 and C_2. Eq. (3.65) yields a *cumulant expansion* that, to second order and for $D_1 = C_1/C_0$ and $D_2 = C_2/C_0$, gives the following approximate option price

$$\frac{C(t_0, t_*, T, K)}{B(t_0, t_*)} \simeq C_0 \int_{Q,\eta} e^{-i\eta Q} (Q - \tilde{K})_+ \exp\left\{ i\eta D_1 - \frac{1}{2} \eta^2 (D_2 - D_1^2) \right\}$$

$$= C_0 \int_{Q,\eta} e^{-i\eta Q} (Q + D_1 - \tilde{K})_+ \exp\left\{ -\frac{1}{2} \eta^2 (D_2 - D_1^2) \right\}$$

$$\Rightarrow \frac{C(t_0, t_*, T, K)}{B(t_0, t_*)} = \frac{1}{\sqrt{2\pi}} C_0 I(X) \sqrt{D_2 - D_1^2} + O(\sigma^3) \tag{3.69}$$

The function $I(X)$ is given in terms of the error function $N(u)$ as follows

$$X = \frac{\tilde{K} - D_1}{\sqrt{D_2 - D_1^2}}; \quad \tilde{K} = K - S(t_0)$$

$$I(X) = \int_{-\infty}^{+\infty} dQ (Q - X)_+ e^{-\frac{1}{2}Q^2}$$

$$= e^{-\frac{1}{2}X^2} + \sqrt{2\pi} X (N(X) - 1); \quad N(u) = \frac{1}{\sqrt{2\pi}} \int_{-\infty}^{u} dQ e^{-\frac{1}{2}Q^2} \tag{3.70}$$

The asymptotic behavior of the error function $N(u)$ yields the following limits

$$I(X) = \begin{cases} 1 + O(X^2) & X \approx 0 \\ e^{-\frac{1}{2}X^2} + O(e^{-X^2}) & X >> 0 \end{cases} \tag{3.71}$$

The option's price, for $X \approx 0$, is the following

$$C(t_0, t_*, K) \approx B(t_0, t_*)C_0 \left[\frac{1}{\sqrt{2\pi}} \sqrt{D_2 - D_1^2} - \frac{1}{2}(\tilde{K} - D_1) \right] + O(X^2) \tag{3.72}$$

The coefficients that determine the option price $C(t_0, t_*, K)$ have the following intuitive interpretation.

- The coefficient $C_0 = E[1]$ is a measure of the paths that contribute to the option. For the barrier option, C_0 is a function of the barrier and has a nontrivial value.
- The coefficient $D_1 = C_1/C_0$ with $C_1 = E[V]$ the expectation value of the effective potential V, normalized by the allowed paths.
- The coefficient $D_2 = E[V^2]/C_0$ appears in the option price through the combination $D_2 - D_1^2$ and is a measure of the standard deviation of V.

The volatility expansion is of far-reaching significance since it will be used in later chapters to find the approximate option price for a variety of cases, including the European, Asian, and barrier options for coupon bonds and interest rate swaptions.

The volatility of a security, be it a stock price or a forward interest rate, is usually a small quantity. The volatility expansion defines the price of the option as a power series in the volatility of the security. This expansion requires the evaluation of higher and higher moments of a modified form of the payoff function, showing in an explicit manner how the payoff function determines the option price.

3.15 Derivatives and the real economy

Financial derivatives, which include options, have vastly expanded the domain of finance and created many new opportunities for return on capital and economic growth. Derivatives play a key role in optimizing the utilization of assets, minimizing risks and maximizing returns on investments.

There is an old adage which says that 'one should do what one is good at', and an equally well-known proverb states that 'don't put all your eggs in one basket'. In the language of finance, there seems to be a conflict between pursuing a line of economic activity in which one has a comparative advantage versus efficient diversification. The adage points to concentrating, as an individual or as a country, on doing what one is good at – on doing things where one has a comparative advantage. The proverb, on the other hand, asks one not to overspecialize and, instead, to efficiently diversify one's risks. These two imperatives, to specialize and to diversify, seem to be irreconcilable. Surprisingly enough, derivative instruments can reconcile these apparently contradictory positions.

A *specific* financial derivative, namely a swap, where one exchanges two income streams, can resolve these apparently conflicting economic priorities.[11] A country producing only one product, say garments, can enter a swap with one arm of the swap being payments at the world average rate from the garments industry, say G, and the other arm of the swap being payments, W, at the world average return on all industries. This swap ensures that the country retains its comparative advantage while efficiently diversifying.

Since it has comparative advantage, the garment producer has a rate of return higher than the world average on garments, say $G + \Delta$. By entering the swap, the garment producer exchanges payments G for W and thus ends up with a return of $W + \Delta$. Hence, the garment producer gets a minimum return at the world average rate of return W plus Δ; the garment producer has diversified the risk of a global downturn in garments while retaining the comparative advantage Δ.

A swap derivative is noninvasive and needs no permanent construction or facilities; rather, it is a reversible financial arrangement that can be switched on and off without interfering with the real economy. Swaps, properly employed, add substantial amounts of value to the underlying real economy. It has been shown by empirical studies that if developing countries enter the swap discussed above, they would increase their income from exports by 60% every year; historical data show that the increase in income would have been about 500% over the last 30 years [80]. This example shows that the effect of financial derivatives on the returns on economic activity can be large, of 'first' order, and not just yield marginal gains.

Derivatives facilitate risk diversification and mediate the efficient *transfer of risk*; in the example of a country producing only garments, it spreads the country's risk to the rest of the world. In general, a financial instrument has many forms of risk that are inherent in it, such as credit risk, liquidity risk, foreign exchange risk, interest rate risk, and so on. A particular financial institution may be best prepared to handle a specific form of risk, such as credit risk. In this case, derivative instruments provide the means for un-bundling, factoring out, and hedging other forms of risk that the institution is not fully prepared for, and leaves it to focus on the forms of risk where expertise gives it a comparative advantage. In general, derivatives allow practitioners to hive off all forms of risks that are inherent in a financial

---

[11] Interest rate swaps are discussed in Section 4.2.

instrument – but extraneous to their expertise – and leave them to focus on managing those forms of risk at which they can specialize.

With the advent of derivatives, the traditional role of banks of specializing in the borrowing and lending of money is being transformed. Derivatives can hive off the multifarious risks inherent in a financial contract and the bank can hedge away all risks except the one it is prepared to manage. For example, a bank can concentrate on credit risk for a loan and hedge away other risks, such as liquidity risk, accident risk, interest rate risks, and so on. The role of banks in the twenty-first century will be that of risk management, and, in particular, adding value to the economy by their expertise in handling various forms of specialized risk.

The role of derivatives in risk management is essential, since the very concept of hedging one's risk can only be realized if one has derivative instruments that are negatively correlated with the asset that is being hedged. There have been a lot of criticisms on the role of derivatives in contributing to a financial crisis. These criticisms are off the mark. An analysis of the US subprime crisis shows, as discussed in Appendix B.2, that derivatives by no means create a financial crisis; rather, it is the flawed policies and priorities of financial institutions that are at the root of all financial crises.

An analogy of the effect of derivatives on the real economy is the invention of the safety belt for automobiles. Clearly, all things remaining constant, the use of seat belts should lower the risk of accident injuries. However, paradoxically, statistics showed that accident injuries *increased* after the introduction of seat belts. The reason, it was found, is that instead of driving at the speeds that were considered to be safe before the introduction of seat belts, drivers were now driving at higher speeds with the resulting higher accident rates and hence more injuries.

The case of derivatives is similar; they can be used either for hedging and hence reducing risk or for taking higher risks with the purpose of reaping greater profits. One can buy a call option on a stock to hedge a portfolio, or one can buy the call option to speculate on the future price of the stock. The same call option allows for both possibilities. Since the price of a call option is only about 3% to 5% of the stock's price, a speculator enjoys tremendous leverage by buying an option on the stock's future value. A call option is bought by a speculator solely for making profit. In contrast, an informed trader has a fixed future purchase in mind, such as procuring a thousand tons of steel a year in the future, and enters a forward contract based on a real expected future transaction.

Derivatives expand the financial sphere since they provide instruments that can be used for: (a) hedging to reduce risks, or (b) for deriving benefits that come with taking higher risks. Since the drive for profits is relentless, one can expect that economic agents will try and maximize their returns and employ derivatives both for speculation as well as for reducing risk, depending on the circumstances. Hence, derivatives cannot be said to be creating a financial system with lower risks. Instead, by increasing the reach and domain of finance, derivatives create new opportunities for maximizing the efficient utilization of financial and economic resources and assets.

3.16 Summary

Derivatives, including options, have revolutionized mathematical and practical finance. The 1973 seminal paper of F. Black and M. Scholes has created, within a few decades, what is literally a multi-trillion dollar derivatives market. The concept of hedging is central to pricing options; if one can perfectly hedge a derivative instrument, then the price of that derivative should be equal to the cost of hedging it. From this rather deeply intuitive idea arises a natural question, namely: how does one mathematically define perfect hedging? Modeling the future evolution of the underlying security by a stochastic process leads to the following answer: a portfolio, consisting of the instrument and its derivative, is perfectly hedged if it has a deterministic temporal evolution.

One idea was still missing in this line of reasoning. What is the (notional) evolution of the underlying security such that the price of its derivative is free from arbitrage opportunities? It was shown that, for a 'complete' and efficient market, derivative pricing is arbitrage free if the underlying security evolves by a martingale process. The concept of a martingale brought a natural closure to the line of reasoning initiated by Black and Scholes. The mathematical machinery of probability theory could be brought to bear on option pricing and greatly enriched the subject of option theory.

Option theory is far from complete and its final form may take decades to emerge, if at all. There are many open questions regarding option theory, primary amongst these being whether a stochastic process can accurately describe the market evolution of securities, whether the concepts of a complete and efficient market and that of market equilibrium have any empirical support from the capital markets and so on [35, 87].

Notwithstanding the limitations of option theory, one thing is already clear: the truly staggering multi-trillion dollar derivatives market can best be described and understood by quantitative and mathematical models. Option theory, and derivatives in general, provides a fertile ground for the application of mathematics to finance and one can only expect that this trend will continue to grow.

4

Interest rate and coupon bond options

Options on interest rates and coupon bonds share many general properties with equity options, but are far more complex and have a much richer internal structure [34, 83, 84]. There is a great variety of interest rate derivatives [60], which comprise over 50% of the total derivatives' markets. Options on interest rates are primarily based on interest rate caps and interest rate swaptions. Coupon bonds and options are defined and it is shown that swaptions are a special case of coupon bond options. Various put–call parity relations are derived for interest rates and coupon bond options.

The HJM (Heath–Jarrow–Morton) model of interest rates is based on stochastic calculus and is briefly discussed using a path integral formulation. The HJM model serves as an example for demonstrating the point of departure of the quantum finance formulation of interest rates from the one based on stochastic calculus.

4.1 Introduction

Coupon bond options and interest derivatives comprise a major subfield of finance. To convey some of the key features of this subfield, the US credit derivatives' market – being globally the largest – is briefly reviewed. Figure 4.1(a) gives the notional value of outstanding credit derivatives and Figure 4.1(b) gives a break-down of the diverse variety of swap derivatives most frequently used in the US capital markets. Since 2001, the global credit derivatives' market had grown at a phenomenal annual rate of over 100%; from relatively insignificant beginnings at the turn of the new millennium, by 2006 the notional value of credit derivatives had reached US$26 trillion.

Interest rate swaps are the largest component of the credit derivatives' markets. The Bank of International Settlements (Switzerland) estimates that in 2001 the notional value of the swap market was approximately US$40 trillion and that of the combined interest rate caps' and swaptions' market was about US$9 trillion dollars.

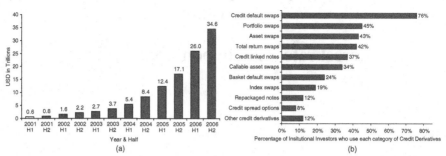

Figure 4.1 (a) US credit derivatives' semi-annual outstanding notional value, from 2001 to 2006. (b) Swap derivatives' usage frequency by North American institutional investors in 2007.

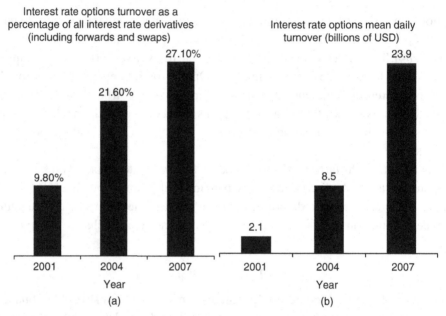

Figure 4.2 (a) Interest rate options' growing share of the global interest rate derivatives' market. (b) The notional value of the daily global turnover of interest rate options.

Figure 4.2(a) shows the rapidly growing importance of interest rate options as a fraction of the total credit derivatives' market; Figure 4.2(b) shows that the daily global turnover of interest rate options has reached close to US$24 billion. One can expect interest rate options to continue to grow and, with this growth, the need for financial engineering techniques for pricing and hedging these instruments.

Interest rate derivatives cover a vast range of financial instruments, from vanilla options, swaps and swaptions defined on underlying interest rate instruments to exotic and hybrid options defined on coupon and zero coupon bonds [36]. The three main forms of interest rate derivatives are swaps, forwards, and options. Interest rate swaps are similar to forward and futures contracts and options consist mostly of interest rate swaptions, caps, floors, and collars. Options on coupon bonds are closely related to interest rate options and are derived from the same underlying interest rates that, in general, drive the entire debt market.

The following derivatives are reviewed [33]:

- Interest rate swaps.
- Interest rate caps, floors, and collars.
- European, American, and barrier coupon bond options.
- Interest rate swaptions.

The general properties of interest rate derivatives' and coupon bond options are reviewed with the purpose of providing the background material for later chapters. The mathematical expressions for interest rate options are defined; model independent features of these options are briefly discussed with particular emphasis on put–call parity that is obeyed by these instruments.

4.2 Interest rate swaps

Interest rate swaps exchange two streams of cash flows. Swaps have many functions, amongst which is to transform the nature of financial liability or assets of a company. One primary utility of swaps is to manage interest rate risks. For example, if a company is assured of a fixed stream of regular payments, it may want to convert this into floating payments using a swap and so on [65].

Interest rate swaps are reviewed and form the basis for analyzing interest rate caps, floors, and options on interest rate swaps.

An interest swap, shown in Figure 4.3 is contracted between two parties in which one party pays at a fixed interest rate and the other pays at a floating interest rate, which is usually taken to be the prevailing three-month Libor rate. For a *floating rate receiver's swap*, namely swap$_L$, the first party receives interest payments, on the notional principal, at the floating interest rate and pays at a fixed interest

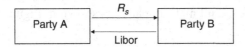

Figure 4.3 Diagram representing a swap in which party A, holder of swap$_L$ receives floating payments fixed by Libor and pays fixed interest rate R_S and party B holder of swap$_R$ recieves payments at fixed interest rate R_S and pays at floating Libor interest rate.

rate R_S. A *fixed rate receiver's swap*, namely swap$_R$, is where the first party receives payments, on the notional principal, at fixed interest rate R_S and pays at the floating Libor rate. Parties entering the swap pay only the net interest due on the notional amount and do not pay or receive the principal amount.[1]

Both swap$_L$ and swap$_R$ are obligatory contracts and hence are not options, but rather should be thought of as forward contracts on interest rates.

Consider a swap defined for fixed and floating interest rates. The simplest forward or deferred swap, called a forward *swaplet*, is entered at time t; the contract has a notional principal ℓV, to be kept in a fixed time deposit from future time T to $T + \ell$, with $\ell = 90$ days. A fixed interest rate R_S is agreed upon; the floating interest rate is taken to be Libor $L(t, T)$. The value of the floating rate receiver swaplet is given by the net outstanding interest difference between the floating and fixed interest payments, that is $L(t_0, T) - R_S$, to be paid out by one of the parties to the other. At time $t = T + \ell$, a payment is made on the principal amount at a rate equal to $L(T, T) - R_S$ and swaplet$_L$ expires.[2]

The value, at time t_0, of a *deferred* or *forward* floating rate receiver swaplet$_L$ is the discounted value of the cash flow at time $T + \ell$. Hence

$$swaplet_L(t_0, T) = \ell V B(t_0, T + \ell)\big[L(t_0, T) - R_S\big] \qquad (4.1)$$

Note the floating rate is fixed by the benchmark three-month Libor $L(t_0, T)$; the bonds discounting the payoff, namely $B(t_0, T + \ell)$, strictly speaking should be obtained from the Libor zero coupon yield curve as given in Eq. (2.27). The TED (Treasury Eurodollar), discussed in Section 2.13, addresses the spread between the zero coupon Treasury Bonds and the Libor zero coupon bonds. The difference between these two zero coupon bonds will be addressed only if necessary.

An interest swap over a longer duration is the sum of individual swaplets with fixed interest rate R_S and notional principal V, which for simplicity is taken to be the same for all the swaplets. The payment dates for the swap coincide with the periods defined by Libor time. Payments are made at fixed intervals, usually $\ell = 90$ or 180 days. The swap has a pre-fixed total duration, starting at time T_0 and with last payment at T_N.

A midcurve forward swap is entered at time t_0 and expires at time t_* before the swap becomes operational at time T_0 and is shown in Figure 4.4(a). A forward swap, entered at time t_0 and maturing at T_0 is shown in Figure 4.4(b).[3]

---

[1] A floating rate payer is equal to the fixed rate receiver's swap, swap$_R$, and a fixed rate payer's swap is equal to a floating rate receiver's swap, swap$_L$. Hence, only the two kinds of receiver's swaps will be discussed as this covers the payer's swaps as well.

[2] From the definition of Libor, $L(t, T) = L(t, T, T + \ell)$ with payments made at calendar time $t = T + \ell$.

[3] A swap that is entered into after the time of the initial payments, that is at time $t_0 > T_0$, can also be priced and is given in [65]; however, for the case of a swaption, this case is not relevant, and will not be discussed.

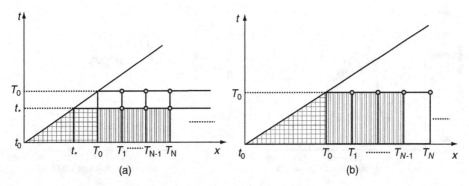

Figure 4.4 The circles signify payment dates; the first payment is at T_1 and the last payment is at T_N. The shaded area inside the rectangles indicate the set of forward interest rates that determine the price of a swap. (a) A midcurve forward swap is entered into at time t_0 and exercised at time t_*, before the interest rate swap becomes operational at time T_0. (b) A forward swap is entered into at time t_0 and exercised at time T_0, when the interest rate swap becomes operational.

Only forward swaps will be considered and the results can be readily extended to midcurve swaps. The forward swap contracted at time t_0, starting at T_0 and ending at T_N is denoted in the market as x by y; which means that the swap matures at $x = T_0 - t_0$ years in the future and the interest payments continue for $y = T_N$ years after the swap matures; hence, the total duration of the swap is $x + y = T_N + T_0 - t_0$ years.

To quantify the value of the swap, let the swap start at Libor time T_0, with payments made at fixed times $T_n = T_0 + n\ell$, with $n = 1, 2, \ldots, N$; the first payment is made at T_1 and the last payment is made at time T_N. In summary, at time t_0, the values of the forward swaplets – corresponding to the interest rate payments made at future times T_n – yield the following forward price for the floating and fixed rate receiver swaps

$$\text{swap}_L(t_0, R_S) = \ell V \sum_{n=0}^{N-1} B(t_0, T_n + \ell)\big[L(t_0, T_n) - R_S\big] \qquad (4.2)$$

$$\text{swap}_R(t_0, R_S) = \ell V \sum_{n=0}^{N-1} B(t_0, T_n + \ell)\big[R_S - L(t_0, T_n)\big] \qquad (4.3)$$

$$\text{swap}_L(t_0, R_S) + \text{swap}_R(t_0, R_S) = 0 \qquad (4.4)$$

Eq. (4.4) shows that a swap contract is a zero sum game, with the gain of one party being exactly equal to the loss of the other party. The value of a swap is taken

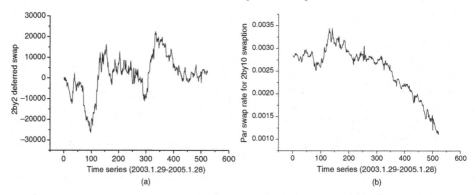

Figure 4.5 (a) Daily market value of a deferred swap$_L(t_0, T)$ on a notional principal of \$1 million with floating taken to be Libor and fixed interest rate $R_S = 2.8\%$. The swap is Libor 2by2: the swap matures two years in the future, that is $T - t_0 = 2$ years and runs for another two years. The market values are given for $T - t_0 \in [29.1.2003{-}28.1.2005]$. (b) Figure shows the time variation of $R_P(t_0)$, the par value for the fixed yearly interest payments, for a 2by10 swap$_L$, with $T - t_0 \in [29.1.2003{-}28.1.2005]$.

to be the difference between the floating and fixed interest rate receiver swaps and is given by

$$V_{\text{swap}} = \text{swap}_L - \text{swap}_R$$

$$= 2\,\text{swap}_L$$

where Eq. (4.4) yields the last line.

The swaplets combine together to form a swap contract that consists of a portfolio of swaplets. A swap that is initiated at time t_0 and runs from T_0 to T_N and is shown in Figure 4.4(b).

In contracts between parties with equally good credit ratings, the value of the swap for both parties, receiving floating or fixed payments, must have equal value.[4] Note from Eq. (4.4) that swap$_L = -$swap$_R$; hence, when the swap contract is initiated, both forward swaps are equal to zero.

It is important to note that although a forward swap contract starts at time t_0 with zero value, swap(t, T_0) has nonzero values during the time interval $t \in [t_0, T_0]$ – that is, until it matures at time T_0. The market value of a swap$_L$ for the Libor market is shown in Figure 4.5(a); the swap matures in two years and payments continue for another two years; the fixed interest rate is $R_S = 2.8\%$.

---

[4] Since, otherwise, the party with the unfavorable price will not enter into the swap contract.

One can simplify the expression for the swaps. The Libor zero coupon yield curve represents Libor in terms of Libor zero coupon bonds, as in Eq. (2.21)

$$L(t,T) = \frac{1}{\ell}\frac{B(t,T) - B(t,T+\ell)}{B(t,T+\ell)}$$

and yields the following

$$\ell V \sum_{n=0}^{N-1} B(t_0, T_n + \ell)L(t_0, T_n) = V \sum_{n=0}^{N-1}\left[B(t_0, T_n) - B(t_0, T_n + \ell)\right]$$

$$= V\left[B(t_0, T_0) - B(t_0, T_N)\right]$$

Hence, from Eq. (4.2)

$$\text{swap}_L(t_0, R_S) = V\left[B(t_0, T_0) - B(t_0, T_N) - \ell R_S \sum_{n=0}^{N-1} B(t_0, T_n + \ell)\right] \quad (4.5)$$

with a similar expression for swap_R.

The floating and receiver swaps are equal for the par value of the fixed rate and this in effect defines the *par value* of the fixed interest rate at time t_0, namely $R_P(t_0)$. Hence

$$\text{swap}_L(t_0, R_P(t_0)) = 0 = \text{swap}_R(t_0, R_P(t_0))$$

$$\Rightarrow \ell R_P(t_0) = \frac{B(t_0, T_0) - B(t_0, T_N)}{\sum_{n=0}^{N-1} B(t_0, T_n + \ell)} = \frac{1 - F(t_0, T_0, T_N)}{\sum_{n=0}^{N-1} F(t_0, T_0, T_n + \ell)} \quad (4.6)$$

where the forward bond prices are given by

$$F(t_0, T_0, T_n + \ell) = \exp\left\{-\int_{T_0}^{T_n+\ell} dx f(t_0, x)\right\} \quad (4.7)$$

The above result shows that the par value $R_P(t_0)$ for the forward swap is fixed by the forward Libor bond prices $F(t_0, T_0, T_n + \ell)$.

The par interest rate $R_P(t)$ changes during the duration of the swap. The empirical value of $R_P(t)$ for a 2by10 swap, at a fixed yearly interest rate of $R_S = 2.6\%$, is given in Figure 4.5(b).

The special case of $t_0 = T_0$, which is the price of a swap on the day that the swap becomes operational, is particularly important as these swaps are widely traded and quoted in the capital markets. Since $B(T_0, T_0) = 1$, the values of the swaps are given by [65]

$$\text{swap}_L(T_0, R_S) = V \left[1 - B(T_0, T_N) - \ell R_S \sum_{n=0}^{N-1} B(T_0, T_n + \ell) \right]$$

$$\text{swap}_R(T_0, R_S) = V \left[\ell R_S \sum_{n=0}^{N-1} B(T_0, T_n + \ell) + B(t, T_N) - 1 \right] \qquad (4.8)$$

with par fixed interest rate at time T_0 given by

$$\ell R_P = \frac{1 - B(T_0, T_N)}{\sum_{n=0}^{N-1} B(T_0, T_n + \ell)} \qquad (4.9)$$

4.3 Interest rate caps and floors

Libor have derivatives written on them, such as caps and floors; these instruments are important interest rate derivatives and have many applications in the financial markets [65]. Interest rate contracts, such as caps and floors, can cover many years and involve a sequence of quarterly payments ranging from one to ten years. Consequently, pricing and hedging such derivatives require modeling of Libor over a long interval of time.

Caps, floors, and collars are interest rate options that are widely used for hedging interest rate risks. Financial companies sometimes have to enter into financial contracts in which they pay or receive cash flows tied to some floating rate, such as Libor. In order to hedge the risks caused by the Libor's variability, participants often enter into options' contracts that guarantee interest payments with a fixed upper limit or lower limit, called a cap or floor respectively. An interest rate collar places a minimum and a maximum limit on the floating rate payments and is a combination of a cap and a floor. Caps and floors can also be used for speculating on the future movements of the interest rates.

One of the most elementary form of an interest rate cap is a caplet, which is an option that puts a maximum upper limit to the floating interest rate that the holder of the caplet will pay for some pre-specified future interval, usually one Libor period $[T, T + \ell]$. Figure 4.6(a) is a graphical representation of an interest rate caplet that limits the maximum floating interest rate to K for $[T, T + \ell]$.

An interest cap of an arbitrarily long duration is composed of a portfolio of caplets, where the caplets can have different strike prices and maturity dates. Typically, the maturity dates for the caplets are on the same cycle as the frequency of the underlying Libor. Since a cap (floor) is a linear sum of independent caplets

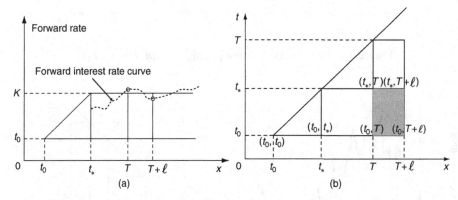

Figure 4.6 (a) Diagram representing a caplet $\ell V B(t_*, T + \ell)[L(t_*, T) - K]_+$. During the time interval $T \le t \le T + \ell$, the holder of a floating receiver *caplet* receives a minimum interest rate of K and, for the case illustrated, will exercise the caplet. (b) Domain for underlying forward interest rates that determines the price of a midcurve caplet maturing at time $t_* < T$. The payoff is defined at time t_*. The shaded portion shows the domain of the forward interest rates that define *caplet*(t_0, t_*, T): the midcurve caplet.

(floorlets); the pricing and hedging of caps (floors) are completely reduced to the analysis of a single caplet (floorlet). It is shown that a caplet and a floorlet obey a put–call parity relation so that the price of a floorlet can be obtained from the price of a caplet. An interest rate collar is an instrument for which the holder pays at a maximum interest rate and receives at a minimum interest rate; it can be shown that a collar is equivalent to buying a cap and selling a floor.

Hence, studying the caplet is sufficient for analyzing interest rate caps, floors, and collars.

Consider a caplet that limits the Libor floating interest rate to a fixed rate K, for the duration of Libor time interval $[T, T + \ell]$. A *midcurve caplet* is defined as an option that is exercised at time $t_* < T$; in other words, the option matures at a time before the caplet becomes operational. Let the caplet price, at time $t_0 < t_*$, be given by caplet(t_0, t_*, T) and $L(t_*, T)$ be the Libor value at time t_*. The payoff for the caplet, similar to a call option on equity given in Eq. (3.1), is given by [11]

$$caplet(t_*, t_*, T) = \ell V B(t_*, T + \ell)\big[L(t_*, T) - K\big]_+ \qquad (4.10)$$

where $B(t_*, T + \ell)$ is a zero coupon bond and V is the principal for which the interest rate caplet is defined.

The domain of the underlying interest rates involved in the pricing of the caplet is given in Figure 4.6(b). The various time intervals that define the interest rate caplet are shown in Figure 4.7.

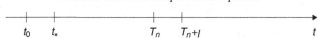

Figure 4.7 Time intervals in the pricing of $caplet(t_0, t_*, T_n)$.

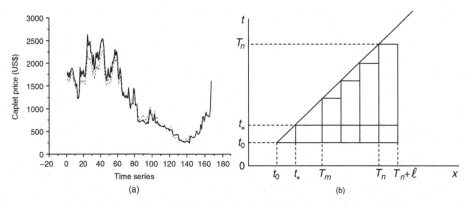

Figure 4.8 (a) A caplet maturing at the time t_* equal to time T when the caplet becomes operational. Caplet prices mature on 12 December 2004 versus time t_0, from 12 September 2003 to 7 May 2004. The unbroken line is the market price of a Libor caplet; the dashed line is the quantum finance model price discussed in Chapter 10. (b) The structure of the payoff function for a midcurve interest rate cap that matures at time t_*. The interest rate cap $cap(t_0, t_*) = \sum_{j=m}^{n} caplet(t_0, t_*, T_j; K_j)$ is defined from future time T_m to time T_n as a portfolio of midcurve caplets.

For the midcurve caplet, the Libor rate $L(t_*, T)$ is determined at time t_*, when the holder of the midcurve caplet exercises the option; hence, the payment is locked-in at time t_* and paid at time $T + \ell$. The market price of a Libor caplet that matures at the time when the caplet becomes operational, that is for $t_* = T$, is given in Figure 4.8(a).

The caplet price $caplet(t_0, t_*, T)$ is given by the martingale condition discussed in Section 3.6; the expectation value of the payoff function, discounted using the money market numeraire, from future (maturity) time t_* to present time t_0, yields the following [12]

$$caplet(t_0, t_*, T) = \ell V E\left[e^{-\int_{t_0}^{t_*} r(t)} B(t_*, T + \ell)[L(t_*, T) - K]_+\right] \quad (4.11)$$

The price of a floorlet is similarly defined by

$$floorlet(t_0, t_*, T) = \ell V E\left[e^{-\int_{t_0}^{t_*} r(t)} B(t_*, T + \ell)[K - L(t_*, T)]_+\right] \quad (4.12)$$

An interest rate cap is made from the sum over caplets spanning the requisite time interval, namely starting from time $T_m = T_0 + m\ell$ and ending at time

$T_n + \ell = T_0 + (n+1)\ell$, with time intervals given by T_j with $j = m, m+1, \ldots, n$; there can be a different fixed interest rate K_j for each time interval. The price of a midcurve cap – the sum of midcurve caplets – is graphically shown in Figure 4.8(b) and is given by

$$cap(t_0, t_*) = \sum_{j=m}^{n} caplet(t_0, t_*, T_j; K_j)$$

$$= \sum_{j=m}^{n} E\left[e^{-\int_{t_0}^{t_*} r(t)} B(t_*, T_j + \ell)\left[L(t_*, T_j) - K_j \right]_+ \right] \quad (4.13)$$

There is a similar expression for an interest rate floor.

4.4 Put–call parity for caplets and floorlets

Put–call parity for caplets and floorlets is fixed by demanding that the prices of two portfolios – having identical cash flows at maturity – must be equal. Failure of the prices to obey the put–call parity relation would lead to arbitrage opportunities.

The payoff function of a caplet can be simplified by using the definition of Libor, given in Eq. (2.21), that $\ell L(t_0, T) = [B(t_0, T) - B(t_0, T + \ell)]/B(t_0, T + \ell)$. Hence the price of a caplet, from Eq. (4.11), is given by

$$caplet(t_0, t_*, T) = \ell V E\left(e^{-\int_{t_0}^{t_*} dt\, r(t)} B(t_*, T + \ell)\left[L(t_*, T) - K \right]_+ \right)$$

$$= V E_M\left(e^{-\int_{t_0}^{t_*} dt\, r(t)} \left[B(t_*, T) - (1 + \ell K) B(t_*, T + \ell) \right]_+ \right)$$

Similarly, a floorlet from Eq. (4.12) is given by

$$floorlet(t_0, t_*, T) = V E\left(e^{-\int_{t_0}^{t_*} dt\, r(t)} \left[(1 + \ell K) B(t_*, T + \ell) - B(t_*, T) \right]_+ \right)$$

The derivation of put–call parity hinges on the identity, which follows from Eq. (A.3), that

$$\Theta(x) + \Theta(-x) = 1 \quad (4.14)$$

and yields

$$[a - b]_+ - [b - a]_+ = (a - b)\Theta(a - b) - (b - a)\Theta(b - a) = a - b \quad (4.15)$$

The difference in the price of a caplet and a floorlet, from Eq. (4.15), yields the following

$$\left[B(t_*,T) - (1+\ell K)B(t_*,T+\ell)\right]_+ - \left[(1+\ell K)B(t_*,T+\ell) - B(t_*,T)\right]_+$$
$$= B(t_*,T) - (1+\ell K)B(t_*,T+\ell)$$

and hence

$$caplet(t_0,t_*,T_n) - floorlet(t_0,t_*,T_n)$$
$$= VE\left(e^{-\int_{t_0}^{t_*} dt r(t)}\left[B(t_*,T) - (1+\ell K)B(t_*,T+\ell)\right]\right)$$

The difference of the caplet and floorlet price does not have any constraint. Taking the expectation value of the zero coupon bonds using the martingale condition given in Eq. (3.4)

$$E\left[e^{-\int_{t_0}^{t_*} dt r(t)}B(t_*,T)\right] = B(t_0,T)$$

yields the following result

$$caplet(t_0,t_*,T_n) - floorlet(t_0,t_*,T_n)$$
$$= V\left[B(t_0,T) - (1+\ell K)B(t_0,T+\ell)\right] \tag{4.16}$$
$$= \ell V B(t_0,T_n+\ell)[L(t_0,T_n) - K] = swaplet_L(t_0,K)$$

The right-hand side of the above equation is the price, at time t_0, of a forward floating receiver swaplet given in Eq. (4.10).

Thus, the floorlet price is given from the caplet price using put–call parity, and an independent derivation of a floorlet's price is not necessary.

A cap or a floor, from Eq. (4.13), is equal to a linear sum of caplets and floorlets. Hence put–call parity for interest rate caplets and floorlets yields the following

$$cap(t_0,t_*) - floor(t_0,t_*)$$
$$= \sum_{n=0}^{N}\left[caplet(t_0,t_*,T_n;K) - floorlet(t_0,t_*,T_n;K)\right]$$
$$= \ell V \sum_{n=0}^{N} B(t_0,T_n+\ell)\left[L(t_0,T_n) - K\right] \tag{4.17}$$
$$= swap_L(t_0,K)$$

4.5 Put–call: empirical Libor caplet and floorlet

Put–call parity for caps and floors is a model-independent result that market prices obey. The prices of interest rate caplets and floorlets for Eurodollar futures contracts – expiring on 13 December 2004 with a fixed interest rate (strike price) of 2% – are analyzed for empirically testing put–call parity. Daily prices, from 12 September 2003 to 7 May 2004, are quoted with the interest rate in basis points (100 basis points = 1% annual interest rate) and need to be multiplied by the notional value of one million Dollars; the caplet price has a fixed maturity date of 13 December 2004.

Using the put–call parity given in Eq. (4.16) – for the case $t_* = T_n = 13$ December 2004 – consider the portfolio

$$\Pi(t_0) = caplet(t_0, T_n) - floorlet(t_0, T_n) - \ell V B(t_0, T_n + \ell)[L(t_0, T_n) - K]$$
(4.18)

The value of the portfolio $\Pi(t_0)$ should be zero if put–call parity holds for the caplet and floorlet prices. The value of the portfolio is directly taken from the market data and is plotted in Figure 4.9; it is seen that the market obeys put–call parity to a high degree of accuracy. The deviations of $\Pi(t_0)$ from zero are negligible compared to the price of a caplet. Hence there are no-arbitrage opportunities in the pricing for caplets and floorlets.

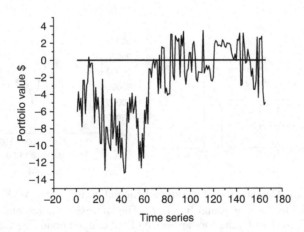

Figure 4.9 Value of portfolio $\Pi(t_0)$ as in Eq. (4.18), which is the difference of the caplet and floorlet prices with notional value one million Dollars, versus time t_0 (12 September 2003–7 May 2004). Put–call parity requires the portfolio value to be zero.

4.6 Coupon bond options

Zero coupon bond options are a special case of the coupon bond options, and hence the analysis is focused on the more general case. Consider the price, at time t_*, of a coupon bond that is given, as in Eq. (2.4), by

$$\mathcal{B}(t_*) = \sum_{i=1}^{N} c_i B(t_*, T_i) \tag{4.19}$$

The payoff function $\mathcal{P}(t_*)$ of a coupon bond *European call option* maturing at time t_*, and with strike price K, is given by

$$\mathcal{P}(t_*) = \left(\sum_{i=1}^{N} c_i B(t_*, T_i) - K \right)_+$$

$$= (\mathcal{B}(t_*) - K)_+ \tag{4.20}$$

A coupon bond European put option and its forward price is shown in Figure 4.10(a).

The price of a European call option at time $t_0 < t_*$ is given by the expectation value of the payoff $\mathcal{P}(t_*)$, discounted from time t_* to time t_0. Using the money

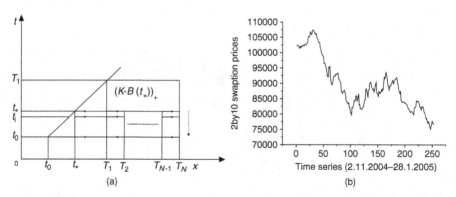

Figure 4.10 (a) The payoff of the coupon bond put option $\mathcal{P}(t_*) = (K - \mathcal{B}(t_*))_+$ is represented by the horizontal line at t_*. The successive horizontal lines with arrows show, for $t < t_*$, the forward price of the coupon bond; the forward zero coupon bond price is given by $F(t_i, T_1, T_2) = B(t_i, T_2)/B(t_i, T_1)$. (b) The market price of a 2by10 Libor swaption $C_L(t, T_0, R_P(t))$ with $T_0 - t = 2$ years. The swaption price is given for fixed interest equal to the par interest rate at time t, namely $R_P(t)$.

market numeraire for discounting the payoff function yields the following

$$C(t_0, t_*, T_N, K) = E\left[e^{-\int_{t_0}^{t_*} dt\, r(t)} \mathcal{P}(t_*)\right]$$

$$= E\left[e^{-\int_{t_0}^{t_*} dt\, r(t)} \left(\sum_{i=1}^{N} c_i B(t_*, T_i) - K\right)_+\right] \quad (4.21)$$

Similarly, the European put option is given by

$$P(t_0, t_*, T, K) = E\left[e^{-\int_{t_0}^{t_*} dt\, r(t)} \left(K - \sum_{i=1}^{N} c_i B(t_*, T_i)\right)_+\right]$$

The (zero) coupon bonds have a martingale evolution for a large class of numeraires. For calculating the price of interest rate options it is very convenient to discount by the forward bond numeraire. The future price of the option is discounted by the zero coupon bond $B(t, t_*)$: t_* fixed, instead of the money market numeraire given by $\exp\{-\int_{t_0}^{t_*} dt\, r(t)\}$: t_0 fixed.

For the forward bond numeraire, one has that $C(t, t_*, T, K)/B(t, t_*)$ is a martingale. From Eq. (3.8), since $B(t_*, t_*) = 1$

$$C(t_0, t_*, T, K) = B(t_0, t_*) E\left[\mathcal{P}(t_*)\right]$$

$$= B(t_0, t_*) E\left[\left(\sum_{i=1}^{N} c_i B(t_*, T_i) - K\right)_+\right] \quad (4.22)$$

Note that the discounting factor $B(t_0, t_*)$ in Eq. (4.22) is determined by the initial value of the forward interest rates $f(t_0, x)$. Unlike $\exp\{-\int_{t_0}^{t_*} dt\, r(t)\}$, $B(t_0, t_*)$ is not a random quantity and, hence, is outside the expectation value $E[\ldots]$; this is the main reason for choosing the forward bond numeraire.

In summary, Eq. (4.22) shows that the price of a (European call) option, at time $t_0 < t_*$, is given by *discounting* the payoff $\mathcal{P}(t_*)$ from time t_* to time t_0 and averaging over all the random (fluctuating) forward interest rates over future calendar time $[t_0, t_*]$ – with the initial conditions specified at time t_0 by $f(t_0, x)$.

4.7 Put–call parity for European bond option

Put–call parity for the coupon bond option, as expected, is model-independent. The martingale property of the zero coupon bonds is the key to the derivation of this section. The difference in the call and put payoff functions for the coupon bond option, using Eq. (4.15), satisfies

$$
\left(\sum_{i=1}^{N} c_i B(t_*, T_i) - K \right)_+ - \left(K - \sum_{i=1}^{N} c_i B(t_*, T_i) \right)_+ = \sum_{i=1}^{N} c_i B(t_*, T_i) - K
$$

Multiplying both sides by $\exp\{-\int_{t_0}^{t_*} dt\, r(t)\}$ and taking the expectation value using the martingale condition $E[e^{-\int_{t_0}^{t_*} dt\, r(t)} B(t_*, T)] = B(t_0, T)$ given in Eq. (3.4), yields the following put–call parity relation

$$
C(t_0, t_*; K) - P(t_0, t_*; K) = E\left[e^{-\int_{t_0}^{t_*} dt\, r(t)} \left(\sum_{i=1}^{N} c_i B(t_*, T_i) - K \right) \right]
$$

$$
= \sum_{i=1}^{N} c_i B(t_0, T_i) - K B(t_0, t_*) = \mathcal{B}(t_0) - K B(t_0, t_*)
$$

Put–call parity yields the expected result that

$$
C(t_0, t_*; K) - P(t_0, t_*; K) = \mathcal{B}(t_0) - K B(t_0, t_*) \quad : \text{put–call parity} \qquad (4.23)
$$

The right-hand side is the difference, at time t_0, between the value of the (underlying) coupon bond and the discounted value of the strike price K. The result for bonds is similar to the earlier result for the equity put–call given in Eq. (3.1).

4.8 American coupon bond option put–call inequalities

The call or put American option on a coupon bond $\mathcal{B}(t_0)$ has the same payoff function as the European option, with the additional feature that the American option can be exercised any time from the time it is sold at t_0 to its maximum possible maturity time t_*. The freedom of an early exercise implies that the price of the American option must always be greater than the corresponding European option since the American option contains the European option as one of its special cases.

It is conjectured in [17], in analogy with Eq. (4.23), that the American call and put coupon bond options satisfy the following put–call inequalities

$$
\mathcal{F}(t_0) - K \leq C_A(t_0, t_*; K) - P_A(t_0, t_*; K) \leq \mathcal{F}(t_0) - K B(t_0, t_*) :
$$

$$
\text{American put–call inequalities} \qquad (4.24)
$$

where $C_A(t_0, t_*; K)$, $P_A(t_0, t_*; K)$ are the price of the American call and put options respectively, K is the strike price, and $B(t_0, t_*)$ is a zero coupon bond. $\mathcal{F}(t_0)$ is the forward price, at time t_0, of the coupon bond $\mathcal{B}(t_*)$ and is given by

$$\mathcal{F}(t_0) \equiv \sum_{i=1}^{N} c_i F(t_0, t_*, T_i) = \frac{\sum_{i=1}^{N} c_i B(t_0, T_i)}{B(t_0, t_*)} = \frac{\mathcal{B}(t_0)}{B(t_0, t_*)}$$

Recall, from Eq. (2.14), that $F(t_0, t_*, T_i)$ is the forward zero coupon bond price of $B(t_*, T_i)$ at t_0. The forward price of the coupon bond is graphically shown in Figure 4.10(a).

4.9 Interest rate swaptions

An interest rate swaption, denoted by C_L and C_R, is an option on a floating or a fixed interest rate receiver swap, $swap_L$, and $swap_R$, respectively.

Consider a swap with N payments dates given by $T_n = T_0 + n\ell$; $n = 1, 2, \ldots, N$; the swap starts at time T_0, the first payment is made at time T_1 and the last payment is made at time T_N. A midcurve swaption, similar to a midcurve caplet, is contracted at time t_0 and matures at time $t_* < T_0$. The payoff function for a midcurve swaption is given in Figure 4.4(a) and is the same as a midcurve forward swap. The swaption is an option on the swap and hence has the same cash flow if it is exercised.

The swaption that will be studied henceforth is the one that matures at $t_* = T_0$, when the swap becomes operational and is shown in Figure 4.4(b). Almost all market data on swaptions are exclusively given for this case and are, consequently, the most important for empirical studies of swaptions.

The swaption, on maturing, will be exercised only if the value of the swap at time T_0 is greater than its initial par value of zero. Hence, the payoff function for the swaption for the floating and fixed receivers swap, from Eqs. (4.2) and (4.3), is given respectively by the following

$$C_L(T_0, T_0; R_S) = \left[swap_L(t_0, R_S) \right]_+$$

$$= \ell V \left[\sum_{n=0}^{N-1} B(t_0, T_n + \ell)(L(t_0, T_n) - R_S) \right]_+ \quad (4.25)$$

$$C_R(t, T_0, R_S) = \left[swap_R(t_0, R_S) \right]_+$$

$$= \ell V \left[\sum_{n=0}^{N-1} B(t_0, T_n + \ell)\big(R_S - L(t_0, T_n)\big) \right]_+ \quad (4.26)$$

In terms of zero coupon bonds, the swaption payoff function, from Eq. (4.5), is given by

$$C_L(T_0, T_0; R_S) = V \left[1 - B(T_0, T_N) - \ell R_S \sum_{n=1}^{N} B(T_0, T_0 + n\ell) \right]_+ \quad (4.27)$$

and a similar expression for C_R. The value of the swaption at an earlier time $t < T_0$ can be obtained by discounting the payoff function using the money market numeraire and yields

$$C_L(t, T_0, R_S) = VE\left[e^{-\int_t^{T_0} r(t')dt'} C_L(T_0; R_S)\right]$$

$$= VE\left[e^{-\int_t^{T_0} r(t')dt'} \left(1 - B(T_0, T_N) - \ell R_S \sum_{n=1}^{N} B(T_0, T_0 + n\ell)\right)\right]_+$$

(4.28)

and similarly for $C_R(t, T_0, R_S)$. One can see that a swap is equivalent to a particular portfolio of coupon bonds, and all techniques that are used for coupon bond options can be used for analyzing swaptions.

Discounting by the forward bond numeraire $B(t, T_0)$, similar to the case of coupon bond options given in Eq. (4.22), makes the swaption price computationally more tractable; the price of the swaption, from Eq. (4.27), is given by

$$\frac{C_L(t, T_0, R_S)}{B(t, T_0)} = VE\left[C_L(T_0; R_S)\right]$$

$$= VE\left[1 - B(T_0, T_N) - \ell R_S \sum_{n=1}^{N} B(T_0, T_0 + n\ell)\right]_+ \quad (4.29)$$

Figure 4.10(b) shows the market price of a 2by10 swaption $C_L(t, T_0, R_P(t))$, with Libor being the floating rate on a notional principal sum for the underlying swap of US$1 million.

Eq. (4.15), together with the martingale property of zero coupon bonds under the money market measure, namely that $E\left[e^{-\int_t^{T_0} r(t')dt'} B(T_0, T_n)\right] = B(t, T_n)$, yields the put–call parity for the swaptions as [11]

$$C_L(t, T_0, R_S) - C_R(t, T_0, R_S)$$

$$= VE\left[e^{-\int_t^{T_0} r(t')dt'} \left[1 - B(T_0, T_0 + N\ell) - \ell R_S \sum_{n=1}^{N} B(T_0, T_0 + n\ell)\right]\right]$$

$$= V\left[B(t, T_0) - B(t, T_0 + N\ell) - \ell R_S \sum_{n=0}^{N-1} B(t, T_0 + n\ell)\right] \quad (4.30)$$

$$= swap_L(t; T_0, R_S)$$

$swap_L(t; T_0, R_S)$ is the price, at time t, of the underlying forward swap that matures at time $T_0 > t$. Eq. (4.23) is the general expression for put–call parity for coupon bond options and the put–call parity for swaptions given in Eq. (4.30) is a special case.

The price of swaption C_R, in which the holder has the option to enter a fixed rate R_S receiver's swap, is given, from Eq. (4.30), by the formula for the call option for a coupon bond. The swaption C_R matures at time T_0; the payoff function on a principal amount V is the following

$$
C_R(T_0, T_0, R_S) = V \left[B(T_0, T_0 + N\ell) + \ell R_S \sum_{n=1}^{N} B(T_0, T_0 + n\ell) - 1 \right]_+
$$

(4.31)

Comparing the payoff for C_R given above with the payoff for the coupon bond call option given in Eq. (4.20), yields the following for the swaption coefficients

$$
c_n = \ell R_S ; \quad n = 1, 2, \ldots, (N-1); \quad \text{payment at time } T_0 + n\ell \qquad (4.32)
$$

$$
c_N = 1 + \ell R_S; \quad \text{payment at time } T_0 + N\ell
$$

$$
K = 1
$$

There are swaptions traded in the market in which the floating rate is paid at $\ell = 90$ day intervals, and the fixed rate payments are paid at $2\ell = 180$ day intervals. For a swaption with fixed rate payments at 90 day intervals – at times $T_0 + n\ell$, with $n = 1, 2, \ldots, N$ – there are N payments. For payments made at 180 day intervals, there are only $N/2$ payments[5] made at times $T_0 + 2n\ell$, $n = 1, 2, \ldots, N/2$, and of amount $2R_S$. The payoff function for the swaption is[6]

$$
C_L(T_0, T_0; R_S) = V \left[1 - B(T_0, T_0 + N\ell) - 2\ell R_S \sum_{n=1}^{N/2} B(T_0, T_0 + 2n\ell) \right]_+
$$

$$
= V \left[1 - \sum_{n=1}^{N/2} \tilde{c}_n B(T_0, T_0 + 2n\ell) \right]_+
$$

(4.33)

The equivalent coupon bond *put* option payoff function is given by

$$
\left(K - \sum_{n=1}^{N/2} \tilde{c}_n B(t_*, T_0 + 2n\ell) \right)_+
$$

(4.34)

---

[5] Suppose the swaption has a duration such that N is even. Note that for $N = 4$ the underlying swap has a duration of one year.

[6] The price of C_R for the case of 90 day floating and 180 day fixed interest payments is given from C_L by using the put–call relation similar to that given in Eq. (4.30).

and has the coefficients and strike price given by

$$\tilde{c}_n = 2\ell R_S; \quad n = 1, 2, \ldots, (N-1)/2; \quad \text{payment at time } T_0 + 2n\ell \quad (4.35)$$

$$\tilde{c}_{N/2} = 1 + 2\ell R_S; \quad \text{payment at time } T_0 + N\ell$$

$$K = 1$$

The par interest rate at time t_0 is fixed by the forward swap contract and is given, similar to Eq. (4.6), by

$$2\ell R_P(t_0) = \frac{B(t_0, T_0) - B(t_0, T_0 + N\ell)}{\sum_{n=1}^{N/2} B(t_0, T_0 + 2n\ell)} \quad (4.36)$$

The par interest rate reduces, at $t_0 = T_0$, to the par value of the fixed interest rate similar to Eq. (4.9) and is given by

$$2\ell R_P = \frac{1 - B(T_0, T_0 + N\ell)}{\sum_{n=1}^{N/2} B(T_0, T_0 + 2n\ell)} \quad (4.37)$$

It is only due to the asymmetric nature of the last coefficient, namely c_N and $\tilde{c}_{N/2}$ for the two cases discussed above, that the swap interest rate R_S does not completely factor out (up to a re-scaling of the strike price) from the swaption price.

Options on $swap_L$ and $swap_R$, namely C_L and C_R, are both *call options* since they give the holder the option to either receive fixed or receive floating payments, respectively. When expressed in terms of coupon bond options, it can be seen from Eqs. (4.28) and (4.31) that the swaption for receiving fixed payments is equivalent to a coupon bond put option, whereas the option to receive floating payments is equivalent to a coupon bond call option.

4.10 Interest rate caps and swaptions

The fundamental ingredient for swaps, caps, and swaptions is the following combination of floating and fixed interest rates, which constitutes a (floating receiver) midcurve caplet at time t_* and is given by

$$B(t_*, T + \ell)\big[L(t_*, T) - K\big]_+ = [swaplet_L(t_*, K)]_+$$

The payment of interest is made at time $T + \ell$; hence the price of the caplet at t_* is given by discounting the payment by the bond $B(t_*, T + \ell)$.

- The interest rate cap price is the following

$$cap(t_0, t_*) = \sum_{j=m}^{n} E\left[e^{-\int_{t_0}^{t_*} r(t)} B(t_*, T_j + \ell)[L(t_*, T_j) - K_j]_+\right]$$

The summation on the caplets is *outside* the expectation value and hence there are no cross-terms between the various caplets.
- In contrast, the swaption price is the following

$$\text{Swaption}_L(t_0, t_*) = E\left[e^{-\int_{t_0}^{t_*} r(t)}\left[\sum_{j=m}^{n} B(t_*, T_j + \ell)[L(t_*, T_j) - K_j]\right]_+\right]$$

The summation on the caplets is now *inside* the expectation value and hence there are cross-terms leading to correlations between all the caplets. These correlations are needed for determining the swaption price, as discussed in Chapter 12.

In summary, the prices of the interest rate cap and swaption show a fundamental difference between the two instruments. The cap price is a *linear sum* of the caplet prices, which are all independent and have no correlation with each other. In contrast, for the swaption the payoff function is a *linear sum* of component caplets. Taking the expectation of the payoff function gives rise to complicated and nontrivial correlations between all the component caplets. This is the fundamental reason why the price of a swaption is much more complex and difficult to evaluate than the price of an interest rate cap.

The price of liquid interest rate options, such as caps and floors, encode all the available market information. The underlying Libor rates are common for these options, and consequently one can extract information on the Libor rates from caps and floors. The main challenge for market participants is to use this information for pricing other (exotic) options. In particular, the market prices of interest caplets are often used for fixing the volatility of Libor. To reduce the number of inputs, volatility parameters in a given Libor time interval are often assumed to be constant and lead to many inaccuracies. Furthermore, longer maturity options require a large number of volatility parameters due to the aggregation of the volatility parameters from each Libor future time interval.

4.11 Heath–Jarrow–Morton path integral

The HJM (Heath–Jarrow–Morton) model [56] is the industry standard for studying interest rates and has been extensively investigated, both analytically and empirically [48, 57, 88]. In particular, an exact expression for coupon bond options has been obtained in the HJM framework using stochastic calculus [65]. The discussion in this section is a preparation for the more general formulation of forward interest rates discussed in Chapter 5.

Eq. (3.5) shows that one can consider the present value of a zero coupon bond as resulting from a stochastic process followed by the spot interest rate $r(t)$. However, in the HJM approach the forward interest rates $f(t, x)$ are considered as fundamental, and the value of the present-day value of the bond $B(t, T)$ is taken as input from the debt market; the spot interest rate $r(t)$ is just one point of the $f(t, x)$ curve, namely $r(t) = f(t, t)$.

The HJM model is re-formulated in the language of path integration [12] as it provides a powerful computational tool for obtaining many nontrivial results of the HJM model. White noise is expressed in terms of a path integral and the concepts of stochastic calculus are seen to be a special case of path integration. The path integral framework provides a natural generalization of white noise to a two-dimensional Gaussian quantum field.[7]

In the one-factor HJM model the time evolution of the forward interest rates is driven by a single white noise $R(t)$ and is given by [56, 63, 84][8]

$$\frac{\partial f}{\partial t}(t, x) = \alpha(t, x) + \sigma(t, x)R(t) \qquad (4.38)$$

$\alpha(t, x)$ is the drift velocity term and $\sigma(t, x)$ is the deterministic volatility of the forward interest rates. For every value of time t, white noise $R(t)$ is an independent Gaussian random variable given by

$$E[R(t)] = 0; \quad E[R(t)R(t')] = \delta(t - t')$$

The forward interest rates $f(t, x)$ are driven by random variables $R(t)$, which give the same random 'shock' at time t to all the future forward rates $f(t, x), x > t$. To bring in the maturity dependence of the random shocks, the volatility function $\sigma(t, x)$, at given time t, weighs this 'shock' differently for each x.

The HJM model evolves an entire curve $f(t, x)$; for the K factor model, at each instant of time t, it is driven by K random variables given by $R_i(t)$, and hence has at most K degrees of freedom.

The HJM combination $[\partial f(t, x)/\partial t - \alpha(t, x)]/\sigma(t, x)$ of the forward interest rates is a generalization of the Sharpe ratio for equity [59], given by $(\mu - r)/\sigma$. μ is the expected rate of return on a stock, r is the spot risk-free spot interest

---

[7] The generalization of the HJM model, which is the subject matter of Chapter 5, makes the theory of forward interest rates mathematically equivalent to a two-dimensional quantum field theory.

[8] The K-factor HJM model is given by

$$\frac{\partial f}{\partial t}(t, x) = \alpha(t, x) + \sum_{i=1}^{K} \sigma_i(t, x)R_i(t)$$

where $\sigma_i(t, x)$ are the deterministic volatilities and $R_i(t)$ is a vector Gaussian white noise. No new insight is offered by the K-factor model and hence only the one-factor model is analyzed.

rate (fixed by the martingale condition), and σ is the stock price's volatility. The Sharpe ratio is an important quantity is assessing the risk premium for a stock. In the quantum finance formulation of the forward interest rates, the quantity $f(t, x)$ always appears in the HJM combination and has far-reaching consequences.

From Eq. (4.38)

$$f(t_*, x) = f(t_0, x) + \int_{t_0}^{t_*} dt \alpha(t, x) + \int_{t_0}^{t_*} dt' \sigma(t, x) R(t) \qquad (4.39)$$

The initial forward interest rate curve $f(t_0, x)$ is determined from the market, and so is the volatility function $\sigma(t, x)$. Similar to the Black–Scholes analysis, the drift term $\alpha(t, x)$ is fixed to ensure that the forward interest rates have a martingale time evolution, which yields [56]

$$\alpha(t, x) = \sigma(t, x) \int_t^x dx' \sigma(t, x')$$

4.12 HJM coupon bond European option price

The coupon bond option price in the one factor HJM model with exponential volatility has been stated in [34, 65] and a path integral derivation is given of this result. The derivation illustrates many key features of path integration in a simple context and serves as an exemplar for more complex derivations.

The payoff function $\mathcal{P}(t_*)$ of a coupon bond European call option, maturing at time t_*, for strike price K, is given, from Eq. (4.20), as follows

$$\mathcal{P}(t_*) = \left(\sum_{i=1}^{N} c_i B(t_*, T_i) - K \right)_+ \qquad (4.40)$$

The European coupon bond option price $C(t_0, t_*, T, K)$, from Eq. (4.22), is given by

$$C(t_0, t_*, T, K) = B(t_0, t_*) E[\mathcal{P}(t_*)]$$

In the HJM model the expectation value is calculated by evaluating the white noise path integral, as discussed in Appendix A.4; more precisely

$$C_{HJM}(t_0, t_*, T, K) = B(t_0, t_*) \int DR \, \mathcal{P}(t_*) \, e^{S_0} \; ; \; S_0 = -\frac{1}{2} \int_{t_0}^{t_*} R^2(t) \qquad (4.41)$$

To explicitly evaluate the path integral, one needs to express the zero coupon bonds in terms of white noise $R(t)$; from Eqs. (2.12) and (4.39)

$$B(t_*, T_i) = \exp\left\{-\int_{t_*}^{T_i} dx f(t, x)\right\}$$

$$= F(t_0, t_*, T_i)e^{-\int_{t_*}^{T_i} dx \int_{t_0}^{t_*} dt[\alpha(t,x)+\sigma(t,x)R(t)]} \qquad (4.42)$$

Choosing the exponential volatility function

$$\sigma(t, x) = \sigma_0 e^{-\lambda(x-t)}$$

leads to the following simplifications

$$\int_{t_*}^{T_i} dx \int_{t_0}^{t_*} dt \sigma(t, x) R(t) = Y_i \sigma_0 \int_{t_0}^{t_*} dt e^{-\lambda(t_*-t)} R(t) \qquad (4.43)$$

$$\int_{t_*}^{T_i} dx \int_{t_0}^{t_*} dt \alpha(t, x) = \frac{1}{2}\sigma_E^2 Y_i^2$$

$$Y_i = Y(t_*, T_i) = \frac{1}{\lambda}[1 - e^{-\lambda(T_i - t_*)}]; \quad \sigma_E^2 = \frac{\sigma_0^2}{2\lambda}[1 - e^{-2\lambda(t_*-t_0)}]$$

All the zero coupon bonds are driven by one random variable, namely

$$W = \sigma_0 \int_{t_0}^{t_*} dt e^{-\lambda(t_*-t)} R(t) = \int_{t_0}^{t_*} dt \sigma(t_*, t) R(t)$$

The payoff function, from Eqs. (4.40), (4.42), and (4.43), is given by

$$\mathcal{P}(t_*) = \left(\sum_{i=1}^{N} c_i F(t_0, t_*, T_i)e^{-\frac{1}{2}\sigma_E^2 Y_i^2 - Y_i W} - K\right)_+$$

Since the payoff function depends on only *one* random variable, namely W, the path integral can be performed exactly. Inserting unity

$$1 = \int_{-\infty}^{+\infty} dW \delta\left[W - \int_{t_0}^{t_*} dt \sigma(t_*, t) R(t)\right]$$

into the path integral for the option price given in Eq. (4.41) yields the following

$$\frac{C_{HJM}(t_0, t_*, T, K)}{B(t_0, t_*)} = \int DR \int_{-\infty}^{+\infty} dW \delta \left[W - \int_{t_0}^{t_*} dt\sigma(t_*,t)R(t) \right] P_* e^{S_0}$$

$$= \int_{-\infty}^{+\infty} dW \left(\sum_{i=1}^{N} c_i F(t_0, t_*, T_i) e^{-\frac{1}{2}\sigma_E^2 Y_i^2 - Y_i W} - K \right)_+ Z(W) \qquad (4.44)$$

where $Z(W) = \int DR \delta \left[W - \int_{t_0}^{t_*} dt\sigma(t_*,t)R(t) \right] e^{S_0}$

To evaluate the path integral, the Dirac-delta function is represented, as in Eq. (A.7), as follows

$$\delta[W - \tilde{W}] = \int_{-\infty}^{+\infty} \frac{d\xi}{2\pi} e^{i\xi(W-\tilde{W})}$$

The path integral for $Z(W)$ is evaluated using Eq. (A.41) and yields

$$Z(W) = \int DR \int_{-\infty}^{+\infty} \frac{d\xi}{2\pi} e^{i\xi(W - \int_{t_0}^{t_*} dt\sigma(t_*,t)R(t))} e^{S_0}$$

$$= \int_{-\infty}^{+\infty} \frac{d\xi}{2\pi} e^{i\xi W} e^{-\frac{1}{2}\xi^2 \int_{t_0}^{t_*} dt\sigma^2(t_*,t)}$$

$$= \frac{1}{\sqrt{2\pi\sigma_E^2}} \exp\left\{ -\frac{1}{2\sigma_E^2} W^2 \right\} \qquad (4.45)$$

since $\int_{t_0}^{t_*} dt\sigma^2(t_*,t) = \sigma_E^2$.

Hence, Eqs. (4.44) and (4.45) yield the following coupon bond option price

$$C_{HJM}(t_0, t_*, T, K) = \frac{B(t_0, t_*)}{\sqrt{2\pi\sigma_E^2}} \int_{-\infty}^{+\infty} dW e^{-\frac{1}{2\sigma_E^2} W^2}$$

$$\times \left(\sum_{i=1}^{N} c_i F(t_0, t_*, T_i) e^{-\frac{1}{2}\sigma_E^2 Y_i^2 - Y_i W} - K \right)_+$$

To further simplify the HJM option price, define quantity w_0 such that[9]

$$\sum_{i=1}^{N} c_i F(t_0, t_*, T_i) e^{-\frac{1}{2}\sigma_E^2 Y_i^2 - Y_i w_0} = K \qquad (4.46)$$

---

[9] The definition of w_0 given in [65] differs from the one given in Eq. (4.46).

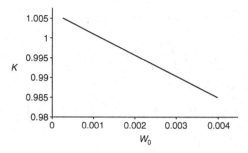

Figure 4.11 The nonlinear relation of K and w_0.

The quantity w_0 is related to the strike price K by a nonlinear transformation that depends on the initial coupon bond price [65]. Figure 4.11 shows a typical dependence of K on w_0.

For values of $W > w_0$ the coupon bond's value is greater than K and the option is not exercised; the option price hence is given by

$$C_{HJM}(t_0, t_*, T, K) = \frac{B(t_0, t_*)}{\sqrt{2\pi\sigma_E^2}} \int_{-\infty}^{w_0} dW e^{-\frac{1}{2\sigma_E^2} W^2}$$

$$\times \left(\sum_{i=1}^{N} c_i F(t_0, t_*, T_i) e^{-\frac{1}{2}\sigma_E^2 Y_i^2 - Y_i W} - K \right)$$

The integration over W yields the following explicit expression for the coupon bond option [34, 57, 65]

$$C_{HJM}(t_0, t_*, K) = \sum_{i=1}^{N} c_i B(t_0, T_i) N(d_i) - K B(t_0, t_*) N(d) \qquad (4.47)$$

$$d_i \equiv \frac{w_0}{\sigma_E} + Y(t_*, T_i)\sigma_E; \quad d = \frac{w_0}{\sigma_E}$$

where $N(d)$ is the probability integral for the normal distribution defined in Eq. (3.61).

The coupon bond European option is equal to the sum of terms that refer to options on its constituent zero coupon bonds; all correlations between the different zero coupon bonds are absent in the HJM coupon bond option price. It will be shown in Chapter 11 that in the quantum finance model of forward interest rates the coupon bond option price has cross terms of the constituent zero coupon bonds of arbitrarily high order.

4.12.1 Small volatility limit of HJM coupon bond option

The small volatility limit of the HJM coupon bond option price is obtained. In Chapter 11, a quantum finance derivation is given of the coupon bond option price; a limiting case of which is shown to be the HJM small volatility price.

The HJM option price is expanded as a power series in the volatility constant σ_0, which is taken to be small. The value of w_0 is taken to be such that w_0/σ_0 is small, which in turn yields that all the d_i, d are small. Using the expansion

$$N(d) = \sqrt{\frac{1}{2\pi}} \int_{-\infty}^{d} dz e^{-\frac{1}{2}z^2} \simeq \frac{1}{2} + \sqrt{\frac{1}{2\pi}} d + O(d^2)$$

Eq. (4.47) yields the following approximate HJM bond option price

$$C_{\text{HJM}}(t_0, t_*, K) \simeq \sum_{i=1}^{N} c_i B(t_0, T_i) \left[\frac{1}{2} + \sqrt{\frac{1}{2\pi}} d_i \right]$$

$$- K B(t_0, t_*) \left[\frac{1}{2} + \sqrt{\frac{1}{2\pi}} d \right] + O(d_i^2, d^2)$$

$$= \frac{1}{2} \left[\sum_{i=1}^{N} c_i B(t_0, T_i) - K B(t_0, t_*) \right] + \sqrt{\frac{1}{2\pi}} \sum_{i=1}^{N} c_i B(t_0, T_i) Y(t_*, T_i) \sigma_R$$

$$+ \sqrt{\frac{1}{2\pi}} \frac{w_0}{\sigma_R} \left[\sum_{i=1}^{N} c_i B(t_0, T_i) - K B(t_0, t_*) \right] + O(d_i^2, d^2)$$

$$= \frac{1}{2} B(t_0, t_*)[F - K] + \sqrt{\frac{1}{2\pi}} B(t_0, t_*) \sigma_R \sum_{i=1}^{N} J_i Y(t_*, T_i)$$

$$+ O\left(\frac{w_0}{\sigma_0} (F - K), (F - K)^2 \right) \tag{4.48}$$

where $J_i = c_i B(t_0, T_i)/B(t_0, t_*)$ and $F = \sum_{i=1}^{N} J_i$.

4.13 Summary

Interest rate derivatives, namely interest rate swaps, caps, and swaptions were briefly discussed. Coupon bond options were also discussed and it was shown that swaptions can be viewed as a special case of coupon bond options. Put–call parity for coupon bond options were derived and were shown to follow from the definition of the payoff functions.

The main focus of this chapter was on the mathematical formulation of the debt instruments. In particular, the precise expression of the options' payoff functions and the formulas for determining the option price were obtained and analyzed.

The HJM model was given a path integral formulation. The forward interest rates were shown, in the HJM model, to be driven by a single white noise. The coupon bond option HJM price was evaluated exactly, for an exponential volatility function, and was seen to have no correlation terms between the constituent zero coupon bonds.

The combination $[\partial f(t,x)/\partial t - \alpha(t,x)]/\sigma(t,x)$ is the most important feature of the HJM model. This combination is reflected in the stochastic differential equation that defines the HJM evolution of $f(t,x)$ and carries over to all forms of generalizations, including the quantum finance model for $f(t,x)$ as well as the quantum Libor Market Model.

5
Quantum field theory of bond forward interest rates

A quantum field theory of forward interest rates is developed as a natural general-ization of the HJM model: the forward interest rates are allowed to have *independent fluctuations* for each future time. The forward interest rates are modeled as a two-dimensional Gaussian quantum field, leading to forward interest rates that have a finite probability of being negative. The model is consistent not for the interest rate sector but only for the bond sector and is consequently called the *bond forward interest rates*. The concept of a quantum field is briefly discussed in Appendix A.7. The 'stiff' quasi-Gaussian model, together with the concept of market time, describes the forward interest rates. A differential formulation of forward interest rates' dynamics is obtained. Using a singular property of the forward interest rates' quantum field, a generalization of Ito calculus follows from the Wilson expansion. A derivation of a risk-neutral measure for zero coupon bonds is obtained based on the differential martingale condition.

5.1 Introduction

The complexity of the forward interest rates is far greater than that encountered in the study of stocks and their derivatives. A stock, at a given instant in time, is descri-bed by only one random variable (degree of freedom) $S(t)$ and which is usually modeled using stochastic differential equations. In the case of interest rates, it is the entire interest rates yield curve $f(t, x)$ that undergoes a random evolution. Clearly, at each instant, the most general random evolution is that the forward interest rates $f(t, x)$, for each of future time x, should be an independent random variable.

In the industry standard HJM model, as discussed in Section 4.11, the forward interest rate evolution equation is similar to a stock $S(t)$; this fact leads to the following major limitations of the HJM model.

- Forward interest rates are not directly observed, in contrast to Libor and Euribor, but instead are derived from traded instruments such as the Treasury Bond zero coupon yield

91

curve. This shortcoming is addressed by the Libor Market Model, discussed in some detail in Chapter 6.

- The forward interest rates in the HJM model are defined by the stochastic differential equation given in Eq. (4.38), which is driven by one random variable $R(t)$, similar to Eq. (3.10) for $S(t)$. For this reason, all the forward rates are exactly correlated, leading, for instance, to the unreasonable possibility of hedging a 30-year Treasury Bond with a six-month Treasury Bill.
- Empirical studies of the debt market show that the forward interest rates have *nontrivial correlation* in the future time direction. These correlations yield observable effects in the pricing and hedging of interest rate instruments.

One needs to look beyond the HJM model to describe the behavior of the forward interest rates. The limitations of the HJM model are redressed quite naturally in the framework of quantum finance. Forward interest rate models, based on quantum field theory, are able to incorporate correlations between forward interest rates with different maturities in a parsimonious and minimal manner. These models are computationally tractable and well suited for empirical implementation. This is the main motivation for studying forward interest rates from the point of view of quantum field theory.

Treating all the forward interest rates as independent random variables has been studied in [52, 67, 85]. In references [52] and [67] a correlation between forward interest rates with different maturities was introduced. In [85] the forward interest rates were modeled as a stochastic string, and a stochastic partial differential equation in infinitely many variables was obtained. A detailed discussion of the various generalizations of the HJM model, and their relation to the quantum field theory model of the forward rates, is given in [92].

In the quantum finance approach, the prices of all interest rate instruments are formally given as a path (functional) integral and hence it is complementary to the approach based on stochastic partial differential equations.

Quantum field theory models of interest rates are based on taking the interest rates as a strongly correlated system, with independent fluctuations for all maturities [14]. It is shown in [12] that the well-known results of the HJM model [56] can be obtained as a limiting case of the quantum finance model of the forward interest rates.

5.2 Bond forward interest rates: a quantum field

Recall, from Eq. (4.38), that in the HJM model the forward interest rates are defined by

$$\frac{\partial f}{\partial t}(t,x) = \alpha(t,x) + \sigma(t,x)R(t)$$

As discussed earlier, in the HJM model the fluctuations in the forward interest rates at a given time t are given by white noise $R(t)$ that delivers 'shocks' to the entire curve $f(t, x)$; white noise does not depend on the maturity direction x.

Figure 2.3 shows the forward interest rates obtained from the Libor and Euribor futures markets. One can see from the figure that each forward interest rate is evolving randomly in time. From the data, it is clear that, for each future time x, the forward interest rates $f(t, x)$ are evolving under the impact of independent random shocks.

It is natural to make a quantum finance generalization of the HJM model. The forward interest rates are *defined* by the following

$$\frac{\partial f}{\partial t}(t, x) = \alpha(t, x) + \sigma(t, x)\mathcal{A}(t, x) \tag{5.1}$$

$$f(t_*, x) = f(t_0, x) + \int_{t_0}^{t_*} dt\alpha(t, x) + \int_{t_0}^{t_*} dt\sigma(t, x)\mathcal{A}(t, x) \tag{5.2}$$

where $\mathcal{A}(t, x)$ is a generalization of white noise $R(t)$.

For the case where $\alpha(t, x)$ and $\sigma(t, x)$ are deterministic, $f(t, x)$ is called the *bond forward interest rates*. The quantum finance model proposed for $f(t, x)$ in Eq. (5.1) is appropriate for studying the bond sector of the debt market. For studying the interest rate component of the debt market, *Libor forward interest rates* is another collection of rates that is more suitable. The Libor case is nonlinear and has stochastic drift and volatility that are very different from the bond case.

The bond forward interest rates will be denoted by $f(t, x)$ for the rest of the book. In contrast, the Libor forward interest rates will be denoted by $f_L(t, x)$. Only the bond forward interest rates are discussed in this chapter and hence, unless necessary, they will be referred to only as forward interest rates. It should be noted that the terms bond and Libor forward interest rates refer to *models* that have been constructed to explain the market's behavior. The empirical forward interest rates are neither bond nor Libor forward interest rates, but, rather, are essentially the same for both the bond and interest rate sectors of the debt markets. It will be clear from the context what forward interest rates are being discussed.

The quantity $\mathcal{A}(t, x)$ is a (classical) stochastic field that delivers, for each future time x, independent 'shocks' to $f(t, x)$. Eq. (5.1) is very different from Eq. (4.38) of the HJM model, since now *both* the stochastic noise term $\mathcal{A}(t, x)$ and the forward interest rates $f(t, x)$ are on par. In fact, Eq. (5.1) is a change of variables from stochastic field $\mathcal{A}(t, x)$ to stochastic forward interest rates $f(t, x)$. Both fields are equally good for describing the interest rates, although, as one can imagine, the choice of which one to use depends on the financial instrument one is studying.

In quantum finance, all financial instruments, such as interest rate options and other derivatives, are defined by averaging the stochastic forward interest rates over all possible values; the averaging over the stochastic field is mathematically identical to the averaging in Euclidean quantum field theory.

The random evolution of the instantaneous forward interest rates implies that $f(t, x)$ is an *independent* random variable for *each x* and *each t*. Or equivalently, $f(t, x)$, for *each x and t*, is an *independent* integration variable. As discussed in Appendix A.7, the generic quantity describing such a system is a *quantum field* [95]. The forward interest rate is mathematically equivalent to a two-dimensional quantum field. In quantum finance, the techniques of quantum field theory are employed for modeling interest rates.

The theory of quantum fields [95] is a vast and complex subject that is at the leading edge of theoretical physics. Quantum field theory has been developed precisely to study problems involving infinitely many degrees of freedom and so one is naturally led to its techniques in the study of the interest yield curve.

For notational simplicity, both t and x are taken to be continuous. In Chapter 16, the lattice theory of the forward interest rates is defined by discretizing both t and x so that they take integer values in a finite set.

5.3 Forward interest rates: Lagrangian and action

The market price of an interest rate instrument, denoted by $F[A]$, is equal to its expectation value $E[F[A]]$ – obtained by performing an average over the two-dimensional quantum (random) field $A(t, x)$. Similar to the case of white noise – given in Eqs. (A.38), (A.39), and (A.40) – to evaluate $E[F[A]]$ one has the generic Feynman path integral given by

$$E[F[A]] = \frac{1}{Z} \int DA\, F[A]\, e^{S[A]}; \quad Z = \int DA e^{S[A]} \tag{5.3}$$

where $\int DA$ stands for integrating over all possible values of $A(t, x)$ – weighted by the probability measure e^S/Z. S is the 'action' for the quantum field $A(t, x)$, and Z is the 'partition function'; see Appendix A.7. One can equivalently do all the calculations directly using the forward interest rates since Eq. (5.1) defines, as follows, a change of variables from $A(t, x)$ to $f(t, x)$ with a constant Jacobian

$$A(t, x) = \frac{\partial f(t, x)/\partial t - \alpha(t, x)}{\sigma(t, x)} \quad \Rightarrow \quad \int DA = \text{constant} \int Df$$

$$\Rightarrow E[F[A]] = E[F[f]] = \frac{1}{Z} \int Df\, F[f]\, e^{S[f]}; \quad Z = \int Df e^{S[f]}$$

To choose an action $S[f]$ for $f(t, x)$, the domain over which t, x take values needs to be specified. For the sake of concreteness, consider the forward interest rates starting from initial calendar time T_i to a future calendar time T_f. Forward interest rates $f(t, x)$ only exist for the future, which yields $x > t$. The quantum field $f(t, x)$ is defined on the domain in the shape of a trapezoid $\mathcal{T}$ that is bounded by parallel lines $x = t$ and $x = T_{FR} + t$ in the maturity direction, and by the lines $t = T_i$ and $t = T_f$ in the time direction, as shown in Figure 5.1(a). *Every point inside the domain $\mathcal{T}$ represents an independent integration variable* $f(t, x)$, and shows the enormous increase over the HJM random variable $R(t)$ given in Figure A.1.

For a financial instrument that matures at some future time T_f, its behavior at earlier time $T_i < T_f$ is determined by the action

$$S[f] = \int_{T_i}^{T_f} dt \int_t^{t+T_{FR}} dx \mathcal{L}[f] \tag{5.4}$$

$$\equiv \int_{\mathcal{T}} \mathcal{L}[f] \tag{5.5}$$

where $\mathcal{L}(t, x)$ is the *Lagrangian* density for the forward interest rates.

What should terms should $\mathcal{L}(t, x)$ contain? What should be the form of $\mathcal{L}(t, x)$?

The HJM model shows that the forward interest rates have a drift velocity $\alpha(t, x)$ and volatility $\sigma(t, x)$; hence, these have to appear directly in the Lagrangian. To be well defined, the Lagrangian needs a kinetic term – denoted by $\mathcal{L}_{kinetic}$ – that is necessary to have forward interest rates' time evolution similar to a stock price.

The important insight of HJM [56] is that it is the combination $[\partial f(t, x)/\partial t - \alpha(t, x)]/\sigma(t, x)$ of the forward interest rates that occurs in finance. The dynamics

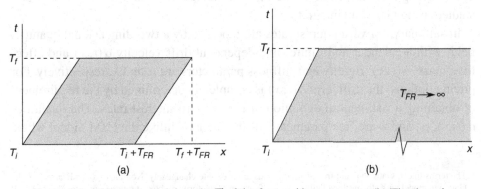

Figure 5.1 (a) Trapezoidal domain $\mathcal{T}$ of the forward interest rates. (b) The domain of forward interest rates for the limit of $T_{FR} \rightarrow \infty$.

of the forward interest rate, in particular its Lagrangian, are functions of the HJM combination.

There needs to be a term in the Lagrangian that constrains the change of shape of the forward interest rates in the future time direction x. Sharp changes in the shape of the forward interest rates need to be attenuated because the interest yield curve is not expected to change suddenly. The existence of a risk-neutral measure requires that the forward interest rates' Lagrangian contains only derivative terms in future time x [12]. Such systems have been studied in [81] and are said to be strings with finite rigidity.

The forward interest rates' Lagrangian has a piece $\mathcal{L}_{\text{rigidity}}$ that is of the form $(\partial^2 f/\partial x \partial t)^2$, with rigidity parameter μ quantifying the strength of this term. One has to further include yet another term in the Lagrangian, namely the stiffness term $\mathcal{L}_{\text{stiffness}}$, which has the form $(\partial^3 f/\partial x^2 \partial t)^2$; this term is needed for suppressing fluctuations that cause discontinuities in the *slope* of a correlation of forward interest rates [16].

Keeping in mind the considerations discussed above, the 'stiff' Lagrangian density $\mathcal{L}[f]$ for the bond forward rates is given by [12, 16]

$$\mathcal{L}[f] = \mathcal{L}_{\text{kinetic}}[f] + \mathcal{L}_{\text{rigidity}}[f] + \mathcal{L}_{\text{stiffness}}[f] \tag{5.6}$$

$$= -\frac{1}{2}\left[\left(\frac{\frac{\partial f(t,x)}{\partial t} - \alpha(t,x)}{\sigma(t,x)}\right)^2 + \frac{1}{\mu^2}\left\{\frac{\partial}{\partial x}\left(\frac{\frac{\partial f(t,x)}{\partial t} - \alpha(t,x)}{\sigma(t,x)}\right)\right\}^2\right.$$

$$\left. + \frac{1}{\lambda^4}\left\{\frac{\partial^2}{\partial x^2}\left(\frac{\frac{\partial f(t,x)}{\partial t} - \alpha(t,x)}{\sigma(t,x)}\right)\right\}^2\right]$$

$$-\infty \leq f(t,x) \leq +\infty \tag{5.7}$$

The Lagrangian is quadratic in the forward interest rates and can be analytically studied using Gaussian integrations.[1]

In summary, forward interest rates are modeled by a two-dimensional quantum field, with a calendar and future time dependent drift velocity $\alpha(t,x)$ and effective 'mass' $\sigma(t,x)$; rigidity and stiffness parameters are μ and λ, respectively. For pricing options, the drift term $\alpha(t,x)$ is completely determined by the requirement of obtaining a risk-neutral evolution of the forward interest rates. The quantities $\sigma(t,x)$, μ, and λ are free parameters of the theory. Unlike the HJM model where

---

[1] The terms linear, Gaussian, and free quantum fields are used interchangeably. Nonlinear generalizations of the Gaussian model with stochastic volatility that is a function of the forward interest rates are discussed in [12, 13]; the formulation can be further generalized to the case of linear forward rates with (nonlinear) stochastic volatility being another independent quantum field.

a functional form is usually assumed for the volatility function $\sigma(t,x)$, in the quantum finance approach the volatility function is determined from the market. In the limit of $\mu \to 0$, the HJM model is recovered (up to a rescaling), which corresponds to an infinitely rigid interest yield curve.

A remarkable fact about Eq. (5.6) is that only the time derivative of the forward interest rates, that is $\partial f(t,x)/\partial t$, appears in the Lagrangian. It will be shown in Chapter 15 that this translates into a Hamiltonian that has only kinetic terms and no (potential) terms (that depend on $f(t,x)$).

Since the field theory is defined on a finite domain $\mathcal{T}$, as shown in Figure 5.1(a), the boundary conditions need to be specified on all the four boundaries of the finite parallelogram $\mathcal{T}$.

- *Fixed (Dirichlet) initial and final conditions*
  The initial and final (Dirichlet) conditions in the time direction are given by

$$t = T_i \; : \; T_i < x < T_i + T_{FR} \; : \; f(T_i, x) \tag{5.8}$$

$$: \text{specified initial forward rate curve}$$

$$t = T_f \; : \; T_f < x < T_f + T_{FR} \; : \; f(T_f, x) \tag{5.9}$$

$$: \text{specified final forward rate curve}$$

- *Free (Neumann) boundary conditions*
  To specify the boundary condition in the maturity direction, one needs to analyze the action given in Eq. (5.4) and impose the condition that there are no surface terms in the action. The absence of surface terms is a necessary condition for the existence of a Hamiltonian. A straightforward analysis yields the following Neumann conditions

$$T_i < t < T_f, \quad \frac{\partial}{\partial x}\left(\frac{\frac{\partial f(t,x)}{\partial t} - \alpha(t,x)}{\sigma(t,x)} \right) = 0 \tag{5.10}$$

$$: x = t \text{ or } x = t + T_{FR} \tag{5.11}$$

The quantum field theory of the bond forward interest rates, for the finite domain $\mathcal{T}$, is defined by the Feynman path integral

$$Z = \int Df \, e^{S[f]} \tag{5.12}$$

$$\int Df \equiv \prod_{(t,x)\in\mathcal{T}} \int_{-\infty}^{+\infty} df(t,x) \tag{5.13}$$

$e^{S[f]}/Z$ is the probability for different field configurations to occur when the functional integral over $f(t,x)$ is performed.

The forward interest rates, starting from some initial calendar time $t = t_0$ can, in principle, be defined into the infinite future calendar time, that is to $t = \infty$. Since all options mature at some finite future calendar time, the limit of $t = \infty$ will not be necessary. On the other hand, the maximum future time $t + T_{FR}$ is so far in the future that the limit of $T_{FR} \rightarrow \infty$ is taken for most calculations. Hence, the domain $\mathcal{T}$ of the forward rates is extended, as shown in Figure 5.1(b), to a semi-infinite parallelogram that is bounded by parallel lines $t = T_i$ and $t = T_f$ and by the straight line $x = t$ (since forward interest rates exist only for future time $x > t$).

5.4 Velocity quantum field $\mathcal{A}(t, x)$

The action $S[f]$ expressed in terms of the forward interest rates, given in Eq. (5.4) is suitable for studying many properties of the debt market. In particular, in Chapter 15 the Hamiltonian of the interest rates is derived from $S[f]$; in Chapter 16 the algorithm for evaluating the coupon bond American option price is based on $S[f]$. It is, however, simpler for many computational purposes to change variables from quantum field $f(t, x)$ to quantum field $\mathcal{A}(t, x)$, which is the drift adjusted velocity of the forward interest rates.

Recall from Eq. (5.1) that $\mathcal{A}(t, x)$ is related to $f(t, x)$ by the transformation[2]

$$\frac{\partial f}{\partial t}(t, x) = \alpha(t, x) + \sigma(t, x)\mathcal{A}(t, x)$$

The quantum field theory is defined by a functional integral over all configurations of $\mathcal{A}(t, x)$ with the partition function given by

$$Z = \int D\mathcal{A}\, e^{S[\mathcal{A}]} \tag{5.14}$$

The action, in terms of the $\mathcal{A}(t, x)$ field, is given by

$$S[A] = \int_{\mathcal{T}} \mathcal{L}[\mathcal{A}] \tag{5.15}$$

$$= -\frac{1}{2}\int_{t_0}^{\infty} dt \int_{t}^{t+T_{FR}} dx \left\{ \mathcal{A}^2(t, x) + \frac{1}{\mu^2}\left(\frac{\partial \mathcal{A}(t, x)}{\partial x}\right)^2 + \frac{1}{\lambda^4}\left(\frac{\partial^2 \mathcal{A}(t, x)}{\partial^2 x}\right)^2 \right\}$$

The action $S[\mathcal{A}]$ given in Eq. (5.15) has *no time derivative couplings*. For each instant t, the $\mathcal{A}(t, x)$ is equivalent to a quantum mechanical system. This fact has far-reaching consequences. It is shown in Section 5.10 that absence of time derivatives

---

[2] The Jacobian of the above transformation is a constant.

leads to the Wilson expansion for $\mathcal{A}(t,x)$. This expansion is the mathematical basis for a generalization of Ito calculus; in particular, the Wilson expansion yields a differential derivation of the martingale condition in Section 5.12 as well as of Libor's drift in Section 6.7.

The field $\mathcal{A}(t,x)$, from Eq. (5.10), satisfies Neumann boundary conditions given by

$$\frac{\partial \mathcal{A}(t,x)}{\partial x}\Big|_{x=t} = 0 = \frac{\partial \mathcal{A}(t,x)}{\partial x}\Big|_{x=t+T_{FR}} \qquad (5.16)$$

The quantum field variables at the boundary $x=t$ and $x=t+T_{FR}$, namely $A(t,t)$ and $A(t, t+T_{FR})$ take all possible values, and result in the Neumann boundary conditions given above. In other words, the values of $\mathcal{A}(t,x)$ on the boundary of $\mathcal{T}$ are integration variables [12].

On integrating the future time variable in the action $S[A]$ by parts, the Neumann boundary conditions ensure that there are no surface terms and yield

$$S[A] = -\frac{1}{2}\int_{\mathcal{T}} \mathcal{A}(t,x)\left(1 - \frac{1}{\mu^2}\frac{\partial^2}{\partial x^2} + \frac{1}{\lambda^4}\frac{\partial^4}{\partial x^4}\right)\mathcal{A}(t,x) \qquad (5.17)$$

A more general Gaussian Lagrangian, which will be useful in studying the empirical behavior of interest rate instruments, is nonlocal in future time x and has the form

$$\mathcal{L}(A) = -\frac{1}{2}\mathcal{A}(t,x)\mathcal{N}^{-1}(t,x,x')\mathcal{A}(t,x') \qquad (5.18)$$

The HJM model, obtained in the limit of $\mu, \lambda \to 0$, leads to a drastic truncation of the full quantum field theory. The HJM model considers only the fluctuations of the average value of the quantum field $\mathcal{A}(t,x)$ and, in effect, 'freezes-out' all the other fluctuations of $\mathcal{A}(t,x)$. It is shown in [12] that, in the limit of $\mu \to 0$, the HJM model emerges from the field theory model in the following manner

$$\frac{\partial f}{\partial t}(t,x)\Big|_{HJM} = \alpha(t,x) + \sigma(t,x) \times \frac{1}{T_{FR}}\int_t^{t+T_{FR}} dx\,\mathcal{A}(t,x)$$

$$= \alpha(t,x) + \sigma(t,x)R(t)$$

where $R(t) = \int_t^{t+T_{FR}} dx\,\mathcal{A}(t,x)/T_{FR}$ is white noise.

One can think of the field $\mathcal{A}(t_0,x)$, at some instant t_0, as giving the position of a 'string' [85], as shown in Figure A.2. The action $S[A]$ given in Eq. (5.15) allows *all* points x of the field $\mathcal{A}(t_0,x)$ to fluctuate independently and can be thought of as a 'string' with rigidity equal to $1/\mu^2$. In the 'string' language, the HJM model of forward interest rates is a string with infinite rigidity.

5.5 Generating functional for $\mathcal{A}(t,x)$: propagator

The action for $\mathcal{A}(t,x)$, from Eq. (5.17) is given by

$$S[A] = -\frac{1}{2}\int_T \mathcal{A}(t,x)\mathcal{D}^{-1}(x,x';t)\mathcal{A}(t,x) \qquad (5.19)$$

$$\left(1 - \frac{1}{\mu^2}\frac{\partial^2}{\partial x^2} + \frac{1}{\lambda^4}\frac{\partial^4}{\partial x^4}\right)\mathcal{D}(x,x';t) = \delta(x-x') + \text{Neumann B.C.} \qquad (5.20)$$

The complete content of a quantum field is contained in its generating functional defined by

$$Z[h] = E\left[\exp\left\{\int_{t_0}^{\infty} dt \int_t^{\infty} dxh(t,x)\mathcal{A}(t,x)\right\}\right]$$

Functional differentiation of $Z[h]$ by $h(t,x)$, discussed in Section A.5, and setting $h(t,x) = 0$ yields all the correlation functions of $\mathcal{A}(t,x)$.

Quantum field $\mathcal{A}$ has a quadratic Lagrangian, as given in Eq. (5.20). Hence, the generating functional is evaluated exactly by Gaussian integration, reviewed in Section A.3 and yields the following

$$Z[h] = \frac{1}{Z}\int D\mathcal{A}\, e^{S[A]+\int_{t_0}^{\infty} dt \int_0^{\infty} dzh(t,z)\mathcal{A}(t,z)}$$

$$= \exp\left(\frac{1}{2}\int_{t_0}^{\infty} dt \int_t^{\infty} dxdx'h(t,z)\mathcal{D}(x,x';t)h(t,x')\right) \qquad (5.21)$$

The *propagator* (connected quadratic correlator) of the $\mathcal{A}(t,x)$ quantum field is given by

$$E\left[\mathcal{A}(t,z)\mathcal{A}(t',z')\right] = \frac{1}{Z}\int D\mathcal{A}e^{S[A]}\mathcal{A}(t,z)\mathcal{A}(t',z')$$

$$= \frac{\delta^2}{\delta h(t,x)\delta h(t',x')}Z[h]\Big|_{h=0}$$

$$= \delta(t-t')\mathcal{D}(x,x';t) \qquad (5.22)$$

The propagator $\delta(t-t')\mathcal{D}(x,x';t)$ is of central importance in the study of quantum fields. In fact, for Gaussian quantum fields it can be shown that the propagator encodes the full content of its quantum field theory [95].

Recall from Eq. (5.1) that $\mathcal{A}(t,x)$ is related to $f(t,x)$ by

$$\frac{\partial f}{\partial t}(t,x) = \alpha(t,x) + \sigma(t,x)\mathcal{A}(t,x)$$

Hence[3]

$$E\left[\frac{\partial f}{\partial t}(t,x)\right] = E[\alpha(t,x)] = \alpha(t,x)$$

$$E\left[\frac{\partial f}{\partial t}(t,x)\frac{\partial f}{\partial t'}(t',x')\right]_c = \sigma(t,x)\sigma(t',x')E[A(t,x)A(t',x')] \quad (5.23)$$

$$= \delta(t-t')\sigma(t,x)\mathcal{D}(x,x';t)\sigma(t,x') \quad (5.24)$$

showing that $\partial f(t,x)/\partial t$ has nontrivial correlations for future time x.

5.6 Future market time

The empirical analysis of forward interest rates leads to a further generalization of the interest rates' Lagrangian. Remaining future time $\theta = x - t$ is not what the market traders and practitioners perceive; instead, a modified form of time, called z: *future market time*, is what determines the price of debt instrument options.

The defining equation for future market time $z(\theta)$ is given by

$$\frac{\partial f}{\partial t}(t,t+\theta) = \alpha(t,z(\theta)) + \sigma(t,z(\theta))A(t,z(\theta)); \quad \theta = x - t$$

and which yields the following representation for zero coupon bonds

$$B(t,T) = \exp\left\{-\int_0^{T-t} d\theta f(t,t+\theta)\right\}$$

In Chapter 7 it is shown that market data imply a future time of the form $z = \theta^\eta$, where η is a dimensionless number with $\eta < 1$. The constants $\tilde{\mu}, \tilde{\lambda}$, which are more natural for the z variable, are defined as follows

$$z = (x-t)^\eta = \theta^\eta$$

$$\lambda z = [\tilde{\lambda}z]^\eta \; ; \; \mu z = [\tilde{\mu}z]^\eta \; : \; \text{dimensionless}$$

$$\text{dimension of } \tilde{\mu} = \text{dimension of } \tilde{\lambda} = 1/\text{time}$$

The stiff Lagrangian for the velocity field $A(t,x)$ is given by

$$\mathcal{L}[A] = -\frac{1}{2}\left\{A^2(t,z) + \frac{1}{\mu^2}\left(\frac{\partial A(t,z)}{\partial z}\right)^2 + \frac{1}{\lambda^4}\left(\frac{\partial^2 A(t,z)}{\partial^2 z}\right)^2\right\} \quad (5.25)$$

The action $S[A]$ of the Lagrangian, for $T_{FR} \to \infty$ yields the following

$$S[A] = \int_{T_i}^{T_f} dt \int_0^\infty dz \mathcal{L}[A] \quad (5.26)$$

---

[3] Note the connected correlation of AB is given by $E[AB]_c = E[AB] - E[A]E[B]$.

Suppose one were to write a Lagrangian in remaining future time $x - t$ such that it produces the results of the Lagrangian with market time given in Eq. (5.26). Such a Lagrangian would need to be nonlinear, having new terms in addition to ones given in Eq. (5.17). By introducing market time one is, in effect, describing the forward interest rates with a nonlinear Lagrangian; all the nonlinearities are encoded in market time, with the quantum field $\mathcal{A}(t, z)$ remaining Gaussian. Hence, the Lagrangian given in Eq. (5.26) is pseudo-Gaussian, being nonlinear when expressed in terms of the θ-variable.

5.7 Stiff propagator

Consider the special case of the stiff Lagrangian with $\eta = 1$, that is $z = x - t$; the case of $\eta \neq 1$ will be derived by a change of variables from the simpler case. The propagator is given by[4]

$$G(x, x'; t) = \lambda^4 < x | \frac{1}{\lambda^4 + (\lambda^2/\mu)^2 p^2 + p^4} | x' > \qquad (5.27)$$

$$\text{where} \quad p^2 \equiv -\frac{\partial^2}{\partial x^2}$$

Define new variables

$$\theta_\pm = \theta \pm \theta'; \quad z(\theta) = \theta^\eta; \quad z' = z(\theta')$$

$$\theta = x - t; \quad \theta' = x' - t$$

$$\alpha_\pm = \frac{\lambda^4}{2\mu^2} \left[1 \pm \sqrt{1 - 4 \left(\frac{\mu}{\lambda}\right)^4} \right]$$

Eq. (5.27) yields [12]

$$G(\theta_+; \theta_-) = \left(\frac{\lambda^4}{\alpha_+ - \alpha_-}\right) \left[\frac{1}{\alpha_-} D_-(\theta_+; \theta_-) - \frac{1}{\alpha_+} D_+(\theta_+; \theta_-) \right] \qquad (5.28)$$

where

$$D_\pm(\theta_+; \theta_-) = \frac{\sqrt{\alpha_\pm}}{2} \left[e^{-\sqrt{\alpha_\pm}\theta_+} + e^{-\sqrt{\alpha_\pm}|\theta_-|} \right] \qquad (5.29)$$

The stiff propagator for $\eta \neq 1$ is given by

$$\mathcal{D}(z, z'; t) = G(z(\theta_+); z(\theta_-)); \quad z(\theta_\pm) = z(\theta) \pm z(\theta') \qquad (5.30)$$

---

[4] Henceforth $T_{FR} \to \infty$.

All the pricing formulas for interest rate options depend on the following:

- The volatility function $\sigma(t, x)$
- Parameters μ, λ, η in the Lagrangian and
- The initial term structure $f(t_0, x)$

For notational simplicity and unless it is necessary otherwise, only the case of $\eta = 1$ will be considered; in other words, all integrations over z are replaced with those over future time x. For $\eta = 1$ the dimension of the quantum field $\mathcal{A}(t, x)$ is 1/time and volatility $\sigma(t, x)$ of the forward interest rates also has dimension of 1/time.

In many cases, where an empirical analysis of an interest option is carried out, the explicit value of the propagator $\mathcal{D}(z, z'; t)$ is not used. Instead, only the Gaussian property of the Lagrangian is used. In particular, the effective propagator that describes the market behavior of interest rate options is given by $M(x, x'; t) = \sigma(t, x)\mathcal{D}(x, x'; t)\sigma(t, x')$. In the empirical studies of swaptions carried out in Chapter 12, $M(x, x'; t)$ is evaluated directly from market data.

The expression for $\mathcal{D}(z, z'; t)$ given in Eq. (5.30) is discussed in Chapter 7 and is shown to provide a very accurate description of the correlation of the forward interest rates for both the Libor and Euribor forward interest rates. The stiff propagator's prediction on the correlation function of the changes in the forward interest rates matches market data to an accuracy of over 99%.

5.8 Integral condition for interest rates' martingale

The integral martingale condition for zero coupon bonds is given in Eq. (3.4) as follows

$$B(t_0, T) = E[e^{-\int_{t_0}^{t_*} r(t)dt} B(t_*, T)] \qquad (5.31)$$

and has the following functional integral representation in quantum field theory

$$B(t_0, T) = \frac{1}{Z} \int Df e^{-\int_{t_0}^{t_*} r(t)dt} B(t_*, T) e^{S[f]} \qquad (5.32)$$

A change of variables from $f(t, x)$ to $\mathcal{A}(t, x)$ is given, from Eq. (5.1), by

$$f(t, x) = f(t_0, x) + \int_{t_0}^{t'} dt' \alpha(t', x) + \int_{t_0}^{t'} dt' \sigma(t', x)\mathcal{A}(t', x)$$

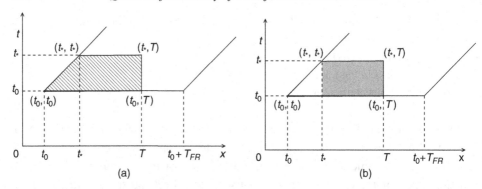

Figure 5.2 (a) Shaded trapezoidal domain $\mathcal{T}$ of the forward interest rates required for money market numeraire martingale condition. (b) Shaded rectangular domain $\mathcal{R}$ of the forward interest rates required for forward numeraire martingale condition.

Eqs. (5.1), (5.31), and (5.32) yield[5]

$$\exp \int_{\mathcal{T}} \alpha(t,x) = \frac{1}{Z} \int D\mathcal{A} e^{-\int_{\mathcal{T}} \sigma(t,x)\mathcal{A}(t,x)} e^{S[\mathcal{A}]} \qquad (5.33)$$

$$= \exp \frac{1}{2} \int_{t_0}^{t_*} dt \int_t^T dx dx' \sigma(t,x)D(x,x';t,T_{FR})\sigma(t,x') \qquad (5.34)$$

where the last equation follows from the generating functional given in Eq. (5.21). The trapezoidal domain $\mathcal{T}$, as shown in Figure 5.2(a), is the domain of the forward interest rates required for carrying out the calculation.

Dropping the time integration in Eq. (5.34) yields

$$\int_t^T dx \alpha(t,x) = \frac{1}{2} \int_t^T dx dx' \sigma(t,x)D(x,x';t,T_{FR})\sigma(t,x') \qquad (5.35)$$

Differentiating the above expression with respect to T yields the drift velocity

$$\alpha(t,x) = \sigma(t,x) \int_t^x dx' D(x,x';t,T_{FR})\sigma(t,x') \qquad (5.36)$$

The drift $\alpha(t,x)$ obtained in Eq. (5.36) is for the money market numeraire. One can, instead of the money market numeraire, choose the forward bond numeraire given by the zero coupon bond $B(t,t_*)$; t_*: fixed. For the forward bond numeraire,

---

[5] Integration over the trapezoidal domain $\mathcal{T}$ is defined by $\int_{\mathcal{T}} = \int_{t_0}^{t_*} dt \int_t^T dx$.

the drift, given by $\alpha_F(t, x)$, is fixed by the martingale condition similar to Eq. (3.8), namely that

$$\frac{B(t_0, T)}{B(t_0, t_*)} = E_F\left[\frac{B(t_*, T)}{B(t_*, t_*)}\right] = E_F[B(t_*, T)]$$

$$\Rightarrow B(t_0, T) = B(t_0, t_*)E_F[B(t_*, T)]$$

A calculation similar, but simpler, to the one for obtaining the drift for the money market numeraire gives the drift $\alpha_F(t, x)$. The domain required for the forward numeraire calculation is given in Figure 5.2(b) and yields

$$\alpha_F(t, x) = \sigma(t, x)\int_{t_*}^{x} dx' D(x, x'; t, T_{FR})\sigma(t, x') \qquad (5.37)$$

For most cases, the subscript F in the drift and expectation value will be omitted as it will be clear from the context which numeraire is being used.

In summary, the martingale condition determines the drift $\alpha(t, x)$ in the action $S[\mathcal{A}]$.[6]

5.9 Pricing kernel and path integration

Consider a European option on the forward interest rates that matures at calendar time t_* and has a payoff function $\mathcal{P}[f(t_*, \cdot)]$.[7] The price of the option at time t_0, using the forward numeraire, is given from Eq. (3.8), by the following

$$C[t_0, t_*; f(t_0, \cdot)] = B(t_0, t_*)\int Df_* \mathcal{K}[f(t_0, \cdot); f(t_*, \cdot)]\mathcal{P}[f(t_*, \cdot)]$$

$$\int Df_* \equiv \prod_{x=t_*}^{+\infty}\int_{-\infty}^{+\infty} df(t_*, x)$$

The pricing kernel $\mathcal{K}[f(t_0, \cdot); f(t_*, \cdot)]$, expressed as a Feynman path integral, provides an exemplar of the path integral as a computational tool. The pricing kernel is the conditional probability of the occurrence of the initial forward interest rates $f(t_0, \cdot)$, given the occurrence of the final values $f(t_*, \cdot)$. The path integral is a sum over all possible configurations of the forward interest rates that start with $f(t_0, \cdot)$

---

[6] The result obtained here has been obtained using path integration. A Hamiltonian derivation is given of the drift in Section 9.3.

[7] The notation $f(t, \cdot)$ means that the quantity in question depends on all the forward interest rates at time t, namely on $f(t, x)$; $x \in [t, t + T_{FR}]$.

and end at $f(t_*, \cdot)$. It is convenient to evaluate the path integral using the velocity field $\mathcal{A}(t, x)$; from Eq. (5.1)

$$f(t_*, x) = f(t_0, x) + \int_{t_0}^{t_*} dt\, \alpha(t, x) + \int_{t_0}^{t_*} dt\, \sigma(t, x)\mathcal{A}(t, x)$$

In the equation above, all configurations of $\mathcal{A}(t, x)$ have the initial condition $f(t_0, \cdot)$ built into the change of variables. The final condition has to be imposed on the path integral, which is implemented by a Dirac-delta function.

Hence, the pricing kernel is given by

$$K[f(t_0, \cdot); f(t_*, \cdot)] = \frac{1}{Z} \int D\mathcal{A} \prod_{t_* \leq x \leq \infty} \delta\left[F(x) - \int_{t_0}^{t_*} dt\, \sigma(t, x)\mathcal{A}(t, x) \right] e^{S[\mathcal{A}]}$$

$$F(x) = f(t_*, x) - f(t_0, x) - \int_{t_0}^{t_*} dt\, \alpha(t, x)$$

$$S[\mathcal{A}] = -\frac{1}{2} \int_{t_0}^{t_*} dt \int_{t}^{\infty} dx\, \mathcal{A}(t, x)\mathcal{D}^{-1}(x, x'; t)\mathcal{A}(t, x')$$

where the action is given in Eq. (5.19). The Dirac-delta function is represented as follows

$$\prod_{t_* \leq x \leq \infty} \delta\left[F(x) - \int_{t_0}^{t_*} dt\, \sigma(t, x)\mathcal{A}(t, x) \right] = \int DK e^{i \int_x k(x) \left[F(x) - \int_{t_0}^{t_*} dt\, \sigma(t, x)\mathcal{A}(t, x) \right]}$$

$$\int DK = C \prod_{t_* \leq x \leq \infty} \int_{-\infty}^{+\infty} dk(x); \quad \int_{t_*}^{\infty} dx \equiv \int_x$$

where C is a normalization constant.

The Gaussian path integral $\int D\mathcal{A}$ can be performed exactly using the generating functional given in Eq. (5.21) and yields

$$K[f(t_0, \cdot); f(t_*, \cdot)] = \int DK e^{i \int_x k(x) F(x)} e^{-\frac{1}{2} \int_{x,x'} k(x) M(x, x') k(x')}$$

$$= \mathcal{N} \exp\left\{ -\frac{1}{2} \int_{x,x'} F(x) M^{-1}(x, x') F(x') \right\}$$

$$M(x, x') = \int_{t_0}^{t_*} dt\, \sigma(t, x)\mathcal{D}(x, x'; t)\sigma(t, x') \tag{5.38}$$

The normalization constant $\mathcal{N} = \mathcal{N}(t_0, t_*, \sigma(t, x), \mathcal{D})$ can be evaluated by a detailed calculation; it can also be obtained by demanding that the pricing kernel obeys the composition law for conditional probabilities [47].

5.9.1 Infinitesimal limit

Consider an infinitesimal transition from time t to time $t_* = t + \epsilon$. This yields, from Eq. (5.38)

$$M(x, x') \rightarrow \epsilon \sigma(t, x) \mathcal{D}(x, x'; t) \sigma(t, x')$$

$$\Rightarrow M^{-1}(x, x') \rightarrow \frac{1}{\epsilon} \frac{1}{\sigma(t, x)} \mathcal{D}^{-1}(x, x'; t) \frac{1}{\sigma(t, x)}$$

$$F(x) \rightarrow F_\epsilon(x) = f(t + \epsilon, x) - f(t, x) - \epsilon \alpha(t, x)$$

$$\mathcal{K}[f(t, \cdot); f(t + \epsilon, \cdot)] \rightarrow$$

$$\mathcal{N} \exp \left\{ -\frac{1}{2\epsilon} \int_{x, x'} \frac{F_\epsilon(x)}{\sigma(t, x)} \mathcal{D}^{-1}(x, x'; t) \frac{F_\epsilon(x')}{\sigma(t, x')} \right\} \quad (5.39)$$

Consider the action for infinitesimal time; to simplify the notation, define the following

$$\frac{\partial f(t, x)}{\partial t} = \frac{1}{\epsilon}[f(t + \epsilon, x) - f(t, x)]$$

$$\mathcal{A}_\epsilon(t, x) = \frac{1}{\epsilon}[f(t + \epsilon, x) - f(t, x) - \epsilon \alpha(t, x)] = \frac{1}{\epsilon} F_\epsilon(x)$$

From Eq. (5.19), the time integral in the action reduces to ϵ and yields

$$S_\epsilon[f] = \epsilon \int_t^\infty dx \mathcal{L}[f(t, x); f(t + \epsilon, x)]$$

$$= -\frac{\epsilon}{2} \int_t^\infty dx dx' \mathcal{A}_\epsilon(t, x) \mathcal{D}^{-1}(x, x'; t) \mathcal{A}_\epsilon(t, x')$$

$$\simeq -\frac{1}{2\epsilon} \int_t^\infty dx dx' \frac{F_\epsilon(x)}{\sigma(t, x)} \mathcal{D}^{-1}(x, x') \frac{F_\epsilon(x')}{\sigma(t, x')} \quad (5.40)$$

On comparing Eq. (5.39) with Eq. (5.40), it is seen that the pricing kernel for ϵ time evolution is simply the action for infinitesimal time, namely that

$$\mathcal{K}[f(t, \cdot); f(t + \epsilon, \cdot)] = \mathcal{N} \exp \left\{ S_\epsilon[f] \right\}$$

$$= \mathcal{N} \exp \left\{ \epsilon \int_t^\infty dx \mathcal{L}[f(t, x); f(t + \epsilon, x)] \right\} \quad (5.41)$$

The pricing kernel will be given a Hamiltonian interpretation in Chapter 15. The American option's pricing will be seen, in Sections 16.2 and 16.5, to hinge on the properties of the pricing kernel of the forward interest rates.

Eq. (5.41) has been derived in the context of Gaussian quantum fields for which the path integral can be explicitly evaluated. However, Eq. (5.41) is of great generality and holds true for any action for which a Lagrangian exists; Eq. (5.41) states that the pricing kernel, or more technically, the conditional probability, for infinitesimal time is, up to a normalization, given by the exponential of the action for infinitesimal time. In physics, Eq. (5.41) is called the Dirac–Feynman relation.

5.10 Wilson expansion of quantum field $\mathcal{A}(t,x)$

Modeling in finance widely uses the concept of stochastic differential equations and of Ito calculus. The Wilson expansion of quantum fields is a very general technique that allows one to isolate the singularities in the product of quantum fields [93, 94]. In the context of mathematical finance, the Wilson expansion provides a generalization of Ito calculus to the case where the stochastic phenomenon is driven by the two-dimensional Gaussian quantum field $\mathcal{A}(t, x)$ [3].

The time derivative of various quantities like the underlying security $S(t)$ or stochastic volatility are generically expressed as follows

$$\frac{dS(t)}{dt} = \mu(t) + \sigma(t)R(t)$$

Ito's stochastic calculus, for discrete time $t = n\epsilon$, is a result of the following identity [12]

$$E[R(t)R(t)] = \delta(t - t') \quad \Rightarrow \quad R^2(t) = \frac{1}{\epsilon} + O(1) \qquad (5.42)$$

The singular piece of $R^2(t)$ is *deterministic*, namely equal to $1/\epsilon$; all the random terms that occur for $R^2(t)$ are finite as $\epsilon \to 0$.

Interest rate models need to incorporate future time x and both the HJM [56] and BGM–Jamshidian [32, 62] models are expressed as functions of white noise, given by

$$\frac{\partial f(t,x)}{\partial t} = \alpha(t,x) + \sigma(t,x)R(t) \qquad (5.43)$$

$$\frac{1}{L(t, T_k)}\frac{\partial L(t, T_k)}{\partial t} = \zeta_k(t) + \gamma_k(t)R(t); \quad T_k = \ell k \qquad (5.44)$$

Note that future time x has been introduced in the HJM and BGM models only in the drift and volatility of the interest rates term structure, and the same single white noise $R(t)$ drives the entire forward interest rates' curve.

The two-dimensional quantum field $\mathcal{A}(t, x)$ is an integration variable for each t and each x. For Gaussian quantum fields such as $\mathcal{A}(t, x)$ that have a quadratic action, one can give differential formulation of the theory of forward interest rates similar to HJM and BGM. This is possible because the full content of a Gaussian (free) quantum field, as discussed in Eq. (5.22), is encoded in its *propagator*.

The quantum finance *differential formulation* of Libor and the bond forward interest rates – generalizing the HJM and BGM–Jamshidian models – is given by 'promoting' white noise $R(t)$ to a two-dimensional quantum field $\mathcal{A}(t, x)$ and yields the following

$$\frac{\partial f(t, x)}{\partial t} = \alpha(t, x) + \sigma(t, x)\mathcal{A}(t, x) \tag{5.45}$$

$$f(t, x) = f(t_0, x) + \int_{t_0}^{t} dt\alpha(t, x) + \int_{t_0}^{t} dt\sigma(t, x)\mathcal{A}(t, x)$$

$$\frac{1}{L(t, T_k)}\frac{\partial L(t, T_k)}{\partial t} = \zeta_k(t) + \int_{T_k}^{T_{k+1}} dx\gamma(t, x)\mathcal{A}_L(t, x)$$

$$E[\mathcal{A}(t, x)\mathcal{A}(t', x')] = \delta(t - t')\mathcal{D}(x, x'; t)$$

Libor $L(t, T_k)$ and quantum field $\mathcal{A}_L(t, x)$ are discussed in detail in Chapter 6. An empirical analysis of Libor data, carried out in Chapter 7, shows that the quantum field $\mathcal{A}_L(t, x)$ driving Libor is also defined by a stiff Lagrangian, but with parameters and volatilities that are different from $\mathcal{A}(t, x)$.

Similar to white noise, the correlation function $E[\mathcal{A}(t, x)\mathcal{A}(t', x')]$ is infinite for $t = t'$ (equal calendar time). The product of nonlinear (non-Gaussian) quantum fields is the subject matter of what is called the short 'distance' Wilson expansion [93]. The singular product of two Gaussian quantum fields is the simplest case of the Wilson expansion. The singularity of the correlation function in the product of the quantum field $\mathcal{A}(t, x)$, similar to Eq. (5.42), is expressed as follows

$$\mathcal{A}(t, x)\mathcal{A}(t, x') = \frac{1}{\epsilon}\mathcal{D}(x, x'; t) + O(1) \tag{5.46}$$

All the fluctuating components, which are contained in $\mathcal{A}(t, x)\mathcal{A}(t, x')$, are regular and finite as $\epsilon \to 0$. The correlation of $\mathcal{A}(t, x)$ is singular for $t = t'$ – very much like the singularity of white noise $R(t)$.

Since $\mathcal{A}(t, x)$ is an integration variable for each x and each t, one may question as to how one can assign it a deterministic numerical value as in Eq. (5.46)? What Eq. (5.46) means is that in any correlation function, wherever a product of fields is at the same time, namely $\mathcal{A}(t, x)\mathcal{A}(t, x')$, then – to leading order in ϵ – the product can be replaced by the deterministic quantity $\mathcal{D}(x, x'; t)/\epsilon$. In terms of symbols, Eq. (5.46) states the following

$$E[\mathcal{A}(t_1, x_1)\mathcal{A}(t_2, x_2)\ldots\mathcal{A}(t, x_n)\mathcal{A}(t, x_{n+1})\ldots\mathcal{A}(t_N, x_N)]$$

$$= \frac{1}{\epsilon}E[\mathcal{A}(t_1, x_1)\mathcal{A}(t_2, x_2)\ldots\mathcal{A}(t_{n-1}, x_{n-1})\mathcal{D}(x_n, x_{n+1}; t)$$

$$\times \mathcal{A}(t_{n+2}, x_{n+2})\ldots\mathcal{A}(t_N, x_N)] + O(1)$$

As discussed in Section 7.3, one can choose the normalization of $\sigma(t, x)$ so that $\mathcal{D}(x, x; t) = \frac{1}{\epsilon}$ and which yields from Eq. (5.46)

$$\mathcal{A}^2(t, x) = \frac{1}{\epsilon^2} + O(1) \tag{5.47}$$

showing even more clearly the similarity of Eqs. (5.42) and (5.47).

The HJM model is a special case of quantum finance, given by taking the limit

$$\mathcal{A}(t, x) \to R(t); \quad \mathcal{D}(x, x'; t) \to 1 \tag{5.48}$$

$$\Rightarrow \quad E[\mathcal{A}(t, x)\mathcal{A}(t', x')] \to E[R(t)R(t)] = \delta(t - t')$$

Since Eqs. (5.42) and (5.46) have a similar singularity structure, one expects that there should be a natural generalization of Ito calculus for Gaussian quantum fields. The singularity of the equal time quadratic product of the quantum field, in particular, leads to a differential formulation of the martingale condition for discounted zero coupon bonds and is discussed in Section 5.12.

5.11 Time evolution of a bond

To illustrate the content of the singularity in the equal time quadratic product of the quantum field, namely $\mathcal{A}(t, x)\mathcal{A}(t, x')$ given in Eq. (5.46), a concrete analysis is carried out of the evolution of zero coupon bond – with and without discounting.

Consider a zero coupon bond. Eq. (5.2) yields the following

$$B(t_*, T) = \exp\left\{ -\int_{t_*}^{T} dx f(t_*, x) \right\}; \quad F(t_0, t_*, T) = \exp\left\{ -\int_{t_*}^{T} dx f(t_0, x) \right\}$$

$$B(t_*, T) = F(t_0, t_*, T)$$

$$\times \exp\left\{ -\int_{t_0}^{t_*} dt \int_{t_*}^{T} dx \alpha(t, x) - \int_{t_0}^{t_*} dt \int_{t_*}^{T} dx \sigma(t, x)\mathcal{A}(t, x) \right\} \tag{5.49}$$

The domain of integration is a rectangle $\mathcal{R}$, equal to $[t_0, t_*] \times [t_*, T]$ and shown in Figure 5.2(b).

To calculate the time evolution of the bond one needs to compute the time derivative of $\exp\{-\int_{t_0}^{t_*} dt \int_{t_*}^{T} dx \sigma(t, x)\mathcal{A}(t, x)\}$; to do this computation one needs to take account of the fact that the quadratic power of $\mathcal{A}(t, x)$ is singular. Consider[8]

$$\exp\left\{ +\int_{t_0}^{t_*} dt \int_{t_*}^{T} dx \sigma(t, x)\mathcal{A}(t, x) \right\} \frac{\partial}{\partial t_*} \exp\left\{ -\int_{t_0}^{t_*} dt \int_{t_*}^{T} dx \sigma(t, x)\mathcal{A}(t, x) \right\}$$

$$= \frac{1}{\epsilon}\left[\exp\left\{ -\epsilon \int_{t_*}^{T} dx \sigma(t_*, x)\mathcal{A}(t_*, x) + \epsilon \int_{t_0}^{t_*} dt \sigma(t, t_*)\mathcal{A}(t, t_*) \right\} - 1 \right]$$

---

[8] Note, on the right-hand side of the equation, the positive sign of the second term in the exponent.

Expanding the exponential to second order yields all the nontrivial terms, as follows[9]

$$\frac{1}{\epsilon}\left[\exp\left\{-\epsilon\int_{t_*}^{T}dx\sigma(t_*,x)\mathcal{A}(t_*,x)+\epsilon\int_{t_0}^{t_*}dt\sigma(t,t_*)\mathcal{A}(t,t_*)\right\}-1\right]$$

$$=-\int_{t_*}^{T}dx\sigma(t_*,x)\mathcal{A}(t_*,x)+\int_{t_0}^{t_*}dt\sigma(t,t_*)\mathcal{A}(t,t_*)$$

$$+\frac{\epsilon}{2}\left(\int_{t_*}^{T}dx\sigma(t_*,x)\mathcal{A}(t_*,x)\right)^2+O(\epsilon)$$

$$=-\int_{t_*}^{T}dx\sigma(t_*,x)\mathcal{A}(t_*,x)+\int_{t_0}^{t_*}dt\sigma(t,t_*)\mathcal{A}(t,t_*)$$

$$+\frac{1}{2}\int_{t_*}^{T}dx\int_{t_*}^{T}dx'\mathcal{M}(x,x';t_*)$$

since, from Eq. (5.46)

$$\epsilon\left(\int_{t_*}^{T}dx\sigma(t_*,x)\mathcal{A}(t_*,x)\right)^2=\int_{t_*}^{T}dx\int_{t_*}^{T}dx'\mathcal{M}(x,x';t)$$

$$\mathcal{M}(x,x';t_*)\equiv\sigma(t_*,x)\sigma(t_*,x')\mathcal{D}(x,x';t)$$

Collecting all the results yields

$$\exp\left\{+\int_{t_0}^{t_*}dt\int_{t_*}^{T}dx\sigma(t,x)\mathcal{A}(t,x)\right\}$$

$$\times\frac{\partial}{\partial t_*}\exp\left\{-\int_{t_0}^{t_*}dt\int_{t_*}^{T}dx\sigma(t,x)\mathcal{A}(t,x)\right\}$$

$$=-\int_{t_*}^{T}dx\sigma(t_*,x)\mathcal{A}(t_*,x)+\int_{t_0}^{t_*}dt\sigma(t,t_*)\mathcal{A}(t,t_*)$$

$$+\frac{1}{2}\int_{t_*}^{T}dx\int_{t_*}^{T}dx'\mathcal{M}(x,x';t_*)$$

The only term in the zero coupon bond that needs to be examined carefully is the one involving $\mathcal{A}(t,x)$; the other terms all obey the usual rules of calculus. Hence,

---

[9] The second term in the exponential, namely $\int_{t_0}^{t_*}dt\sigma(t,t_*)\mathcal{A}(t,t_*)$, does not contribute to the singular piece; when this term is squared the singular term coming from the quadratic product of $\mathcal{A}(t,x)$ at equal time is canceled by the vanishing integration measure.

since $\partial F(t_0, t_*, T)/\partial t_* = +f(t_0, t_*) F(t_0, t_*, T)$

$$\frac{1}{B(t_*, T)} \frac{\partial B(t_*, T)}{\partial t_*} = f(t_0, t_*) - \int_{t_*}^{T} dx\, \alpha(t_*, x) + \int_{t_0}^{t_*} dt\, \alpha(t, t_*)$$

$$- \int_{t_*}^{T} dx\, \sigma(t_*, x) \mathcal{A}(t_*, x) + \int_{t_0}^{t_*} dt\, \sigma(t, t_*) \mathcal{A}(t, t_*)$$

$$+ \frac{1}{2} \int_{t_*}^{T} dx \int_{t_*}^{T} dx'\, M(x, x'; t_*)$$

Using the no-arbitrage condition on $\alpha(t, x)$ given in Eq. (5.36) and combining terms using Eq. (5.2) yields the result

$$\frac{\partial B(t_*, T)}{\partial t_*} = \left[f(t_*, t_*) - \int_{t_*}^{T} dx\, \sigma(t_*, x) \mathcal{A}(t_*, x) \right] B(t_*, T) \qquad (5.50)$$

$$\Rightarrow E\left[\frac{\partial B(t_*, T)}{\partial t_*} \right] = f(t_*, t_*) B(t_*, T) \neq 0$$

The zero coupon bond is not a martingale since it has nonzero expectation value and hence does not satisfy Eq. (A.17).

5.12 Differential martingale condition for bonds

A fundamental theorem of finance states that for a derivative instrument to have a price that does not allow for any arbitrage opportunities, the instrument must have a martingale evolution. The martingale condition, in turn, uniquely fixes the drift term $\alpha(t, x)$. The drift term has been obtained in Eq. (5.36) using Gaussian integration and is now re-derived using the differential formulation of the forward interest rates given in Eq. (5.1). A derivation, similar to Section 5.11, involving the ϵ approach is carried out to further illustrate and clarify the behavior of the quantum field $\mathcal{A}(t, x)$.

The *discounted* zero coupon bond is given by

$$D(t_*, T) \equiv \exp\left\{ -\int_{t_0}^{t_*} dt\, f(t, t) \right\} B(t_*, T)$$

From Eq. (A.17), the differential formulation of the martingale condition requires the following

$$E\left[\frac{\partial D(t_*, T)}{\partial t_*} \right] = 0 \qquad (5.51)$$

It can be shown, similar to Eq. (5.49), that

$$D(t_*, T) = B(t_0, T)$$

$$\times \exp\left\{-\int_{t_0}^{t_*} dt \int_t^T dx\alpha(t, x) - \int_{t_0}^{t_*} dt \int_t^T dx\sigma(t, x)A(t, x)\right\} \quad (5.52)$$

The domain of integration of $f(t, x)$ is over a trapezoid $\mathcal{T}$ shown in Figure 5.2(a). In contrast, the domain of integration of $f(t, x)$ for the (nondiscounted) zero coupon bond, as given in Eq. (5.49), was over a rectangular domain $\mathcal{R}$ shown in Figure 5.2(b).

The rate of change of the discounted zero coupon bond is given by

$$\frac{1}{D(t_*, T)} \frac{\partial D(t_*, T)}{\partial t_*} \equiv \frac{1}{\epsilon} \frac{[D(t_* + \epsilon, T) - D(t_*, T)]}{D(t_*, T)}$$

$$= \frac{1}{\epsilon} \left[\exp\left\{-\epsilon \int_{t_*}^T dx\alpha(t_*, x) - \epsilon \int_{t_*}^T dx\sigma(t_*, x)A(t_*, x)\right\} - 1\right]$$

$$= -\int_{t_*}^T dx\alpha(t, x) - \int_{t_*}^T dx\sigma(t_*, x)A(t_*, x) + \frac{\epsilon}{2}\left(\int_{t_*}^T dx\sigma(t_*, x)A(t_*, x)\right)^2$$

and hence

$$\frac{1}{D(t_*, T)} \frac{\partial D(t_*, T)}{\partial t_*} \quad (5.53)$$

$$= -\int_{t_*}^T dx\alpha(t_*, x) - \int_{t_*}^T dx\sigma(t_*, x)A(t_*, x) + \frac{1}{2}\int_{t_*}^T dxdx'\mathcal{M}(x, x'; t_*)$$

The martingale condition given in Eq. (5.51) requires that the drift be zero and hence, from $E[A(t, x)] = 0$, yields

$$0 = \int_{t_*}^T dx\alpha(t_*, x) - \frac{1}{2}\int_{t_*}^T dxdx'\mathcal{M}(x, x'; t_*)$$

$$\Rightarrow \alpha(t, x) = \int_t^x dx'\mathcal{M}(x, x'; t) = \sigma(t, x)\int_t^x dx'\mathcal{D}(x, x'; t)\sigma(t, x') \quad (5.54)$$

This is the result obtained earlier, in Section 5.8, using Gaussian path integration. Hence, from Eqs. (5.53) and (5.54)

$$\frac{\partial D(t_*, T)}{\partial t_*} \equiv \frac{\partial}{\partial t_*}\left[\exp\left\{-\int_{t_0}^{t_*} dtf(t, t)\right\}B(t_*, T)\right]$$

$$= -\left[\int_{t_*}^T dx\sigma(t_*, x)A(t_*, x)\right]D(t_*, T) \quad (5.55)$$

The evolution of the discounted zero coupon bond satisfies the martingale condition given in Eq. (5.51).

Due to the simple dependence on t_* of $\mathcal{T}$, the domain of integration of $f(t,x)$ for $D(t_*, T)$ as shown in Figure 5.2(a), one can derive the martingale condition for the drift $\alpha(t,x)$ by directly differentiating on t_*; more precisely

$$
\exp\left\{+\int_{t_0}^{t_*} dt \int_{t}^{T} dx\sigma(t,x)\mathcal{A}(t,x)\right\} \frac{\partial}{\partial t_*} \exp\left\{-\int_{t_0}^{t_*} dt \int_{t}^{T} dx\sigma(t,x)\mathcal{A}(t,x)\right\}
$$

$$
= -\int_{t_*}^{T} dx\sigma(t_*,x)\mathcal{A}(t_*,x) + \frac{1}{2}\int_{t_*}^{T} dx \int_{t_*}^{T} dx' \mathcal{M}(x,x';t_*)
$$

Directly differentiating the discounted bond given in Eq. (5.52) yields

$$
\frac{1}{D(t_*,T)} \frac{\partial D(t_*,T)}{\partial t_*} = -\int_{t_*}^{T} dx\alpha(t_*,x)
$$

$$
- \int_{t_*}^{T} dx\sigma(t_*,x)\mathcal{A}(t_*,x) + \frac{1}{2}\int_{t_*}^{T} dx \int_{t_*}^{T} dx' \mathcal{M}(x,x';t_*)
$$

Requiring that the drift be zero in the equation above for $D(t_*,T)$ yields the value for $\alpha(t,x)$ given in Eq. (5.54) and one recovers Eq. (5.55).

The differential derivation of the martingale condition for zero coupon bonds given in Eq. (5.54) is an example of the generalization of Ito calculus for the case of the quantum field $\mathcal{A}(t,x)$.

5.13 HJM limit of forward interest rates

All the formulas for coupon and zero coupon bond options and derivatives have expressions in quantum finance that are similar to the HJM model – when the instruments are expressed in terms of the underlying forward interest rates $f(t,x)$. The difference emerges when one computes any expectation value – due to the dissimilarity between the two-dimensional quantum field $\mathcal{A}(t,x)$ and white noise $R(t)$.

The crucial difference is that, unlike the case in quantum finance, the forwards interest rates in the HJM model are *exactly* correlated (in the future time direction). To see this note[10]

$$
E\left[\left\{\frac{\partial f(t,x)}{\partial t} - \alpha(t,x)\right\}\left\{\frac{\partial f(t',x')}{\partial t'} - \alpha(t',x')\right\}\right]\bigg|_{HJM}
$$

$$
= \delta(t-t')\sigma(t,x)\sigma(t',x')
$$

---

[10] The result is not changed for the N-factor model, in which case the right-hand side is given by $\delta(t-t')\sum_i \sigma_i(t,x)\sigma_i(t',x')$.

In other words, the quantum finance model reduces to the HJM model for the following limits

$$\Rightarrow M(x,x';t)\Big|_{HJM} = \sigma(x-t)\mathcal{D}(x,x';t)\Big|_{HJM}\sigma(x'-t) = \sigma(x-t)\sigma(x'-t)$$

The HJM limit can be taken of all the formulas derived in quantum finance by the following prescription

$$\mathcal{A}(t_*,x)\Big|_{HJM} \rightarrow R(t)$$

$$\mathcal{D}(x,x';t)\Big|_{HJM} \rightarrow 1$$

Hence, in the HJM model, the time evolution of the zero coupon bonds, from Eq. (5.50), is given by

$$\frac{\partial B(t_*,T)}{\partial t_*}\Big|_{HJM} = \left[f(t_*,t_*) - R(t) \int_{t_*}^{T} dx\sigma(t_*,x) \right] B(t_*,T) \qquad (5.56)$$

and, from Eq. (5.55), the time evolution of the discounted bond is given by

$$\frac{\partial D(t_*,T)}{\partial t_*}\Big|_{HJM} = -\left[R(t) \int_{t_*}^{T} dx\sigma(t_*,x) \right] D(t_*,T)$$

The HJM martingale condition, from Eq. (5.54), is similarly given by

$$\alpha_{HJM}(t,x) = \sigma(t,x) \int_{t}^{x} dx'\sigma(t,x')$$

5.14 Summary

The quantum field theory of forward interest rates offers a different perspective on the debt markets as well as providing a variety of analytical and computational tools. It will be seen in later chapters that interest rate options are modeled by nonlinear terms in the action $S[\mathcal{A}]$ that can be incorporated into the framework of quantum field theory in a fairly straightforward manner.

The quantum finance formulation of the bond forward interest rates is based on the concept of the Lagrangian, action, and path integrals. The instantaneous forward interest rates $f(t,x)$ are modeled as a two-dimensional quantum field defined on a trapezoidal domain given by $x \geq t$. A velocity field $\mathcal{A}(t,x)$ was introduced that is also a two-dimensional quantum field and is the generalization of the concept of white noise. A stiff Gaussian Lagrangian was written for determining the dynamics of $f(t,x)$. Introducing the concept of market time leads to a pseudo-Gaussian stiff Lagrangian and will be seen to accurately describe the empirical behavior of $f(t,x)$.

The path integral formulation of $f(t,x)$ is a powerful tool for many computations; to illustrate this, the martingale condition for zero coupon bonds as well as the pricing kernel for $f(t,x)$ were derived using a path integral derivation. For an infinitesimal time interval, the pricing kernel was shown to have a direct connection with the infinitesimal action. The path integral and Lagrangian framework are useful for studying path independent coupon bond European options as well as path dependent coupon bond options such as the Asian and American options.

The 'short distance' Wilson expansion was defined for $\mathcal{A}(t,x)$; the equal time singularity in the correlation function of $\mathcal{A}(t,x)$, provided a generalization of Ito calculus for the quantum field $\mathcal{A}(t,x)$. The Wilson expansion was used for calculating the evolution of zero coupon bonds and for deriving the drift of the forward interest rates using the martingale condition.

The quantum formulation encodes the imperfect correlation of the bond forward interest rates in an efficient and transparent manner. As will be demonstrated in the chapters that follow, numerous calculations for a variety of derivative instruments can be efficiently carried out using the quantum finance formulation of the bond forward interest rates.

6

Libor Market Model of interest rates

Libor $L(t, T)$ is one of the primary interest rate instruments in the capital markets, the other being Euribor. The term Libor will be used generically for all interest rates on fixed deposits. The Libor Market Model (LMM) is defined in the framework of quantum finance and leads to a key generalization: the Libors, for different future times, are imperfectly correlated. A major difference between a forward interest rates' model and the LMM lies in the fact that the LMM is calibrated directly from the observed market values for $L(t, T)$. The short distance Wilson expansion of the Gaussian quantum field $\mathcal{A}(t, x)$ driving the Libors yields a derivation of the Libor drift term that incorporates imperfect correlations of the different Libors [3]. The logarithm of Libor $\phi(t, x)$ is defined and leads to a quantum field theory of Libor. The Lagrangian and Feynman path integral are obtained for the log Libor quantum field $\phi(t, x)$.

6.1 Introduction

Interest rates can be modeled using either the zero coupon bonds $B(t, T)$ or the simple interest Libor $L(t, T)$. Both these approaches are, in principle, equivalent but are quite different from an empirical, computational, and analytical point of view.

One can take the view that there exists a single set of underlying forward interest rates $f(t, x)$ that can be used for modeling both $L(t, T)$ and $B(t, T)$. The HJM [56] approach, in fact, takes this view and the HJM's quantum finance generalization goes a long way in accurately modeling interest rate instruments [12]. However, the HJM model and its quantum finance generalization have one serious shortcoming. From Eq. (2.10), forward interest rates are strictly positive, that is $f(t, x) \geq 0$. The positivity of $f(t, x)$ is intuitively obvious – and is also required by absence of arbitrage. In contrast, both the HJM model and its quantum finance generalization, given in Eqs. (4.38) and (5.1) respectively, allow $f(t, x)$ to be *negative* with a finite

117

probability – which implies, from Eqs. (2.15) and (2.16), that the simple interest rate $L(t; T_1, T_2)$ has a finite probability of being negative.

Giving up $f(t, x) \geq 0$ does not pose a very serious problem for the bond sector since $B(t, T)$ is strictly positive even for those configurations for which $f(t, x) \leq 0$. However, for the interest rates sector of the debt market, a model that allows Libor to be negative can yield results that allow for arbitrage – and hence are not permissible as a consistent model for interest rate instruments. One needs to go beyond Gaussian modeling of $f(t, x)$ and instead develop a model based directly on Libor $L(t, T_n)$.

The Libor Market Model (LMM) approach was pioneered by Bruce–Gatarek–Musiela (BGM) [32] and Jamshidian [62], with many of its subsequent developments discussed by Rebonata [83, 84]. The LMM aims at modeling interest rates in terms of debt instruments that are *directly* traded in the financial markets. In particular, forward interest rates are not directly traded, but, instead, what are traded are (a) Libor and Euribor for fixed time cash deposits and (b) zero coupon bonds $B(t, T)$ as well as coupon bonds. LMM takes the traded values of Libor $L(t, T)$ to be the main ingredient in modeling interest rates – instead of deriving Libor from an underlying Libor forward interest rates' model. In the LMM, all Libors are *strictly positive*: $L(t, T) > 0$. Zero coupon bonds and the Libor forward interest rates are derived from Libor. Strictly positive Libor has the added advantage that all zero coupon bonds, and hence coupon bonds as well, are all strictly positive.

One of the biggest achievements of the LMM is a derivation of Black's formula for pricing interest rate caplets from an arbitrage free model – something that many experts thought was not possible. Various extensions of the LMM have been made: Anderson and Andresean [1] and Joshi and Rebonata [66] incorporate stochastic volatility into the LMM, whereas Henry-Labordere [58] combines LMM with the SABR model [54]. The calibration and applications of the BGM–Jamshidian model have been extensively studied [33, 83].

In the BGM–Jamshidian approach, similar to the HJM modeling of the forward interest rates, all Libors for different future times are exactly correlated. In contrast, in the quantum finance formulation, Libors are driven not by white noise, but rather, by the two-dimensional stochastic field $\mathcal{A}_L(t, x)$. Libor velocity quantum field $\mathcal{A}_L(t, x)$ has the same stiff Lagrangian $\mathcal{L}[\mathcal{A}_L]$ as the bond velocity quantum field $\mathcal{A}(t, x)$ given in Eq. (5.25), except that the parameters μ_L, λ_L, and ν_L have different empirical values. As was the case for the bond forward interest rates in Chapter 5, the value of all Libor instruments are given by averaging $\mathcal{A}_L(t, x)$ over all its possible values. Hence, $\mathcal{A}_L(t, x)$ is mathematically equivalent to a two-dimensional quantum field.

The quantum finance generalization of the LMM contains crucial correlation terms reflecting the imperfect correlation of the different Libors and avoids systematic errors that arise from the assumption of perfectly correlated Libors.

The key link in deriving the quantum finance version of the LMM, and, in particular, of the Libor drift, is the singular property of the bilinear product of the Gaussian quantum field $\mathcal{A}_L(t,x)$. The equal time Wilson expansion of the bilinear product of the quantum field $\mathcal{A}_L(t,x)$, as discussed in Section 5.10, provides a generalization of Ito calculus.

The LMM is driven by $f_L(t,x)$, the *Libor forward interest rates*, which are distinct from both the empirical forward interest rates and the bond forward interest rates. It will be shown that $f_L(t,x)$ has a nonlinear evolution equation with both, its drift and volatility being stochastic. Libor forward interest rates are strictly positive and nonsingular, being finite for all calendar and future time. It is more efficient for describing Libor instruments to do a nonlinear change of independent variables from $f_L(t,x)$ to $L(t,T_n)$ and then to log Libor $\phi(t,x)$.

It is shown that, when the limit of perfectly correlated Libor is taken, the quantum finance LMM reduces to the BGM–Jamshidian model, which, in turn, yields – in the limit of zero Libor tenor ($\ell \to 0$) – the HJM model for the bond forward interest rates.

6.2 Libor and zero coupon bonds

Libor zero coupon bonds $B_L(t,T)$, derived from the Libor ZCYC are given in Eq. (2.27). In terms of Libor forward interest rates $f_L(t,x)$

$$B_L(t,T) = \exp\left\{ -\int_t^T dx f_L(t,x) \right\} \tag{6.1}$$

Libor zero coupon bonds $B_L(t,T)$ are not actual instruments traded in the market but rather a way of encoding the Libor ZCYC. As discussed in Section 2.13, the price of a traded zero coupon Treasury Bond $B(t,T)$ is not equal to a Libor bond $B_L(t,T)$ due to TED among other reasons, but these differences will be ignored. The subscript on $B_L(t,T)$ will be dropped henceforth.

Libor forward interest rates, from Eqs. (2.20) and (2.21), are given by

$$L(t,T_n) = \frac{1}{\ell}\left[\exp\left\{ \int_{T_n}^{T_n+\ell} dx f_L(t,x) \right\} - 1 \right] \tag{6.2}$$

$$= \frac{B(t,T_n) - B(t,T_{n+1})}{\ell B(t,T_{n+1})} \tag{6.3}$$

The Libor rates are related to the zero coupon bonds by Eq. (6.3), namely that

$$B(t,T+\ell) = \frac{B(t,T)}{1 + \ell L(t,T)} \tag{6.4}$$

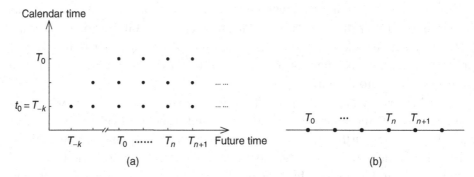

Figure 6.1 (a) Libor future and calendar time lattice $T_n = T_0 + \ell n$; the tenor (Libor time lattice spacing) is given by $\ell = 90$ days. (b) Libor future time lattice.

Eq. (6.4) provides a recursion equation that allows one to express $B(t, T)$ solely in terms of $L(t, T)$. Note that Libors are only defined for the discrete future time given by *Libor future time* $T = T_n = n\ell$, $n = 0, \pm 1, \pm 2, \ldots, \pm\infty$. Calendar time as well as the future Libor time lattice is shown in Figure 6.1(a) and the Libor future time lattice is shown in Figure 6.1(b).

Hence, from Eq. (6.4)

$$B(t, T_{k+1}) = \frac{B(t, T_k)}{1 + \ell L_k(t)} = B(t, T_0) \prod_{n=0}^{k} \frac{1}{1 + \ell L_n(t)}$$

where $L(t, T_n) \equiv L_n(t)$

Bonds $B(t, T_0)$ that have time t not at a Libor time ℓk cannot be expressed solely in terms of Libor rates. Zero coupon bonds that are issued at Libor time, say T_0, and mature at another Libor time T_{k+1}, since $B(T_0, T_0) = 1$, can be expressed entirely in terms of Libor as follows

$$B(T_0, T_{k+1}) = \prod_{n=0}^{k} \frac{1}{1 + \ell L(T_0, T_n)} = \prod_{n=0}^{k} \frac{1}{1 + \ell L_n(T_0)} \qquad (6.5)$$

6.2.1 Forward bond price and Libor

Let present time be $t_0 = T_{-k}$. Suppose a zero coupon bond $B(T_0, T_n + \ell)$ is going to be issued at some future time $T_0 > T_{-k}$, with expiry at time $T_n + \ell$; the zero coupon bond and its forward price are defined for Libor time and shown in Figure 6.2. From

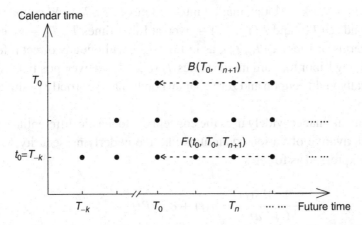

Figure 6.2 The zero coupon bond $B(T_0, T_{n+1})$ is issued at T_0 and expires at $T_n + \ell$. Its forward Libor bond price $F(t_0, T_0, T_{n+1})$ is given at (present) Libor time $t_0 = T_{-k}$.

Eqs. (2.14) and (6.5), the forward bond price is given by

$$F(t_0, T_0, T_n + \ell) = \frac{B(t_0, T_{n+1})}{B(t_0, T_0)}$$

$$= \left\{ \prod_{i=-k}^{n} \frac{1}{1 + \ell L(t_0, T_i)} \right\} \Big/ \left\{ \prod_{i=-k}^{-1} \frac{1}{1 + \ell L(t_0, T_i)} \right\}$$

$$= \prod_{i=0}^{n} \frac{1}{1 + \ell L(t_0, T_i)} \qquad (6.6)$$

6.3 Libor Market Model and quantum finance

The Libor Market Model is defined in the framework of quantum finance. The differential formulation of Libor is similar to the modeling of the bond forward interest rates; the Libor Market Model is defined using the time evolution of the Libor rates $L(t, T)$.

Recall that the quantum finance model of the bond forward interest rates $f(t, x)$ given in Eq. (5.1) – and its HJM limit – have a major unavoidable side effect: there is a finite probability that $f(t, x)$ can take negative values. Since, empirical forward interest rates can never be negative, the Libor Market Model takes the view that for the debt market one should *replace* the bond forward interest rates $f(t, x)$ by strictly positive Libor forward interest rates $f_L(t, x)$. These rates are used for modeling all interest rate instruments and, in particular, yield all $L(t, T)$ as always being positive.

In the Libor Market Model, market interest rates $L(T_0, T_n)$ and coupon and zero coupon bonds $\mathcal{B}(T_N)$ and $B(T_n, T_N)$ – given at Libor times T_n, T_N – are expressed solely in terms of Libor $L(T_0, T_n)$, as in Eq. (6.5), without any direct reference to the underlying Libor forward interest rates $f_L(t, x)$. Moreover, positive Libor rates automatically yield coupon and zero coupon bonds that are strictly positive, as seen in Eq. (6.5).

Modeling in finance widely uses the concept of stochastic differential equations; the time derivative of various quantities like the underlying security $S(t)$ is, for example, expressed as follows

$$\frac{1}{S(t)} \frac{dS(t)}{dt} = a(t) + \sigma(t) R(t)$$

$$E[R(t)] = 0; \quad E[R(t) R(t')] = \delta(t - t')$$

where $R(t)$ is Gaussian white noise, $a(t)$ is the drift, and $\sigma(t)$ is the volatility.

The HJM and BGM–Jamshidian models of interest rates' models are both expressed as functions of white noise, given by

$$\frac{\partial f(t, x)}{\partial t} = \alpha(t, x) + \sigma(t, x) R(t) \quad : \quad \text{HJM model} \tag{6.7}$$

$$\frac{1}{L_k(t)} \frac{\partial L_k(t)}{\partial t} = \zeta_k(t) + \gamma_k(t) R(t) \quad : \quad \text{BGM–Jamshidian model} \tag{6.8}$$

where the volatility functions $\sigma(t, x), \gamma_k(t)$ are deterministic. The drift $\alpha(t, x)$ is deterministic in the HJM model, whereas $\zeta_k(t)$ is stochastic and depends on Libors $L_k(t)$ for the BGM–Jamshidian model.

Future time x and T_k have been introduced in both the HJM and BGM–Jamshidian models only in the drift and volatility of the interest rates' term structure. A *single* white noise $R(t)$ drives the entire forward interest rates' curve and leads, as follows, to the following perfectly correlated rates in both the HJM and BGM–Jamshidian models

$$E\left[\frac{\partial f(t, x)/\partial t - \alpha(t, x)}{\sigma(t, x)} \frac{\partial f(t', x')/\partial t' - \alpha(t', x')}{\sigma(t', x')}\right] = \delta(t - t')$$

$$E\left[\frac{L_k^{-1}(t)\partial L_k(t)/\partial t - \zeta_k(t)}{\gamma_k(t)} \frac{L_{k'}^{-1}(t')\partial L_{k'}(t')/\partial t' - \zeta_{k'}(t')}{\gamma_{k'}(t')}\right] = \delta(t - t')' \tag{6.9}$$

Note that the right-hand sides of the above equations are *independent* of x, x' and $T_k, T_{k'}$ respectively, showing perfect correlation in future time.

The quantum finance formulation of Libor forward interest rates $f_L(t, x)$ – with drift $\mu(t, x)$ and volatility $v(t, x)$ that depend on $f_L(t, x)$ – is given by the following

$$\frac{\partial f_L(t, x)}{\partial t} = \mu(t, x) + v(t, x)\mathcal{A}_L(t, x) \tag{6.10}$$

$$f_L(t, x) = f_L(t_0, x) + \int_{t_0}^{t} dt \mu(t, x) + \int_{t_0}^{t} dt v(t, x)\mathcal{A}_L(t, x) \tag{6.11}$$

It will turn out that volatility $v(t, x) \propto [1 - \exp\{-\ell f_L(t, x)\}]$. From Eq. (5.25)

$$\mathcal{L}[\mathcal{A}_L] = -\frac{1}{2} \left\{ \mathcal{A}_L^2(t, z) + \frac{1}{\mu_L^2} \left(\frac{\partial \mathcal{A}_L(t, z)}{\partial z} \right)^2 + \frac{1}{\lambda_L^4} \left(\frac{\partial^2 \mathcal{A}_L(t, z)}{\partial^2 z} \right)^2 \right\}$$

$$z = (x - t)^{v_L}$$

The correlation function and covariance, similar to Eq. (5.22), are given by

$$E[\mathcal{A}_L(t, x)\mathcal{A}_L(t', x')] = \delta(t - t')\mathcal{D}_L(x, x'; t)$$

$$M_v(x, x'; t) = v(t, x)\mathcal{D}_L(x, x'; t)v(t, x') \tag{6.12}$$

As expected, the Libor forward interest rates are imperfectly correlated

$$E\left[\frac{\partial f_L(t, x)/\partial t - \mu(t, x)}{v(t, x)} \frac{\partial f_L(t', x')/\partial t' - \mu(t', x')}{v(t', x')} \right] = \delta(t - t')\mathcal{D}_L(x, x'; t)$$

$$: \text{imperfectly correlated}$$

6.4 Libor Martingale: forward bond numeraire

A wide class of numeraires can be used to render all traded assets into martingales. The combination $L(t, T_n)B(t, T_{n+1})$, from Eq. (6.3), is equivalent to a portfolio of zero coupon bonds and hence is a traded asset. By a suitable choice of the drift, *all* traded assets $L(t, T_n)B(t, T_{n+1})$ – discounted by the numeraire – can be made into martingales. Choose the zero coupon bond $B(t, T_{I+1})$ as the forward bond numeraire. The instruments $\mathcal{X}_n(t)$

$$\mathcal{X}_n(t) \equiv \frac{L(t, T_n)B(t, T_{n+1})}{B(t, T_{I+1})} : \text{martingale} \tag{6.13}$$

are martingales for *all* values of n [53].

Note that, for $n = I$, the portfolio $\mathcal{X}_I(t) = L_I(t) \equiv L(t, T_I)$; hence the Libor $L(t, T_I)$ is a martingale for the forward bond numeraire given by zero coupon bond

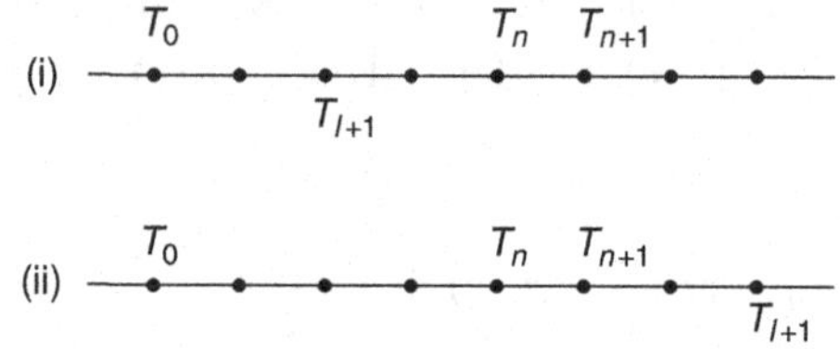

Figure 6.3 Libor time lattice for the forward bond numeraire $B(t, T_{I+1})$ with (i) $T_{I+1} < T_n$ and (ii) $T_{I+1} > T_n$.

$B(t, T_{I+1})$. As shown in Figures 6.3 (i) and (ii), time T_n can be either less than, equal to, or greater than T_I.

In terms of the Libor forward interest rates the martingale is

$$X_n(t) = \frac{1}{\ell}\left[\exp\left\{-\int_{T_{I+1}}^{T_n} dx f_L(t, x)\right\} - \exp\left\{-\int_{T_{I+1}}^{T_{n+1}} dx f_L(t, x)\right\}\right] \quad (6.14)$$

Differentiating portfolio $X_n(t)$ using Eq. (6.10) and the rules derived in Section 5.11 yields

$$\ell\frac{\partial X_n(t)}{\partial t} = \quad (6.15)$$

$$\left[-\int_{T_{I+1}}^{T_n} dx\,\mu(t, x) + \frac{1}{2}\int_{T_{I+1}}^{T_n} dx dx'\, M_v(x, x'; t) - \int_{T_{I+1}}^{T_n} dx\, v(t, x)\mathcal{A}_L(t, x)\right]$$

$$\times e^{-\int_{T_{I+1}}^{T_n} dx f_L(t, x)}$$

$$+\left[\int_{T_{I+1}}^{T_{n+1}} dx\,\mu(t, x) - \frac{1}{2}\int_{T_{I+1}}^{T_{n+1}} dx dx'\, M_v(x, x'; t) + \int_{T_{I+1}}^{T_{n+1}} dx\, v(t, x)\mathcal{A}_L(t, x)\right]$$

$$\times e^{-\int_{T_{I+1}}^{T_{n+1}} dx f_L(t, x)}$$

Note that in obtaining $\partial X_n(t)/\partial t$, *no condition* has been placed on drift $\mu(t, x)$ and volatility $v(t, x)$, which can be arbitrary nonlinear functions of $f_L(t, x)$.

The (discounted) bond portfolio X_n is a martingale, as discussed in Eq. (A.17), if and only if

$$E\left[\frac{\partial X_n(t)}{\partial t}\right] = 0 \quad (6.16)$$

The random terms in Eq. (6.15) are proportional to $\mathcal{A}_L(t, x)$. Since $E[\mathcal{A}_L(t, x)] = 0$, the martingale condition given in Eq. (6.16) requires that the drift – namely, terms independent of $A_L(t, x)$) – must be zero and yield

$$\int_{T_{I+1}}^{T_n} dx \mu_I(t, x) = \frac{1}{2} \int_{T_{I+1}}^{T_n} dx dx' M_v(x, x'; t)$$

The martingale condition given above is satisfied by choosing the following value for drift

$$\mu_I(t, x) = \int_{T_{I+1}}^{x} dx' M_v(x, x'; t) \qquad (6.17)$$

The result is similar to an earlier result given in Eq. (5.37), except that in Eq. (6.17) the volatility $v(t, x)$ is stochastic.

6.5 Time evolution of Libor

The main motivation for introducing the Libor Market Model is to have manifestly positive interest rates and bonds. To ensure that the Libor rates $L(t, T_I)$ are always positive, it is sufficient to show that they are the exponential of real variables. To obtain positive Libor rates requires a nontrivial drift; a quantum finance derivation of the drift term is given in this section and generalizes earlier results of the BGM–Jamshidian approach.

The drift term, as given in Eq. (6.17), is expressed in terms of the Libor forward interest rates volatility function $v(t, x)$. The main theoretical objective of the Libor Market Model is to completely remove $v(t, x)$ from the Libor evolution equation. In particular, to express the drift of the Libor rates in terms of deterministic Libor volatility $\gamma(t, x)$ (defined later in Eq. (6.21)).

From the definition of Libor given in Eq. (6.2), choosing μ given in Eq. (6.10) to be equal to the μ_I given in Eq. (6.17), and using the Wilson expansion for $A_L(t, x)$ yields

$$L(t, T_n) = \frac{1}{\ell} \left[\exp \left\{ \int_{T_n}^{T_{n+1}} dx f_L(t, x) \right\} - 1 \right]$$

$$\frac{\partial L(t, T_n)}{\partial t} = \frac{1}{\ell} \left[\int_{T_n}^{T_{n+1}} dx \mu_I(t, x) + \frac{1}{2} \int_{T_n}^{T_{n+1}} dx M_v(x, x'; t) \right.$$

$$\left. + \int_{T_n}^{T_{n+1}} dx v(t, x) A_L(t, x) \right] \exp \left\{ \int_{T_n}^{T_{n+1}} dx f_L(t, x) \right\} \qquad (6.18)$$

The drift for $\partial L(t, T_n)/\partial t$, from Eq. (6.17), has the following simplification

$$\int_{T_n}^{T_{n+1}} dx \mu_I(t, x) + \frac{1}{2} \int_{T_n}^{T_{n+1}} dx dx' M_v(x, x'; t)$$

$$= \int_{T_n}^{T_{n+1}} dx \left[\int_{T_{I+1}}^{T_n} dx' + \int_{T_n}^{x} dx' \right] M_v(x, x'; t) + \int_{T_n}^{T_{n+1}} dx \int_{T_n}^{x} dx' M_v(x, x'; t)$$

$$= \int_{T_{I+1}}^{T_n} dx \int_{T_n}^{T_{n+1}} dx' M_v(x, x'; t) + \int_{T_n}^{T_{n+1}} dx \int_{T_n}^{T_{n+1}} dx' M_v(x, x'; t)$$

$$= \int_{T_{I+1}}^{T_{n+1}} dx \int_{T_n}^{T_{n+1}} dx' M_v(x, x'; t) \qquad (6.19)$$

and yields, from Eqs. (6.2), (6.18), and (6.19), the following

$$\frac{\partial L(t, T_n)}{\partial t} = \left[\int_{T_{I+1}}^{T_{n+1}} dx \int_{T_n}^{T_{n+1}} dx' M_v(x, x'; t) \right.$$

$$\left. + \int_{T_n}^{T_{n+1}} dx v(t, x) \mathcal{A}_L(t, x) \right] \frac{[1 + \ell L(t, T_n)]}{\ell} \qquad (6.20)$$

Note that, as expected, for $n = I$ the drift is zero, making $\mathcal{X}_I(t) = L(t, T_I)$ a martingale. Libor drift $\zeta(t, T_n)$ and volatility $\gamma(t, x)$ are *defined* as follows

$$\frac{1}{L(t, T_n)} \frac{\partial L(t, T_n)}{\partial t} = \zeta(t, T_n) + \int_{T_n}^{T_{n+1}} dx \gamma(t, x) \mathcal{A}_L(t, x) \qquad (6.21)$$

Volatility $\gamma(t, x)$ is a *deterministic* function – independent of $L(t, T_n)$. The drift $\zeta(t, T_n)$ is determined by the martingale condition and, in particular, is a nonlinear function of $L(t, T_n)$. Volatility $\gamma(t, x)$ and drift $\zeta(t, T_n)$ are discussed, respectively, in Sections 6.6 and 6.7.

6.6 Volatility $\gamma(t, x)$ for positive Libor

A key *assumption* of the Libor Market Model is that the Libor volatility function $\gamma(t, x)$ is a deterministic function that is independent of the Libor. Market data for Libors or for interest rates' caplets can be used for determining the empirical value of $\gamma(t, x)$ [83].

As it stands, Eq. (6.20) for $\partial L(t, T_n)/\partial t$ does not imply that the Libor interest rates $L(t, T_n)$ are strictly positive. Libors are strictly positive only if Eq. (6.21) holds; namely, if there exists a Libor volatility function $\gamma(t, x)$ such that, from

Eqs. (6.20) and (6.21)

$$\frac{[1 + \ell L(t, T_n)]}{\ell} \int_{T_n}^{T_{n+1}} dx v(t, x) \mathcal{A}_L(t, x) = L(t, T_n) \int_{T_n}^{T_{n+1}} dx \gamma(t, x) \mathcal{A}_L(t, x)$$

$$(6.22)$$

$$\Rightarrow \int_{T_n}^{T_{n+1}} dx v(t, x) \mathcal{A}_L(t, x) = \frac{\ell L(t, T_n)}{1 + \ell L(t, T_n)} \int_{T_n}^{T_{n+1}} dx \gamma(t, x) \mathcal{A}_L(t, x) \quad (6.23)$$

$$\Rightarrow v(t, x) = \frac{\ell L(t, T_n)}{1 + \ell L(t, T_n)} \gamma(t, x); \quad x \in [T_n, T_{n+1}] \quad\quad (6.24)$$

In the Libor Market Model, $v(t, x)$ yields a model of the Libor forward interest rates with stochastic volatility. Eq. (6.23) can be viewed as fixing the volatility function $v(t, x)$ of the forward interest rates $f_L(t, x)$ so as to ensure that all Libors are strictly positive.

To have a better understanding of $v(t, x)$ consider the limit of $\ell \to 0$, which yields $\int_{T_n}^{T_{n+1}} dx f_L(t, x) \simeq \ell f_L(t, x)$. From Eqs. (6.2) and (6.24)

$$v(t, x) \simeq [1 - e^{-\ell f_L(t, x)}] \gamma(t, x)$$

The following are the two limiting cases

$$v(t, x) = \begin{cases} \ell \gamma(t, x) f_L(t, x); & \ell f_L(t, x) << 1 \\ \\ \gamma(t, x); & \ell f_L(t, x) >> 1 \end{cases} \quad\quad (6.25)$$

For small values of $f_L(t, x)$, the volatility $v(t, x)$ is proportional to $f_L(t, x)$. It is known [88] that Libor forward interest rates $f_L(t, x)$ with volatility $v(t, x) \simeq f_L(t, x)$ are unstable and diverge after a finite time. However, in the Libor Market Model, when the Libor forward rates become large, that is $\ell f_L(t, x) >> 1$, the volatility $v(t, x)$ becomes deterministic and equal to $\gamma(t, x)$. It is shown in Section 6.11 that Libor forward interest rates $f_L(t, x)$ are never divergent and Libor dynamics yields finite $f_L(t, x)$ for all future calendar time.

6.7 Forward bond numeraire: Libor drift $\zeta(t, T_n)$

Libor drift $\zeta(t, T_n)$ is chosen so that all discounted instruments $\chi_n(t)$ are martingales. The main challenge of the Libor Market Model is to express Libor drift

$\zeta(t, T_n)$ solely in terms of Libor volatility function $\gamma(t, x)$. The Libor drift term $\zeta(t, T_n)$ is defined, from Eq. (6.20), as follows

$$\frac{[1 + \ell L(t, T_n)]}{\ell} \int_{T_{l+1}}^{T_{n+1}} dx \int_{T_n}^{T_{n+1}} dx' M_v(x, x'; t) = L(t, T_n)\zeta(t, T_n)$$

$$\zeta(t, T_n) = \frac{[1 + \ell L(t, T_n)]}{\ell L(t, T_n)} \int_{T_{l+1}}^{T_{n+1}} dx v(t, x) \int_{T_n}^{T_{n+1}} dx' \mathcal{D}_L(x, x'; t) v(t, x')$$

$$(6.26)$$

The Libor forward interest rates' volatility function $v(t, x)$ needs to be expressed in terms of the Libor volatility function $\gamma(t, x)$. To do so, a recursion equation is obtained from Eq. (6.23) in the following manner. Multiply both sides of Eq. (6.23) by $\mathcal{A}_L(t, x') v(t, x')$ and use Eq. (5.46) to obtain

$$\mathcal{A}_L(t, x)\mathcal{A}_L(t, x') = \frac{1}{\epsilon}\mathcal{D}_L(x, x'; t)$$

This removes the quantum field from Eq. (6.23) and, by equating the $1/\epsilon$ term from both sides of the resulting equation, one obtains

$$\int_{T_n}^{T_{n+1}} dx v(t, x)\mathcal{D}_L(x, x'; t) v(t, x')$$

$$= \frac{\ell L(t, T_n)}{1 + \ell L(t, T_n)} \int_{T_n}^{T_{n+1}} dx \gamma(t, x)\mathcal{D}_L(x, x'; t) v(t, x') \qquad (6.27)$$

Since the dynamics of $L(t, T_n)$ is being analyzed, integrate variable x' from T_n to T_{n+1} and obtain

$$\int_{T_n}^{T_{n+1}} dx \int_{T_n}^{T_{n+1}} dx' v(t, x)\mathcal{D}_L(x, x'; t) v(t, x')$$

$$= \frac{\ell L(t, T_n)}{1 + \ell L(t, T_n)} \int_{T_n}^{T_{n+1}} dx v(t, x)\omega_n(t, x) \qquad (6.28)$$

where $\omega_n(t, x)$ is defined by

$$\omega_n(t, x) = \int_{T_n}^{T_{n+1}} dx' \mathcal{D}_L(x, x'; t)\gamma(t, x') \qquad (6.29)$$

Hence, from Eqs. (6.28) and (6.26)

$$\zeta(t, T_n) = \int_{T_{l+1}}^{T_{n+1}} dx v(t, x)\omega_n(t, x) \qquad (6.30)$$

The drift obtained in Eq. (6.30) still depends on the volatility function $v(t, x)$. To express this integral solely in terms of the volatility function $\gamma(t, x)$ one has to carry out a calculation similar to the one used in obtaining Eq. (6.30).

Multiplying both sides of Eq. (6.23), this time by $\mathcal{A}_L(t, x')\gamma(t, x')$ and using Eq. (5.46)

$$\mathcal{A}_L(t, x)\mathcal{A}_L(t, x') = \frac{1}{\epsilon}\mathcal{D}_L(x, x'; t)$$

yields the following

$$\int_{T_n}^{T_{n+1}} dx v(t, x)\mathcal{D}_L(x, x'; t)\gamma(t, x')$$

$$= \frac{\ell L(t, T_n)}{1 + \ell L(t, T_n)} \int_{T_n}^{T_{n+1}} dx \gamma(t, x)\mathcal{D}_L(x, x'; t)\gamma(t, x')$$

Integrating x' from T_n to T_{n+1} yields

$$\int_{T_n}^{T_{n+1}} dx v(t, x)\omega_n(t, x) = \frac{\ell L(t, T_n)}{1 + \ell L(t, T_n)} \int_{T_n}^{T_{n+1}} dx dx' M_\gamma(x, x'; t) \qquad (6.31)$$

where

$$M_\gamma(x, x'; t) \equiv \gamma(t, x)\mathcal{D}_L(x, x'; t)\gamma(t, x') \qquad (6.32)$$

Since

$$\int_t^{T_{n+1}} dx v(t, x)\omega_n(t, x)$$

$$= \int_t^{T_n} dx v(t, x)\omega_n(t, x) + \int_{T_n}^{T_{n+1}} dx v(t, x)\omega_n(t, x) \qquad (6.33)$$

Eqs. (6.31) and (6.33) yield the recursion equation

$$\int_t^{T_{n+1}} dx v(t, x)\omega_n(t, x)$$

$$= \int_t^{T_n} dx v(t, x)\omega_n(t, x) + \frac{\ell L(t, T_n)}{1 + \ell L(t, T_n)} \int_{T_n}^{T_{n+1}} dx \gamma(t, x)\omega_n(t, x)$$

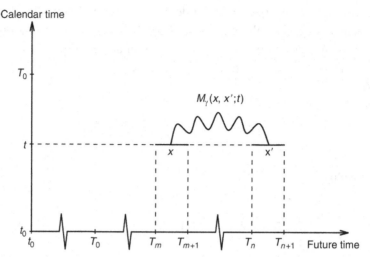

Figure 6.4 Libor propagator $\Lambda_{mn}(t) = \int_{T_m}^{T_{m+1}} dx \int_{T_n}^{T_{n+1}} dx' M_\gamma(x,x';t) = \int_{T_m}^{T_{m+1}} dx \int_{T_n}^{T_{n+1}} dx' \gamma(t,x) \mathcal{D}_L(x,x';t) \gamma(t,x')$ that yields nontrivial and imperfect correlation between the different Libors.

For simplicity, let time $t = T_0$; recursing the above equation yields, using Eq. (6.29), the following

$$\int_{T_0}^{T_{n+1}} dx v(t,x) \omega_n(t,x) = \sum_{m=1}^{n} \frac{\ell L(t,T_m)}{1+\ell L(t,T_m)} \int_{T_m}^{T_{m+1}} dx \gamma(t,x) \omega_n(t,x)$$

$$= \sum_{m=1}^{n} \frac{\ell L(t,T_m)}{1+\ell L(t,T_m)} \Lambda_{mn}(t) \qquad (6.34)$$

where, as shown in Figure 6.4, the Libor propagator is given by

$$\Lambda_{mn}(t) \equiv \int_{T_m}^{T_{m+1}} dx \gamma(t,x) \omega_n(t,x)$$

$$= \int_{T_m}^{T_{m+1}} dx \int_{T_n}^{T_{n+1}} dx' \gamma(t,x) \mathcal{D}_L(x,x';t) \gamma(t,x') \qquad (6.35)$$

$$= \int_{T_m}^{T_{m+1}} dx \int_{T_n}^{T_{n+1}} dx' M_\gamma(x,x';t)$$

There are three cases for T_n, as shown in Figure 6.3, namely $T_n = T_I$, $T_n > T_I$, and $T_n < T_I$.

Case (i) $T_n = T_I$. From Eq. (6.30)

$$\zeta(t, T_I) = 0$$

Case (ii) $T_n > T_I$. Eq. (6.34) yields the following

$$\zeta(t, T_n) = \int_{T_{I+1}}^{T_{n+1}} dx \, v(t, x) \omega_n(t, x)$$

$$= \int_{T_0}^{T_{n+1}} dx \, v(t, x) \omega_n(t, x) - \int_{T_0}^{T_{I+1}} dx \, v(t, x) \omega_n(t, x)$$

$$= \sum_{m=I+1}^{n} \frac{\ell L(t, T_m)}{1 + \ell L(t, T_m)} \Lambda_{mn}(t)$$

Case (iii) $T_n < T_I$. One has, from Eq. (6.30), the following

$$\zeta(t, T_n) = \int_{T_{I+1}}^{T_{n+1}} dx \, v(t, x) \omega_n(t, x) = - \int_{T_{n+1}}^{T_{I+1}} dx \, v(t, x) \omega_n(t, x)$$

$$= - \left[\int_{T_0}^{T_{I+1}} dx \, v(t, x) \omega_n(t, x) - \int_{T_0}^{T_{n+1}} dx \, v(t, x) \omega_n(t, x) \right]$$

$$= - \sum_{m=n+1}^{I} \frac{\ell L(t, T_m)}{1 + \ell L(t, T_m)} \Lambda_{mn}(t)$$

Collecting the results from above yields [53, 55]

$$\zeta(t, T_n) = \begin{cases} \sum_{m=I+1}^{n} \frac{\ell L(t, T_m)}{1+\ell L(t, T_m)} \Lambda_{mn}(t) & T_n > T_I \\[2mm] 0 & T_n = T_I \\[2mm] -\sum_{m=n+1}^{I} \frac{\ell L(t, T_m)}{1+\ell L(t, T_m)} \Lambda_{mn}(t) & T_n < T_I \end{cases} \qquad (6.36)$$

Recall from Eq. (6.35), the Libor correlator $\Lambda_{mn}(t)$ is given by

$$\Lambda_{mn}(t) = \int_{T_m}^{T_{m+1}} dx \int_{T_n}^{T_{n+1}} dx' \gamma(t, x) \mathcal{D}_L(x, x'; t) \gamma(t, x')$$

In summary, the forward bond numeraire $B(t, T_I)$ fixes the drift $\zeta(t, T_n)$ of the Libor $L(t, T_n)$ for *all* n. The drift $\zeta(t, T_n)$ is nonlinear of the Libors and nonlocal, depending on all the Libors from Libor time T_I to T_n.

6.8 Libor dynamics and correlations

As stated in Eq. (6.21), Libor dynamics are given by

$$\frac{1}{L(t,T_n)}\frac{\partial L(t,T_n)}{\partial t} = \zeta(t,T_n) + \int_{T_n}^{T_{n+1}} dx \gamma(t,x) \mathcal{A}_L(t,x) \tag{6.37}$$

In particular, since $\zeta(t,T_I) = 0$, Libor $L(t,T_I)$ has a martingale evolution given by

$$\frac{\partial L(t,T_I)}{\partial t} = L(t,T_I) \int_{T_I}^{T_{I+1}} dx \gamma(t,x) \mathcal{A}_L(t,x) \tag{6.38}$$

The results obtained express the time evolution of Libor completely in terms of volatility $\gamma(t,x)$, which is a function that is empirically determined in Chapter 7 and given in Figure 7.7. Libor drift $\zeta(t,T_n)$ is fixed by Eq. (6.36) and is a nonlinear and nonlocal function of all Libors. An expansion of the drift term $\zeta(t,T_n)$ about its leading value is generated in Section 8.3.

In the Libor evolution equations given in Eq. (6.37), all references to the volatility function $v(t,x)$ of the Libor forward interest rates $f_L(t,x)$ have been removed – as indeed was the whole purpose of the derivations of the previous section – with the drift being completely expressed in terms of Libor $L(t,T_n)$ and its volatility $\gamma(t,x)$.

Eq. (6.37) needs to be integrated to confirm that Libor dynamics yield positive valued Libors. Let $T_0 > t_0$ be two points on the Libor time lattice. From Eqs. (5.46) and (6.37), the differential of log Libor are given by

$$\frac{\partial \ln L(t,T_n)}{\partial t} = \lim_{\epsilon \to 0} \frac{1}{\epsilon} \Big[\ln L(t+\epsilon,T_n) - \ln L(t,T_n) \Big] \tag{6.39}$$

$$= \frac{1}{L(t,T_n)}\frac{\partial L(t,T_n)}{\partial t} - \frac{\epsilon}{2}\left[\frac{1}{L(t,T_n)}\frac{\partial L(t,T_n)}{\partial t}\right]^2 + O(\epsilon)$$

$$\Rightarrow \frac{\partial \ln L(t,T_n)}{\partial t} = \zeta(t,T_n) + \int_{T_n}^{T_{n+1}} dx \gamma(t,x) \mathcal{A}_L(t,x) - \frac{1}{2}\Lambda_{nn}(t) \tag{6.40}$$

Integrating the above equation over time yields

$$L(T_0,T_n) = L(t_0,T_n) e^{\beta(t_0,T_0,T_n) - \frac{1}{2}q_n^2 + \int_{t_0}^{T_0} dt \int_{T_n}^{T_{n+1}} dx \gamma(t,x)\mathcal{A}_L(t,x)}$$

$$= L(t_0,T_n) e^{\beta(t_0,T_0,T_n) + W_n} \tag{6.41}$$

$$\beta(t_0, T_0, T_n) = \int_{t_0}^{T_0} dt \zeta(t, T_n); \quad q_n^2 = \int_{t_0}^{T_0} dt \Lambda_{nn}(t) \qquad (6.42)$$

$$W_n = -\frac{1}{2}q_n^2 + \int_{t_0}^{T_0} dt \int_{T_n}^{T_{n+1}} dx \gamma(t, x) \mathcal{A}_L(t, x) \qquad (6.43)$$

Libor dynamics lead to positive Libor, as given in Eq. (6.41); Libor is proportional to the exponential of real quantities, namely a real drift $\zeta(t, T_n) - q_n^2/2$ and a real valued (Gaussian) quantum field $\mathcal{A}_L(t, x)$.

The action for $\mathcal{A}_L$ from Eq. (5.17) is given by

$$S[\mathcal{A}_L] = -\frac{1}{2}\int_{t_0}^{T_0} dt \int_{t}^{\infty} dx dx' \mathcal{A}_L(t, x) \mathcal{D}_L^{-1}(x, x'; t) \mathcal{A}_L(t, x) \qquad (6.44)$$

The explicit expression for the propagator $\mathcal{D}_L(x, x'; t)$ is given in Eq. (5.20). The partition function is given by

$$Z = \int D\mathcal{A}_L e^{S[\mathcal{A}_L]}$$

The generating functional given in Eq. (5.21) yields[1]

$$E[e^{W_n}] = \frac{1}{Z}\int D\mathcal{A}_L e^{W_n} e^{S[\mathcal{A}_L]}$$

$$= \exp\left\{-\frac{1}{2}q_n^2 + \int_{t_0}^{T_0} dt \int_{T_n}^{T_{n+1}} dx dx' M_\gamma(x, x'; t)\right\} = 1$$

$$E[(e^{W_m} - 1)(e^{W_n} - 1)] = \frac{1}{Z}\int D\mathcal{A}_L (e^{W_m} - 1)(e^{W_n} - 1)e^{S[\mathcal{A}_L]}$$

$$= e^{\Delta_{mn}} - 1$$

$$\Delta_{nm} \equiv \int_{t_0}^{T_0} dt \int_{T_n}^{T_{n+1}} dx \int_{T_m}^{T_{m+1}} dx' M_\gamma(x, x'; t) = \int_{t_0}^{T_0} dt \Lambda_{mn}(t)$$

Figure 6.5 graphically represents the Libor correlator Δ_{mn}.

Two results from above that will be useful in Chapter 8 are

$$E[e^{W_n} - 1] = 0 \qquad (6.45)$$

$$E[(e^{W_m} - 1)(e^{W_n} - 1)] \simeq \Delta_{mn} + O(\Delta_{mn}^2) \qquad (6.46)$$

---

[1] Field theorists will recognize that e^{W_n} is the normal ordered product : $\exp\{\int_{t_0}^{T_0} dt \int_{T_n}^{T_{n+1}} dx \gamma(t, x) \mathcal{A}_L(t, x)\}$:.

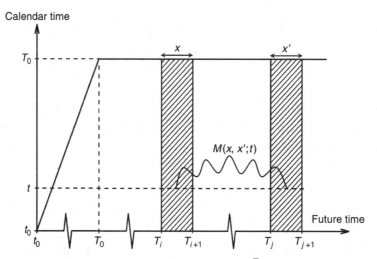

Figure 6.5 Libor correlator $\Delta_{ij} = \int_{t_0}^{T_0} dt \int_{T_i}^{T_{i+1}} dx \int_{T_j}^{T_{j+1}} dx' M_\gamma(x, x'; t)$ for the Libor European swaption.

6.9 Logarithmic Libor rates $\phi(t, x)$

Since $\gamma(t, x)$, the volatility of $L(t, T_n)$, is deterministic it is convenient to change variables from $f_L(t, x)$ to $L(t, T_n)$. Eq. (6.41) shows that Libor $L(t, T_n)$ is a positive random variable. A change of variables to logarithmic coordinates shows the structure of the Libor Market Model more clearly. Let $\phi(t, x)$ be a two-dimensional quantum field; similar to the definition of forward interest rates from zero coupon bonds given in Eq. (2.12), define a change of variables by

$$\ell L(t, T_n) = \exp \left\{ \int_{T_n}^{T_{n+1}} dx \phi(t, x) \right\} \equiv e^{\phi_n(t)} \qquad (6.47)$$

From its definition, $\phi(t, x)$ has dimensions of 1/time and can be thought of as the effective *logarithmic Libor* interest rates.

Consider, at some time t, a contract for a deposit to be made from T_0 to T_2; the principal plus simple interest earned is given, at time T_2, by $1 + (T_2 - T_0)L(t, T_0, T_2)$. This amount must be equal to that earned by first depositing the principal at time T_0 and then rolling over, at $T_0 + \ell = T_1$, the deposit and interest earned and collecting the principal and interest at time $T_2 = T_1 + \ell$; see Figure 6.6, which is the inverse of Figure 2.1. For there to be no-arbitrage opportunities the two procedures must

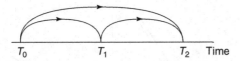

Figure 6.6 Simple interest earned over Libor time interval T_0 to T_2. Simple interest earned over the two sub-intervals T_0 to T_1 and from T_1 to T_2 must be equal to the interest earned from T_0 to T_2.

be equal, namely[2]

$$1 + (T_2 - T_0)L(t, T_0, T_2) = [1 + \ell L(t, T_0)][1 + \ell L(t, T_1)]$$
$$\Rightarrow (T_2 - T_0)L(t, T_0, T_2) = e^{\phi_0(t) + \phi_1(t)} + e^{\phi_0(t)} + e^{\phi_1(t)}$$
$$\text{where } e^{\phi_0(t) + \phi_1(t)} = e^{\int_{T_0}^{T_2} dx \phi(t,x)} = \ell L(t, T_0) \ell L(t, T_1) \quad (6.48)$$

Similarly, the integral of $\phi(t, x)$ over many Libor future time intervals yields the following

$$\exp\left\{\int_{T_n}^{T_m + \ell} dx \phi(t, x)\right\} = \prod_{i=n}^{m} [\ell L(t, T_i)]$$

$\exp\{\int_{T_n}^{T_m + \ell} dx \phi(t, x)\}$, similar to Eq. (6.48), is related to the future Libor rate $L(t, T_n, T_{m+1})$.

The 90-day benchmark Libor forward interest rates $f_L(t, x)$ are related to logarithmic Libor by Eq. (2.20)

$$\exp\left\{\int_{T_n}^{T_{n+1}} dx f_L(t, x)\right\} = 1 + \ell L(t, T_n)$$
$$\Rightarrow \exp\left\{\int_{T_n}^{T_n + \ell} dx f_L(t, x)\right\} = 1 + \exp\left\{\int_{T_n}^{T_n + \ell} dx \phi(t, x)\right\} \quad (6.49)$$

The definition of $\phi(t, x)$ depends on the tenor, and for the benchmark case is taken to be $\ell = 90$-days (three months). For Libor, the tenor is always finite, being a minimum of overnight (24 hours). The logarithm Libor $\phi(t, x)$ is well defined for

---

[2] Recall, from Eq. (2.19), that $L(t, T) \equiv L(t, T, T + \ell)$.

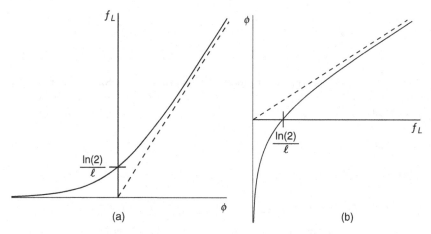

Figure 6.7 The dependence of $f_L(t,x)$ on $\phi(t,x)$ and visa versa.

any nonzero tenor ℓ. For the limit of the zero tenor, let $\ell = \epsilon \to 0$; from the defining Eq. (6.49), it follows that, since $f_L(t,x)$ is always finite

$$1 + \epsilon f_L(t,x) = 2 + \epsilon\phi(t,x)$$

$$\Rightarrow \phi(t,x) = \frac{1}{\epsilon}[f_L(t,x) - 1] \to -\infty$$

In other words, the zero tenor limit is singular for $\phi(t,x)$; however, for finite tenor $\ell \neq 0$, the field $\phi(t,x)$ is always well defined,

Since the interest derivative market is based on the three-month Libor, let $\ell = 1/4$ year; one can approximately evaluate the integral and obtain the following

$$\exp\{\ell f_L(t,x)\} \simeq 1 + \exp\{\ell\phi(t,x)\} \qquad (6.50)$$

Eq. (6.50) is plotted in Figure 6.7. For $f_L(t,x) << \ln(2)/\ell \sim 400\%$/year, the value of $\phi(t,x) \simeq -\infty$; only when both the rates $f_L(t,x)$ and $\phi(t,x)$ are large are they approximately equal. Hence, there is no domain where *both* the quantum fields $f_L(t,x)$ and $\phi(t,x)$ take *small* values and, consequently, there is no consistent scheme for *simultaneously* defining a perturbation expansion, in powers of $\phi(t,x)$ and $f_L(t,x)$, for both the quantum fields. In summary, one can define a perturbation expansion for either $f_L(t,x)$ or for $\phi(t,x)$, but not simultaneously for both the fields.

Furthermore, as can be seen from Figure 6.7, for Eq. (6.50) to have real values for both $f_L(t,x)$ and $\phi(t,x)$, the following is required

$$0 \leq f_L(t,x) \leq +\infty; \quad -\infty \leq \phi(t,x) \leq +\infty$$

The dynamics of Libor $L(t,T_n)$ are specified in Eq. (6.37) and yield, from Eq. (6.40), the following *defining equation* for $\phi(t,x)$

$$\frac{\partial}{\partial t}\int_{T_n}^{T_{n+1}} dx\phi(t,x) = \frac{\partial \ln(\ell L(t,T_n))}{\partial t}$$

$$= \zeta(t,T_n) + \int_{T_n}^{T_{n+1}} dx\gamma(t,x)\mathcal{A}_L(t,x) - \frac{1}{2}\Lambda_{nn} \qquad (6.51)$$

The drift $\zeta(t,T_n)$ for forward bond numeraire $B(t,T_{I+1})$ is given by Eqs. (6.35) and (6.36). Integrating Eq. (6.51) from calendar Libor time t_0 to T_0 yields

$$\int_{T_n}^{T_{n+1}} dx\phi(T_0,x) = \int_{T_n}^{T_{n+1}} dx\phi(t_0,x) + \int_{t_0}^{T_0} dt\left[\zeta(t,T_n)\right.$$

$$\left. + \int_{T_n}^{T_{n+1}} dx\gamma(t,x)\mathcal{A}_L(t,x) - \frac{1}{2}\Lambda_{nn}\right] \qquad (6.52)$$

Exponentiating Eq. (6.52) yields, as expected, Eq. (6.41).

Dropping the $\int_{T_n}^{T_{n+1}} dx$ integration from both sides of Eq. (6.51) yields the following time evolution for logarithmic Libor

$$\frac{\partial\phi(t,x)}{\partial t} = -\frac{1}{2}\Lambda_n(t,x) + \rho_n(t,x) + \gamma(t,x)\mathcal{A}_L(t,x) \qquad (6.53)$$

$$T_n \le x < T_{n+1}; \quad \Lambda_n(t,x) = \int_{T_n}^{T_{n+1}} dx' M_\gamma(x,x';t) \qquad (6.54)$$

The function $\rho_n(t,x)$ is defined as follows

$$\zeta(t,T_n) = \int_{T_n}^{T_{n+1}} dx\rho_n(t,x) \qquad (6.55)$$

Hence, from Eqs. (6.36) and (6.55)

$$\rho_n(t,x) = \begin{cases} \sum_{m=I+1}^{n} \frac{e^{\phi_m(t)}}{1+e^{\phi_m(t)}}\Lambda_m(t,x) & n > I \\[2mm] 0 & n = I \\[2mm] -\sum_{m=n+1}^{I} \frac{e^{\phi_m(t)}}{1+e^{\phi_m(t)}}\Lambda_m(t,x) & n < I \end{cases} \qquad (6.56)$$

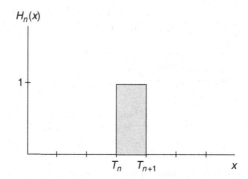

Figure 6.8 The characteristic function $H_n(x)$ for the Libor interval $[T_n, T_{n+1}]$.

To write Eq. (6.53) in a more compact form, define the characteristic function $H_n(x)$ for the Libor time interval $[T_n, T_{n+1}]$ given by

$$H_n(x) = \begin{cases} 1 & T_n \le x < T_{n+1} \\ 0 & x \notin [T_n, T_{n+1}] \end{cases} \tag{6.57}$$

and is shown in Figure 6.8. The characteristic function has the following important properties

$$f(x) = \sum_{n=0}^{\infty} H_n(x) f_n(x); \quad f(x) = f_n(x) \text{ for } T_n \le x < T_{n+1}$$

$$\int_{T_n}^{T_{n+1}} dx\, H_m(x) = \ell \delta_{m-n}$$

From Eqs. (6.53) and (6.57)

$$\frac{\partial \phi(t,x)}{\partial t} = -\frac{1}{2}\Lambda(t,x) + \rho(t,x) + \gamma(t,x)\mathcal{A}_L(t,x) \tag{6.58}$$

$$\Lambda(t,x) = \sum_{n=0}^{\infty} H_n(x)\Lambda_n(t,x); \quad \rho(t,x) = \sum_{n=0}^{\infty} H_n(x)\rho_n(t,x) \tag{6.59}$$

It is convenient to separate out a 'kinetic' drift $-\frac{1}{2}\Lambda(t,x)$ that does not depend on the Libors, with the remaining drift $\rho(t,x)$ being a nonlinear and nonlocal function of the Libors.

6.10 Lagrangian and path integral for $\phi(t,x)$

Logarithmic Libor rates $\phi(t,x)$ are the natural variables for expressing the Lagrangian for the Libor dynamics. A remarkable result is that the $\phi(t,x)$ variables are 'flat' integration variables for the path integral in spite of having a nonlinear relation to Libor $L(t,T_n)$. The Lagrangian itself is a highly nonlinear and nonlocal function of the $\phi(t,x)$ variables.

The Lagrangian and action for the Gaussian quantum field $\mathcal{A}_L(t,x)$ are given from Eqs. (5.25) and (5.15) by the following

$$\mathcal{L}[\mathcal{A}_L] = -\frac{1}{2}\mathcal{A}_L(t,x)\mathcal{D}_L^{-1}(t,x,x')\mathcal{A}_L(t,x')$$

$$S[\mathcal{A}_L] = \int_T \mathcal{L}[\mathcal{A}_L]$$

with the semi-infinite trapezoidal domain T given in Figure 5.1. The partition function, from Eq. (5.14), is given by

$$Z = \int D\mathcal{A}_L e^{S[\mathcal{A}_L]}$$

Eq. (6.58) encodes a change of variables relating two quantum fields $\phi(t,x)$ and $\mathcal{A}_L(t,x)$ and is given by

$$\mathcal{A}_L(t,x) = \frac{\partial\phi(t,x)/\partial t - \tilde{\rho}(t,x)}{\gamma(t,x)} \tag{6.60}$$

$$\tilde{\rho}(t,x) = -\frac{1}{2}\Lambda(t,x) + \rho(t,x)$$

The Lagrangian and action for logarithmic Libor quantum field $\phi(t,x)$ are given by

$$\mathcal{L}[\phi] = -\frac{1}{2}\left[\frac{\partial\phi(t,x)/\partial t - \tilde{\rho}(t,x)}{\gamma(t,x)}\right]\mathcal{D}_L^{-1}(t,x,x')\left[\frac{\partial\phi(t,x')/\partial t - \tilde{\rho}(t,x')}{\gamma(t,x')}\right]$$

$$S[\phi] = \int_{t_0}^{\infty} dt \int_t^{\infty} dx dx' \mathcal{L}[\phi] \tag{6.61}$$

The Neumann boundary conditions $\mathcal{A}_L(t,x)$ given in Eq. (5.16) yield the following boundary conditions on $\phi(t,x)$

$$\frac{\partial}{\partial x}\left[\frac{\partial\phi(t,x)/\partial t - \tilde{\rho}(t,x)}{\gamma(t,x)}\right]\Bigg|_{x=t} = 0 \tag{6.62}$$

It is shown in Section 6.14 that the Jacobian of the transformation given in Eq. (6.60) is a constant, independent of $\phi(t, x)$; note that the Jacobian is a constant in spite of the fact that the transformation in Eq. (6.60) is nonlinear due to the nonlinearity of Libor drift $\rho(t, x)$. A constant Jacobian leads to $\phi(t, x)$ being *flat variables*, with no measure term in the path integral. Flat variables have a well-defined leading order Gaussian path integral that generates a Feynman perturbation expansion for all financial instruments, thus greatly simplifying all calculations that are based on $\phi(t, x)$.

In summary, up to an irrelevant constant, the log Libor path integral measure is given by

$$\int DA_L = \int D\phi = \prod_{t=t_0}^{\infty} \prod_{x=t}^{\infty} \int_{-\infty}^{+\infty} d\phi(t, x)$$

The partition function for ϕ is given by

$$Z = \int D\phi e^{S[\phi]} = \int DA_L e^{S[A_L]}$$

The expectation value of a financial instrument $\mathcal{O}$ is given by

$$E[\mathcal{O}] = \frac{1}{Z} \int DA_L \mathcal{O}[A_L] e^{S[A_L]} = \frac{1}{Z} \int D\phi \mathcal{O}[\phi] e^{S[\phi]} \qquad (6.63)$$

6.10.1 Path integral for Libor martingale

Consider the special case of the portfolio from Eq. (6.13), of the martingale instrument $\mathcal{X}_I(t) = L(t, T_I)$. The integral formulation of the martingale condition for the forward bond numeraire, from Section 5.8, is given by

$$L(t_0, T_I) = E[L(T_0, T_I)] ; \quad T_0 > t_0$$

The path integral for the right-hand side, from Eq. (6.63), is given by the expectation value of a financial instrument $\mathcal{O} = L(T_0, T_I)$; hence

$$E[L(T_0, T_I)] = \frac{1}{Z} \int D\phi e^{S[\phi]} L(T_0, T_I) \qquad (6.64)$$

For Libor $L(t, T_I)$, the drift is zero, that is $\zeta(t, T_I) = 0$; hence, from Eq. (6.52)

$$\ln \ell L(T_0, T_I) = \int_{T_I}^{T_{I+1}} dx \phi(T_0, x)$$

$$= \int_{T_I}^{T_{I+1}} dx \phi(t_0, x) + \int_{t_0}^{T_0} dt \left\{ \int_{T_I}^{T_{I+1}} dx \gamma(t, x) A_L(t, x) - \frac{1}{2} \Lambda_{II}(t) \right\}$$

Changing path integration variables from $\phi(t,x)$ to $\mathcal{A}_L(t,x)$ and using the generating functional given in Eq. (5.21) and Eq. (6.45) yields

$$E[\ell L(T_0, T_I)] = \ell L(t_0, T_I)\frac{1}{Z}\int D\mathcal{A}_L e^{\int_{t_0}^{T_0} dt\{\int_{T_I}^{T_I+1} dx\gamma(t,x)\mathcal{A}_L(t,x)-\frac{1}{2}\Lambda_{II}(t)\}}e^{S[\mathcal{A}_L]}$$

$$= \ell L(t_0, T_I) \quad : \quad \text{martingale}$$

6.11 Libor forward interest rates $f_L(t,x)$

The dynamics of the log Libor given in Eq. (6.58) also defines the dynamics of the Libor forward interest rates quantum field $f_L(t,x)$. Differentiating Eq. (6.49) and substituting Eq. (6.58) yields the following

$$\int_{T_n}^{T_{n+1}} dx\frac{\partial f_L(t,x)}{\partial t} = \frac{e^{\phi_n(t)}}{1+e^{\phi_n(t)}}\int_{T_n}^{T_{n+1}} dx\frac{\partial\phi(t,x)}{\partial t} \tag{6.65}$$

$$= \frac{e^{\phi_n(t)}}{1+e^{\phi_n(t)}}\int_{T_n}^{T_{n+1}} dx\left[-\frac{1}{2}\Lambda(t,x)+\rho(t,x)+\gamma(t,x)\mathcal{A}_L(t,x)\right] \tag{6.66}$$

From Eqs. (6.10) and (6.66)

$$\frac{\partial f_L(t,x)}{\partial t} = \mu(t,x) + v(t,x)\mathcal{A}_L(t,x)$$

$$\int_{T_n}^{T_{n+1}} dx\mu(t,x) = \frac{e^{\phi_n(t)}}{1+e^{\phi_n(t)}}\int_{T_n}^{T_{n+1}} dx\left[-\frac{1}{2}\Lambda(t,x)+\rho(t,x)\right] \tag{6.67}$$

$$\int_{T_n}^{T_{n+1}} dxv(t,x) = \frac{e^{\phi_n(t)}}{1+e^{\phi_n(t)}}\int_{T_n}^{T_{n+1}} \gamma(t,x) \tag{6.68}$$

The result for $v(t,x)$ has been obtained earlier in Eq. (6.24); the value of μ is a new result. Writing the drift and volatility in terms of $f_L(t,x)$ yields, from Eqs. (6.49) and (6.59), the following

$$f_n(t) \equiv \int_{T_n}^{T_{n+1}} dxf_L(t,x); \quad \frac{e^{\phi_n(t)}}{1+e^{\phi_n(t)}} = 1 - e^{-f_n(t)} \tag{6.69}$$

$$\mu(t,x) = u(t,x)\left[-\frac{1}{2}\Lambda(t,x)+\rho(t,x)\right] \tag{6.70}$$

$$v(t,x) = u(t,x)\gamma(t,x)$$

$$\text{where} \quad u(t,x) = \sum_{n=0}^{\infty} H_n(x)[1 - e^{-f_n(t)}] \tag{6.71}$$

The drift $\mu(t, x)$ and volatility $v(t, x)$ are both functions of only $\exp\{-f_n(t)\}$, which, in turn, is the forward price of a zero coupon bond.

6.11.1 Nonsingular Libor forward interest rates

The underlying Libor forward interest rates driving all the Libors, from Eq. (6.10), are given by the following

$$\frac{\partial f_L(t, x)}{\partial t} = \mu(t, x) + v(t, x)\mathcal{A}_L(t, x)$$

The theory is nonlinear due to the dependence of the volatility $v(t, x)$ and drift $\mu(t, x)$ on the underlying Libor forward interest rates $f_L(t, x)$. A drift that renders Libor to be a martingale apparently implies that the underlying Libor forward interest rates are singular [88]. This aspect of the Libor forward interest rates is analyzed.

An approximation of Eq. (6.71) that is adequate for the analysis of this section is

$$v(t, x) \simeq [1 - e^{-\ell f_L(t, x)}]\gamma(t, x)$$

The following are the two limiting cases

$$v(t, x) = \begin{cases} \ell\gamma(t, x)f_L(t, x); & \ell f_L(t, x) << 1 \\ \gamma(t, x); & \ell f_L(t, x) >> 1 \end{cases} \tag{6.72}$$

For $v(t, x) = \ell f_L(t, x)\gamma(t, x) \sim 0$, ignoring the integration that does not qualitatively change the results, from Eq. (6.17), the drift is

$$\mu(t, x) \simeq v(t, x) \int_{T_I}^{x} dx' \mathcal{D}_L(x, x'; t)v(t, x') + O(\ell)$$

$$\simeq [\ell f_L(t, x)\gamma(t, x)]^2 + O(\ell) \tag{6.73}$$

The limiting cases for $\mu(t, x)$, from Eqs. (6.70) and (6.73), are the following[3]

$$\mu(t, x)$$
$$\simeq \begin{cases} [\ell f_L(t, x)\gamma(t, x)]^2 + O(\ell); & \ell f_L(t, x) << 1 \\ \mu_0 = -\frac{1}{2}\int_{T_n}^{T_{n+1}} dx' M_\gamma(x, x'; t) + \int_{T_{I+1}}^{T_{n+1}} dx' M_\gamma(x, x'; t); & \ell f_L(t, x) >> 1 \end{cases} \tag{6.74}$$

---

[3] Note that, for $\ell \to 0$, one has the limit $\ell \sum_{m=I+1}^{n} \to (T_n - T_I)$: constant.

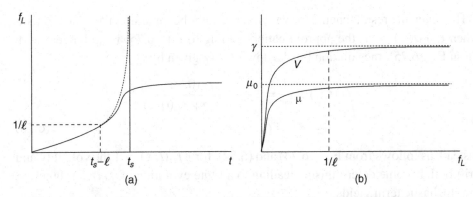

Figure 6.9 (a) The value of $f_L(t,x)$ is finite for all t. The singularity at t_s is spurious, since, at time $t_s - \ell$ when $\ell f_L(t,x) \simeq 1$, nonlinear effects take over. (b) The behavior of volatility $v(t,x)$ and drift $\mu(t,x)$ as a function of $f_L(t,x)$.

In the limit $\ell f_L(t,x) >> 1$, from Eq. (6.72), volatility $v(t,x) \to \gamma(t,x)$; the results obtained in Eqs. (6.74) and (6.72) for this limit are consistent with the earlier result given in Eq. (6.17). The two results agree only in the limit of $\ell \to 0$; for the case when $\ell f_L(t,x)\gamma(t,x)$ is independent of ℓ, the extra term $\int_{T_n}^{T_{n+1}} M_\gamma(x,x';t)/2$ in Eq. (6.74) is of $O(\ell)$ and goes to zero.

For small values of $\ell f_L(t,x)$, $\mu(t,x) \simeq (\ell\gamma(t,x)f_L(t,x))^2$ follows from Eq. (6.74). The stochastic term can be ignored as these do not qualitatively change the impact of the quadratic term on the evolution of $f_L(t,x)$. The simplified dynamics for the Libor forward interest rates, from Eq. (6.10), are the following[4]

$$\frac{\partial f_L(t,x)}{\partial t} \simeq \ell^2\gamma^2 f_L^2(t,x) + \text{random terms}$$

$$\Rightarrow f_L(t,x) \simeq \ell\gamma \frac{f_L(0,x)}{1 - t\ell\gamma f_L(0,x)} + \text{random terms} \qquad (6.75)$$

$$\Rightarrow f_L(t,x) \text{ infinite for } t_s = \frac{1}{\ell\gamma f_L(0,x)}; \; f_L(0,x) > 0$$

From Eq. (6.75), all the Libor forward interest rates become infinite as $t \to t_s = 1/[\ell\gamma f_L(0,x)] > 0$, as shown by the dotted curve in Figure 6.9(a). If $f_L(t,x)$, in fact, becomes singular, then, on including the stochastic term, the singularity is even more severe with the Libor forward interest rates becoming singular almost instantaneously [88].

---

[4] The dependence of γ and μ on t,x is, henceforth, ignored since it simplifies the calculation and does not change the main conclusions.

However, the result that the forward interest rates become singular is not correct; when $\ell f_L(t, x) \sim 1$, the approximation leading to Eq. (6.75) is no longer valid. From Eq. (6.75), the solution breaks down for t_f given by

$$f_L(t, x) \simeq \frac{1}{\ell} = \ell \gamma \frac{f_L(0, x)}{1 - t_f \ell \gamma f_L(0, x)}$$

$$\Rightarrow t_f = t_s - \ell < t_s \tag{6.76}$$

Instead, as follows from Eqs. (6.74) and (6.72), for $\ell f_L(t, x) \sim 1$ the volatility and drift both become deterministic, leading to a finite evolution of $f_L(t, x)$. Ignoring the stochastic term yields

$$f_L(t, x) \sim f_L(0, x) e^{\mu_0 t}; \quad t > t_f$$

The different domains for $f_L(t, x)$ are shown in Figure 6.9(a); $f_L(t, x)$ grows slowly for $t > t_s$ since the coefficient $\mu_0 \sim \gamma^2 << 1$.

Recall that the drift $\mu(t, x)$ and volatility $v(t, x)$ are both functions of only $\exp\{-f_n(t)\} \simeq \exp\{-\ell f_L(t, T_n)\}$. As $f_L(t, x)$ grows large, both $\mu(t, x)$ and $v(t, x)$ rapidly become deterministic and independent of $f_L(t, x)$, as shown in Figure 6.9(b). This in turn means that $f_L(t, x)$ can be described by a linear Gaussian quantum field as discussed in Chapter 5. Gaussian fields have configurations where $f_L(t, x)$ takes small values and can, hence, revert to the regime for its nonlinear evolution. In this manner, the Libor forward interest rates are driven by the exact evolution equation between the nonlinear and linear domains.

6.12 Summary

Quantum finance provides a natural and mathematically tractable formulation of the Libor Market Model. Gaussian white noise $R(t)$, in effect, is 'promoted' by quantum finance to a quantum field $\mathcal{A}_L(t, x)$: the evolution equation of Libor is driven by $\mathcal{A}_L(t, x)$.

In an economy where Libor rates are perfectly correlated across different maturities, a single volatility function is sufficient. However, the Libor term structure that is imperfectly correlated introduces many new features. One way of accounting for imperfect correlations is to extract Libor volatility from many caplets and then aggregate them – leading to a proliferation of parameters. In the quantum finance approach, due to the specific properties of Gaussian quantum fields, the entire Libor volatility function can be taken directly from the market and thus incorporates many key features of the market in a parameter-free manner.

The Gaussian quantum field $\mathcal{A}_L(t, x)$ has a quadratic action and hence one can obtain a differential formulation of the Libor Market Model. The singular quadratic product $\mathcal{A}_L(t, x)$ has a 'short distance' Wilson expansion that generalizes the results of Ito calculus and yields a derivation of Libor drift. It is seen that the underlying Libor forward interest rates $f_L(t, x)$ of the Libor Market Model are nonlinear and nonsingular.

The Libor forward interest rates $f_L(t, x)$ are related to logarithmic Libor by Eq. (6.49)

$$\exp\left\{\int_{T_n}^{T_n+\ell} dx f_L(t, x)\right\} = 1 + \exp\left\{\int_{T_n}^{T_n+\ell} dx \phi(t, x)\right\} \qquad (6.77)$$

The transformation, in Eq. (6.49), from $\phi(t, x)$ to $f_L(t, x)$ is nonlinear and nonlocal and is well defined only for strictly positive $f_L(t, x)$, that is for $f_L(t, x) \geq 0$. In particular, this means that $f_L(t, x)$ *cannot* be a Gaussian quantum field. The linear approximation developed in Chapter 5 treats $f(t, x)$ as a Gaussian quantum field and needs to be carefully examined to ascertain whether it can be applied to the interest rates (Libor and Euribor) market. Nonlinear models for $f(t, x)$, in general, are fairly intractable [12, 13].

Note the transformation given in Eq. (6.77) is completely general and only requires that $f_L(t, x) \geq 0$; for example, one can take $f_L(t, x) \sim e^{\chi(t, x)}$, where $\chi(t, x)$ is any real quantum field and this yields strictly positive Libors. The Libor Market Model makes a very *specific* choice for $f_L(t, x)$, namely one for which the log Libor quantum field $\phi(t, x)$ has deterministic volatility given $\gamma(t, x)$. This choice for $f_L(t, x)$ leads to the field $\phi(t, x)$ being a 'flat' quantum field that takes values on the entire real line. The leading order effects of $\phi(t, x)$ are described by a Gaussian quantum field; the nonlinear terms are all contained in the drift and can be treated perturbatively using Feynman diagrams.

Bond forward interest rates modeled, in Chapter 5, $f(t, x)$ as a Gaussian quantum field lead to strictly positive (zero) coupon bonds but negative Libors. In contrast, log Libor variables $\phi(t, x)$, yield strictly positive coupon bonds and strictly positive Libors. A nonlinear quantum field theory for flat quantum field $\phi(t, x)$ is the most appropriate formalism for analyzing coupon bond and Libor instruments.

The study of interest rates shows that the complexity and nonlinearity of the interest rates partly stems from Eq. (6.77): Libor can be described by $\phi(t, x)$ variables that are nonlinear, whereas the bond sector can be consistently described by bond forward interest rates $f(t, x)$ that are Gaussian. Hybrid instruments combine coupon bonds with Libor [79]. A consistent model of *both* $\phi(t, x)$ and $f(t, x)$, which are ingredients for hybrid instruments, necessarily requires that *either* $f(t, x)$ or $\phi(t, x)$ be modeled as a nonlinear two-dimensional quantum field – thus greatly increasing the difficulty of the analysis.

6.13 Appendix: Limits of the Libor Market Model

The following three different limits of the Libor Market Model evolution equations are taken:

- The limit of zero tenor, namely $\ell \to 0$
- The BGM–Jamshidian limit
- The HJM limit

6.13.1 Tenor $\ell \to 0$

Consider the limit of the tenor $\ell \to 0$. Let $x = T_n$, $x' = T_m$, and $x, x' > T_I$. From Eq. (6.24), $v(t, x) \simeq \ell f_L(t, x) \gamma(t, x)$; hence, from Eq. (6.70)

$$x \in [T_n, T_{n+1}]; \quad L(t, T_n) \sim f_L(t, x) + O(\ell); \quad e^{\phi(t)} \sim \ell f_L(t, x)$$
$$1 - e^{-f_n(t)} \simeq \ell f_L(t, x); \quad \Lambda_n \simeq \ell M_\gamma(x, x; t) = \ell \gamma(t, x) \mathcal{D}_L(x, x; t) \gamma(t, x)$$

$$\mu(t, x) = \sum_{j=0}^{\infty} H_j(x)[1 - e^{-f_j(t)}] \left[-\frac{1}{2} \Lambda_j(t, x) + \rho_j(t, x) \right]$$

$$\sim [\ell f_L(t, x)] \left[-\frac{\ell}{2} \gamma(t, x) \mathcal{D}_L(x, x; t) \gamma(t, x) \right.$$

$$\left. + \sum_{m=I+1}^{n} \frac{\ell f_L(t, T_m)}{1 + \ell f_L(t, x)} \ell \gamma(t, x) \mathcal{D}_L(x, T_m; t) \gamma(t, T_m) \right]$$

The term $\sum_m H_m(x)$ collapses to one term since $x \in [T_n, T_{n+1})$. The limit $\ell \to 0$ is taken holding $v(t, x) = \ell f_L(t, x) \gamma(t, x)$ fixed; hence, in this limit

$$\mu(t, x) \simeq \ell f_L(t, x) \ell^2 \sum_{m=I+1}^{n} \left[f_L(t, T_m) \gamma(t, x) \mathcal{D}_L(x, T_m; t) \gamma(t, T_m) \right]$$

$$- \frac{\ell}{2} [\ell f_L(t, x) \gamma(t, x)] \mathcal{D}_L(x, x; t) \gamma(t, x)$$

$$\simeq v(t, x) \int_{T_I}^{x} dx' \mathcal{D}_L(x, x'; t) v(t, x') + O(\ell) \tag{6.78}$$

Hence, from Eq. (6.10)

$$\frac{\partial f_L(t, x)}{\partial t} \simeq v(t, x) \int_{T_I}^{x} dx' \mathcal{D}_L(x, x'; t) v(t, x') + v(t, x) \mathcal{A}_L(t, x)$$

The limit of tenor $\ell \to 0$ yields an evolution equation for the Libor forward interest rates $f_L(t, x)$ that looks similar to the quantum HJM model, except

$v(t,x) = \ell f_L(t,x)\gamma(t,x)$ is stochastic. Recall the drift results from requiring a martingale evolution of the Libor forward interest rates $f_L(t,x)$ with the forward bond numeraire being the zero coupon bond $B(t, T_I)$, with T_I fixed.

6.13.2 BGM–Jamshidian limit

The BGM–Jamshidian limit of the quantum finance results for the Libor Market Model can be obtained using the following prescription

$$\mathcal{D}_L(x,x';t)\Big|_{BGM} \to 1$$

$$\mathcal{A}_L(t,x)\Big|_{BGM} \to R(t); \quad E[R(t)R(t')] = \delta(t-t') \tag{6.79}$$

where, recall, $R(t)$ is Gaussian white noise.

The Libor evolution for $T_I < T_n$, given in Eq. (6.21), yields the BGM–Jamshidian limit. From Eq. (6.35)

$$\Lambda_{mn}(t) = \int_{T_m}^{T_{m+1}} dx\gamma(t,x) \int_{T_n}^{T_{n+1}} dx' \mathcal{D}_L(x,x';t)\gamma(t,x') \to \gamma_n(t)\gamma_n(t)$$

$$\text{where} \quad \gamma_n(t) = \int_{T_n}^{T_{n+1}} dx\gamma(t,x)$$

Hence, the BGM–Jamshidian limit of the Libor Market Model evolution equation, from Eqs. (6.21), (6.56), and (6.79), is given by

$$\frac{1}{L(t,T_n)}\frac{\partial L(t,T_n)}{\partial t} = \zeta_n(t) + \gamma_n(t)R(t); \quad n > I \tag{6.80}$$

$$\zeta_n(t) = \gamma_n(t)\sum_{m=I+1}^{n}\frac{\ell L(t,T_m)}{1+\ell L(t,T_m)}\gamma_m(t)$$

6.13.3 HJM limit

The HJM limit requires the following three conditions.

- The zero tenor limit $\ell \to 0$ is taken.
- It assumed that $v(t,x) = \ell f_L(t,x)\gamma(t,x)$ is a deterministic function equal to the HJM volatility function; that is $v(t,x) \to \sigma(t,x)$: the HJM bond forward interest rates' volatility that is independent of $f_L(t,x)$.
- $\mathcal{D}_L(x,x';t)\Big|_{HJM} \to 1$.

With these assumptions Eqs. (6.79) and (6.80) yield

$$\frac{\partial f_L(t,x)}{\partial t} = \sigma(t,x) \int_{T_I}^{x} dx' \sigma(t,x') + \sigma(t,x) R(t)$$

which is the expected HJM equation; the drift is fixed by the forward bond numeraire $B(t, T_I)$.

6.14 Appendix: Jacobian of $A_L(t,x) \rightarrow \phi(t,x)$

The change of variables from quantum field $A_L(t,x)$ to $\phi(t,x)$ is, from Eq. (6.58), given by

$$\frac{\partial \phi(t,x)}{\partial t} = \rho(t,x) + \gamma(t,x) A_L(t,x); \quad t \in [t_0, t_*] \tag{6.81}$$

Eq. (6.81) is a nonlinear change of variables since drift $\rho(t,x)$ depends on $\phi(t,x)$; the transformation can, in principle, have a nontrivial Jacobian. Taking the differential of Eq. (6.81) yields

$$\int_{t_0}^{t_*} dt \left[\delta(t'-t) \frac{\partial}{\partial t} - \frac{\delta \rho(t',x)}{\delta \phi(t,x)} \right] d\phi(t,x) = \gamma(t',x) dA_L(t',x) \tag{6.82}$$

$$\Rightarrow \det[\mathcal{J}] D\phi = \text{const} \times DA_L \tag{6.83}$$

The change of variables factorizes for the x variable; hence, for notational simplicity, the x coordinate is suppressed and only the time variable t is displayed. The Jacobian is equal to $\det[\mathcal{J}]$, where the matrix of transformation $\mathcal{J}$ is given by

$$\mathcal{J}(t',t) = \delta(t'-t) \frac{\partial}{\partial t} - \frac{\delta \rho(t',x)}{\delta \phi(t,x)} \equiv \delta(t'-t) \frac{\partial}{\partial t} - J(t',t)$$

$$\Rightarrow \mathcal{J} = \mathcal{U} \frac{\partial}{\partial t} \mathcal{U}^{-1}; \quad J(t',t) = \frac{\delta \rho(t',x)}{\delta \phi(t,x)} \tag{6.84}$$

In Eq. (6.84) matrix multiplication is an integration over t given by $\int_{t_0}^{t_*} dt$. Time is discretized $t \rightarrow t_n = t' + (n-1)\epsilon$, $n = 1, 2, 3, \ldots, M = (t - t')/\epsilon$ to explicitly write out the matrix elements of $\mathcal{U}$ as follows

$$\mathcal{U}(t',t) = \left[\prod_{n=1}^{M-1} \int_{t_0}^{t_*} dt_n \right] \prod_{n=0}^{M-1} \exp\{\epsilon J(t_n, t_{n+1})\}; \quad t_1 = t'; t_M = t$$

From Eq. (6.84) the Jacobian is given by

$$\det[\mathcal{J}] = \det \left[\mathcal{U} \frac{\partial}{\partial t} \mathcal{U}^{-1} \right] = \det \left[\delta(t'-t) \frac{\partial}{\partial t} \right] = \text{constant}$$

The Jacobian of the transformation in going from $\mathcal{A}_L(t,x) \rightarrow \phi(t,x)$ is a constant; all constants involved in going from $\mathcal{A}_L(t,x) \rightarrow \phi(t,x)$ cancel due to division by the partition function Z. Henceforth, the path integration measure will taken to be invariant, namely

$$\int D\phi = \int D\mathcal{A}_L$$

The Jacobian being a constant will be essential in deriving the Libor Hamiltonian in Section 15.7.

7

Empirical analysis of forward interest rates

Empirical forward interest rates drive the debt markets. The quantum finance bond and Libor forward interest rates' models of the empirical rates are analyzed using market data [6]. In particular, the models are calibrated to the market and empirically tested, with all the model parameters, including interest rate volatilities, being obtained from market data.[1]

The bond forward interest rates, from Eq. (5.1), are given by

$$\frac{\partial f(t,x)}{\partial t} = \alpha(t,x) + \sigma(t,x)\mathcal{A}(t,x)$$

$$E[\mathcal{A}(t,x)] = 0; \quad E[\mathcal{A}(t,x)\mathcal{A}(t',x')] = \delta(t-t')\mathcal{D}(x,x';t)$$

In studying bonds, the volatility $\sigma(t,x)$ is taken to be deterministic, with drift $\alpha(t,x)$ being fixed by a martingale condition. The Libor Market Model of the Libor forward interest rates $f_L(t,x)$ and log Libor $\phi(t,x)$ are given by Eqs. (6.66) and (6.67) as follows

$$\frac{\partial f_L(t,x)}{\partial t} = \mu(t,x) + v(t,x)\mathcal{A}_L(t,x)$$

$$\frac{\partial \phi(t,x)}{\partial t} = \rho(t,x) + \gamma(t,x)\mathcal{A}_L(t,x)$$

$$E[\mathcal{A}_L(t,x)] = 0; \quad E[\mathcal{A}_L(t,x)\mathcal{A}_L(t',x')] = \delta(t-t')\mathcal{D}_L(x,x';t)$$

Libor volatility $\gamma(t,x)$ is deterministic, whereas $v(t,x)$ is stochastic. Libor drifts $\mu(t,x)$ and $\rho(t,x)$ are both stochastic.

Empirical forward interest rates can be taken from either the bond market, as realized in the bond ZCYC (zero coupon yield curve), or from the Libor and Euribor

---

[1] The term 'Libor' is generic and stands for Libor and Euribor.

market. The forward interest rates $f(t,x)$ are assumed to be approximately equal to Libor; that is[2]

$$L(t,T) \simeq f(t,T) + O(\ell)$$

In analyzing the Libor and Euribor data for determining the parameters of the Libor Market Model, it will be assumed that the Libor interest rates $f_L(t,x)$ are related to Libor by Eq. (6.2)

$$1 + \ell L(t,T) = \exp\left\{ \int_T^{T+\ell} dx f_L(t,x) \right\}$$

Empirical Libor data are used for analyzing *both* the bond forward interest rates and the Libor Market Model. The empirical study will fix the parameters of both $\mathcal{A}(t,x)$ that drives $f(t,x)$ and $\mathcal{A}_L(t,x)$ that drives $\phi(t,x)$.

7.1 Introduction

Libor futures data from 17 April 2002 to 29 April 2003, consisting of 261 trading days, are used for the empirical analysis. The Treasury Bond market is empirically studied using ZCYC data for 523 trading days, from 29 January 2003 to 28 January 2005. The Libor ZCYC is empirically studied using 261 days of data from 10 October 2007 to 8 August 2008.

Both bond and Libor forward interest rates are empirically studied using various approximation schemes. The bond forward interest rates are taken to be equal to empirical Libor and the volatility $\sigma(t,x)$ and correlation of changes in the forward interest rates are evaluated. The stiff Lagrangians for $\mathcal{A}(t,x)$ and $\mathcal{A}_L(t,x)$ are seen to provide an excellent fit to the market data. The Libor forward interest rates $f_L(t,x)$ are empirically studied in the framework of the Libor Market Model and, in particular, the relation between deterministic volatility $\gamma(t,x)$ and stochastic $v(t,x)$ is analyzed.

The market data for the ZCYC are analyzed for both the zero coupon bonds and Libor markets. It is seen that empirical ZCYC for both markets could not be satisfactorily explained by the bond and Libor forward interest rates.

---

[2] Market prices of zero coupon bonds yield empirical values for the bond forward interest rates since, from Eq. (2.12), $B(t,T) = \exp\{-\int_t^T dx f(t,x)\}$. The forward interest yield curves defined by the bond and Libor ZCYC differ by the TED (Treasury Eurodollar) spread. The difference will be ignored; errors that are much larger than those due to TED are introduced by the spline and other fits – required for fitting the various parameters of the models to discrete market Libor and bond data.

7.2 Interest rate correlation functions

The market provides data on interest rates as a time series, given at discrete moments. Hence one needs to discretize both calendar and future time to empirically study interest rate models. For notational convenience, let $f(t,x)$ denote both the bond forward interest rates as well as the log Libor interest rates $\phi(t,x)$. The volatility of $f(t,x)$ and $\phi(t,x)$ is given by $\sigma(t,x)$ and $\gamma(t,x)$, respectively; both the volatilities are deterministic and the basis of many of the results of this section. A more specific analysis for Libor forward interest rates $f_L(t,x)$, having stochastic volatility $v(t,x)$, is carried out in Section 7.8.

In principle, the drift terms $\alpha(t,x)$, $\rho(t,x)$ can depend on the forward interest rates, as is the case for the Libor Market Model, and are taken to be completely general.[3]

Discretize time into a lattice of points $t = n\epsilon$, with spacing $\epsilon = 1$ day. Hence

$$\delta f(t,x) = \epsilon\alpha(t,x) + \epsilon\sigma(t,x)\mathcal{A}(t,x) \tag{7.1}$$

$$\text{where } \delta f(t,x) \equiv f(t+\epsilon,x) - f(t,x)$$

Recall from Eq. (5.22) that

$$E[\mathcal{A}(t,x)] = 0 \tag{7.2}$$

$$E[\mathcal{A}(t,x)\mathcal{A}(t',x')] = \delta(t-t')\mathcal{D}(x,x';t) \tag{7.3}$$

Henceforth, all correlation functions will be expressed as functions of only remaining future time, namely $\theta = x - t, \theta' = x' - t$, as these are the coordinates appropriate for the empirical study. Define $f(t,x) = f(t,t+\theta)$ as shown in Figure 2.4(a); the notation $f(t,\theta)$ is sometimes used for $f(t,t+\theta)$ since it simply re-labels what one means by θ.

On discretizing time, the equal time expectation value of the fields at two future times is singular. From Eq. (A.10), $\delta(0) = 1/\epsilon$; hence, from Eq. (5.22)

$$E[\mathcal{A}(t,\theta)\mathcal{A}(t,\theta')] \equiv < \mathcal{A}(t,\theta)\mathcal{A}(t,\theta') > = \frac{1}{\epsilon}\mathcal{D}(\theta,\theta') \tag{7.4}$$

From Eq. (5.23)

$$< \delta f(t,\theta) > = \epsilon < \alpha(\theta) > \tag{7.5}$$

---

[3] The drift term is fixed using the martingale condition. For the bond forward interest rates, drift is deterministic; for the Libor Market Model, the drift depends on log Libor $\phi(t,x)$.

The drift velocity is fixed by the martingale condition and the drift in the market is not the one given by the martingale measure; hence martingale drift $\alpha(t, x)$ cannot be determined from the forward interest rates' market data.

7.2.1 *Forward interest rates' covariance*

Central to the empirical analysis is the following *covariance*

$$< \delta f(t,\theta)\delta f(t,\theta') >_c \equiv\; < \delta f(t,\theta)\delta f(t,\theta') > - < \delta f(t,\theta) >< \delta f(t,\theta') >$$
$$= \epsilon^2 \sigma(\theta)\sigma(\theta') < \mathcal{A}(t,\theta)\mathcal{A}(t,\theta') >$$
$$\Rightarrow\; < \delta f(t,\theta)\delta f(t,\theta') >_c = \epsilon\sigma(\theta)\sigma(\theta')\mathcal{D}(\theta,\theta') \tag{7.6}$$

The model's volatility $\sigma(t, x) = \sigma(\theta)$ is given in terms of empirical volatility $\sigma_E(\theta)$ as follows

$$\sigma_E^2(\theta) \equiv\; < [\delta f(t,\theta)]^2 >_c\; = \epsilon\sigma^2(\theta)\mathcal{D}(\theta,\theta)$$
$$\sigma(\theta) = \frac{\sigma_E(\theta)}{\sqrt{\epsilon\mathcal{D}(\theta,\theta)}} \tag{7.7}$$

Hence, from Eqs. (7.6) and (7.7)

$$< \delta f(t,\theta)\delta f(t,\theta') >_c = \sigma_E(\theta)\frac{\mathcal{D}(\theta,\theta')}{\sqrt{\mathcal{D}(\theta,\theta)}\sqrt{\mathcal{D}(\theta',\theta')}}\sigma_E(\theta') \tag{7.8}$$

Eq. (7.8) shows that the covariance is uniquely fixed by the empirical volatility $\sigma_E(\theta)$ and the model's normalized propagator.

7.3 Interest rate volatility

For many calculations, it is convenient to have an explicit expression for the volatility $\sigma(t, x)$ that appears in the definition of the model. However, $\sigma(t, x)$ is not uniquely specified in the model; one can rescale $\sigma(t, x) \rightarrow \tilde{\sigma}(t, x) = \kappa(\theta)\sigma(t, x)$ and rescale $\mathcal{A}(t, x) \rightarrow \tilde{\mathcal{A}}(t, x) = \mathcal{A}(t, x)/\kappa(\theta)$, leaving the defining equation Eq. (5.1) unchanged. The scaling factor is chosen to make $\sigma(t, x)$ equal to the empirical volatility $\sigma_E(t, x)$ and yields, from Eq. (7.7)

$$\kappa(\theta) = \sqrt{\epsilon\mathcal{D}(\theta,\theta)} \tag{7.9}$$
$$\tilde{\sigma}(\theta) = \kappa(\theta)\sigma(t, x) = \sigma_E(\theta)$$

The model's volatility, from Eq. (7.7), is hence given by

$$\tilde{\sigma}^2(\theta) = \sigma_E^2(\theta) =< [\delta f(t,\theta)]^2 >_c \tag{7.10}$$

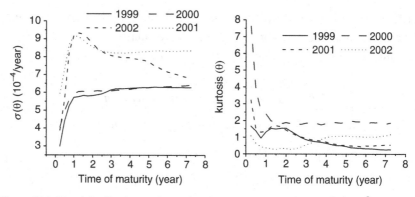

Figure 7.1 Empirically determined volatility function $\sigma(\theta) = \sqrt{< \delta f^2(t,\theta) >_c}$ and kurtosis $\kappa(t,\theta) =< [\delta f(t,\theta)]^4 > /\sigma^4(t,\theta) - 3$ for Libor forward interest rates. The functions are given for four distinct time periods showing a significant change in their values.

The empirical volatility and kurtosis for four different periods is shown in Figure 7.1.

The rescaled field $\tilde{\mathcal{A}}(t,\theta) = \mathcal{A}(t,\theta)/\kappa(\theta)$ has propagator given from Eq. (7.4)

$$\frac{1}{\epsilon}\tilde{\mathcal{D}}(\theta,\theta') = E[\tilde{\mathcal{A}}(t,\theta)\tilde{\mathcal{A}}(t,\theta')] = \frac{E[\mathcal{A}(t,\theta)\mathcal{A}(t,\theta')]}{\kappa(\theta)\kappa(\theta')}$$

$$\Rightarrow \tilde{\mathcal{D}}(\theta,\theta') = \frac{\mathcal{D}(\theta,\theta')}{\kappa(\theta)\kappa(\theta')} = \frac{1}{\epsilon}\frac{\mathcal{D}(\theta,\theta')}{\sqrt{\mathcal{D}(\theta,\theta)}\sqrt{\mathcal{D}(\theta',\theta')}} \qquad (7.11)$$

$$\tilde{\mathcal{D}}(\theta,\theta) = \frac{1}{\epsilon} \qquad (7.12)$$

The rescaling is consistent; in particular, note from Eqs. (7.6), (7.9), and (7.12)

$$< \delta f(t,\theta)\delta f(t,\theta') >_c= \epsilon\sigma(\theta)\sigma(\theta')\mathcal{D}(\theta,\theta') = \epsilon\tilde{\sigma}(\theta)\tilde{\sigma}(\theta')\tilde{\mathcal{D}}(\theta,\theta')$$

$$= \tilde{\sigma}(\theta)\frac{\tilde{\mathcal{D}}(\theta,\theta')}{\sqrt{\tilde{\mathcal{D}}(\theta,\theta)}\sqrt{\tilde{\mathcal{D}}(\theta',\theta')}}\tilde{\sigma}(\theta') = \sigma_E(\theta)\frac{\mathcal{D}(\theta,\theta')}{\sqrt{\mathcal{D}(\theta,\theta)}\sqrt{\mathcal{D}(\theta',\theta')}}\sigma_E(\theta')$$

The result agrees with Eq. (7.8), as indeed it must, for consistency. The normalization chosen in Eq. (7.9) will be used from now on and the tildes on $\tilde{\sigma}$ and $\tilde{\mathcal{D}}$ are henceforth dropped. Note the value of the covariance given by $< \delta f(t,\theta)\delta f(t,\theta') >_c$ is independent of the choice of the scaling factor. Choosing a form for $\kappa(\theta)$ fixes a particular frame; the specific choice was made so that volatility $\sigma(t,x)$, which appears in the model, is equal to the empirical volatility $\sigma_E(t,x)$.

All prices of traded instruments are independent of the scaling factor $\kappa(\theta)$ and depend only on the covariance $< \delta f(t,\theta)\delta f(t,\theta') >_c$.

7.4 Empirical normalized propagators

The connected correlation function $< \delta f(t,\theta)\delta f(t,\theta') >_c$ is independent of drift velocity $\alpha(\theta)$. In quantum finance models, *deterministic* volatility, as expressed in Eqs. (7.6) and (7.7), can be completely factorized out of the correlation functions. The empirical volatility, as in Eq. (7.10), determines the value of $\sigma(\theta)$ and reflects information encoded in the interest rates. Determining volatility $\sigma(\theta)$ from the market greatly improves the applicability and accuracy of the quantum finance model.

The *normalized propagator* is given by

$$\mathcal{C}(\theta,\theta') = \frac{\mathcal{D}(\theta,\theta')}{\sqrt{\mathcal{D}(\theta,\theta)\mathcal{D}(\theta',\theta')}} \tag{7.13}$$

The normalized propagator, for $\theta \neq \theta'$, is used for testing quantum finance models by comparing the models' predictions with the observed market behavior of the forward interest rates.

The observed market values of the forward interest rates are assumed to be the possible outcomes (sample values) of the random values of $f(t,x) = f(t,t + \theta)$; $\theta = x - t$. A fundamental assumption in the empirical analysis is to treat expectation values of the various financial instruments as being equal to the time average value of these instruments, taken over the time series of the forward interest rates. This assumption is called the ergodic hypothesis in statistical physics.

The expectation value determining the correlation functions are obtained by summing over historical data of the forward interest rates. Suppose historical data are given for L days denoted by t_i; the stochastic averages for all financial instruments are taken to be equal to its average over historical data. Since the correlation functions are assumed to depend only on remaining future time $\theta = x - t$, one holds θ fixed and sums over the L historical values of $f(t_i, t_i + \theta)$ for calendar time $t_i = i\epsilon$; hence, in all empirical analysis $x = t + \theta$, the new set of coordinate variables are (t, θ).

From Eqs. (7.6) and (7.10), the empirical values of the correlation function are determined as follows

$$\sigma^2(\theta) = < [\delta f(t,t + \theta)]^2 >_c = \frac{1}{L}\sum_{i=1}^{L}[\delta f(t_i, t_i + \theta)]^2\Big|_c \tag{7.14}$$

$$\equiv \frac{1}{L}\sum_{i=1}^{L}[\delta f(t_i, t_i + \theta)]^2 - \left[\frac{1}{L}\sum_{i=1}^{L}\delta f(t_i, t_i + \theta)\right]^2 \tag{7.15}$$

From Eq. (7.6)

$$\epsilon\sigma(\theta)\sigma(\theta')\mathcal{D}(\theta,\theta') = \frac{1}{L}\sum_{i=1}^{L}\delta f(t_i, t_i + \theta)\delta f(t_i, t_i + \theta')\Big|_c$$

As shown in Figure 2.4(a), as time t runs over historical data for $f(t_i, t_i + \theta)$, one moves along the line $\theta = $ constant, at a slope of $45°$ in the xt-plane.

The covariance and normalized correlation between $\delta f(t,\theta)$ and $\delta f(t,\theta')$, required for evaluating the volatility and propagator of the interest rates, are given by

$$\langle\delta f(t,\theta)\delta f(t,\theta')\rangle_c : \text{ covariance}$$

$$C(\theta,\theta') = \frac{\langle\delta f(t,\theta)\delta f(t,\theta')\rangle_c}{\sqrt{\langle[\delta f(t,\theta')]^2\rangle_c}\sqrt{\langle[\delta f(t,\theta)]^2\rangle_c}} : \text{ normalized covariance} \quad (7.16)$$

Comparing the above result with Eq. (7.13) yields

$$\frac{\langle\delta f(t,\theta)\delta f(t,\theta')\rangle_c}{\sqrt{\langle[\delta f(t,\theta')]^2\rangle_c}\sqrt{\langle[\delta f(t,\theta)]^2\rangle_c}} = \frac{\mathcal{D}(\theta,\theta')}{\sqrt{\mathcal{D}(\theta,\theta)\mathcal{D}(\theta',\theta')}} \quad (7.17)$$

Defining the normalized propagator to be equal to the normalized covariance makes it *independent* of $\sigma(\theta)$. No assumption needs to be made regarding the form of the volatility. This is the reason for using the normalized propagator, rather than the covariance itself, for modeling forward interest rates. In particular, parameters such as η, μ, λ, and so on, which need calibration in quantum finance models, are fitted from market data independent of the value of $\sigma(\theta)$.

Eq. (7.17) provides the link between market correlations and the predictions made by the model. The calibration of the model's parameters are based on this equation.

The empirical value of the correlation functions for the bond forward interest rates $f(t,x)$ are estimated from the market Libor and Euribor futures' data for $L(t,T)$ using the approximation

$$L(t,T) \simeq f(t,T)$$

The result of the empirical evaluation of the covariance and normalized covariance for Libor is shown in Figures 7.2(a) and 7.2(b) respectively; the empirical Euribor normalized covariance is given in Figure 7.3.

The normalized covariance, for all values of its arguments, is always greater than about 0.55, showing that all the forward interest rates are *highly correlated*. Any

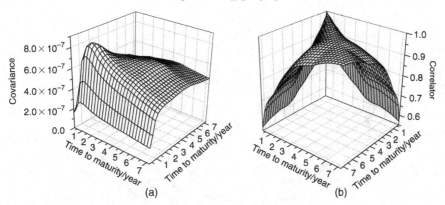

Figure 7.2 Libor data from 17 April 2002 to 29 April 2003, consisting of 261 trading days, are used for evaluating all the correlation functions. It is assumed that $L(t, T) \simeq f(t, T)$. (a) Covariance of $< \delta L(\theta)\delta L(\theta') >$. (b) The normalized covariance is equal to $< \delta L(\theta)\delta L(\theta') > /\sigma(\theta)\sigma(\theta')$.

two forward rates are strongly correlated – no matter how large is their separation in maturity time.

7.5 Empirical stiff propagator

Figures 7.2 and 7.3 clearly show that both the Libor and Euribor normalized coviariance have extremely smooth surfaces with no discontinuities or 'kinks' along the diagonal that appears in all models without the stiffness term in the Lagrangian [12]. It is to explain the highly correlated behavior of the forward interest rates that the stiff Lagrangian, given in Eq. (5.6), has been introduced in [16]. The normalized propagator is given by

$$C(\theta, \theta') = \frac{G(\theta, \theta')}{\sqrt{G(\theta, \theta)G(\theta', \theta')}} \qquad (7.18)$$

The stiff propagator has three branches and the real branch, which is relevant to the empirical analyis is given, from the results of Section 5.7, as follows [12]

$$G(\theta_+; \theta_-) \equiv \frac{\lambda}{2\sinh(2b)}[g_+(\theta_+) + g_-(\theta_-)] \qquad (7.19)$$

where

$$g_+(\theta_+) = e^{-\lambda\theta_+ \cosh(b)} \sinh\{b + \lambda\theta_+ \sinh(b)\} \qquad (7.20)$$

$$g_-(\theta_-) = e^{-\lambda|\theta_-| \cosh(b)} \sinh\{b + \lambda|\theta_-| \sinh(b)\} \qquad (7.21)$$

$$\theta_\pm = \theta \pm \theta' \qquad (7.22)$$

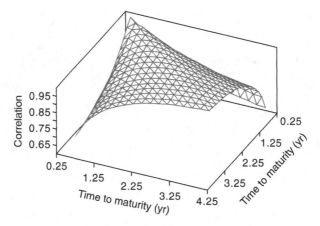

Figure 7.3 It is assumed that $L(t, T) \simeq f(t, T)$. Covariance is given by $C(\theta, \theta') = \langle \delta f(t, \theta) \delta f(t, \theta') \rangle_c / \sqrt{\langle [\delta f(t, \theta)]^2 \rangle_c} \sqrt{\langle [\delta f(t, \theta')]^2 \rangle_c}$; the correlation for Euribor forward interest rates is based on daily data from 26 May 1999 to 17 May 2004.

In this representation

$$C(\theta_+; \theta_-) = \frac{g_+(\theta_+) + g_-(\theta_-)}{\sqrt{[g_+(\theta_+ + \theta_-) + g_-(0)][g_+(\theta_+ - \theta_-) + g_-(0)]}} \tag{7.23}$$

The diagonal axis is a line of maxima for the normalized propagator since

$$\left. \frac{\partial C(\theta_+; \theta_-)}{\partial \theta_-} \right|_{(\theta_-=0)} \equiv \frac{\partial C(\theta_+; 0)}{\partial \theta_-} = 0 \tag{7.24}$$

The propagator $G(\theta_+; \theta_-)$ has a finite curvature perpendicular to the diagonal and hence one can compare it with the curvature of $C(\theta, \theta')$ given by the market data. The curvature orthogonal to the diagonal axis is defined as follows

$$R(\theta_+) = -\left. \frac{\partial^2 C(\theta_+; \theta_-)}{\partial \theta_-^2} \right|_{(\theta_-=0)} \equiv -\frac{\partial^2 C_Q(\theta_+; 0)}{\partial \theta_-^2} \tag{7.25}$$

It can be shown that [12]

$$R(\theta_+) = \frac{|g_-''(0)|}{g_+(\theta_+) + g_-(0)} - \frac{|g_+''(\theta_+)|[g_+(\theta_+) + g_-(0)] + [g_+'(\theta_+)]^2}{[g_+(\theta_+) + g_-(0)]^2} \tag{7.26}$$

Eq. (7.26) shows that $R(\theta_+)$, as θ_+ increases – which in effect means that one is moving on the diagonal axis away from the origin (having $\theta_+ = 0$) – the curvature (slowly) *increases*. The denominator of the first term decreases while at the same

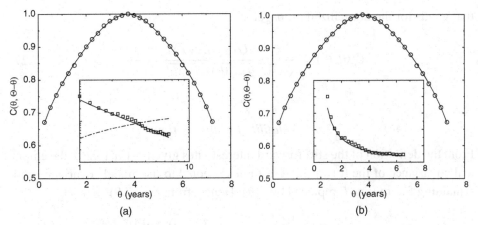

Figure 7.4 Empirical values of $\mathcal{C}(\theta_+;\theta_-)$ are shown by empty circles and squares, whereas the model's values with market time $z = \theta^\eta$ are shown by unbroken lines. $\mathcal{C}(\theta_+;\theta_-)$ is empirically determined for Libor forward interest rates (from 1990 to 1996). (a) The inset shows a plot of with log (θ_+) as the horizontal axis. The curvature $\log(R(\theta_+))$, shown as the dashed line, has an (incorrect) upwards slope. Market time $z = \theta^\eta$ has curvature $\log[z'(\theta_+)]^2 R(2z(\theta_+/2))$ – shown by an unbroken line – which correctly slopes downwards. (b) Figure shows the propagator fitted with market time. The inset shows curvature $[z'(\theta_+)]^2 R(2z(\theta_+/2))$ versus θ_+ as the horizontal axis.

time the second (negative) term becomes smaller; this behavior of the curvature, shown by the dotted line in the inset of Figure 7.4(a), is seen to slope upwards.

The curvature calculation predicts that the model's normalized propagator should fall off more rapidly as one moves on the diagonal away from the origin. If one looks carefully at Figures 7.2(a) and 7.3, one can see that the empirical normalized propagator shows the *opposite* behavior. As one moves away from the origin on the diagonal axis the curve *flattens* out, showing that the curvature is *decreasing* as θ_+ increases. Hence, as it stands, the stiff Lagrangian cannot explain the empirical behavior of the forward interest rates.

7.6 Empirical stiff propagator: future market time

The curvature for the normalized stiff propagator increases very slowly. One could try and rectify the problem by multiplying the propagator with a pre-factor that cancels the gradual rise in curvature and, instead, makes it fall off with a power law. Future market time variable $z(\theta)$ plays precisely this role.

The defining equation for market time $z(\theta)$, from Eq. (5.25), is given by

$$\frac{\partial f}{\partial t}(t, t + \theta) = \alpha(t, z(\theta)) + \sigma(t, z(\theta))\mathcal{A}(t, z(\theta)); \quad \theta = x - t$$

that yields for the normalized propagator

$$C_z(\theta, \theta') = \frac{G(z(\theta), z(\theta'))}{\sqrt{G(z(\theta), z(\theta))G(z(\theta'), z(\theta'))}} \tag{7.27}$$

7.6.1 Volatility for market time

From the definition of the stiff forward interest rates given in Eq. (5.25), the empirical covariance of the forward interest rates is equal to the model covariance with remaining future time θ replaced by $z(\theta)$. Hence, from Eq. (7.8)

$$< \delta f(t, \theta) \delta f(t, \theta') >_c = \sigma(z(\theta)) \frac{\mathcal{D}(z(\theta), z(\theta'))}{\sqrt{\mathcal{D}(z(\theta), z(\theta))} \sqrt{\mathcal{D}(z(\theta'), z(\theta'))}} \sigma(z(\theta'))$$

Hence, for $\theta = \theta'$

$$< [\delta f(t, \theta)]^2 >_c = \sigma_E^2(\theta) = \sigma^2(z(\theta))$$
$$\Rightarrow \sigma(z(\theta)) = \sigma_E(\theta)$$

In other words, no separate calculation is required for $\sigma(z(\theta))$, but, rather, volatility for market time is simply a re-labeling of the empirical volatility $\sigma_E(\theta)$.

7.6.2 Stiff propagator for market time

The empirical value of the normalized propagator of the forward interest rates $C_z(\theta, \theta')$ does not change when going to market future time. Instead, the description of this normalized propagator by the quantum finance model changes, and, consequently, the left-hand side of the above equation depends only on the remaining future time variables θ, θ', whereas the right-hand side depends only on the market time variables $z(\theta), z(\theta')$. Writing the normalized propagator more explicitly, similar to Eq. (7.23), yields

$$C_z(\theta+; \theta-) = \frac{g_+(z_+) + g_-(z_-)}{\sqrt{[g_+(z_+ + z_-) + g_-(0)][g_+(z_+ - z_-) + g_-(0)]}} \tag{7.28}$$

$$z_\pm(\theta+; \theta-) \equiv z(\theta) \pm z(\theta') \tag{7.29}$$

The curvature in the nonlinear variable $z(\theta)$ is [12]

$$\frac{\partial^2 C_z(\theta+; 0)}{\partial \theta_-^2} = [z'(\theta_+)]^2 R(2z(\theta_+/2)) \tag{7.30}$$

On empirically studying the curvature, one finds a power law fall-off for the curvature that is given by $C_Q(\theta_+) \simeq 1/\theta_+^{1.3}$. The ansatz $z(\theta) = \theta^\eta$ is used for fitting the data. Using the fact that $R_Q(2z(\theta_+/2))$ varies very slowly as a function of θ_+, one can make the following approximation

$$[z'(\theta_+)]^2 \propto \frac{1}{\theta_+^{1.3}}$$

$$\Rightarrow \theta_+^{2\eta-2} \propto \frac{1}{\theta_+^{1.3}} \Rightarrow \eta \simeq 0.35 \tag{7.31}$$

The best fit for Libor yields $\eta = 0.34$ showing that the market time variable almost completely dominates the curvature of the normalized propagator.

The units for λ and μ are fixed so that λz and μz are dimensionless; since $z = \theta^\eta$, define $\lambda z = [\tilde{\lambda}\theta]^\eta$ so that new constant $\tilde{\lambda}$ always has dimensions of $(\text{time})^{-1}$. η is a scaling exponent and is always dimensionless. In a unit where θ is measured in years, the result of the empirical study is summarized below. The parameter b is defined by

$$\mu = \frac{\lambda}{\sqrt{2\cosh(b)}} \quad \Rightarrow \quad \tilde{\mu} = \frac{\tilde{\lambda}}{(2\cosh(b))^{0.5/\eta}}$$

The stiff propagator with nonlinear maturity time $z(\theta)$ has an almost perfect match with Libor data, with a root mean square error of only 0.4%. Figure 7.4 shows a plot of the model's propagator on the diagonal line that is orthogonal to the $\theta_- = 0$ diagonal – since this is the longest stretch for the normalized propagator; the agreement with data is almost exact. What is noteworthy is that, even though the nonlinear maturity variable $z(\theta)$ was introduced to address the behavior of the propagator in the neighborhood of the diagonal axis, it continues to give the correct behavior for the propagator even far from the diagonal region.

The market Euribor normalized covariance, given in Figure 7.3, can also be fitted with the stiff propagator. Figure 7.5(a) shows the fit for Euribor along the diagonal orthogonal to $\theta = \theta'$ axis. The fit is almost perfect. Varying η as well as all the other parameters yields the best fit, given in Table 7.1; the effect of market time is shown in Figure 7.5(b). The fit, of the model, for Euribor is even better than for Libor, with both fits having an overall accuracy of over 99.6%.

The parametric fit for volatility $\sigma(t, x)$ given in Table 7.4 in Section 7.81, where $\sigma(\theta) = v(\theta)$, leads to an error of 5.04% for covariance $\sigma(t, x)\mathcal{D}(x, x'; t)\sigma(t, x')$, where $\mathcal{D}(x, x'; t)$ is the value of the stiff propagator; the error is almost entirely due to the errors in the fit for $\sigma(t, x)$.

Table 7.1 *The parameters for the stiff Lagrangian derived from Libor and Euribor data; the best fit was obtained by varying η.*

	$\tilde{\lambda}$	$\tilde{\mu}$	b	η	rms error for the entire fit
Libor	1.79/year	0.40/year	0.85	0.34	0.40%
Euribor	4.48/year	0.06/year	0.99	0.13	0.37%

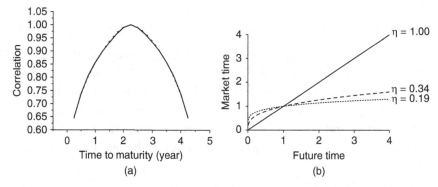

Figure 7.5 (a) Figure showing the fitted propagator for Euribor data from 26 May 1999 to 17 May 2004. (b) Figure shows Libor market time $[\lambda(x - t)]^{\eta}$ compared with future time $\lambda(x - t)$.

7.6.3 Euribor and Libor propagators: market and model

For many interest rate options – such as a swap for which the floating interest rates are paid in US Dollars and the fixed interest rates are paid in Euros – one has to simultaneously model both the Libor and Euribor forward interest rates. For an instrument that combines the US Dollars and Euros all market participants should have the same subjective view of what constitutes future time.

The η parameter quantifying market future time should be equal for such common instruments and, in general, for the Libor and Euribor markets. The reason for choosing η to be the same is mathematical as well as empirical; for a Lagrangian to exist on the same domain of future time the variable $z = \theta^{\eta}$ has to be common to both the forward interest rates.

The best fit was made based on both Libor and Euribor having the same $\eta = 0.19$. Data from the period from 26 May 1999 to 17 May 2004 yield the parameters given in Table 7.2 that fit data to better than 99% accuracy. If one takes $\eta = 0.34$ from the Libor market to be the common market time exponent, then the best fit, obtained for both the Libor and Euribor data, yields the following: $\tilde{\lambda} = 3.15$; $b = 0.57$ with root mean square error = 1.78%.

Table 7.2 *The parameters for the stiff Lagrangian. Best fit for Libor and Euribor with common* $\eta = 0.19$.

	$\tilde{\lambda}$	$\tilde{\mu}$	b	η	rms error for the entire fit
Libor	2.273/year	0.07/year	1.245	0.19	0.82%
Euribor	2.831/year	0.21/year	0.816	0.19	0.69%

Market future time for Libor and Euro is given by $(x - t)^{0.19}$. For $x - t = 2$ years the dimensionless Libor market time is $[\tilde{\lambda}(x - t)]^{\eta} = 1.33$, in contrast to 4.54 for $\eta = 1$, as shown in Figure 7.5(b).

Market future time $z(\theta) = \theta^{\eta}$ is a result of far-reaching significance. It shows that future time in the financial markets, as proposed in [23], is significantly different (slower) than calendar time, and influences all financial instruments. Market future time index η may vary over time, similar to volatility, in that it slowly changes over a long period of calendar time as well as being affected by market sentiment.

The parameters of the stiff Lagrangian, in particular the volatility $\sigma(t, x)$ and μ, λ, and η, depend on the market and the interest rate instrument one is fitting. Libor and Euribor give quite distinct values for the parameters; interest caplets yield parameters different than those obtained from the Libor data or from swaption data. For hybrid instruments that straddle many instruments and markets, one needs to further develop the models considered so far.

7.7 Empirical analysis of the Libor Market Model

The LMM (Libor Market Model) is studied empirically for calibrating the model as well as for comparing the behavior of Libor forward interest rates $f_L(t, x)$ with the log Libor field $\phi(t, x)$ [6]. A stiff propagator, with not necessarily the same parameters as the bond forward interest rates, is assumed to drive both the Libor forward interest rates $f_L(t, x)$ and log Libor field $\phi(t, x)$.

The defining equations of the Libor Market Model are Eqs. (6.10), (6.47), and (6.58) and yield the following

$$\frac{\partial f_L(t, x)}{\partial t} = \mu(t, x) + v(t, x) \mathcal{A}_L(t, x)$$

$$\ell L(t, T_n) = \exp \left\{ \int_{T_n}^{T_{n+1}} dx \phi(t, x) \right\} \equiv e^{\phi_n(t)}$$

$$\frac{\partial \phi(t, x)}{\partial t} = \rho(t, x) + \gamma(t, x) \mathcal{A}_L(t, x)$$

The correlations of $\partial\phi(t,x)/\partial t$, for $\theta = x - t$ and $\theta' = x' - t$, are the following

$$E\left[\frac{\partial\phi(t,x)}{\partial t}\right] = E[\rho(t,x)]$$

$$E\left[\frac{\partial\phi(t,x)}{\partial t}\frac{\partial\phi(t',x')}{\partial t'}\right]_c = \delta(t-t')\gamma(t,x)\gamma(t,x')\mathcal{D}_L(t;x,x')$$

$$= \delta(t-t')\gamma(\theta)\gamma(\theta')\mathcal{D}_L(\theta,\theta')$$

Eq. (6.51) relates the changes in $\phi(t,x)$ to Libor

$$\frac{\partial\ln(\ell L(t,T_n))}{\partial t} = \int_{T_n}^{T_{n+1}} dx \frac{\partial\phi(t,x)}{\partial t} \simeq \ell\frac{\partial\phi(t,T_n)}{\partial t}$$

Discretize time $t \to t_i = i\epsilon$ with $\epsilon = 1$ day; define

$$\delta\ln(\ell L(t_i,T_n)) = \ln(\ell L(t_i + \epsilon,T_n)) - \ln(\ell L(t_i,T_n)) = \ln\left[\frac{L(t_i+\epsilon,T_n)}{L(t_i,T_n)}\right]$$

Including the effects of market time, Eqs. (7.8) and (7.10) yield

$$E\left[\delta\ln(\ell L(t_i,\theta))\delta\ln(\ell L(t_i,\theta))\right]_c^2 = [\ell\gamma(z(\theta))]^2$$

$$E\left[\delta\ln(\ell L(t_i,\theta))\delta\ln(\ell L(t_i,\theta'))\right]_c = \frac{\ell^2\gamma(z(\theta))\mathcal{D}_L(z(\theta),z(\theta'))\gamma(z(\theta'))}{\sqrt{\mathcal{D}_L(z(\theta),z(\theta))}\sqrt{\mathcal{D}_L(z(\theta'),z(\theta'))}}$$

Libor forward interest rates $f_L(t,x)$ have stochastic volatility and are defined by Eqs. (6.21), (6.66), (6.67), and (6.47)

$$\int_{T_n}^{T_{n+1}} dx f_L(t,x) = \ln[1 + \ell L(t,T_n)]$$

$$\delta f_L(t,x) = \epsilon\mu(t,x) + \epsilon v(t,x)\mathcal{A}_L(t,x)$$

$$v(t,x) = \frac{e^{\phi_n(t)}}{1 + e^{\phi_n(t)}}\gamma(t,x) = \frac{\ell L(t,T_n)}{1 + \ell L(t,T_n)}\gamma(t,x); \quad T_n \le x < T_{n+1} \qquad (7.32)$$

Note that in the Libor Market Model, the *same* Gaussian quantum field $\mathcal{A}_L(t,x)$ drives both log Libor $\phi(t,x)$ and Libor forward interest rate $f_L(t,x)$.

Assume for now that $v(t,x)$ is deterministic; this assumption, in effect, makes $v(t,x)$ identical to the volatility $\sigma(t,x)$ of the bond forward interest rates defined

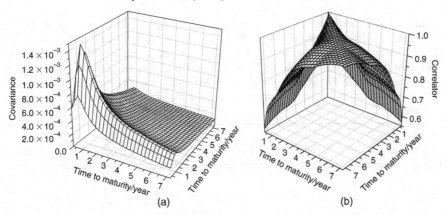

Figure 7.6 (a) Covariance of $< \delta \ln(\ell L)(\theta)\delta \ln(\ell L)(\theta') >_c$. (b) The normalized propagator $< \delta \ln(\ell L)(\theta)\delta \ln(\ell L)(\theta') >_c /\gamma(\theta)\gamma(\theta')$.

in Eq. (5.1). The same Libor data are used for analyzing both volatilities.[4] The covariance of δL yields the following

$$E\left[\delta L(t_i,\theta)\delta L(t_i,\theta')\right]_c = \frac{v(z(\theta))\mathcal{D}_L(z(\theta), z(\theta'))v(z(\theta'))}{\sqrt{\mathcal{D}_L(z(\theta), z(\theta))}\sqrt{\mathcal{D}_L(z(\theta'), z(\theta'))}}$$

The best fit for the parameters of the stiff propagator is obtained from the empirical propagator $\mathcal{D}_L(z(\theta), z(\theta'))$ for $< \delta L\delta L >_c$ and $< \delta \ln(\ell L)\delta \ln(\ell L) >_c$, with η_L having values 0.058 and 0.074 for the two cases, respectively. However, it is intuitively more appropriate to fix the future market time index η_L to be equal for both cases, since both normalized propagators result from the same market and the same instrument. The best fit for market time common to Libor and Euribor yields $\eta_L = 0.07$ with 99% accuracy.

The same data set is used for evaluating $< \delta \ln(\ell L)\delta \ln(\ell L) >_c$, the covariance of log Libor, as was used in Section 7.4 for evaluating the Libor covariance $< \delta L\delta L >_c$.

The empirical covariance and normalized covariance for $< \delta L\delta L >_c$ are shown in Figures 7.2(a) and 7.2(b) and those for $< \delta \ln(\ell L)\delta \ln(\ell L) >_c$ are shown in Figures 7.6(a) and 7.6(b). The result is fairly robust and convergence is stable.

The parameters obtained by fitting the stiff propagators to the two covariances are given in Table 7.3. The covariance of the instantaneous change in Libor δL as well as the instantaneous change in logarithm of Libor $\delta \ln(\ell L)$, in the Libor Market Model, are *both* driven by the same quantum field $\mathcal{A}_L(t,x)$. Table 7.3 shows that, to within the rms error of 1%, both the covariances yield almost the

---

[4] Note that, in the context of the Libor Market Model, in Section 7.8 $v(t,x)$ will be analyzed for its stochastic behavior.

Table 7.3 *The parameters for the Lagrangian of $A_L(t,x)$ for a common η. The normalized propagator from $< \delta f_L \delta f_L >_c \equiv < \delta L \delta L >_c / v(\theta) v(\theta')$ and from the log Libor case $\ell^2 < \delta \phi \delta \phi >_c \equiv < \delta \ln (\ell L) \delta \ln (\ell L) >_c / \gamma(\theta) \gamma(\theta')$ are both fitted for the parameters of the stiff propagator. The rms (root mean square) error is for the entire fit. Parameters for both fits are equal to within the 1% rms error.*

Covariance	λ_L	μ_L	b_L	η_L	rms error
$< \delta f_L \delta f_L >_c$	9.95	4.95	1.33	0.07	1.07%
$< \delta \phi \delta \phi >_c$	10.33	5.02	1.38	0.07	1.21%

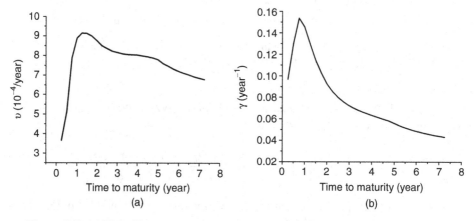

Figure 7.7 (a) Volatility $v(\theta)$ driving Libor forward interest rates $f_L(t,x)$. (b) Volatility $\gamma(\theta)$ driving log Libor interest rates $\phi(t,x)$.

same parameters for $A_L(t,x)$. This result is consistent with the quantum finance Libor Market Model.

7.8 Stochastic volatility $v(t,x)$

The empirical volatilities $v(\theta)$ and $\gamma(\theta)$ are plotted in Figures 7.7(a) and 7.7(b) respectively. Note that $v(\theta)$ is about two orders of magnitude smaller than $\gamma(\theta)$. The reason is because the daily changes in Libor δL and log Libor $\delta L/L$ differ by 10^{-2}; more precisely

$$L \simeq 0.01; \ \delta L \simeq 10^{-4}; \ \frac{\delta L}{L} \simeq 0.01$$

$$E[\delta L \delta L] \simeq 10^{-8} \simeq v^2 \Rightarrow v \sim 10^{-4}$$

$$E\left[\frac{\delta L}{L} \frac{\delta L}{L}\right] \simeq 10^{-4} \simeq \gamma^2 \Rightarrow \gamma \sim 10^{-2}$$

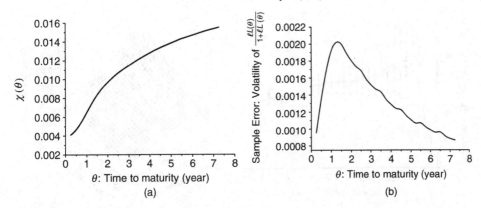

Figure 7.8 (a) The expectation value $\chi(\theta) = E[\ell L(\theta)/(1 + \ell L(\theta))]$. (b) The volatility of $\ell L(\theta)/(1 + \ell L(\theta))$ shows a peak around 1.5 years for the remaining future time.

Although, in Section 7.7, volatility $v(\theta)$ was assumed to be deterministic, Eq. (7.32) shows that it is, in fact, a *derived* stochastic quantity given by

$$v(t, x) = \frac{e^{\phi_n(t)}}{1 + e^{\phi_n(t)}} \gamma(t, x) = \frac{\ell L(t, T_n)}{1 + \ell L(t, T_n)} \gamma(t, x) \qquad (7.33)$$

A measure of the error in replacing stochastic volatility $v(t, x)$ by a deterministic function is given by the variance of $\ell L(\theta)/(1 + \ell L(\theta))$, which is the stochastic quantity that makes $v(t, x)$ stochastic. More precisely

$$\frac{\ell L(t, T_n)}{1 + \ell L(t, T_n)} \simeq E\left[\frac{\ell L(t, T_n)}{1 + \ell L(t, T_n)}\right] \pm \text{ volatility} \equiv \chi(\theta) \pm \text{ volatility}$$

$\chi(\theta)$ is plotted in Figure 7.8(a).

Figure 7.8(b) plots the volatility of $\ell L(\theta)/(1 + \ell L(\theta))$, which shows an expected peak around 1.25 years for remaining future time: the most volatile period for Libor. From Figure 7.8 the empirical values yield the following

$$\chi(\theta) = E\left[\frac{\ell L(t, T_n)}{1 + \ell L(t, T_n)}\right] \simeq 10^{-2} - 10^{-3}$$

$$\text{volatility of } \left[\frac{\ell L(t, T_n)}{1 + \ell L(t, T_n)}\right] \simeq 10^{-3}$$

Hence, $\ell L(\theta)/(1 + \ell L(\theta))$ has volatility comparable to $\chi(\theta)$ and replacing it by $\chi(\theta)$ will lead to substantial errors.

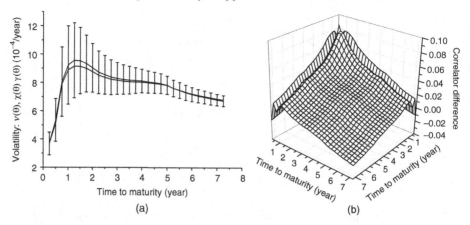

Figure 7.9 (a) The upper line is the effective volatility $v(\theta)$ and the lower line is $\chi(\theta)\gamma(\theta)$; the fluctuations of stochastic $v(\theta)$ about the value of $\chi(\theta)$ are indicated by error bars. (b) The graph shows the empirical value of $\Delta\mathcal{D}(\theta,\theta')$.

The empirical volatilities $\gamma(\theta)$ and $v(\theta)$ can be taken to be approximately related by the following

$$v(t,x) = v(\theta) \simeq \chi(\theta)\gamma(\theta)$$

$v(\theta)$ and $\chi(\theta)\gamma(\theta)$ are plotted in Figure 7.9(a). The total error in the fit of $v(\theta)$ with $\chi(\theta)\gamma(\theta)$ is about 2.1% of the value of $v(\theta)$, with the error close to 4.5% near the remaining future time around 1.5 years.

The difference in the normalized propagators of L and $\ln(\ell L)$ is a measure of the error (for different maturities) made by the approximation $v(\theta) \simeq \chi(\theta)\gamma(\theta)$; the difference should be zero if the approximation has no error. Define the difference of the normalized propagators by

$$\Delta\mathcal{D}(\theta,\theta') = \frac{< \delta \ln(\ell L(\theta))\delta \ln(\ell L(\theta')) >_c}{\ell^2\gamma(\theta)\gamma(\theta')} - \frac{< \delta L(\theta)\delta L(\theta') >_c}{\chi(\theta)\gamma(\theta)\chi(\theta')\gamma(\theta')}$$

The empirical value of $\Delta\mathcal{D}(\theta,\theta')$ is shown in Figure 7.9(b). The errors are substantial due to the volatility of $\ell L(\theta)/(1 + \ell L(\theta))$; in particular, there is an error of almost 10% for the region near remaining future time of about 1.5 years, for which the volatility is maximum.

In conclusion, stochastic volatility $v(t,x)$ of the Libor forward interest rates $f_L(t,x)$ cannot be treated as a deterministic function. Errors of about 10% are the result of the volatility of Libor. A more productive approach seems to be to focus on the log Libor quantum field $\phi(t,x)$ and develop efficient numerical algorithms based on the deterministic volatility $\gamma(t,x)$.

Table 7.4 *The parametric fit of interest rates volatility. Note the coefficient c_3 determines the exponential fall off of the volatility. The rms (root mean square) error is for the entire fit.*

Volatility	c_1	c_2	c_3	c_4	rms error
$\gamma(\theta)$	0.051	0.038	1.360	0.279	7.73%
$v(\theta)$	0.001	0.000	1.047	0.001	4.70%

7.8.1 Interest rate volatility in the Libor Market Model

The empirical volatility function $v(\theta)$ for Libor forward interest rates $f_L(t,x)$ and volatility $\gamma(\theta)$ of log Libor $\phi(t,x)$ are fitted with analytic expressions. The best fit parameters are given in Table 7.4, based on the formula below

$$c_1 + c_2 \exp\{-c_3(\theta - 0.25)\} + c_4(\theta - 0.25)\exp\{-c_3(\theta - 0.25)\}$$

The linearly increasing term – given by coefficient c_4 – results from the market's projection of the anticipated trends of the spot rate [27]. The relative error is computed by

$$\text{rms error} = \sqrt{\frac{1}{N}\sum_{n=1}^{N}\left[\frac{\gamma_n(\text{theory}) - \gamma_n(\text{market})}{\gamma_n(\text{market})}\right]^2}$$

where the sum is taken over all the data points. Figure 7.10 shows that, as expected, volatility $\gamma(\theta)$ is much higher, by two orders of magnitude, than volatility $v(\theta)$.

The empirical values of $v(\theta)$ and $\gamma(\theta)$, together with the best fits, are plotted in Figures 7.10(a) and (b).

For volatility $\gamma(t,x)$, the covariance $\gamma(t,x)\mathcal{D}_L(x,x';t)\gamma(t,x')$ has an error of 11.28%. The errors are largely due to the errors in fitting the volatilities. If one uses the empirical value for the volatility and the stiff propagator, the error in the covariance is about 1%.

7.9 Zero coupon yield curve and covariance

All the discussion so far has concentrated on Libor and Euribor futures data. It was assumed that *both* the bond and Libor forward interest rates can be calibrated using Libor data. The results obtained so far are consistent with the quantum finance models for bond forward interest rates and with the Libor Market Model's quantum finance generalization.

Treasury Bond and Libor ZCYC data have been discussed in Sections 2.13 and 2.11. One can evaluate correlation functions by averaging over historical ZCYC

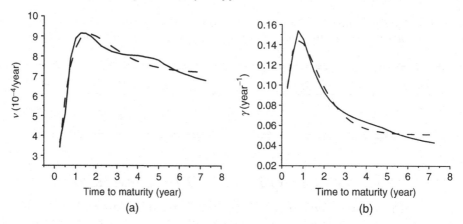

Figure 7.10 The empirical volatility of $v(\theta)$ and $\gamma(\theta)$ are taken to be equal to the average over 260 days of historical data. The unbroken line is market volatility and the dashed line is the best fit. (a) The best fit for $v(\theta)$, with relative rms error of 4.70%. (b) The best fit for $\gamma(\theta)$, with relative rms error of 7.73%.

market data. The correlation functions can be written directly in terms of the ZCYC and it is seen below that the ZCYC provides an estimate of the integrated covariance of the underlying forward interest rates driving the ZCYC.

From Eq. (2.28), the ZCYC, which is compounded c times a year, has the following relation with zero coupon bonds and forward interest rates

$$B(t,T) = \frac{1}{(1 + \frac{1}{c}Z(t,T))^{c[T-t]}} = \exp\left\{-\int_t^T dx f(t,x)\right\}$$

$$\Rightarrow \int_t^T dx f(t,x) = c[T-t]\ln\left(1 + \frac{1}{c}Z(t,T)\right)$$

Consider the ZCYC for the forward interest rates integrated over a fixed interval $[t_*, T]$; this yields the following integrated covariance of the forward interest rates

$$\int_{t_*}^T dx \frac{\partial f}{\partial t}(t,x) = \frac{\partial}{\partial t}\int_{t_*}^T dx f(t,x) \tag{7.34}$$

$$\Rightarrow \int_{t_*}^T dx \int_{t_*}^{T'} dx' < \delta f(t,x)\delta f(t,x') >$$

$$= \left\langle \delta\left[\int_{t_*}^T dx f(t,x)\right] \delta\left[\int_{t_*}^T dx' f(t,x')\right] \right\rangle$$

The effects of the spot rate $r(t) = f(t,t)$ (boundary term) can be studied by letting the lower limit be equal to time t. This yields the following integrated forward interest rates

$$\frac{\partial}{\partial t} \int_t^T dx f(t,x) = -f(t,t) + \int_t^T dx \frac{\partial f}{\partial t}(t,x)$$

$$\int_t^T dx [\delta f(t,x)] = \epsilon f(t,t) + \delta \int_t^T dx f(t,x)$$

$$\Rightarrow \int_t^T dx \int_t^{T'} dx' < \delta f(t,x)\delta f(t,x') > \tag{7.35}$$

$$= \left\langle \left[\epsilon f(t,t) + \delta \int_t^T dx f(t,x) \right] \left[\epsilon f(t,t) + \delta \int_t^T dx' f(t,x') \right] \right\rangle$$

All terms on the right-hand side can be evaluated using the ZCYC.

7.9.1 Empirical US Treasury Bond ZCYC covariance

The bond market is studied to ascertain whether it can be explained by the stiff propagator. The empirical result for the Treasury Bond market is shown in Figure 7.11(a) for Eq. (7.34). The result for Eq. (7.35) is shown in Figure 7.11(b) and looks very similar to the earlier result given in Figure 7.11(a), showing that including the boundary term due to the spot rate $r(t) = f(t,t)$ in the covariance does not make much of a difference.

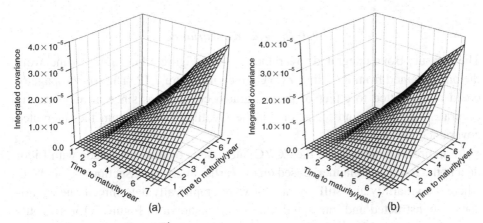

Figure 7.11 (a) $\int_{t_*}^T dx \int_{t_*}^{T'} dx' < \delta f(t,x)\delta f(t,x') >$: Treasury Bond ZCYC integrated covariance from fixed $t_* = 0.25$ years and T and T' range from 0.5 years to 7.25 years. (b) $\int_t^T dx \int_t^{T'} dx' < \delta f(t,x)\delta f(t,x') >$: Treasury Bond ZCYC integrated covariance; contains boundary terms $f(t,t)$. Note $t = 0$ and T and T' range from 0.25 years to 7.25 years.

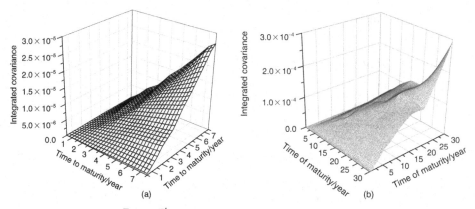

Figure 7.12 (a) $\int_{t_*}^{T} dx \int_{t_*}^{T'} dx' < \delta f(t,x)\delta f(t,x') >$: Libor ZCYC integrated covariance from fixed $t_* = 0.25$ years and for maturity time T and T' range from 0.5 to 7.5 years. (b) $\int_{t}^{T} dx \int_{t_*}^{T'} dx' < \delta f(t,x)\delta f(t,x') >$: Libor ZCYC integrated covariance from fixed $t = 0$; maturity time T and T' range from 0.25 to 30 years.

7.9.2 Empirical Libor ZCYC covariance

The Libor ZCYC is used to evaluate the integrated covariance. The data are taken from calendar dates 10 August 2007 to 8 August 2008, totaling 261 trading days; the average is taken over 260 trading days. Figure 7.12(a) shows the integrated covariance out to 7.5 years and Figure 7.12(b) shows it out to 30 years.

7.9.3 Integrated covariance

The Treasury Bond integrated covariance given in Figure 7.11 is quite distinct from the one obtained from the Libor ZCYC given in Figure 7.12. The stiff propagator could not be fitted to either of the integrated covariances since no numerically accurate way was found to factor out the volatility $\sigma(t,x)$ from the integrated covariance.

An indirect comparison of the ZCYC integrated covariance is made with Libor data in the following manner. Based on the earlier analysis in Sections 7.5 and 7.6, where the volatility and stiff propagator were empirically determined, the covariance is constructed and integrated; the result is shown in Figure 7.13(a) for the case of parameters found from the covariance of δL and in Figure 7.13(b) from the covariance of $\delta \ln(\ell L)$. The result shown in Figure 7.13(a) has a surface quite different from the one generated by the Treasury Bond ZCYC covariance given in Figure 7.11(a); there is some similarity with the surface generated by the Libor ZCYC covariance given in Figure 7.12(a), with an error of about 21.4%.

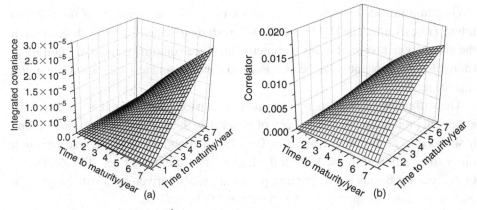

Figure 7.13 (a) $\int_{t_*}^{T} dx \int_{t_*}^{T'} dx' < \delta L(t,x)\delta L(t,x') >_c$: integration of *Libor* covariance. (b) $\int_{t_*}^{T} dx \int_{t_*}^{T'} dx' < \delta(\ln L(t,x))\delta(\ln L(t,x')) >_c$: integration of *log Libor* covariance.

One does not expect the covariance of log Libor to reproduce the ZCYC covariance but the covariance of Libor should – since empirical Libor $L(t,x)$ is approximately equal to the forward interest rates $f(t,x)$. However, the surfaces in Figures 7.11 and 7.13 are significantly different, leading to the conclusion that forward interest rates from Libor are not the same as the forward interest rates obtained from the Treasury Bond ZCYC.

A possible explanation for the difference in the behavior of Libor and ZCYC is the TED (Treasury Eurodollar) spread. The zero coupon Treasury Bond ZCYC is constructed from a risk-free instrument. Libor, on the other hand, carries an element of risk and the spread of TED is taken as an indicator of credit risk, reflecting the default possibility of corporate borrowers. As the spread increases, so does the risk. Another possible reason for the discrepancy of the two ZCYCs is that Libor contains a fundamental scale, namely the nonzero tenor for simple interest rate period ℓ, whereas there is no such scale in the zero coupon bond market since there is almost instantaneous discounting.

Hence, one may conclude that the difference between Libor and bond forward interest rates reflects the two major components of the debt market, namely the (Libor) interest rates and bond markets.

7.10 Summary

Quantum finance models of the forward interest rates were empirically analyzed. The models were simple to calibrate and made many empirically testable predictions. On the balance of the results, the quantum finance model for both the bond and Libor markets give excellent results.

Quantum finance provides a framework in which the volatility of the forward interest rates, unlike the HJM and BGM–Jamshidian models, is taken directly from the market with no need for any parametric fit. Once volatility is fixed, the empirical (normalized) propagator can be evaluated to calibrate and test the various quantum finance models.

The stiff propagator of quantum finance is seen to provide an excellent fit for the empirical propagator, to an accuracy of better than 99%. The index for market time η shows a large variation from $\eta = 0.34$ for the bond forward interest rates to a fairly small value of $\eta_L = 0.07$ for the Libor forward interest rates. The ZCYC for both Treasury Bonds and Libor are, presumably, also driven by a stiff propagator and the parameters appropriate for them need to be studied.

The quantum finance generalization of white noise $R(t)$ to a two-dimensional quasi-Gaussian quantum field $\mathcal{A}(t, x)$ was empirically studied. The results of the empirical study point to a very general and fundamental role of the stiff Lagrangian in describing the random processes that drive the debt markets. The stiff Lagrangian is a pseudo-Gaussian (free) quantum field and one may wonder why it can so accurately describe forward interest rates that one expects to be nontrivial and nonlinear. A possible answer lies in the concept of market future time and, in particular, the index η. To generate market future time from a Lagrangian – instead of directly putting it into the Lagrangian 'by hand' – would require nonlinear interactions. Presumably, the index for future market time is like a critical exponent that appears in phase transitions. η is the result of strongly correlated and, at present unknown, nonlinear interactions of the underlying fundamental theory, which is defined for future market time $x - t$ with no reference to η.

The Libor Market Model shows that the volatility $v(t, x)$ of the Libor forward interest rates $f_L(t, x)$ is a stochastic quantity derived from deterministic volatility $\gamma(t, x)$ of log Libor $\phi(t, x)$. It remains an open question whether volatility $\gamma(t, x)$, which drives the bond forward interest rates $f(t, x)$, is the fundamental interest rate from which one can derive the volatility $\gamma(t, x)$ of log Libor rates $\phi(t, x)$.

The bonds' and the interest rates' markets are two sectors of the debt markets. Both the bond and interest rates' markets are driven by underlying interest rates and their inter-relationship is an empirical question. The introduction of tenor ℓ into the debt market creates a nonlinear relationship between zero coupon bonds and time deposits. One of the main conclusions of the empirical study is that the models of bond and Libor forward interest rates are distinct and different. The difference of the two can be attributed to the TED spread and to the minimum tenor $\ell = 90$ days for the three-month benchmark Libor.

Parameters μ, λ, and the market time index η need to be fixed from market data. The velocity field $\mathcal{A}(t, x)$ that drives $f(t, x)$ and $\mathcal{A}_L(t, x)$ that drives $\phi(t, x)$ are determined by the same stiff action but their parameters are different.

Furthermore, volatility $\sigma(t, x)$ and $\gamma(t, x)$ are different by two orders of magnitude; both volatilities have a nontrivial structure and, in quantum finance, are determined from market data.

The parameters, including volatility, required for specifying the behavior of the debt market point to a major difference between the theories of physics and those of finance. In physics, all physical constants, such as Planck's constant, speed of light, mass and charge of an electron, and so on, are determined by nature. In contrast, in finance all parameters are the summation of political and economic activities. In particular, the capital markets are the result of the motivation and psychology of the market practitioners that reflect their social environment and cultural priorities. The market index of time η is a case in point: it is the subjective view of the traders that replaces, in the pricing of instruments, future time by market future time. η shows that, unlike equations in the natural sciences that relate quantities which are entirely independent of human subjectivity, the equations in finance seem to incorporate the presence of human intervention and manipulation in the defining laws of finance.

It might be possible to change the parameters of finance through human intervention, such as herd behavior in the market, or by government policies. The explanation of the parameters determining the characteristics of the financial markets is thought to be found in mathematical behavioral finance [87, 77] and which has been shown, in some cases, to lead to quantitative results [41]. A major challenge of theoretical finance is to obtain the market values of the various parameters from underlying principles of finance.

8

Libor Market Model of interest rate options

The prices of Libor options are obtained for the quantum finance Libor Market Model [4]. The option prices show new features of the Libor Market Model arising from the fact that, in the quantum finance formulation, all the different Libor payments are coupled and (imperfectly) correlated.

Black's caplet formula for quantum finance is given an exact derivation. The coupon and zero coupon bond options as well as the Libor European and Asian swaptions are derived for the quantum finance Libor Market Model. The approximate Libor option prices are derived using the volatility expansion developed in Section 3.14.

The BGM–Jamshidian expression for the Libor interest rate caplet and swaption prices is obtained as the limiting case when all the Libors are exactly correlated.

8.1 Introduction

The Libor option prices are obtained from the Libor zero coupon bonds $B_L(t, T)$ – obtained from the Libor ZCYC curve $Z_L(t, T)$ discussed in Section 7.9 – and the benchmark three-month Libor $L(t, T)$. For notational convenience, Libor zero coupon bonds $B_L(t, T)$ will be denoted by $B(t, T)$.

All the options are defined to mature at future calendar time T_0, with present time given by $t_0 = T_{-k}$; the notation of present being denoted by t_0 is used to simplify the notation. It is natural for these options to choose $B(t, T_0)$ as the forward bond numeraire. In other words, the forward bond numeraire is $B(t, T_{I+1})$ with $I = -1$ and Libor drift is calculated for this numeraire. Libor calendar and future time are shown in Figure 6.1 and Libor times $t_0 = T_{-k}$, T_0, and T_n are shown in Figure 8.1.

The option price is governed by the defining equations of the LMM. From Eqs. (6.41), (6.42), and (6.43), Libor is given by the following

$$t_0 = T_{-k} \qquad T_0 \qquad T_n$$

Figure 8.1 The Libor lattice is defined by $T_n = \ell n$; the payoff is defined at Libor time T_0. The option price is evaluated at present Libor calendar time $t_0 = T_{-k}$.

$$L(T_0, T_n) = L(t_0, T_n) e^{\beta(t_0, T_0, T_n) - \frac{1}{2}q_n^2 + \int_{t_0}^{T_0} dt \int_{T_n}^{T_{n+1}} dx \gamma(t,x) \mathcal{A}_L(t,x)}$$

$$\equiv L(t_0, T_n) e^{\beta_n + W_n} \tag{8.1}$$

where

$$\beta_n = \int_{t_0}^{T_0} dt \zeta(t, T_n); \quad \beta_{-1} = 0 \tag{8.2}$$

$$\zeta(t, T_n) = \sum_{m=0}^{n} \frac{\ell L(t, T_m)}{1 + \ell L(t, T_m)} \Lambda_{mn}(t); \quad q_n^2 = \int_{t_0}^{T_0} dt \Lambda_{nn}(t) \tag{8.3}$$

$$W_n = -\frac{1}{2} q_n^2 + \int_{t_0}^{T_0} dt \int_{T_n}^{T_{n+1}} dx \gamma(t,x) \mathcal{A}_L(t,x)$$

$$\Lambda_{mn}(t) = \int_{T_m}^{T_{m+1}} dx \int_{T_n}^{T_{n+1}} dx' \gamma(t,x) \mathcal{D}_L(x, x'; t) \gamma(t, x')$$

The two expectation values that are required for the option calculations, from Eqs. (6.45) and (6.46), are the following

$$E[e^{W_n} - 1] = 0 \tag{8.4}$$

$$E[(e^{W_m} - 1)(e^{W_n} - 1)] = e^{\Delta_{mn}} - 1 \simeq \Delta_{mn} + O(\gamma^4) \tag{8.5}$$

$$\Delta_{nm} \equiv \Delta_{mn}(T_0) = \int_{t_0}^{T_0} dt \int_{T_n}^{T_{n+1}} dx \int_{T_m}^{T_{m+1}} dx' M_\gamma(x, x'; t)$$

$$= \int_{t_0}^{T_0} dt \Lambda_{mn}(t) \simeq O(\gamma^2) \tag{8.6}$$

Empirical values of $\Delta_{mn}(T_0)$ are given in Figure 8.2 for two well-separated time intervals $T_0 - t_0 = 1$ month and $T_0 - t_0 = 1$ year; the correlator $\Delta_{mn}(T_0)$ has a 'hump' for a short time due to the effects of volatility $\gamma(t,x)$ that smooths out for longer times.

Libor European and Asian swaption prices are evaluated to only $O(\gamma^2)$ as this is sufficient for demonstrating the main features of the calculations. The higher-order terms rapidly proliferate but, nevertheless, can be computed in a straightforward manner. The swaption price BGM–Jamshidian limit is taken to understand the new features that arise due to imperfectly correlated Libors.

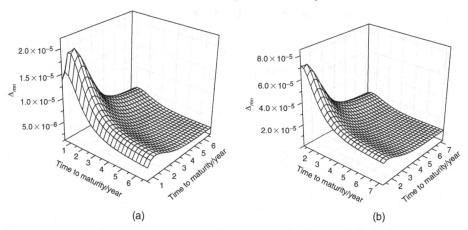

Figure 8.2 Empirical value for Libor correlator $\Lambda_{mn}(T_0) = \int_{t_0}^{T_0} dt\,\Lambda_{mn}(t)$ (a) for $T_0 - t_0 = 1$ month and (b) for $T_0 - t_0 = 1$ year. The graphs are constructed from Libor ZCYC data for 261 days, from 10 October 2007 to 8 August 2008.

Libor options in the BGM–Jamshidian framework have been extensively studied [61, 84]. A review of Libor derivative instruments, priced using the BGM–Jamshidian model, is given in [33].

8.2 Quantum Libor Market Model: Black caplet

The Black caplet price is derived exactly for the quantum finance Libor Market Model. The key simplification is that Libor $L(t, T_I)$ is a martingale for numeraire $B(t, T_{I+1})$; hence the Libor drift $\zeta(t, T_I)$ is zero and this reduces the dynamics of $L(t, T_I)$ to the Gaussian case.

Hence, from Eq. (6.38), $L(t, T_I)$ has a martingale evolution given by

$$\frac{\partial L(t, T_I)}{\partial t} = L(t, T_I) \int_{T_I}^{T_{I+1}} dx\, \gamma(t, x) \mathcal{A}_L(t, x)$$

and from Eq. (6.40)

$$L(t_*, T_I) = L(t_0, T_I) \exp\left\{ -\frac{1}{2}q_I^2 + \int_{t_0}^{t_*} dt \int_{T_I}^{T_{I+1}} dx\, \gamma(t, x) \mathcal{A}_L(t, x) \right\} \quad (8.7)$$

The payoff for a midcurve caplet on Libor $L(t_*, T_I)$, maturing at time $t_* < T_I$, from Eq. (4.10), is given by[1]

$$caplet(t_*, t_*, T_I) = \ell V B(t_*, T_I + \ell) \big[L(t_*, T_I) - K \big]_+$$

where the notional sum of the caplet is V.

---

[1] A caplet that matures when Libor $L(t_*, T_I)$ becomes operational is obtained by setting $t_* = T_I$.

The caplet is a traded instrument and follows a martingale evolution for numeraire $B(t, T_{I+1})$; hence, the price of the caplet at present time t_0 is given by the martingale condition

$$\frac{caplet(t_0, t_*, T_I)}{B(t_0, T_{I+1})} = E\left[\frac{caplet(t_*, t_*, T_I)}{B(t_*, T_{I+1})}\right] = \ell V E\left[L(t_*, T_I) - K\right]_+$$

$$\Rightarrow caplet(t_0, t_*, T_I) = \ell V B(t_0, T_{I+1}) E\left[L(t_*, T_I) - K\right]_+ \qquad (8.8)$$

The payoff can be re-expressed, from Eq. (A.7), as follows

$$\left[L(t_*, T_I) - K\right]_+ = \int_{-\infty}^{+\infty} dQ \frac{d\eta}{2\pi} e^{i\eta\left(\int_{t_0}^{t_*} dt \int_{T_I}^{T_{I+1}} dx \gamma(t,x) \mathcal{A}_L(t,x) + Q\right)}$$

$$\times \left[L(t_0, T_I) e^{-\frac{1}{2}q_I^2 - Q} - K\right]_+ \qquad (8.9)$$

To obtain the caplet price one evaluates Eq. (8.8); from Eq. (5.21)

$$E\left[e^{i\eta \int_{t_0}^{t_*} dt \int_{T_I}^{T_{I+1}} dx \gamma(t,x) \mathcal{A}_L(t,x)}\right] = \frac{1}{Z} \int D\mathcal{A}_L e^S e^{i\eta \int_{t_0}^{t_*} dt \int_{T_I}^{T_{I+1}} dx \gamma(t,x) \mathcal{A}_L(t,x)}$$

$$= \exp\left\{-\frac{1}{2}\eta^2 \int_{t_0}^{t_*} dt \int_{T_I}^{T_{I+1}} dx dx' \gamma(t,x) \mathcal{D}_L(x,x';t) \gamma(t,x')\right\}$$

$$= \exp\left\{-\frac{1}{2}q_I^2 \eta^2\right\} \text{ where } q_I^2 = \int_{t_0}^{t_*} dt \int_{T_I}^{T_{I+1}} dx \int_{T_I}^{T_{I+1}} dx' M_\gamma(x,x';t) \quad (8.10)$$

Hence, from Eqs. (8.8), (8.9), and (8.10), the caplet price is given by

$$\frac{caplet(t_0, t_*, T_I)}{B(t_0, T_{I+1})}$$

$$= \ell V \int_{-\infty}^{+\infty} dQ \frac{d\eta}{2\pi} e^{-\frac{1}{2}q_I^2 \eta^2} e^{i\eta Q} \left[L(t_0, T_I) e^{-\frac{1}{2}q_I^2 - Q} - K\right]_+$$

$$= \ell V \int_{-\infty}^{+\infty} \frac{dQ}{\sqrt{2\pi q_I^2}} e^{-\frac{1}{2q_I^2} Q^2} \left[L(t_0, T_I) e^{-\frac{1}{2}q_I^2 - Q} - K\right]_+ \qquad (8.11)$$

$$= \ell V \left[L(t_0, T_I) N(d_+) - K N(d_-)\right] \qquad (8.12)$$

$$d_\pm = \frac{1}{q_I} \ln\left[\frac{L(t_0, T_I)}{K}\right] \pm \frac{q_I}{2}$$

Eq. (8.12) is the well-known Black's formula for a Libor caplet [34, 59]. The additional information that the quantum LMM yields is that Black's volatility σ_B^2

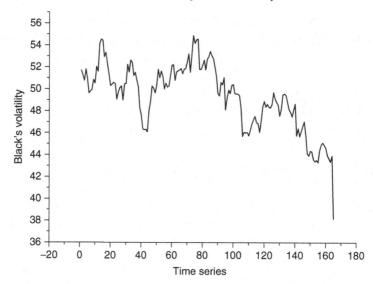

Figure 8.3 Market implied Black volatility σ_B for a caplet that matures on 12 December 2004 versus time t_0, from 12 September 2003 to 7 May 2004.

is in fact equal to q_I^2, which in turn is given by Libor volatility $\gamma(t, x)$ and the correlator $\mathcal{D}_L(x, x'; t)$. More precisely

$$\sigma_B^2 = \frac{q_I^2}{t_* - t_0} = \frac{1}{t_* - t_0} \int_{t_0}^{t_*} dt \int_{T_I}^{T_{I+1}} dx \int_{T_I}^{T_{I+1}} dx' M_\gamma(x, x'; t) \quad (8.13)$$

Figure 8.3 shows the implied Black's volatility σ_B for the market price of a Libor caplet.

One can either choose to calibrate the quantum LMM from caplet data and ascertain $\gamma(t, x)$ or else obtain $\gamma(t, x)$ independently from the correlation of Libor rates as shown in Figure 7.7(b). Knowing $\gamma(t, x)$ and the Libor propagator allows one to predict the value of the caplet [84].

The BGM–Jamshidian limit is obtained when all the Libors are exactly correlated, namely that $\mathcal{D}_L(x, x'; t) \to 1$, which yields $M_\gamma(x, x'; t) \to \gamma(t, x)\gamma(t, x')$; hence

$$\sigma_B^2 \to \sigma_B^2 \Big|_{BGM} = \frac{1}{t_* - t_0} \int_{t_0}^{t_*} dt \left[\int_{T_I}^{T_{I+1}} dx \gamma(x, t) \right]^2$$

8.3 Volatility expansion for Libor drift

Libor drift, defined for Libor time $T_n = T_0 + \ell n$, can be expressed completely in terms of Libors with no reference to the Libor forward interest rates $f_L(t, x)$.

Due to the nontrivial Libor drift β_n given in Eqs. (8.2) and (8.3), the expression for Libor interest rates is a nonlinear, nonlocal, and fairly intractable function of $\mathcal{A}_L(t,x)$. A natural expansion for Libor drift is a perturbation series in Libor volatility $\gamma(t,x)$. The empirical value of γ, from Figure 7.10(b), is about 10^{-1}/year; the expansion parameter is the dimensionless quantity $\ell\gamma$, which is about 10^{-2}/year. The volatility expansion, as a power series in $\gamma(t,x)$, generates a convergent expression for the drift and other quantities.

All future time is taken to be on the Libor time lattice given by Libor time $T_n = \ell n$.

$$L(t, T_n) = L(t_0, T_n)e^{\beta_n + W_n}; \quad t > t_0$$

$$\beta_n = \int_{t_0}^{t} dt' \zeta(t', T_n); \quad \beta_{-1} = 0$$

$$\zeta(t, T_n) = \sum_{m=0}^{n} \frac{\ell L(t, T_m)}{1 + \ell L(t, T_m)} \Lambda_{mn}(t)$$

where $\beta_n = \beta_n(t)$ and $W_n = W_n(t)$.

To obtain the volatility expansion, the leading term in β_n is isolated and a recursion equation then generates the expansion. Eq. (8.1) yields

$$\frac{\ell L(t, T_m)}{1 + \ell L(t, T_m)} = \frac{\ell L(t_0, T_m)e^{\beta_m + W_m}}{1 + \ell L(t_0, T_m)e^{\beta_m + W_m}}$$

$$= \frac{\ell L(t_0, T_m)}{1 + \ell L(t_0, T_m)} + \frac{\ell L(t_0, T_m)(e^{\beta_m + W_m} - 1)}{(1 + \ell L(t_0, T_m))(1 + \ell L(t_0, T_m)e^{\beta_m + W_m})}$$

$$= O(1) + O(\gamma)$$

Hence, from above and Eqs. (8.2) and (8.3) one obtains the following implicit equation

$$\beta_n = \int_{t_0}^{t} dt' \sum_{m=0}^{n} \Lambda_{mn}(t') \left[\frac{\ell L(t_0, T_m)}{1 + \ell L(t_0, T_m)} \right.$$

$$\left. + \frac{\ell L(t_0, T_m)(e^{\beta_m + W_m} - 1)}{(1 + \ell L(t_0, T_m))(1 + \ell L(t_0, T_m)e^{\beta_m + W_m})} \right]$$

$$= \beta_n^{(0)} + \Psi_n(\beta) \tag{8.14}$$

where

$$\Psi_n(\beta) = \int_{t_0}^{t} dt' \sum_{m=0}^{n} \Lambda_{mn}(t') \left[\frac{\ell L(t_0, T_m)(e^{\beta_m + W_m} - 1)}{(1 + \ell L(t_0, T_m))(1 + \ell L(t_0, T_m)e^{\beta_m + W_m})} \right]$$

$$= O(\gamma^3)$$

$$\beta_n^{(0)} = \sum_{m=0}^{n} \Delta_{mn} \frac{\ell L(t_0, T_m)}{1 + \ell L(t_0, T_m)} = O(\gamma^2) : \text{ deterministic} \qquad (8.15)$$

$$\Delta_{mn} = \int_{t_0}^{t} dt' \Lambda_{mn}(t') = O(\gamma^2)$$

In Eq. (8.14), the term $\beta_n^{(0)}$ is $O(\gamma^2)$ and $\Psi_n(\beta)$ is $O(\gamma^3)$; hence, it provides the following recursion equation for evaluating drift β_n as a power series in γ

$$\begin{aligned}
\beta_n &= \beta_n^{(0)} + \Psi_n(\beta) \\
&= \beta_n^{(0)} + \Psi_n\big(\beta^{(0)} + \Psi(\beta)\big) \\
&= \beta_n^{(0)} + \Psi_n\Big(\beta^{(0)} + \Psi\big(\beta^{(0)} + \Psi(\beta)\big)\Big) \\
&= \ldots \\
&= \beta_n^{(0)} + \beta_n^{(1)} + \ldots + \beta_n^{(\ell)} + O(\gamma^{\ell+2})
\end{aligned}$$

Carrying out the above expansion to the first nontrivial order yields

$$\beta_n^{(1)} = \Psi_n(\beta^{(0)}) = O(\gamma^3)$$

$$= \int_{t_0}^{t} dt' \sum_{m=0}^{n} \Lambda_{mn}(t') \left[\frac{\ell L(t_0, T_m)(e^{\beta_m^{(0)} + W_m(t')} - 1)}{(1 + \ell L(t_0, T_m))\big(1 + \ell L(t_0, T_m)e^{\beta_m^{(0)} + W_m(t')}\big)} \right]$$

The option price will be evaluated to only $O(\gamma^2)$ and hence the drift will be taken as follows

$$\beta_n = \beta_n^{(0)} + O(\gamma^3) : \text{deterministic} \qquad (8.16)$$

8.4 Zero coupon bond option

The zero coupon Libor bond option price is derived based on the volatility expansion and is analyzed to illustrate certain key features of the volatility expansion. The results of this section provide the necessary ingredients for the more complex derivation of the coupon bond option price in the Libor Market Model.

8.4.1 Zero coupon bond volatility expansion

Consider the Libor future time lattice of points given by $T_n = \ell n$. The zero coupon bond $B(T_0, T_n)$ issued at Libor time T_0 has a forward bond price $F(t_0, T_0, T_n)$ at earlier Libor time $t_0 = T_{-k}$. The three Libor times $t_0 = T_{-k}$, T_0, and T_n are shown in Figure 8.1.

The relatively simple formula relating the zero coupon bond to its forward price, based on the instantaneous forward interest rates and given in Eq. (2.14), is re-expressed in terms of Libors. A volatility expansion is also developed as this is required for generating the approximate option price in the Libor Market Model.

Since $\beta_n^{(0)} W_n(t) = O(\gamma^3)$, Libor has the following volatility expansion

$$L(t, T_n) = L(t_0, T_n)e^{\beta_n(t) + W_n(t)} \simeq L(t_0, T_n)e^{\beta_n^{(0)}(t) + W_n(t)}$$

$$= \tilde{L}(t_0, T_n) + L(t_0, T_n)(e^{W_n} - 1) + O(\gamma^3) \qquad (8.17)$$

where $\tilde{L}(t_0, T_n) = L(t_0, T_n) \exp\{\beta_n^{(0)}\}$: deterministic

The zero coupon bond, from Eq. (6.5) is given by

$$B(T_0, T_n) = \prod_{i=0}^{n-1} \frac{1}{[1 + \ell L(T_0, T_i)]}$$

and from Eq. (8.17) has the following volatility expansion

$$B(T_0, T_n) = \prod_{i=0}^{n-1} \frac{1}{[1 + \ell L(T_0, T_i)]}$$

$$= \prod_{i=0}^{n-1} \frac{1}{[1 + \ell \tilde{L}(t_0, T_i) + \ell L(t_0, T_i)(e^{W_i} - 1)]}$$

$$= \tilde{F}(t_0, T_0, T_n) \exp\left\{-\sum_{i=0}^{n-1} \ln(1 + a_i(e^{W_i} - 1))\right\}$$

$$= \tilde{F}(t_0, T_0, T_n)[1 + A_n] + O(\gamma^3) \qquad (8.18)$$

where

$$\tilde{F}(t_0, T_0, T_n) = \prod_{i=0}^{n-1} \frac{1}{[1 + \ell \tilde{L}(t_0, T_i)]}; \quad a_i = \frac{\ell L(t_0, T_i)}{1 + \ell L(t_0, T_i)} \qquad (8.19)$$

$$A_n = -1 + \exp\left\{-\sum_{i=0}^{n-1} \ln(1 + a_i(e^{W_i} - 1))\right\}$$

$$= -\sum_{i=0}^{n-1} a_i(e^{W_i} - 1) + \frac{1}{2}\sum_{i=0}^{n-1} a_i^2(e^{W_i} - 1)^2$$

$$+ \frac{1}{2}\sum_{i,j=0}^{n-1} a_i a_j(e^{W_i} - 1)(e^{W_j} - 1) + O(\gamma^3)$$

$$A_n^2 = \sum_{i,j=0}^{n-1} a_i a_j (e^{W_i} - 1)(e^{W_j} - 1) + O(\gamma^3) \tag{8.20}$$

To leading order in γ, the forward price of the zero coupon bond $B(T_0, T_n + \ell)$ is $\tilde{F}(t_0, T_0, T_n)$. Since $t_0 = T_{-K}$, T_0, and T_n are all on the Libor time lattice, the forward price of a zero coupon bond, from Eq. (6.6), is given by the following

$$F(t_0, T_0, T_n) = \prod_{i=0}^{n-1} \frac{1}{[1 + \ell L(t_0, T_i)]} \tag{8.21}$$

Hence, if one replaces $\tilde{L}(t_0, T_i)$ by $L(t_0, T_i)$ in Eq. (8.19), then $\tilde{F}(t_0, T_0, T_n)$ is equal to $F(t_0, T_0, T_n)$.

8.4.2 Zero coupon bond option price

Consider a zero coupon bond, $B(T_0, T_n)$, issued at Libor time T_0 and maturing at T_n. A European call option, maturing at T_0, has a payoff given by

$$\mathcal{P} = \Big[B(T_0, T_n) - K \Big]_+ \tag{8.22}$$

Let the option price, at $t_0 < T_0$, be denoted by $C(t_0, T_0, K)$. The three Libor times $t_0 = T_{-k}$, T_0, and T_n are shown in Figure 8.1. Forward bond numeraire $B(t, T_0)$ yields the following martingale

$$\frac{C(t_0, T_0, K)}{B(t_0, T_0)} = E\left[\frac{\mathcal{P}}{B(T_0, T_0)} \right] = E\,[\mathcal{P}]$$

$$\Rightarrow C(t_0, T_0, K) = B(t_0, T_0) E\Big[B(T_0, T_n) - K \Big]_+$$

Zero coupon bond volatility expansion, given in Eq. (8.18), yields

$$C(t_0, T_0, K) = B(t_0, T_0) E\Big[\tilde{F}(t_0, T_0, T_n)\big(1 + A_n\big) - K \Big]_+ + O(\gamma^3)$$

$$= B(t_0, T_0) \tilde{F}(t_0, T_0, T_n) E\big[A_n - \tilde{K} \big]_+ + O(\gamma^3)$$

$$\tilde{K} = \frac{K}{\tilde{F}(t_0, T_0, T_n)} - 1$$

All stochastic terms are contained in A_n.

From Section 3.14, the call option's volatility expansion requires the evaluation of $E[A_n]$ and $E[A_n^2]$. From Eqs. (8.4), (8.6), and (8.20)

$$E[A_n] = \frac{1}{2}\sum_{i=0}^{n-1} a_i^2 E[(e^{W_i} - 1)^2] + \frac{1}{2}\sum_{i,j=0}^{n-1} a_i a_j E\big[(e^{W_i} - 1)(e^{W_j} - 1)\big]$$

$$= \frac{1}{2}\sum_{i=0}^{n-1} a_i^2 \Delta_{ii} + \frac{1}{2}\sum_{i,j=0}^{n-1} a_i a_j \Delta_{ij} \tag{8.23}$$

$$E[A_n^2] = \sum_{i,j=0}^{n-1} a_i a_j \Delta_{ij} \tag{8.24}$$

Hence, from Eq. (3.69), since $E[1] = 1$, the zero coupon bond call option price is given by

$$C(t_0, T_0, T, K) = \frac{1}{\sqrt{2\pi}} B(t_0, T_0) \tilde{F}(t_0, T_0, T_n) I(X) \sqrt{E[A_n^2] - E[A_n]^2} + O(\gamma^3)$$

$$X = \frac{\tilde{K} - E[A_n]}{\sqrt{E[A_n^2] - E[A_n]^2}}; \quad \tilde{K} = \frac{K}{\tilde{F}(t_0, T_0, T_n)} - 1$$

The Libor price for the zero coupon bond call option is approximate. In contrast, in the quantum finance model for bond forward interest rates, the zero coupon bond call option price is evaluated exactly in [12] and given in Eq. (11.49).

Whether the Libor or bond forward interest rate pricing formula is more accurate is an empirical question and needs to be studied further.

8.5 Libor Market Model coupon bond option price

The payoff function $\mathcal{P}$ of a coupon bond European call option maturing at Libor time T_0 and with strike price $\mathcal{K}$ is given, from Eq. (4.20), by

$$\mathcal{P}(T_0) = \left(\sum_{I=1}^{N} c_I B(T_0, T_I) - \mathcal{K}\right)_+$$

$$= (\mathcal{B}(T_0) - \mathcal{K})_+ \tag{8.25}$$

The coupon bond option price at time $t_0 = T_{-K}$, for the forward bond numeraire, is given by Eq. (4.22) as follows

$$C(t_0, T_0, \mathcal{K}) = B(t_0, T_0) E\big[\mathcal{P}(T_0)\big]$$

$$= B(t_0, T_0) E \left[\left(\sum_{I=1}^{N} c_I B(T_0, T_I) - \mathcal{K} \right)_+ \right]$$

Libor times $t_0 = T_{-k}$, T_0, and T_n are shown in Figure 8.1.

Similar to the case of the zero coupon bond option analyzed in Section 8.4, a volatility expansion is developed for the coupon bond option price. One expands the payoff about the forward bond prices as in Eq. (8.18); to do this, the payoff is written by isolating the leading order effect in the following manner.

$$\mathcal{P}(T_0) = \left(\sum_{I=1}^{N} c_I [B(T_0, T_I) - F(t_0, T_0, T_I)] + c_I F(t_0, T_0, T_I) - \mathcal{K} \right)_+$$

$$= (V + F - \mathcal{K})_+$$

$$F = \sum_{I=1}^{N} J_I; \quad J_I = c_I F(t_0, T_0, T_I)$$

All the random terms in the payoff are in the 'potential' term V given by

$$V = \sum_{I=1}^{N} c_I [B(T_0, T_I) - F(t_0, T_0, T_I)]$$

$$= \sum_{I=1}^{N} J_I \left[\frac{B(T_0, T_I)}{F(t_0, T_0, T_I)} - 1 \right] \tag{8.26}$$

The Libor expression of zero coupon bonds and its forward price, given in Eqs. (8.18) and (8.21), yields the following

$$\frac{B(T_0, T_I)}{F(t_0, T_0, T_I)} = \prod_{i=0}^{I-1} \left[\frac{1 + \ell L(t_0, T_i)}{1 + \ell L(T_0, T_i)} \right] \tag{8.27}$$

The stochastic Libors $L(T_0, T_i)$ have a volatility expansion given by the following. From Eqs. (6.43), (8.15), and (8.17)

$$L(T_0, T_n) = L(t_0, T_n) + L(t_0, T_n)(\beta_n^{(0)} + e^{W_n} - 1) + O(\gamma^3)$$

$$W_n = -\frac{1}{2} q_n^2 + \int_{t_0}^{T_0} dt \int_{T_n}^{T_{n+1}} dx \gamma(t, x) \mathcal{A}_L(t, x)$$

$$\beta_n^{(0)} = \sum_{m=0}^{n} \Delta_{mn} \frac{\ell L(t_0, T_m)}{1 + \ell L(t_0, T_m)} \sim O(\gamma^2)$$

Hence, from Eqs. (8.27) and (8.26), the 'potential' term V given by

$$V = \sum_{I=1}^{N} J_I \left[\prod_{i=0}^{I-1} \left\{ \frac{1 + \ell L(t_0, T_i)}{1 + \ell L(T_0, T_i)} \right\} - 1 \right]$$

$$= \sum_{I=1}^{N} J_I \left[-1 + \exp \left\{ -\sum_{i=0}^{I-1} \ln(1 + a_I(\beta_n^{(0)} + e^{W_n} - 1)) \right\} \right]$$

$$= \sum_{I=1}^{N} J_I A_I$$

$$\Rightarrow A_I = -1 + \exp \left\{ -\sum_{i=0}^{I-1} \ln(1 + a_i(\beta_i^{(0)} + e^{W_i} - 1)) \right\} \qquad (8.28)$$

$$a_i = \frac{\ell L(t_0, T_i)}{1 + \ell L(t_0, T_i)}$$

The coefficient has a volatility expansion given by the following

$$A_I = -\sum_{i=0}^{I-1} a_i(\beta_i^{(0)} + e^{W_i} - 1) + \frac{1}{2} \sum_{i=0}^{I-1} [a_i(e^{W_i} - 1)]^2$$

$$+ \frac{1}{2} \sum_{i,j=0}^{I-1} a_i a_j(e^{W_i} - 1)(e^{W_j} - 1) + O(\gamma^3)$$

$$A_I A_J = \sum_{i=0}^{I-1} \sum_{j=0}^{J-1} a_i a_j(e^{W_i} - 1)(e^{W_j} - 1) + O(\gamma^3) \qquad (8.29)$$

From Eqs. (8.4) and (8.6)

$$E[e^{W_n} - 1] = 0$$

$$E[(e^{W_m} - 1)(e^{W_n} - 1)] = \Delta_{mn} + O(\gamma^4)$$

$$\Delta_{nm} \equiv \int_{t_0}^{T_0} dt \int_{T_n}^{T_{n+1}} dx \int_{T_m}^{T_{m+1}} dx' M_\gamma(x, x'; t) \simeq O(\gamma^2)$$

The option price is determined by $E[V]$ and $E[V^2]$; from the results obtained above

$$E[V] = \sum_{I=1}^{N} J_I E[A_I] = \sum_{I=1}^{N} J_I B_I$$

$$B_I = -\sum_{i=0}^{I-1} a_i \beta_i^{(0)} + \frac{1}{2}\sum_{i=0}^{I-1} a_i^2 \Delta_{ii} + \frac{1}{2}\sum_{i,j=0}^{I-1} a_i a_j \Delta_{ij}$$

$$E[V^2] = \sum_{I=1}^{N}\sum_{L=1}^{N} J_I J_L E[A_I A_L] = \sum_{I=1}^{N}\sum_{L=1}^{N} J_I J_L \sum_{i=0}^{I-1}\sum_{j=0}^{L-1} a_i a_j \Delta_{ij}$$

Hence, from Eq. (3.69), the coupon bond option price is given by

$$\frac{C(t_0, T_0, \mathcal{K})}{B(t_0, T_0)} = \frac{1}{\sqrt{2\pi}} I(X)\sqrt{C_2 - C_1^2} + O(\gamma^3)$$

$$C_1 = E[V]; \quad C_2 = E[V^2]; \quad X = \frac{\mathcal{K} - F - C_1}{\sqrt{C_2 - C_1^2}}$$

8.5.1 Libor swaption

The coupon bond option includes interest rate swaptions as a special case. Consider all the zero coupon bonds as being constructed from the Libor forward interest rates $f_L(t, x)$, as given in Eq. (6.1); hence

$$B(T_0, T_i) \rightarrow B_L(T_0, T_i) = \exp\left\{-\int_{T_0}^{T_i} dx f_L(t, x)\right\} \qquad (8.30)$$

The Libor ZCYC data can be used for obtaining the Libor zero coupon bond as given in Eq. (2.27). One can also construct the Libor zero coupon bond, from Eq. (6.5), as follows

$$B_L(T_0, T_n) = \prod_{i=0}^{n-1}\left[\frac{1}{[1 + \ell L(T_0, T_i)]}\right]$$

and the Eurodollar futures data can be used to find the market value of $L(T_0, T_i)$.

The Libor ZCYC and Eurodollar futures should give the same value for the Libor zero coupon bond $B_L(T_0, T_i)$; in practice, one of the two expressions for evaluating $B_L(T_0, T_i)$ may be more useful, depending on the approximations that need to be made for using market data.

From Eq. (4.32) the payoff for a swaption, in which the holder has the option to enter a fixed rate R_S receiver's swap and both floating and fixed payments are made at the same Libor time, yields the following coupon bond coefficients

$$c_n = \ell R_S; \quad n = 1, 2, \ldots, (N-1); \quad \text{payment at time } T_0 + n\ell \qquad (8.31)$$

$$c_N = 1 + \ell R_S; \quad \text{payment at time } T_0 + N\ell$$

$$K = 1$$

8.6 Libor Market Model European swaption price

Consider a Libor swaption maturing at time T_0, with swap payments being made at times $T_n = T_0 + \ell n$; the first payment is made at time T_1 and let final swap payment be made at $T_N = T$. Consider the swaption price at earlier Libor time t_0. The three Libor times $t_0 = T_{-k}$, T_0, and T_n are shown in Figure 8.1. For fixed payments being made at rate R_S, the swaption price for paying fixed rate and receiving floating payments at Libor, from Eq. (4.25), is given by the following expectation value

$$C(t_0, T_0; R_S) = \ell V B(t_0, T_0) E \left[\sum_{n=0}^{N-1} B(T_0, T_n + \ell)(L(T_0, T_n) - R_S) \right]_+ \quad (8.32)$$

$$\frac{C(t_0, T_0; R_S)}{B(t_0, T_0)} = E \left[\mathrm{swap}_L(T_0, R_S) \right]_+$$

where the numeraire (discounting) is given by the zero coupon bond $B(t, T_0)$. Figure 8.4 graphically represents the Libor swaption payoff function.

An approximate price for an interest rate swaption has been obtained in Eq. (8.31) by expressing the zero coupon bond in terms of the underlying Libor forward interest rates $f_L(t, x)$, as given in Eq. (8.30). In this section another expansion for swaption price is generated by using Eq. (6.5), which expresses zero coupon bonds directly in terms of Libors. The swaption price from both approaches should be equal, but the approximate prices may have different realizations depending on the market data being used. Consequently, the derivation of this section is of interest even though Section 8.5 does indeed provide a perturbation expansion for the swaption price.

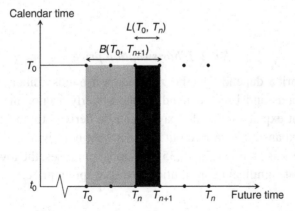

Figure 8.4 $\ell V \left[\sum_{n=0}^{N-1} B(T_0, T_{n+1})(L(T_0, T_n) - R_S) \right]_+$: Libor European swaption payoff.

8.6.1 Interest rate swap's volatility expansion

From Eq. (4.2), at time T_0, the price of the floating receiver swap is given by

$$\text{swap}_L(T_0, R_S) = \ell V \sum_{n=0}^{N-1} B(T_0, T_n + \ell)[L(T_0, T_n) - R_S] \qquad (8.33)$$

Based on the volatility expansion of the Libor zero coupon bonds given in Eq. (8.18)

$$B(T_0, T_n + \ell) = \tilde{F}(t_0, T_0, T_n + \ell)\big[1 + A_n\big] + O(\gamma^3)$$

the interest rate swap has the following volatility expansion

$$\text{swap}_L(T_0, R_S) \approx \ell V \sum_{n=0}^{N-1} \tilde{F}(t_0, T_0, T_n + \ell)\big[1 + A_n\big]$$

$$\times \left\{ \tilde{L}(t_0, T_n) + L(t_0, T_n)(e^{W_n} - 1) - R_S \right\}$$

$$= \ell V \left[\sum_{n=0}^{N-1} \tilde{F}(t_0, T_0, T_n + \ell)[\tilde{L}(t_0, T_n) - R_S] + V + O(\gamma^3) \right] \qquad (8.34)$$

where the 'potential' V, which contains all the stochastic terms, is given by[2]

$$V = \sum_{n=0}^{N-1} \tilde{F}(t_0, T_0, T_n + \ell)\Big\{ L(t_0, T_n)(e^{W_n} - 1) + L(t_0, T_n)A_n(e^{W_n} - 1)$$

$$+ [\tilde{L}(t_0, T_n) - R_S]A_n \Big\} + O(\gamma^3) \qquad (8.35)$$

8.6.2 Libor swaption price

The swaption price depends on the zero coupon bonds, Libor, and the initial Libors; both bonds and Libor depend on the underlying nonlinear drift $\beta_n(t_0)$. A self-consistent expansion for the payoff can be derived to any order in $O(\gamma)$. The volatility expansion is carried out to the lowest nontrivial order of $O(\gamma^2) = O(\Delta_{mn})$, where V is given in Eq. (8.35). For most purposes, the lowest order term contains the most significant contribution to the swaption price.

---

[2] Potential V is not to confused with the principal of the swap given by ℓV.

The swaption price is obtained as an expansion in powers of $\Delta = O(\gamma^2)$, which entails computing $E[\mathcal{V}]$ and $E[\mathcal{V}^2]$ to $O(\Delta)$. Recall from Eqs. (8.4) and (8.6)

$$E[e^{W_n} - 1] = 0; \quad E[(e^{W_m} - 1)(e^{W_n} - 1)] \simeq \Delta_{mn} + O(\gamma^4) \quad (8.36)$$

Using the above formulae and from Eq. (8.35)

$$E[\mathcal{V}] = \sum_{n=0}^{N-1} \tilde{F}(t_0, T_0, T_n + \ell)$$

$$\times \left\{ \tilde{L}(t_0, T_n) E[A_n(e^{W_n} - 1)] + [\tilde{L}(t_0, T_n) - R_S] E[A_n] \right\} + O(\gamma^3)$$

Eqs. (8.20) and (8.36) yield the following

$$E[(e^{W_m} - 1)A_n] = -\sum_{i=0}^{n} a_i \Delta_{mi} + O(\gamma^3)$$

and hence, using the value of $E[A_n]$ given in Eq. (8.23), we have

$$C_1 = E[\mathcal{V}] = \sum_{n=0}^{N-1} \tilde{F}(t_0, T_0, T_n + \ell) \left\{ -\tilde{L}(t_0, T_n) \sum_{i=0}^{n} a_i \Delta_{ni} \right.$$

$$\left. + \frac{1}{2} [\tilde{L}(t_0, T_n) - R_S] \left(\sum_{i=0}^{n} a_i^2 \Delta_{ii} + \sum_{i,j=0}^{n} a_i a_j \Delta_{ij} \right) \right\} + O(\gamma^3) \quad (8.37)$$

In computing $E[\mathcal{V}^2]$, the term $\tilde{F}(t_0, T_0, T_n + \ell)\tilde{L}(t_0, T_n)A_n(e^{W_n} - 1)$ contributes only to $O(\Delta^2)$ since $E[(e^{W_l} - 1)(e^{W_m} - 1)(e^{W_n} - 1)] = O(\Delta^2)$ and, hence, is dropped. This yields

$$E[\mathcal{V}^2] = \sum_{m,n=0}^{N-1} \tilde{F}(t_0, T_0, T_m + \ell)\tilde{F}(t_0, T_0, T_n + \ell)$$

$$\times E\left[\left\{ \tilde{L}(t_0, T_m)(e^{W_m} - 1) + [\tilde{L}(t_0, T_m) - R_S]A_m \right\} \right.$$

$$\left. \times \left\{ \tilde{L}(t_0, T_n)(e^{W_n} - 1) + [\tilde{L}(t_0, T_n) - R_S]A_n \right\} \right] + O(\gamma^3) \quad (8.38)$$

Note that

$$E[A_m A_n] = \sum_{i=0}^{m} \sum_{j=0}^{n} a_i a_j \Delta_{ij} + O(\gamma^3)$$

Hence the expectation value is given by the following

$$C_2 = E[V^2] = \sum_{m,n=0}^{N-1} \tilde{F}(t_0, T_0, T_m + \ell)\tilde{F}(t_0, T_0, T_n + \ell)$$

$$\times \left\{ \tilde{L}(t_0, T_m)\tilde{L}(t_0, T_n)\Delta_{mn} - 2\tilde{L}(t_0, T_m)[\tilde{L}(t_0, T_n) - R_S]\sum_{i=0}^{n} a_i \Delta_{mi} \right.$$

$$\left. + [\tilde{L}(t_0, T_n) - R_S][\tilde{L}(t_0, T_m) - R_S]\sum_{i=0}^{m}\sum_{j=0}^{n} a_i a_j \Delta_{ij} \right\} + O(\gamma^3) \quad (8.39)$$

For the Libor swaption, from (8.37) and (8.39)

$$C_0 = E[1] = 1; \quad C_1 = E[V]; \quad C_2 = E[V^2]$$

Hence, from Eq. (3.69), the swaption price is given by

$$\frac{C(t_0, T_0, T, K)}{B(t_0, T_0)} = E[\text{swap}_L(T_0, R_S)]_+ = \frac{\ell V}{\sqrt{2\pi}}I(X)\sqrt{C_2 - C_1^2}$$

$$X = \frac{\tilde{K} - C_1}{\sqrt{C_2 - C_1^2}}; \quad \tilde{K} = \sum_{n=0}^{N-1}\tilde{F}(t_0, T_0, T_n + \ell)[R_S - \tilde{L}(t_0, T_n)]$$

From Eq. (3.72), the Libor swaption for $X \approx 0$, yields the following approximate price

$$C(t_0, T_0, K) \approx \ell V B(t_0, T_0)\left[\frac{1}{\sqrt{2\pi}}\sqrt{C_2 - C_1^2} - \frac{1}{2}(\tilde{K} - C_1)\right] + O(X^2)$$

8.7 Libor Asian swaption price

Consider an Asian swaption maturing at time T_0. The receiver floating swap holder receives the time average, weighted by function $\rho(t)$, of the difference between N floating payments at Libor and payments at fixed rate R_S. The *Asian payoff function*, shown graphically in Figure 8.5, is given by generalizing Eq. (8.32) and yields the following

$$\mathcal{P}_{Asn} = C_{Asn}(T_0, T_0; R_S)$$

$$= \ell V\left[\frac{1}{T_0 - t_0}\int_{t_0}^{T_0} dt\,\rho(t)\sum_{n=0}^{N-1} F(t, T_0, T_n + \ell)\{L(t, T_n) - R_S\}\right]_+$$

$$F(t, T_0, T_n + \ell) = \prod_{i=0}^{n}\frac{1}{[1 + \ell L(t, T_i)]} \quad (8.40)$$

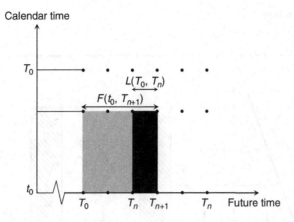

Figure 8.5 Libor swaption payoff for Asian option.

A volatility expansion, similar to the case for the European swaption given in Eqs. (6.41), (6.42), and (6.43) Libor, yields the following for the Asian case[3]

$$L(t, T_n) \simeq \tilde{L}(t_0, T_n) + L(t_0, T_n)(e^{W_n(t)} - 1) + O(\gamma^3)$$

$$W_n(t) = -\frac{1}{2}q_n^2(t) + \int_{t_0}^{t} d\tau \int_{T_n}^{T_{n+1}} dx \gamma(\tau, x) \mathcal{A}_L(\tau, x)$$

$$q_n^2(t) = \int_{t_0}^{t} d\tau \int_{T_n}^{T_{n+1}} dx dx' M_\gamma(x, x'; \tau) = \int_{t_0}^{t} d\tau \Lambda_{nn}(\tau)$$

$$\Lambda_{mn}(t) = \int_{t_0}^{t} d\tau \int_{T_m}^{T_{m+1}} dx \int_{T_n}^{T_{n+1}} dx' M_\gamma(x, x'; \tau) = \int_{t_0}^{t} d\tau \Lambda_{mn}(\tau)$$

Figure 8.6 provides a graphical representation of the Libor correlator $\Lambda_{mn}(t)$ that appears in the Asian Libor swaption pricing. Figure 8.7 shows the empirical values of $\Lambda_{mn}(t)$ for two well-separated time periods, namely $t = 1$ month and $t = 5$ years. Similar to results shown in Figure 8.2, the correlator $\Lambda_{mn}(t)$ has a 'hump' for a short time due to the effects of volatility $\gamma(t, x)$, which smooths out for longer times.

---

[3] The quantities defined earlier for the European swaption, in Eqs. (6.43) and (6.45), are special cases of the Asian case and given by
$$W_n = W_n(T_0); \quad \Lambda_{mn} = \Lambda_{mn}(T_0).$$

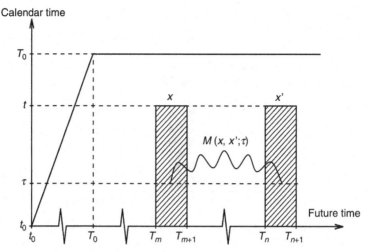

Figure 8.6 Libor correlator $\Delta_{mn}(t) = \int_{t_0}^{t} d\tau \int_{T_m}^{T_{m+1}} dx \int_{T_n}^{T_{n+1}} dx' M_\gamma(x, x'; \tau)$ for the Libor Asian swaption.

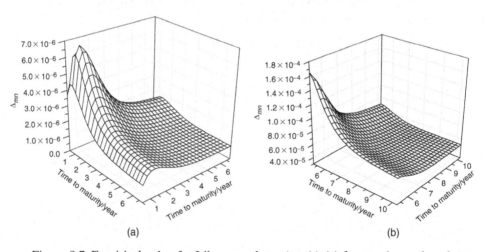

Figure 8.7 Empirical value for Libor correlator $\Delta_{mn}(t)$ (a) for $t = 1$ month and (b) for $t = 5$ years. The graphs are constructed from Libor ZCYC data for 261 days, from 10 October 2007 to 8 August 2008.

The forward price of the zero coupon bond, similar to Eq. (8.18), has the following expansion

$$F(t, T_0, T_n + \ell) = \tilde{F}(t_0, T_0, T_n + \ell)\left[1 + A_n(t)\right]$$

where, similar to Eq. (8.19)

$$A_n(t) \approx -\sum_{i=0}^{n} a_i(e^{W_i(t)} - 1) + \frac{1}{2}\sum_{i=0}^{n} a_i^2(e^{W_i(t)} - 1)^2$$

$$+ \frac{1}{2} \sum_{i,j=0}^{n} a_i a_j (e^{W_i(t)} - 1)(e^{W_j(t)} - 1)$$

Similar to the European case given in Eqs. (8.34) and (8.35), the Asian swap has the following expansion

$$\frac{1}{T_0 - t_0} \int_{t_0}^{T_0} dt \rho(t) \sum_{n=0}^{N-1} F(t, T_0, T_n + \ell)\{L(t, T_n) - R_S\}$$

$$= \tilde{\rho} \sum_{n=0}^{N-1} \tilde{F}(t_0, T_0, T_n + \ell)[\tilde{L}(t_0, T_n) - R_S] + \mathcal{V}_{Asn}; \quad \tilde{\rho} = \frac{1}{T_0 - t_0} \int_{t_0}^{T_0} dt \rho(t)$$

$$\Rightarrow \mathcal{V}_{Asn} = \frac{1}{T_0 - t_0} \int_{t_0}^{T_0} dt \rho(t) \sum_{n=0}^{N-1} \tilde{F}(t_0, T_0, T_n + \ell) \left\{ \tilde{L}(t_0, T_n)(e^{W_n(t)} - 1) \right.$$

$$\left. + \tilde{L}(t_0, T_n) A_n(t)(e^{W_n(t)} - 1) + [\tilde{L}(t_0, T_n) - R_S] A_n(t) \right\} \quad (8.41)$$

To find the swaption price one needs to evaluate $E[\mathcal{V}_{Asn}]$ and $E[\mathcal{V}^2_{Asn}]$, for which one needs to evaluate the following correlator

$$E[e^{W_m(t)} e^{W_n(t')}] = \exp\left\{ \int_{t_0}^{t} d\tau \int_{t_0}^{t'} d\tau' \delta(\tau - \tau') \right.$$

$$\times \int_{T_m}^{T_{m+1}} dx \int_{T_n}^{T_{n+1}} dx' M_\gamma(x, x'; \tau) \right\}$$

$$= \exp\left\{ \int_{t_0}^{t} d\tau \theta(t' - \tau) \Lambda_{mn}(\tau) \right\}$$

since

$$\int_{t_0}^{t'} d\tau' \delta(\tau - \tau') = \theta(t' - \tau)$$

Note from above that $E[e^{W_m(t)} e^{W_n(t)}] = \exp\{\Lambda_{mn}(t)\}$ and hence, similar to Eq. (8.37), the Asian case is given by

$$E[\mathcal{V}_{Asn}] = \frac{1}{T_0 - t_0} \sum_{n=0}^{N-1} \tilde{F}(t_0, T_0, T_n + \ell) \int_{t_0}^{T_0} dt \rho(t) \left\{ -\tilde{L}(t_0, T_n) \sum_{i=0}^{n} a_i \Delta_{ni}(t) \right.$$

$$\left. + \frac{1}{2}[\tilde{L}(t_0, T_n) - R_S] \left(\sum_{i=0}^{n} a_i^2 \Delta_{ii}(t) + \sum_{i,j=0}^{n} a_i a_j \Delta_{ij}(t) \right) \right\} + O(\Delta^2)$$

Similar to Eq. (8.38)

$$E[V_{Asn}^2] = \frac{1}{(T_0 - t_0)^2} \sum_{m,n=0}^{N-1} \tilde{F}(t_0, T_0, T_m + \ell)\tilde{F}(t_0, T_0, T_n + \ell) \int_{t_0}^{T_0} dt\,\rho(t)$$

$$\times \int_{t_0}^{T_0} dt'\rho(t')E\Big[\{\tilde{L}(t_0, T_m)(e^{W_m(t)} - 1) + [\tilde{L}(t_0, T_m) - R_S]A_m(t)\}$$

$$\times \{\tilde{L}(t_0, T_n)(e^{W_n(t')} - 1) + [\tilde{L}(t_0, T_n) - R_S]A_n(t')\}\Big] + O(\Delta^2)$$

$$(8.42)$$

and yields, similar to Eq. (8.39)

$$E[V_{Asn}^2] = \sum_{m,n=0}^{N-1} \tilde{F}(t_0, T_0, T_m + \ell)\tilde{F}(t_0, T_0, T_n + \ell)$$

$$\times \Bigg\{ \tilde{L}(t_0, T_m)\tilde{L}(t_0, T_n)\Gamma_{mn} - 2\tilde{L}(t_0, T_m)\big(\tilde{L}(t_0, T_n) - R_S\big)\sum_{i=0}^{n} a_i\Gamma_{mi}$$

$$+ \big(\tilde{L}(t_0, T_n) - R_S\big)\big(\tilde{L}(t_0, T_m) - R_S\big)\sum_{i=0}^{m}\sum_{j=0}^{n} a_i a_j\Gamma_{ij} \Bigg\} + O(\Delta^2)$$

where

$$\Gamma_{mn} = \frac{1}{(T_0 - t_0)^2}\int_{t_0}^{T_0} dt \int_{t_0}^{T_0} dt'\,\rho(t)\rho(t')\int_{t_0}^{t} d\tau\,\theta(t' - \tau)\Lambda_{mn}(\tau)$$

For the special case of $\rho(t) = 1$ it can be shown that[4]

$$\Gamma_{mn} = \frac{1}{(T_0 - t_0)^2}\int_{t_0}^{T_0} dt(t - t_0)(T_0 - t)\Lambda_{mn}(t) \qquad (8.43)$$

$$< \frac{1}{4}\int_{t_0}^{T_0} dt\,\Lambda_{mn}(t) = \frac{1}{4}\Delta_{mn}$$

Note that $\Gamma_{mn} < \Delta_{mn}$ due to the time average in the Asian payoff function that 'irons' out large fluctuations. This is the reason that the Asian option is always cheaper than the European option.

---

[4] Note one has the inequality

$$\frac{(t - t_0)(T_0 - t)}{(T_0 - t_0)^2} \leq \frac{1}{4}; \quad t_0 \leq t \leq T_0.$$

The Asian swaption price is given, from Eq. (3.69), by

$$X_{Asn} = \frac{\tilde{K}_{Asn} - C_1}{\sqrt{C_2 - C_1^2}}; \quad E[1] = 1; \quad C_1 = E[\mathcal{V}_{Asn}]; \quad C_2 = E[\mathcal{V}_{Asn}^2]$$

$$\tilde{K}_{Asn} = \tilde{\rho} \sum_{n=0}^{N-1} \tilde{F}(t_0, T_0, T_n + \ell)[R_S - \tilde{L}(t_0, T_n)]$$

$$\frac{C_{Asn}(t_0, T_0, T, K)}{B(t_0, T_0)} = \frac{\ell V}{\sqrt{2\pi}} I(X_{Asn}) \sqrt{C_2 - C_1^2} \approx \frac{\ell V}{\sqrt{2\pi}} \sqrt{C_2 - C_1^2} + O(X_{Asn}^2)$$

8.8 BGM–Jamshidian swaption price

The BGM–Jamshidian swaption price is widely discussed in the finance litera-ture and the usual derivation is obtained by using a 'rolling' forward measure and techniques of stochastic calculus [34]. An independent derivation of the swaption price is given by performing a path integral over the Gaussian white noise that drives the BGM–Jamshidian version of the Libor Market Model. The derivation is based on two assumptions, namely that (a) the Libor drift is a constant and (b) that the volatility functions are de-correlated. The result obtained illustrates the differences between the quantum finance and BGM–Jamshidian formulations of the Libor Market Model.

Consider the payoff function for the receiver floating swaption that matures at time T_0 and for which the swap holder receives N payments at Libor and pays at fixed rate R_S. The payoff is given, from Eq. (4.25), by the following

$$\mathcal{P} = C(T_0, T_0; R_S) = \ell V \left[\sum_{n=0}^{N-1} B(T_0, T_n + \ell)\{L(t_0, T_n) - R_S\} \right]_+$$

The swaption price at earlier time t_0 is evaluated using the forward bond numeraire $B(t, T_0)$. The martingale conditon yields

$$\frac{C(t_0, T_0; R_S)}{B(t_0, T_0)} = E\left[\frac{C(T_0, T_0; R_S)}{B(T_0, T_0)}\right] = E[\mathcal{P}]$$

The BGM–Jamshidian limit of $\mathcal{D}_L(x, x'; t) \to 1$, is given in Eq. (6.79), and yields from Eqs. (6.40) and (6.42) the following

$$\int_{t_0}^{T_0} dt \int_{T_n}^{T_{n+1}} dx \gamma(t, x) \mathcal{A}_L(t, x) \to \int_{t_0}^{T_0} dt \gamma_n(t) R(t)$$

$$\gamma_n(t) \equiv \int_{T_n}^{T_{n+1}} dx \gamma(t, x)$$

Hence the drift is given, by Eq. (8.15), as follows

$$\zeta_n \equiv \int_{t_0}^{T_0} dt \zeta_{BGM}(t, T_n) = \int_{t_0}^{T_0} dt \sum_{m=1}^{n} \frac{\ell L(t, T_m)}{[1 + \ell L(t, T_m)]} \gamma_m(t) \gamma_n(t)$$

The *first approximation*, the first of two in the BGM–Jamshidian approach, is the following

$$\frac{\ell L(t, T_m)}{[1 + \ell L(t, T_m)]} \approx \frac{\ell L(t_0, T_m)}{[1 + \ell L(t_0, T_m)]} \tag{8.44}$$

The drift, in this approximation, depends only on the Libor $L(t_0, T_n)$ at initial time t_0, which is given by the market. In particular, the drift no longer depends on the random values of Libor $L(t, T_n)$, with $t_0 \le t \le T_0$ and the approximation *linearizes* the expression for Libor given in Eq. (6.40). The BGM–Jamshidian approximation given in Eq. (8.44) is, in fact, the leading term – as a function of $\gamma(t, x)$ – of the quantum finance approximation for Libor drift given in Eq. (8.16).

Hence, from Eq. (8.44)

$$\zeta_n \approx \sum_{m=1}^{n} \frac{\ell L(t_0, T_m)}{[1 + \ell L(t_0, T_m)]} \int_{t_0}^{T_0} dt \gamma_n(t) \gamma_m(t)$$

Collecting the results yields Libor given by

$$\ell L(t, T_i) = \ell L(t_0, T_i) e^{\zeta_i - \frac{1}{2} q_i^2 + \int_{t_0}^{T_0} dt \gamma_i(t) R(t)} \tag{8.45}$$

where, in the BGM–Jamshidian limit, Eq. (6.35) yields

$$q_i^2 \rightarrow -\frac{1}{2} \int_{t_0}^{T_0} dt \gamma_i^2(t)$$

The swaption price, in the BGM–Jamshidian approximate scheme, is given by

$$\frac{C(t_0, T_0; R_S)}{\ell V B(t_0, T_0)} = E \left[\sum_{n=0}^{N-1} B(T_0, T_n + \ell) \{ L(T_0, T_n) - R_S \} \right]_+ \equiv E[\mathcal{P}]$$

$$\mathcal{P} = \left[\sum_{n=0}^{N-1} B(T_0, T_n + \ell) \{ L(T_0, T_n) - R_S \} \right]_+$$

$$= \left[\sum_{n=0}^{N-1} B(T_0, T_n + \ell) \{ L(t_0, T_n) e^{\zeta_n - \frac{1}{2} q_n^2 + \int_{t_0}^{T_0} dt \gamma_n(t) R(t)} - R_S \} \right]_+ \tag{8.46}$$

$$\text{with } B(T_0, T_n + \ell) = \prod_{i=0}^{n} \frac{1}{\left[1 + \ell L(t_0, T_i) e^{\zeta_i - \frac{1}{2} q_i^2 + \int_{t_0}^{T_0} dt \gamma_i(t) R(t)}\right]} \qquad (8.47)$$

Note the BGM–Jamshidian swaption price depends on white noise $R(t)$ only through the combination $\int_{t_0}^{T_0} dt \gamma_n(t) R(t)$. The expectation value over white noise $R(t)$ is evaluated using a path integral.

Using the Dirac-delta function's representation given in Eq. (A.7), consider the following representation of unity

$$1 = \prod_{n=0}^{N-1} \int_{-\infty}^{+\infty} d\xi_n \delta\left[\xi_n - \int_{t_0}^{T_0} dt \gamma_n(t) R(t)\right]$$

$$= \prod_{n=0}^{N-1} \int_{-\infty}^{+\infty} d\xi_n \int_{-\infty}^{+\infty} \frac{d\eta_n}{2\pi} e^{i\eta_n(\xi_n - \int_{t_0}^{T_0} dt \gamma_n(t) R(t))}$$

$$\equiv \int_{\xi, \eta} e^{i \sum_{i=0}^{n} \eta_n (\xi_n - \int_{t_0}^{T_0} dt \gamma_n(t) R(t))}$$

Inserting the above resolution of unity into the path integral for the swaption price yields

$$E[\mathcal{P}] = \int DR \mathcal{P}[R(t)] e^{S_0}; \quad S_0 = -\frac{1}{2} \int_{t_0}^{T_0} dt R^2(t)$$

$$= \int DR \int_{\xi, \eta} e^{i \sum_{n=0}^{N-1} \eta_n (\xi_n - \int_{t_0}^{T_0} dt \gamma_n(t) R(t))} \mathcal{P}[R(t)] e^{S_0}$$

$$= \int_{\xi, \eta} \mathcal{P}(\xi) e^{i \sum_{n=0}^{N-1} \eta_n \xi_n} Z(\eta) \qquad (8.48)$$

where the payoff function is given by

$$\mathcal{P}(\xi) = \left[\sum_{n=0}^{N-1} B(T_0, T_n + \ell)\{L(t_0, T_n) e^{\zeta_n - \frac{1}{2} q_n^2 + \xi_n} - R_S\}\right]_+$$

$$\text{with } B(T_0, T_n + \ell) = \prod_{i=0}^{n} \frac{1}{\left[1 + \ell L(t_0, T_i) e^{\zeta_i - \frac{1}{2} q_i^2 + \xi_i}\right]}$$

The white noise path integral $\int DR$ can be performed exactly, and yields the following partition function

$$Z(\eta) = \int DR e^{-i \sum_{n=0}^{N-1} \eta_n \int_{t_0}^{T_0} dt \gamma_n(t) R(t)} e^{S_0}$$

$$= \exp\left\{ -\frac{1}{2}\sum_{m,n=0}^{N-1} \eta_i \eta_j \int_{t_0}^{T_0} dt \gamma_m(t)\gamma_n(t) \right\} \tag{8.49}$$

The BGM–Jamshidian derivation makes a *second assumption*, namely that the γ_ns are de-correlated, and yields the following

$$\int_{t_0}^{T_0} dt \gamma_m(t)\gamma_n(t) \approx \Gamma_m \Gamma_n \tag{8.50}$$

$$\Rightarrow \quad q_n^2 = -\frac{1}{2}\int_{t_0}^{T_0} dt \gamma_n(t)\gamma_n(t) \approx -\frac{1}{2}\Gamma_n^2 \tag{8.51}$$

$$\zeta_n \approx \Gamma_n d_n; \quad d_0 = 0; \quad d_n = \sum_{m=1}^{n} \frac{\ell L(t_0, T_m)}{[1 + \ell L(t_0, T_m)]}\Gamma_m \tag{8.52}$$

Numerical studies provide some evidence in support of the de-correlation given in Eq. (8.50) [32]. Eqs. (8.49) and (8.50) yield

$$Z(\eta) = \exp\left\{ -\frac{1}{2}\left[\sum_{n=0}^{N-1} \eta_n \Gamma_n\right]^2 \right\}$$

$$= \int_{-\infty}^{+\infty} \frac{dW}{\sqrt{2\pi}} \exp\left\{ -iW\sum_{n=0}^{N-1} \eta_n \Gamma_n - \frac{1}{2}W^2 \right\} \tag{8.53}$$

Hence, from Eqs. (8.48) and (8.53), the swaption price is given by

$$\frac{C(t_0, T_0; R_S)}{\ell V B(t_0, T_0)} = \int_{\xi,\eta} e^{i\sum_{n=0}^{N-1}\eta_i\xi_i}\mathcal{P}(\xi) \int_{-\infty}^{+\infty} \frac{dW}{\sqrt{2\pi}} \exp\left\{ -iW\sum_{n=0}^{N-1} \eta_n \Gamma_n - \frac{1}{2}W^2 \right\}$$

$$= \int_{-\infty}^{+\infty} \frac{dW}{\sqrt{2\pi}} e^{-\frac{1}{2}W^2} \int_{\xi}\prod_{n=0}^{N-1} \delta(\xi_n - W\Gamma_n)\mathcal{P}(\xi)$$

$$= \int_{-\infty}^{+\infty} \frac{dW}{\sqrt{2\pi}} e^{-\frac{1}{2}W^2}\mathcal{P}(W\Gamma) \tag{8.54}$$

The assumption that the γ_ns are de-correlated yields a major simplification, namely that $\int_{t_0}^{T_0} dt R(t)\gamma_n(t) \to W\Gamma_n$, where W is a $N(0,1)$ normal random variable; in other words, the *infinite collection* of random variables $R(t)$ – one for each $t \in [t_0, T_0]$ – is replaced by a *single* random variable W.

Hence, from Eqs. (8.46) and (8.49), the BGM–Jamshidian swaption price is given by

$$\frac{C(t_0, T_0; R_S)}{\ell V B(t_0, T_0)} = \int_{-\infty}^{+\infty} \frac{dW}{\sqrt{2\pi}} e^{-\frac{1}{2}W^2}$$

$$\times \left[\sum_{n=0}^{N-1} B(T_0, T_n + \ell) \{ L(t_0, T_n) e^{(d_n - \frac{1}{2}\Gamma_n - W)\Gamma_n} - R_S \} \right]_+ \quad (8.55)$$

$$B(T_0, T_n + \ell) = \prod_{i=0}^{n} \frac{1}{\left[1 + \ell L(t_0, T_i) e^{(d_i - \frac{1}{2}\Gamma_i - W)\Gamma_i} \right]} \quad (8.56)$$

Let w_0 be defined by

$$\sum_{n=0}^{N-1} B(T_0, T_n + \ell) \{ L(t_0, T_n) e^{(d_n - \frac{1}{2}\Gamma_n - w_0)\Gamma_n} - R_S \} = 0 \quad (8.57)$$

The BGM–Jamshidian swaption price is given by

$$C(t_0, T_0; R_S) = \ell V B(t_0, T_0) \sum_{n=0}^{N-1} \int_{-\infty}^{w_0} \frac{dW}{\sqrt{2\pi}} e^{-\frac{1}{2}W^2} B(T_0, T_n + \ell)$$

$$\times \{ L(t_0, T_n) e^{(d_n - \frac{1}{2}\Gamma_n - W)\Gamma_n} - R_S \} \quad (8.58)$$

Due to the de-correlation of Libor volatility $\gamma(t, x)$, the BGM–Jamshidian approximation for the swaption price completely *factorizes* into an independent sum over the individual payments. This factorization leads to systematic errors in the BGM–Jamshidian swaption price since crucial correlations are being neglected.

A further simplification can be made by assuming that $B(T_0, T_n + \ell)$ is approximately equal to its forward price, namely

$$B(T_0, T_n + \ell) \simeq F(t_0, T_0, T_n + \ell) + O(\gamma^2)$$

This, in turn, allows the BGM–Jamshidian swaption price to be expressed in terms of the normal cumulative distribution $N(d)$. Eq. (8.57) yields the following approximate equation for determining w_0

$$\sum_{n=0}^{N-1} F(t_0, T_0, T_n + \ell) \{ L(t_0, T_n) e^{(d_n - \frac{1}{2}\Gamma_n - w_0)\Gamma_n} - R_S \} = 0$$

Hence, since $B(t_0, T_0)F(t_0, T_0, T_n + \ell) = B(t_0, T_n + \ell)$, the BGM–Jamshidian swaption price is given by [34, 61]

$$C(t_0, T_0; R_S) = \ell V \sum_{n=0}^{N-1} B(t_0, T_n + \ell)$$

$$\times \int_{-\infty}^{w_0} \frac{dW}{\sqrt{2\pi}} e^{-\frac{1}{2}W^2} \left\{ L(t_0, T_n) e^{(d_n - \frac{1}{2}\Gamma_n - W)\Gamma_n} - R_S \right\}$$

$$= \sum_{n=0}^{N-1} B(t_0, T_n + \ell) \left\{ L(t_0, T_n) e^{d_n \Gamma_n} N(w_0 + \Gamma_n) - R_S N(w_0) \right\}$$

8.9 Summary

Imperfect and nontrivial correlations between the different Libors are parsimoniously captured in the Libor Market Model by the correlations of $\mathcal{A}_L(t, x)$. All Libor option payoffs are written in terms of Libor zero coupon bonds $B_L(t, T_n)$ and Libor $L(t, T_n)$; no reference was made to the bond forward interest rates in defining the Libor options.

Libor drift is a nonlinear function of Libor and a volatility expansion was developed that can be used to evaluate the drift to any degree of accuracy. A volatility expansion was developed for the payoff of the zero coupon bond and swaption. Libor options prices were analytically evaluated as a perturbation expansion in powers of $\gamma(t, x)$, the volatility of log Libor.

The caplet price was exactly evaluated and provides a quantum finance generalization of Black's formula. Determining the zero coupon bond option price in the Libor Market Model is a nonlinear problem and could only be approximately evaluated. In contrast, the zero coupon bond option price can be exactly evaluated in the bond forward interest rates, as discussed in Section 11.6.

Swaption price is a nonlinear problem for the quantum finance Libor Market Model as well as for bond forward interest rates, discussed in detail in Chapter 11. The Libor European and Asian swaption prices were obtained as a perturbation expansion in the log Libor volatility function $\gamma(t, x)$ and which demonstrates the crucial role of the Libor correlator in pricing these instruments.

Two different approximate Libor Market Model swaption prices, based on representing the swaption either in terms of the Libor forward interest rates $f_L(t, x)$ or in terms of Libors $L(t, T_n)$, have been obtained in Sections 8.5.1 and 8.6. In principle, these two prices are equal but in practice this may not be the case, given the nature of the available data and the calibration of the Libor Market Model. These two prices need to be empirically studied to decide on which one is the best suited for applications.

The well-known BGM–Jamshidian result was obtained by taking the limiting case of perfectly correlated Libors. The path integral over the white noise driving the Libors in the BGM–Jamshidian model was exactly evaluated, and the assumption of the de-correlation of Libor volatility led to a complete factorization of the swaption price.

The pricing of hybrid instruments that combine bond and Libor forward interest rates has yet to be addressed. The quantum finance formalism needs to be extended for analyzing these instruments.

9

Numeraires for bond forward interest rates

Various numeraires are defined in the framework of the bond forward interest rates $f(t, x)$ discussed in Chapter 5. Eqs. (5.1) and (2.20) yield the following

$$\frac{\partial f}{\partial t}(t, x) = \alpha(t, x) + \sigma(t, x)\mathcal{A}(t, x); \quad -\infty \leq f(t, x) \leq +\infty$$

$$1 + \ell L(t, T_n) = \exp\left\{\int_{T_n}^{T_n+\ell} dx f(t, x)\right\}$$

where $L(t, T_n)$ is the three-month benchmark Libor with tenor denoted by ℓ.

The main result of this chapter is that a numeraire, called the forward numeraire, can be chosen for the bond forward interest rates, such that all forward bond prices for future Libor time $T_n = T_0 + \ell n$ with tenor ℓ have a martingale evolution. In other words, the numeraire is chosen such that all Libor tenor forward bond prices are martingales; hence

$$F(t, T_n) = \exp\left\{-\int_{T_n}^{T_n+\ell} dx f(t, x)\right\} \quad : \quad \text{martingale for all } n$$

As an academic exercise, Libor is expressed in terms of the bond forward interest rates and a numeraire is chosen so that all the three-month tenor Libor interest rates have a martingale evolution; that is, the numeraire makes all three-month tenor Libor into martingales

$$L(t, T_n) \quad : \quad \text{martingale for all } n$$

In quantum finance the interest rates, at each instant, are driven by infinitely many independent random variables. This feature allows one to consistently vary the choice numeraire so that all forward bond prices and three-month tenor Libor are martingales. A common numeraire, with its concomitant drift, is seen to emerge naturally.

For comparison, the money market numeraire is also analyzed. The forward numeraire is similar to the money market numeraire for which, from Eq. (3.4), all zero coupon bonds are martingales; namely

$$B(t, T) = E\left[e^{-\int_t^{t*} dt' r(t')} B(t_*, T)\right]$$

The price of an interest rate caplet is computed as a test case for all three numeraires and it is shown that the price is numeraire invariant. Put–call parity is discussed in some detail and shown to emerge due to nontrivial properties of the numeraires. Some properties of swaps, and their relation to caps and floors, are briefly discussed. The focus of this chapter is on the choice of numeraire and is not geared towards applications.

9.1 Introduction

The main focus of this chapter is on the properties of bond forward interest rates, and in particular on finding a common numeraire (measure) that yields a martingale evolution for all forward bonds. Two other numeraires for bond forward interest rates are also considered, namely the money market numeraire and a common numeraire for Libors.

All calculations are performed using the quantum field theory of bond forward interest rates. It is shown that a numeraire can be chosen so that *all* the forward bond prices simultaneously have a martingale evolution. One of the effects of this numeraire is that all Libors, written in terms of the bond forward interest rates, are martingales. This outcome is in contrast with the result obtained in Section 6.4 for the Libor Market Model, where only a *single* three-month tenor Libor, namely $L(t, T_I)$, was rendered into a martingale by the choice of the zero coupon bond $B(t, T_{I+1})$ as the forward numeraire; all the other three-month tenor Libors $L(t, T_n)$ are not martingales, having a nonzero drift given by $\zeta(t, T_n)$ as in Eq. (6.40).

The (future) payoff of a financial instrument has to be discounted by a numeraire to obtain its current price. It has been shown by Geman *et al.* [45, 49] that any positive valued security can be used for discounting the payoff function. In particular, one can use other zero coupon bonds with different maturities as a discounting factor instead of using the money market numeraire. The forward numeraire is first discussed; then the money market numeraire and lastly Libor market measure is discussed. The main purpose is to elaborate different choices for the numeraire of bond forward interest rates.

The martingale condition for different numeraires leads to a change in the drift term for the bond forward interest rates $\alpha(t, x)$ [49]. The freedom of choosing a numeraire results from the fact that, for every numeraire, the corresponding drift makes the price of all traded instrument independent of the numeraire. *Numeraire invariance* is an important tool in creating models for the pricing of financial instruments [49, 50]. To concretely illustrate numeraire invariance, interest rate caplet prices are evaluated for three different numeraires. It is verified that, as expected, all three numeraires yield the same price: the price of the caplet is numeraire invariant.

9.2 Money market numeraire

The money market defines martingale measure by a numeraire $M(t, t_*)$ given by

$$M(t, t_*) = e^{\int_t^{t_*} r(t')dt'}; \quad t : \text{fixed}$$

$r(t) = f(t, t)$ is the spot interest rate. The quantity $B(t, T)/M(t, t)$ is defined to be a martingale

$$\frac{B(t, T)}{M(t, t)} = E_M \left[\frac{B(t_*, T)}{M(t, t_*)} \right]$$

$$\Rightarrow B(t, T) = E_M \left[e^{-\int_t^{t_*} r(t')dt'} B(t_*, T) \right] \tag{9.1}$$

where $E_M[\ldots]$ denotes expectation values taken with respect to the money market measure. The martingale condition can be solved for its corresponding drift velocity, which is given by Eq. (5.36)

$$\alpha_M(t, x) = \sigma(t, x) \int_t^x dx' \mathcal{D}(x, x'; t) \sigma(t, x') \tag{9.2}$$

9.3 Forward bond numeraire

Choose numeraire $B(t, T_I)$, with T_I: fixed. The martingale condition for zero coupon bonds $B(t, T)$ is the following

$$\frac{B(t, T)}{B(t, T_I)} = E_I \left[\frac{B(T_I, T)}{B(T_I, T_I)} \right]$$

$$\Rightarrow B(t, T) = B(t, T_I) E_I [B(T_I, T)]$$

The drift is given by the well-known result that [12]

$$\alpha_I(t, x) = \sigma(t, x) \int_{T_I}^x dx' \mathcal{D}(x, x'; t) \sigma(t, x') \tag{9.3}$$

where $E_I[\ldots]$ denotes taking the expectation value with respect to the forward risk-neutral measure. In fact, the derivation in Section 6.4 shows that Eq. (9.3) for the drift of the money market numeraire continues to hold even if $\sigma(t, x)$ is stochastic.

9.4 Change of numeraire

The effect of changing the numeraire is investigated in the framework of quantum finance. In particular, it is shown that changing the numeraire changes the action S that determines the probability distribution e^S/Z in the path integral.

The money market numeraire yields the following martingale condition

$$B(t_0, T) = E_M\left[e^{-\int_{t_0}^{t_*} r(t)dt} B(t_*, T)\right]$$

$$\equiv \frac{1}{Z}\int DA \, e^{-\int_{t_0}^{t_*} r(t)dt} e^{S_M} B(t_*, T) \tag{9.4}$$

Similarly, the martingale condition for the forward numeraire is defined by

$$B(t_0, T) = B(t_0, T_I)E_I[B(T_I, T)] \tag{9.5}$$

$$\equiv B(t_0, T_I)\frac{1}{Z}\int DA \, B(T_I, T)e^{S_I} \tag{9.6}$$

where $E_I[\ldots]$ denotes taking the expectation value with respect to the risk-neutral measure e^{S_I}/Z.

The relation of the risk-neutral probability measures e^{S_M}/Z and e^{S_I}/Z can be explicitly obtained for Gaussian forward interest rates.

From Eqs. (5.1) and (5.25)

$$S_M = S[\alpha_M] = -\frac{1}{2}\int_T \frac{f(t,x) - \alpha_M(t,x)}{\sigma(t,x)}\mathcal{N}^{-1}(t,x,x')\frac{f(t,x') - \alpha_M(t,x')}{\sigma(t,x')}$$

$$S_I = S[\alpha_I] = -\frac{1}{2}\int_T \frac{f(t,x) - \alpha_I(t,x)}{\sigma(t,x)}\mathcal{N}^{-1}(t,x,x')\frac{f(t,x') - \alpha_I(t,x')}{\sigma(t,x')}$$

where the drifts are given in Eqs. (9.2) and (9.3).

It is shown in [12] that

$$e^{S_I} = \frac{e^{-\int_{t_0}^{T_I} r(t)dt}}{B(t_0, T_I)}e^{S_M} \tag{9.7}$$

The factor $\exp\{-\int_{t_0}^{T_I} r(t)dt\}/B(t_0, T_I)$, relating the two actions, is evaluated in the finance literature using the Radon–Nikodyn derivative [49, 45].

Eq. (9.7) is particularly useful in evaluating European options for zero coupon bonds. From Eq. (3.8), for $t_0 < t_*$ the call option price $C(t_0, T_I, T, K)$

is given by

$$C(t_0, T_I, T, K) = E_M[e^{-\int_{t_0}^{T_I} dt\, r(t)}(B(T_I, T) - K)_+] \tag{9.8}$$

$$= B(t_0, T_I)E_I[(B(T_I, T) - K)_+] \tag{9.9}$$

where Eq. (9.7) has been used in obtaining Eq. (9.9) above.

To compute the call option using Eq. (9.9) is much simpler than doing the calculation using Eq. (9.8) since the discounting term, after a change in numeraire, is the deterministic function $B(t_0, T_I)$.[1]

9.5 Forward numeraire

Consider a numeraire that renders the futures price of zero coupon bonds, that is $F(t, T_n, T_{n+1})$, into a martingale. Zero coupon bonds are traded instruments; hence, discounting it by the numeraire must yield an instrument that undergoes a martingale evolution [49].

The forward numeraire is fixed by an (infinite) collection of zero coupon bonds defined for Libor time in the following manner, namely

$$B(t, T_0); B(t, T_1), \ldots; B(t, T_n); \; B(t, T_{n+1}); \; B(t, T_{n+2}), \ldots; \; T_n = T_0 + \ell n$$

Consider a zero coupon bond with maturity at Libor time, T_{n+1} namely $B(t, T_{n+1})$; the numeraire is chosen to be $B(t, T_n)$. The forward value of the bond at time T_n is given by

$$F(t_0, T_n, T_{n+1}) = e^{-\int_{T_n}^{T_{n+1}} dx\, f(t_0, x)} = \frac{B(t_0, T_{n+1})}{B(t_0, T_n)} \tag{9.10}$$

The drift is fixed so that the forward bond price

$$F(t, T_n, T_{n+1}) = \frac{B(t, T_{n+1})}{B(t, T_n)}: \; \text{martingale} \tag{9.11}$$

is a martingale. Namely, the expected value of the future price of the bond at time T_n is equal to its present value; hence

$$F(t_0, T_n, T_{n+1}) = E_F\big[F(t_*, T_n, T_{n+1})\big]$$

$$\Rightarrow e^{-\int_{T_n}^{T_{n+1}} dx\, f(t_0, x)} = E_F\big[e^{-\int_{T_n}^{T_{n+1}} dx\, f(t_*, x)}\big] \tag{9.12}$$

As expressed in the equation above, the drift is chosen to make the forward bond price a martingale.

---

[1] The zero coupon bond call option is computed in Section 11.13 using discounting by $B(t_0, T_I)$.

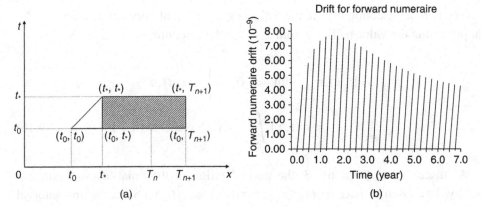

Figure 9.1 (a) The domain of integration $\mathcal{M}$ for evaluating the drift of the three-month tenor Libor market numeraire. (b) The drift velocity $\alpha_F(t,x)$ for the forward numeraire.

To determine the corresponding drift velocity $\alpha_F(t,x)$, the right-hand side of Eq. (9.12) is explicitly evaluated. Note from Eq. (5.2)

$$E_F\left[e^{-\int_{T_n}^{T_{n+1}} dxf(t_*,x)}\right] = e^{-\int_{T_n}^{T_{n+1}} dxf(t_0,x) - \int_{\mathcal{M}} \alpha_F(t,x)} \int D\mathcal{A} e^{-\int_{\mathcal{M}} \sigma(t,x)\mathcal{A}(t,x)} e^{S[\mathcal{A}]}$$

where the integration domain $\mathcal{M}$ is given in Figure 9.1(a).

Hence, from Eqs. (5.21) and (9.12)

$$e^{\int_{\mathcal{M}} \alpha_F(t,x)} = \int D\mathcal{A} e^{-\int_{\mathcal{M}} \sigma(t,x)\mathcal{A}(t,x)} e^{S[\mathcal{A}]}$$

$$= \exp\left\{\frac{1}{2}\int_{t_0}^{t_*} dt \int_{T_n}^{T_{n+1}} dxdx' \sigma(t,x)\mathcal{D}(x,x';t)\sigma(t,x')\right\} \quad (9.13)$$

Hence the drift velocity for the forward measure is given by

$$\alpha_F(t,x) = \sigma(t,x)\int_{T_n}^{x} dx'\mathcal{D}(x,x';t)\sigma(t,x'); \quad T_n \le x < T_n + \ell \quad (9.14)$$

The forward numeraire's drift $\alpha_F(t,x)$ is plotted in Figure 9.1(b). The value of $\sigma(t,x)$ is taken from the market [12, 27].

From its definition, the drift at Libor time T_n is zero; namely,

$$\alpha_F(t,T_n) = 0 \quad (9.15)$$

Eq. (9.15) shows that the forward numeraire has zero drift at Libor time T_n. This result has important consequences in the numerical evaluation of the American coupon bond option discussed in Chapter 16.

There is a discontinuity in the value of $\alpha_F(t, x)$ at forward time $x = T_n$; approaching the value $\alpha_L(t, x)$ from $x < T_n$, the discontinuity is given by

$$\Delta \alpha_F(t, x) \equiv \left[\lim_{x \to T_n-} \alpha_F(t, x) \right] - \alpha_F(t, T_n)$$

$$= \sigma(t, x) \int_{T_n - \ell}^{T_n} dx' \mathcal{D}(x, x'; t) \sigma(t, x') \qquad (9.16)$$

As discussed in Section 7.3, the normalization of the volatility function can always be chosen so that $\sigma(t, x) \mathcal{D}(x, x; t) \sigma(t, x) = \sigma^2(t, x)$. Since the time interval for the three-month tenor $\ell = 90$ days is quite small, one can approximate the drift by the following

$$\alpha_F(t, x) = \sigma(t, x) \int_{T_n}^{x} dx' \mathcal{D}(x, x'; t) \sigma(t, x')$$

$$\simeq (x - T_n) \sigma^2(t, x); \quad T_n \leq x < T_n + \ell \qquad (9.17)$$

The value of discontinuity at $x = T_n$, in this approximation, is given by $\ell \sigma^2(t, T_n)$.

One can see from the graph that, in a given three-month interval, the drift velocity is approximately linear in forward time and the maximum drift goes as $\ell \sigma^2(t, x)$, both of which are expected from Eq. (9.17).

9.6 Common Libor numeraire

For the purpose of modeling the Libor term structure, it is worth exploring if one can choose a numeraire such that *all* the Libor rates have a martingale evolution [63].

Consider a numeraire that is fixed by an (infinite) collection of zero coupon bonds, namely

$$B(t, T_0); B(t, T_1), \dots; B(t, T_n); B(t, T_{n+1}); B(t, T_{n+2}), \dots$$

Consider a portfolio, such as a coupon bond $\mathcal{B}(t, T)$ that matures at some time T. The numeraire is chosen in the following manner

$$T_n \leq T < T_{n+1}; \quad \frac{B(t, T)}{B(t, T_n)} : \text{martingale} \qquad (9.18)$$

In terms of the Libor forward interest rate $f(t, x)$, from Eq. (2.20) the three-month tenor Libor is given by

$$L(t, T_n) = \frac{1}{\ell}\left(e^{\int_{T_n}^{T_n+\ell} dx f(t,x)} - 1\right) \tag{9.19}$$

Re-write the Libor as follows

$$\begin{aligned}
L(t, T_n) &= \frac{1}{\ell}\left(e^{\int_{T_n}^{T_n+\ell} dx f(t,x)} - 1\right) \\
&= \frac{1}{\ell}\left[\frac{B(t, T_n) - B(t, T_n + \ell)}{B(t, T_n + \ell)}\right]
\end{aligned} \tag{9.20}$$

$L(t, T_n)B(t, T_n+\ell)$ is a traded portfolio. For the Libor $L(t, T_n)$, choose numeraire $B(t, T_n+\ell)$. Note that the common Libor measure has been defined in such a manner so that $L(t, T_n)B(t, T_n + \ell)$, for *each* T_n, is a martingale; that is, for $t > t_0$

$$\frac{L(t_0, T_n)B(t_0, T_n + \ell)}{B(t_0, T_n + \ell)} = E_L\left[\frac{L(t, T_n)B(t, T_n + \ell)}{B(t, T_n + \ell)}\right]$$
$$\Rightarrow L(t_0, T_n) = E_L[L(t, T_n)] \tag{9.21}$$

An equivalent way of thinking of Eq. (9.21) is to consider the coupon bond portfolio $\mathcal{B}_L(t, T_{n+1}) = [B(t, T_n) - B(t, T_n + \ell)]/\ell$; from the definition of the numeraire, as given in Eq. (9.18), the bond portfolio needs to be discounted by $B(t, T_n + \ell)$; hence

$$L(t, T_n) = \frac{\mathcal{B}_L(t, T_{n+1})}{B(t, T_{n+1})} = \frac{B(t, T_n) - B(t, T_n + \ell)}{\ell B(t, T_n + \ell)}: \text{ martingale}$$

The common Libor market numeraire makes every Libor $L(t, T_n)$ into a martingale. One can also interpret the Libor as being equal to the bond portfolio $\big(B(t, T_n) - B(t, T_n + \ell)\big)/l$, with the discounting factor being equal to $B(t, T_n + \ell)$.

9.6.1 *Bond forward interest rates and Libor*

A detailed discussion in Chapter 6 on the Libor forward interest rates $f_L(t, x)$ shows that, in particular, both its drift and volatility are stochastic. Libor forward interest rates are nonlinear and not amenable to analytical studies. For the purpose of understanding the implications of choosing a numeraire, a drastic simplification is made by ignoring the nonlinearities of the Libor forward interest rates and equating it to the bond forward interest rates. The results that are obtained are of only academic interest since the market Libor is not described by such a simplified model.

In terms of the bond forward interest rates, the Libors are given by the following

$$f_0 \equiv \int_{T_n}^{T_n+l} dx f(t_0, x); \quad f_* \equiv \int_{T_n}^{T_n+l} dx f(t_*, x) \tag{9.22}$$

$$\Rightarrow L(t_0, T_n) = \frac{1}{\ell}(e^{f_0} - 1); \quad L(t_*, T_n) = \frac{1}{\ell}(e^{f_*} - 1) \tag{9.23}$$

and hence from Eqs. (9.21) and (9.23) the martingale condition for Libor market measure can be written as

$$e^{f_0} = E_L[e^{f_*}] \tag{9.24}$$

Denote the drift for the market measure by $\alpha_L(t, x)$, and let $T_n \le x < T_n + \ell$; the evolution equation for the bond forward interest rates is given by Eq. (5.2), namely

$$f(t, x) = f(t_0, x) + \int_{t_0}^{t} dt' \alpha_L(t', x) + \int_{t_0}^{t} dt' \sigma(t', x) \mathcal{A}(t', x) \tag{9.25}$$

Hence

$$E_L[e^{f_*}] = e^{f_0 + \int_{\mathcal{M}} \alpha_L(t', x)} \frac{1}{Z} \int D\mathcal{A} e^{\int_{\mathcal{M}} \sigma(t', x)\mathcal{A}(t', x)} e^{S[\mathcal{A}]} \tag{9.26}$$

where the integration domain $\mathcal{M}$ is given in Figure 9.1(a).

From Eqs. (5.21), (9.24), and (9.26)

$$e^{-\int_{\mathcal{M}} \alpha_L(t, x)} = \int D\mathcal{A} e^{\int_{\mathcal{M}} \sigma(t, x)\mathcal{A}(t, x)} e^{S[\mathcal{A}]}$$

$$= \exp\left\{ \frac{1}{2} \int_{t_0}^{t_*} dt \int_{T_n}^{T_n+\ell} dx dx' \sigma(t, x) \mathcal{D}(x, x'; t) \sigma(t, x') \right\} \tag{9.27}$$

Hence the three-month tenor Libor drift velocity is given by

$$\alpha_L(t, x) = -\sigma(t, x) \int_{T_n}^{x} dx' \mathcal{D}(x, x'; t) \sigma(t, x'); \quad T_n \le x < T_n + \ell \tag{9.28}$$

The Libor drift velocity $\alpha_L(t, x)$ is *negative* for a martingale evolution. The negative drift is required for compensating the growing payments due to the compounding of interest.

The Libor drift velocity $\alpha_L(t, x)$ is the negative of the drift for the forward measure, that is

$$\alpha_L(t, x) = -\alpha_F(t, x)$$

9.7 Linear pricing a mid-curve caplet

To check the consistency of the three numeraires, the price of a mid-curve caplet is evaluated for the three cases. It was shown in Section 8.2 that the Libor Market Model, which is defined by a nonlinear model of the Libor forward interest rates, results in Black's caplet formula. The caplet price can also be evaluated using the bond forward interest rates model and is called the *linear* caplet price to distinguish it from Black's formula. The linear caplet model is a Gaussian model for caplet pricing. It is hence different from Black's formula and is shown in Chapter 10 to be empirically inaccurate as well.

In spite of its limitations, the linear caplet price is a suitable instrument for studying the consistency of the three numeraires. Since the linear caplet price is based on a Gaussian model, the computations for all three numeraires can be performed exactly. The pricing formula for an interest rate caplet is derived for a general volatility function $\sigma(t, x)$ and propagator $\mathcal{D}(x, x'; t)$ of the bond forward interest rates.

Interest rate caplets, floorlets, caps, and floors have been discussed in Section 4.3. A mid-curve caplet can be exercised at any fixed time t_*, that is before the time T_n at which the caplet caps the interest rate. Recall from Section 4.3 that $caplet(t_0, t_*, T_n)$ denotes the price – at time t_0 – of an interest rate European option contract that must be exercised at time $t_* > t_0$ for an interest rate caplet that puts an upper limit on the interest from time T_n to $T_n + \ell$. Let the principal amount be equal to ℓV, and the caplet rate be K. The caplet is exercised at time t_*, with the payment made in arrears at time $T_n + \ell$. Note that, although the payment is made at time $T_n + \ell$, the *amount* that will be paid is fixed by $L(t_*, T_n)$ at time t_*.

The payoff function of an interest rate caplet is the value of the caplet when it matures at calendar t_* and is given, from Eq. (4.10), by

$$caplet(t_*, t_*, T_n) = \ell V B(t_*, T_n + \ell) \big[L(t_*, T_n) - K \big]_+ \tag{9.29}$$

$$= \ell V \left[\frac{B(t_*, T_n) - B(t_*, T_n + \ell)}{\ell} - K B(t_*, T_n + \ell) \right]_+$$

$$= \tilde{V} B(t_*, T_n + \ell) \big(X e^{f_*} - 1 \big)_+ \tag{9.30}$$

The various time intervals that define the interest rate caplet are shown in Figure 4.7. Recall from Eq. (9.22)

$$f_* \equiv \int_T^{T_n + \ell} dx f(t_*, x) \quad \text{and} \quad X = \frac{1}{1 + \ell K}; \quad \tilde{V} = (1 + \ell K) V$$

The payoff for an interest rate floorlet is similarly given by

$$floorlet(t_*, t_*, T_n) = \ell V B(t_*, T_n + \ell)\big[K - L(t_*, T_n)\big]_+$$

$$= \tilde{V} B(t_*, T_n + \ell)\big(1 - X e^{f_*}\big)_+ \qquad (9.31)$$

As shown in Section 4.3, the price of the caplet automatically determines the price of a floorlet due to put–call parity, and hence the price of the floorlet does not need an independent derivation.

9.8 Forward numeraire and caplet price

The forward numeraire is given by the zero coupon bond $B(t, T_n)$. Hence the caplet is a martingale when discounted by $B(t, T_n)$; the price of the caplet at time $t_0 < t_*$ is, consequently, given by

$$\frac{caplet(t_0, t_*, T_n)}{B(t_0, T_n)} = E_F\left[\frac{caplet(t_*, t_*, T_n)}{B(t_*, T_n)}\right]$$

$$= \tilde{V} E_F\big(X - e^{-f_*}\big)_+$$

The price of a caplet is given by

$$caplet(t_0, t_*, T_n) = \tilde{V} B(t_0, T_n) E_F\big(X - e^{-f_*}\big)_+ \qquad (9.32)$$

and yields the payoff function Eq. (9.30) for $t_0 = t_*$.

The payoff function for the caplet given in Eq. (9.32) yields the following price of the caplet [12, 65]

$$caplet(t_0, t_*, T_n) = \tilde{V} B(t_0, T_n) \int_{-\infty}^{+\infty} dG \Psi_F(G)(X - e^{-G})_+ \qquad (9.33)$$

with the pricing kernel $\Psi_F(G) = \Psi_F(G, t_0, t_*, T_n)$ given by

$$\Psi_F(G) = \sqrt{\frac{1}{2\pi q^2}} \exp\left\{-\frac{1}{2q^2}\left(G - \int_{T_n}^{T_n+\ell} dx f(t_0, x) - \frac{q^2}{2}\right)^2\right\} \qquad (9.34)$$

$$q^2 = q^2(t_0, t_*, T_n)$$

$$= \int_{t_0}^{t_*} dt \int_{T_n}^{T_n+\ell} dx dx' \sigma(t, x) \mathcal{D}(x, x'; t) \sigma(t, x') \qquad (9.35)$$

The price of the caplet is given by the following Black–Scholes type formula

$$caplet(t_0, t_*, T_n) = \tilde{V} B(t_0, T) \big[X N(-d_-) - F N(-d_+)\big] \qquad (9.36)$$

where $N(d_\pm)$ is the cumulative distribution for the normal random variable and

$$F = e^{-\int_{T_n}^{T_n+\ell} dx f(t_0,x)} = e^{-f_0}$$

$$d_\pm = \frac{1}{q}\left[\ln\left(\frac{F}{X}\right) \pm \frac{q^2}{2}\right] \tag{9.37}$$

9.9 Common Libor measure and caplet price

The Libor market measure has as its numeraire the zero coupon bond $B(t_*, T_n + \ell)$; the caplet is a martingale when discounted by this numeraire, and hence the price of the caplet at time $t_0 < t_*$ is given by

$$\frac{caplet(t_0, t_*, T_n)}{B(t_0, T_n + \ell)} = E_L\left[\frac{caplet(t_*, t_*, T_n)}{B(t_*, T_n + \ell)}\right]$$

$$= \tilde{V} E_L(Xe^{f_*} - 1)_+$$

$$\Rightarrow caplet(t_0, t_*, T_n) = \tilde{V} B(t_0, T_n + \ell) E_L(Xe^{f_*} - 1)_+ \tag{9.38}$$

The price of the caplet is given by

$$caplet(t_0, t_*, T_n) = \tilde{V} B(t_0, T_n + \ell)\int_{-\infty}^{+\infty} dG\Psi_L(G)(Xe^G - 1)_+ \tag{9.39}$$

where $\Psi_L(G) = \Psi_L(G, t_0, t_*, T_n)$, the pricing kernel, is given by

$$\Psi_L(G) = \sqrt{\frac{1}{2\pi q^2}} \exp\left\{-\frac{1}{2q^2}\left(G - \int_{T_n}^{T_n+\ell} dx f(t_0, x) + \frac{q^2}{2}\right)^2\right\} \tag{9.40}$$

Note $\Psi_L(G)$ differs from the pricing kernel $\Psi_F(G)$ given in Eq. (9.34) by the sign of the q^2 in the exponent.

The price of the caplet obtained from the forward measure is equal to the one obtained using the three-month tenor Libor market measure, since, from Eqs. (9.34) and (9.40), one can prove the following remarkable result

$$B(t, T_n)\Psi_F(G)(X - e^{-G})_+ = B(t, T_n + \ell)\Psi_L(G)(Xe^G - 1)_+ \tag{9.41}$$

The identity above shows how the three factors required in the pricing of an interest rate caplet, namely the numeraires, the pricing kernel, and the payoff functions, all 'conspire' to yield numeraire invariance for the price of the interest rate option.

The payoff function is correctly given by the price of the caplet, since in the limit of $t_0 \to t_*$, Eq. (9.35) yields

$$\lim_{t_0 \to t_*} q^2 = (t_* - t_0)\int_{T_n}^{T_n+\ell} dxdx'\sigma(t, x)\mathcal{D}(x, x'; t)\sigma(t, x')$$

$$= \epsilon C \tag{9.42}$$

where C is a constant, and $\epsilon = t_* - t_0$. Hence, from Eqs. (9.39) and (9.40)

$$
\lim_{t_0 \to t_*} caplet(t_0, t_*, T_n) = \tilde{V} B(t_*, T_n + \ell) \int_{-\infty}^{+\infty} dG \delta(G - f_*)(Xe^G - 1)_+
$$

$$
= \tilde{V} B(t_*, T_n + \ell)(Xe^{f_*} - 1)_+
$$

verifying the payoff function is the one given in Eq. (9.30).

9.10 Money market numeraire and caplet price

The money market numeraire is given by the spot interest rate $M(t_0, t_*) = \exp\{\int_{t_0}^{t_*} dt\, r(t)\}$. Expressed in terms of the money market numeraire, the price of the caplet is given by

$$
\frac{caplet(t_0, t_*, T_n)}{M(t_0, t_0)} = E_M \left[\frac{caplet(t_*, t_*, T_n)}{M(t_0, t_*)} \right]
$$

$$
\Rightarrow caplet(t_0, t_*, T_n) = E_M \left[e^{-\int_{t_0}^{t_*} dt\, r(t)} caplet(t_*, t_*, T_n) \right]
$$

To simplify the calculation, consider the change of numeraire from $M(t_0, t_*) = \exp\left\{\int_{t_0}^{t_*} dt'\, r(t')\right\}$ to discounting by the zero coupon bond $B(t_0, t_*)$; it then follows, from Eq. (9.7), that

$$
e^{-\int_{t_0}^{t_*} dt\, r(t)} e^S = B(t_0, t_*) e^{S_*}
$$

where the drift for the action S_* is given by

$$
\alpha_*(t, x) = \sigma(t, x) \int_{t_*}^{x} dx' \mathcal{D}(x, x'; t) \sigma(t, x') \tag{9.43}
$$

In terms of the money market measure, the price of the caplet is given by

$$
caplet(t_0, t_*, T_n) = E_M \left[e^{-\int_{t_0}^{t_*} dt\, r(t)} caplet(t_*, t_*, T_n) \right] \tag{9.44}
$$

$$
= B(t_0, t_*) E_M^* \left[caplet(t_*, t_*, T_n) \right]
$$

$$
= \tilde{V} B(t_0, t_*) E_M^* \left[B(t_*, T_n + \ell)(Xe^{f_*} - 1)_+ \right] \tag{9.45}
$$

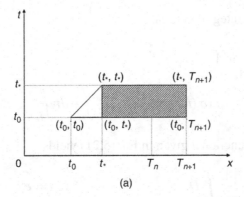

Figure 9.2 Domain of integration $\mathcal{R}$ for evaluating the price of a caplet using the money market numeraire.

From the expression for the bond forward interest rates given in Eqs. (9.25) and (9.45), the price of the caplet can be written out as follows

$$caplet(t_0, t_*, T_n) = \tilde{V}B(t_0, T_n + \ell)e^{-\int_{\mathcal{R}}\alpha_*}\frac{1}{Z}\int D\mathcal{A}e^{-\int_{\mathcal{R}}\sigma\mathcal{A}}e^{S_*}\left(Xe^{f_*} - 1\right)_+$$

(9.46)

where the integration domain $\mathcal{R}$ is given in Figure 9.2.

The payoff can be re-expressed using the Dirac-delta function, given in Eq. (A.7), as follows

$$\left(Xe^{f_*} - 1\right)_+ = \int dG\delta(G - f_*)\left(Xe^G - 1\right)_+$$

$$= \int dG \int \frac{d\xi}{2\pi}e^{i\xi(G - f_*)}\left(Xe^G - 1\right)_+ \qquad (9.47)$$

From Eq. (9.22), and the domain of integration $\mathcal{M}$ given in Figure 9.1(a), one obtains

$$f_* \equiv \int_{T_n}^{T_n+\ell} dx f(t_*, x)$$

$$= \int_{T_n}^{T_n+\ell} dx f(t_0, x) + \int_{\mathcal{M}} \alpha_* + \int_{\mathcal{M}} \sigma\mathcal{A}$$

Hence, from Eqs. (9.46) and (9.47) the price of the caplet, for $f_0 = \int_{T_n}^{T_n+\ell} dx$ $f(t_0, x)$, is given by

$$caplet(t_0, t_*, T_n) = \tilde{V}B(t_0, T_n + \ell)e^{-\int_{\mathcal{R}}\alpha_*}$$

(9.48)

$$\times \int dG \int \frac{d\xi}{2\pi}e^{i\xi(G - f_0 - \int_{\mathcal{M}}\alpha_*)}\left(Xe^G - 1\right)_+\frac{1}{Z}\int D\mathcal{A}e^{-\int_{\mathcal{R}}\sigma\mathcal{A}}e^{-i\xi\int_{\mathcal{M}}\sigma\mathcal{A}}e^{S_*}$$

To perform the path integral note that

$$\int_{\mathcal{R}} \sigma \mathcal{A} + i\xi \int_{\mathcal{M}} \sigma \mathcal{A}$$

$$= \int_{t_0}^{t_*} dt \int_{t_*}^{T_n+\ell} dx \sigma(t,x) \mathcal{A}(t,x) + i\xi \int_{t_0}^{t_*} dt \int_{T_n}^{T_n+\ell} dx \sigma(t,x) \mathcal{A}(t,x)$$

and the generating functional given in Eq. (5.21) yields

$$\frac{1}{Z} \int D\mathcal{A} e^{-\int_{\mathcal{R}} \sigma \mathcal{A} - i\xi \int_{\mathcal{R}_L} \sigma \mathcal{A}} e^{S_*} = e^{\Gamma}$$

where

$$\Gamma = \frac{1}{2} \int_{t_0}^{t_*} dt \int_{t_*}^{T_n+\ell} dx dx' \sigma(t,x) \mathcal{D}(x,x';t) \sigma(t,x')$$

$$- \frac{\xi^2}{2} \int_{t_0}^{t_*} dt \int_{T_n}^{T_n+\ell} dx dx' \sigma(t,x) \mathcal{D}(x,x';t) \sigma(t,x')$$

$$+ i\xi \int_{t_0}^{t_*} dt \int_{t_*}^{T_n+\ell} dx \int_{T_n}^{T_n+\ell} dx' \sigma(t,x) \mathcal{D}(x,x';t) \sigma(t,x')$$

The expression for Γ above, using the definitions of q^2, α_* given in Eqs. (9.35) and (9.43) respectively, can be shown to yield the following

$$\Gamma = \int_{\mathcal{R}} \alpha_* - \frac{\xi^2}{2} q^2 + i\xi \left(\int_{\mathcal{M}} \alpha_* + \frac{1}{2} q^2 \right) \qquad (9.49)$$

Simplifying Eq. (9.48) using Eq. (9.49) yields the price of the caplet as given by

$$caplet(t_0, t_*, T_n) = \tilde{V} B(t_0, T_n + \ell) \int_{-\infty}^{+\infty} dG \Psi_L(G)(Xe^G - 1)_+ \quad (9.50)$$

Hence we see that the money market numeraire yields the same price for the caplet as the one obtained for the Libor market measure, but with a different and longer derivation.

9.11 Numeraire invariance: numerical example

To illustrate the differences in the choice of numeraire, a typical example is fully worked out. Since the final expression for the Libor market and money market numeraire are equal, only the Libor market measure and forward numeraire are considered.

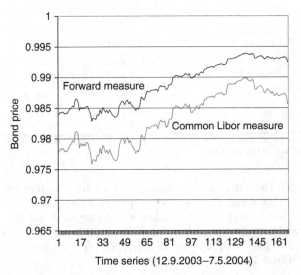

Figure 9.3 The discounting zero coupon bonds $B(t, T_n)$, $B(t, T_n + \ell)$ for the common Libor and forward numeraires, for $T_n = 2004.12.13$ and $t \in [2003.9.12$–$2004.5.7]$.

The integrand required for evaluating the price of a caplet for the common Libor and forward numeraire are given, from Eq. (9.41) by

Forward numeraire: $B(t, T_n)\Psi_F(G)(X - e^{-G})_+$

Common Libor numeraire: $B(t, T_n + \ell)\Psi_L(G)(Xe^G - 1)_+$

$$X = \frac{1}{1 + \ell K}; \quad \ell = 3 \text{ months} \tag{9.51}$$

Consider a caplet that matures at a fixed date, say $T_n = 2004.12.13$ (13 December 2004), with strike price of $K = 0.02\%$ and let $G = 0.01$ in the above formulas. The factors that go into the integrand are evaluated for a range of time interval $t \in [2003.9.12$–$2004.5.7]$, and given in Figures 9.3 and 9.4. Data on US treasury Bonds are taken from the market. The payoff functions $(X - e^{-G})_+$, $(Xe^G - 1)_+$ differ only by a constant scale factor and hence are not plotted. The result for the difference of the integrands – given in Figure 9.4(b) – verifies, within the errors of the computation, the numeraire invariance of caplet pricing.

9.12 Put–call parity for numeraires

In this section a derivation is given for put–call parity for interest rate caps and floors. Note the derivation for put–call parity for a caplet for the money market

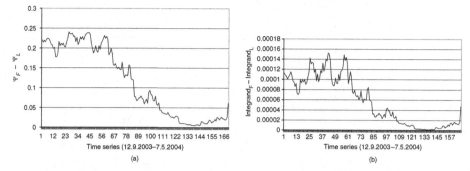

Figure 9.4 (a) The difference in pricing kernel for the forward and three-month tenor Libor numeraires $\Psi_F(G) - \Psi_L(G)$ with $G = 0.01$ and for $t \in [2003.9.12–2004.5.7]$. (b) The difference of the integrands for the three-month tenor Libor and forward numeraires $B(t, T_n)\Psi_F(G)(X - e^{-G})_+ - B(t, T_n + \ell)$ $\Psi_L(G)(Xe^G - 1)_+$ with $T_n = 2004.12.13$, $G = 0.01$ and for $t \in [2003.9.12–2004.5.7]$.

numeraire has been derived in Section 4.4 and given in Eq. (4.16). A derivation is given for the forward and Libor market numeraires and illustrates how their properties are essential for the price of the caplet and floorlet to satisfy put–call parity.

The following simple identity, from Eq. (4.15)

$$(a - b)_+ - (b - a)_+ = (a - b)\Theta(a - b) - (b - a)\Theta(b - a)$$
$$= a - b$$

is required for deriving put–call parity.

9.12.1 Put–call parity for Libor tenor forward numeraire

The price of a caplet and floorlet at time t_0 is given by discounting the payoff functions with the discounting factor of $B(t_0, T_n)$. From Eq. (9.33)

$$caplet(t_0, t_*, T_n) = B(t_0, T_n)E_F\left[\frac{caplet(t_*, t_*, T_n)}{B(t_*, T_n)}\right]$$
$$= \tilde{V}B(t_0, T_n)E_F\left(X - e^{-f_*}\right)_+$$

and the floorlet is given by

$$floorlet(t_0, t_*, T_n) = \tilde{V}B(t_0, T_n)E_F\left(e^{-f_*} - X\right)_+ \qquad (9.52)$$

Consider the expression

$$caplet(t_0, t_*, T_n) - floorlet(t_0, t_*, T_n) \tag{9.53}$$

$$= \tilde{V} B(t_0, T_n)\left[E_F\left(X - e^{-f_*}\right)_+ - E_F\left(e^{-f_*} - X\right)_+\right]$$

$$= \tilde{V} B(t_0, T_n) E_F\left(X - e^{-f_*}\right) \tag{9.54}$$

where Eq. (4.15) has been used to obtain Eq. (9.54).

For the forward measure, from Eq. (9.12)

$$E_F\left[e^{-f_*}\right] = e^{-f_0} \tag{9.55}$$

Hence, since for constant X we have $E_F(X) = X E_F(1) = X$, from the above equation and Eq. (9.54), the price of a caplet and floorlet obeys the put–call relation

$$caplet(t_0, t_*, T_n) - floorlet(t_0, t_*, T_n) = \tilde{V} B(t_0, T_n) E_F\left(X - e^{-f_*}\right)$$

$$= \tilde{V} B(t_0, T_n)(X - e^{-f_0})$$

$$= \ell V B(t_0, T_n + \ell)(L(t_0, T_n) - K) \tag{9.56}$$

and yields Eq. (4.16) as expected.

9.12.2 Put–call for common Libor numeraire

The price of a caplet for the Libor market measure is given from Eq. (9.38) by

$$caplet(t_0, t_*, T_n) = \tilde{V} B(t_0, T_n + \ell) E_L\left(X e^{f_*} - 1\right)_+ \tag{9.57}$$

and the floorlet is given by

$$floorlet(t_0, t_*, T_n) = \tilde{V} B(t_0, T_n + \ell) E_L\left(1 - X e^{f_*}\right)_+ \tag{9.58}$$

Hence, similar to the derivation given in Eq. (9.54), we have

$$caplet(t_0, t_*, T_n) - floorlet(t_0, t_*, T_n) = \tilde{V} B(t_0, T_n + \ell) E_L\left(X e^{f_*} - 1\right) \tag{9.59}$$

For the three-month tenor Libor market measure, from Eq. (9.24)

$$E_L[e^{f_*}] = e^{f_0}$$

Hence the equation above, together with Eq. (9.59), yields the expected put–call parity given in Eq. (4.16), namely that

$$caplet(t_0, t_*, T_n) - floorlet(t_0, t_*, T_n) = \tilde{V}B(t_0, T_n + \ell)(Xe^{f_0} - 1)$$

$$= \ell V B(t_0, T_n + \ell)(L(t_0, T_n) - K)$$

9.13 Summary

A common numeraire was derived for the forward interest rates, and it was shown that a numeraire – consisting of a *collection* of zero coupon bonds – renders *all* forward zero coupon bonds $F(t, T_n, T_{n+1})$ into martingales. The drifts for the forward numeraire and the Libor market measure are rather unusual, having discontinuities at future Libor times. At exactly Libor time, the drift is zero; this fact has major implications in the evaluation of the American coupon bond option in Chapter 16.

Two other numeraires were studied for the forward interest rates, each having its own drift velocity. All the numeraires have their own specific advantages. It was demonstrated by actual computation that all three yield the same price for an interest rate caplet. They also satisfy put–call parity, as is necessary for the prices of interest caps and floors to be free from arbitrage opportunities.

The interest rate caplet payoff function, from Eq. (4.10), is given by the following

$$caplet(t_*, t_*, T_n) = \ell V B(t_*, T_n + \ell)\big[L(t_*, T_n) - K\big]_+$$

To verify the consistency of the three numeraires, the price of the mid-curve caplet was computed using the different numeraires. The caplet price was indeed found to be numeraire invariant due to a remarkable combination of discounting factor, payoff function, and pricing kernel. The caplet prices obey the expected put–call parity, further confirming that the numeraires are consistent.

10

Empirical analysis of interest rate caps

The industry standard for pricing an interest rate caplet is Black's formula, which was derived in Section 8.2 from the Libor Market Model. The underlying Libor forward interest rates $f_L(t, x)$ are known to be nonlinear, as discussed in Section 6.11.1. A different price of the caplet, namely the linear pricing formula, was derived in Section 9.7 using the bond forward interest rates.

An empirical study is carried out of the linear caplet pricing formula [19, 40]. The main purpose is to ascertain how important are the differences in the Libor Market Model and bond forward interest rates – using the pricing of caps and caplets as an example [71]. In particular, the linear caplet price is compared with the market price of caps and caplets to obtain an estimate of the importance of the nonlinear effects that are the hallmark of the Libor Market Model.

Historical volatility and correlation of forward interest rates are used for predicting the linear caplet price; another approach is to predict the linear price from a parametric formula of the effective volatility using market caplet prices. The study shows that bond forward interest rates generate prices of a caplet and cap with fairly large errors, greater than 17%.

10.1 Introduction

The price of a mid-curve caplet has been obtained in Eq. (9.36) based on the bond forward interest rates' model. The result is called the linear caplet price to distinguish it from Black's caplet price, which is an exact result of the Libor Market Model. The linear caplet price is given by the following Black–Scholes type formula. Re-writing Eq. (9.36) for later analysis yields the following[1,2]

---

[1] Note, one recovers the normal caplet by setting $t_* = T_n$.
[2] All the zero coupon bond prices are fixed from the market and are the same for bond forward interest rates $f(t, x)$ or for Libor forward interest rates $f_L(t, x)$.

$$caplet(t_0, t_*, T_n) = V B(t_0, T_{n+1})$$

$$\times \left[(1 + \ell L(t_0, T_n))N(-d_-) - (1 + \ell K)N(-d_+) \right] \qquad (10.1)$$

where $N(d_\pm)$ is the cumulative distribution for the normal random variable with the following definitions

$$d_\pm = \frac{1}{q} \left[\ln \left(\frac{1 + \ell L(t_0, T_n)}{1 + \ell K} \right) \pm \frac{q^2}{2} \right] \qquad (10.2)$$

$$q^2 = q^2(t_0, t_*, T_n)$$

$$= \int_{t_0}^{t_*} dt \int_{T_n}^{T_n + \ell} dx dx' \sigma(t, x) \mathcal{D}(x, x'; t) \sigma(t, x') \qquad (10.3)$$

The domain of integration for evaluating q^2 is given in Figure 4.6. Note that q is the effective volatility for the caplet linear pricing formula and that the propagator for forward interest rates is required for pricing the caplet. The pricing formulas for caplets and floorlets are fixed by the volatility function $\sigma(t, x)$, the correlation parameters μ, λ, η contained in the Lagrangian for the forward interest rates, as well as the initial interest rates term structure.

The Libor Market Model is based on nonlinear Libor forward interest rates and yields, as in Eq. (8.12), Black's caplet formula given by [59, 61]

$$caplet_B(t_0, t_*, T_n) = \ell V B(t_0, T_{n+1}) \left[L(t_0, T_n)N(\tilde{d}_+) - KN(\tilde{d}_-) \right] \qquad (10.4)$$

$$\tilde{d}_\pm = \frac{1}{q_\gamma^2} \left[\ln \left(\frac{L(t_0, T_n)}{K} \right) \pm \frac{q_\gamma^2}{2} \right]$$

Black's volatility σ_B, from Eq. (8.13), is given by

$$\sigma_B^2 = \frac{q_\gamma^2}{t_* - t_0} = \frac{1}{t_* - t_0} \int_{t_0}^{t_*} dt \int_{T_n}^{T_{n+1}} dx \int_{T_n}^{T_{n+1}} dx' \gamma(t, x) \mathcal{D}_L(x, x'; t) \gamma(t, x')$$

The two caplet prices given in Eqs. (10.1) and (10.4) are very different, reflecting the differences in the bond and Libor forward interest rates. The predictions of the linear only caplet price will be tested by comparing it with the market prices, leaving a similar study of Black's formula for the future.

Black's formula, as it is currently used in the financial markets, has no predictive power but instead is simply a convenient way of representing the price of a caplet. The main utility of Black's formula is that implied σ_B is more stable than the price itself and, similar to yield-to-maturity for coupon bonds, can be used for comparing caplets with different maturities, payoffs, and principal amounts.

The linear caplet price and Black's caplet formula are studied empirically using Libor market data. In particular, the *effective volatility q* determining the linear caplet price is computed by a three-dimensional integration on the covariance of the changes in the bond forward interest rates.

The following three different approaches are discussed for fixing the effective volatility q for pricing caplets.

- The volatility function σ_H and parameters of the bond propagator, μ, λ, and η, are all fitted from historical Libor data.
- The market covariance is computed directly from Libor market data.
- A parametric formula for the effective volatility q, and consequently for the implied volatility σ_I for the linear caplet pricing model, is determined from historical caplet prices.[3] The value of σ_I is quite distinct from σ_B since σ_I is a function of future time and can be used for extrapolating the future. In contrast, σ_B is a value that has to be computed every day from caplet prices.

10.2 Linear and Black caplet prices

The pricing formula, *at the money* for a caplet maturing at $t_* = T_n$, is given by

$$1 + \ell K = 1 + \ell L(t_0, T_n); \quad \Rightarrow K = L(t_0, T_n)$$

The linear price of the caplet, at the money, is given by

$$caplet_L(t_0, T_n) = V[1 + \ell L(t_0, T_n)]B(t_0, T_{n+1})[N(d_+) - N(d_-)]$$

Note

$$d_\pm = \pm \frac{q}{2} \tag{10.5}$$

This formula is compared with Black's formula. From Eq. (10.4), Black's caplet formula – at the money – has $K = L(t_0, T_n)$ and yields the price

$$caplet_B(t_0, t_*, T_n) = \ell V L(t_0, T_n)B(t_0, T_{n+1})[N(\tilde{d}_+) - N(\tilde{d}_-)]$$

$$\tilde{d}_\pm = \pm \frac{q_\gamma}{2} = \pm \frac{\sigma_B\sqrt{t_* - t_0}}{2}$$

At the money, since the pre-factors of two pricing formulas are different, the effective volatility q in the linear pricing formula and σ_B in Black's formula are not equal. Black's formula is multiplied by $(1 + \ell L(t_0, T_n))/\ell L(t_0, T_n)$ so that – at

---

[3] Note, in contrast, σ_H is obtained from the historical Libor data.

the money – the two pricing formulas are taken to be exactly equal and this allows q to be equated to $\sigma_B\sqrt{t_* - t_0}$. Hence

$$caplet_B(t_0, t_*, T_n)\Big|_{\text{At the money}} = \frac{\ell L(t_0, T_n)}{1 + \ell L(t_0, T_n)}caplet_L(t_0, t_*, T_n)\Big|_{\text{At the money}}$$

$$\Rightarrow q = \sigma_B\sqrt{t_* - t_0}$$

The linear caplet price is equal to Black's formula multiplied by the factor $\ell L(t_0, T_n)/(1+\ell L(t_0, T_n))$. This is the same factor that relates the volatility $v(t, x)$ of the Libor forward interest rates, which is taken to be a deterministic quantity in the bond forward interest rates, to the deterministic log Libor volatility $\gamma(t, x)$, as in Eq. (7.33) and given below

$$v(t, x) = \frac{\ell L(t, T_n)}{1 + \ell L(t, T_n)}\gamma(t, x)$$

Replacing the stochastic quantity $\ell L(t_0, T_n)/(1 + \ell L(t_0, T_n))$ by the deterministic factor $\ell L(t, T_n)/(1 + \ell L(t, T_n))$, as discussed in Eq. (7.32), results in errors of about 10% for future time of around 1.5 years. Since this pre-factor is not the only source of error in the linear caplet price, we expect errors larger than 10% in fitting the linear price with the market price for caplets; this expectation is borne out by the empirical analysis.

In Figure 10.1(a), the caplet rate K is varied to compare the normalized pricing formula away from the money. The linear caplet price is shown more clearly in Figure 10.1(b). It can be seen that it is only at the money – which for the example is at cap rate $K = 0.02$ – that the two pricing formulas give the same result. The

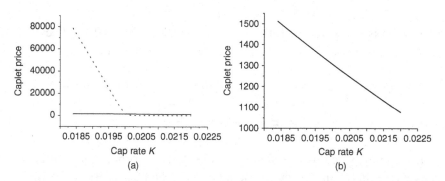

Figure 10.1 (a) Caplet price from normalized Black's formula (dashed line) and linear caplet formula (unbroken line) versus cap rate K. Libor is given at 0.02 and the caplet is at the money for $K = 0.02$. (b) Caplet price from linear formula versus cap rate K. Libor is given at 0.02. The caplet is at the money when $K = 0.02$.

caplet pricing for the two formulas in general is not equal and deviates quite rapidly, specially when K is deep in the money.

Working with q has many advantages over σ_B. Most importantly, the effective volatility q is obtained from the underlying bond forward interest rates, as computed in Eq. (10.3), which are common to all caplets.

10.2.1 Numerical example of Black's caplet price

Black's formula for pricing a caplet is illustrated by working out an example. Consider a contract that caps the interest rate at 2%, on a $1 million loan for three months with Libor as the floating rate. The bond price $B(t_0, t_*)$ is 0.984. The contract is written on t_0=13 September 2003 and matures on $t_* = T_n = $ 12 December 2004, with a cap rate K given by 2% per annum. The Libor $L(t_0, t_*)$ at 13 September 2003 is given by 2.95% per annum for a $\ell = 90$ days Eurodollar deposit – from 12 December 2004 to 12 December 2005. Hence

$$L(t_0, T_n) = 0.0295$$

Black's volatility σ_B is given by $0.5168/\sqrt{\text{year}}$. Hence

$$d_+^B = \frac{1}{0.5168\sqrt{1.25}} \left[\ln \frac{0.0295}{0.02} + \frac{0.5168^2 \times 1.25}{2} \right] = 0.527$$
$$d_-^B = d_+^B - 0.5168\sqrt{1.25} = -0.0508$$

Thus

$$caplet_B(t_0, T_n, 0.02) = \ell V B(t_0, T_{n+1}) \left[L(t_0, T_n) N(d_+) - K N(d_-) \right]$$
$$= \$\frac{1000000 \times 0.25}{1 + 0.25 \times 0.0295} \times 0.984[0.0295 N(0.527) - 0.02 N(-0.0508)]$$
$$= \$1587.655$$

10.3 Linear caplet price: parameters

An empirical study of three different approaches for implementing the linear caplet pricing is carried out and the results are discussed.

The pricing formula for the daily prices of the caplets requires the daily initial interest rates term structure as input as well as the volatility function and parameters μ, λ, and η for the propagator. Daily fit of the volatility function and propagator parameters can be derived by a daily moving average Libor rate history. For the sake of simplicity, from [12, 28] the volatility function is taken to depend only on remaining future time, namely

$$\sigma(t,x) = \sigma(x - t)$$

Although this assumption cannot be indefinitely extended, it is valid for up to three years [28], which is enough for the empirical study. Thus, the parametric fit is done only once, and these parameters are used for the whole data set projected to 1.5 years in the future. It should be noted that one can always do the parametric fit more frequently to get more accurate results.

Since the forward interest rates are defined on a domain $x \geq t$, the propagator satisfies

$$\mathcal{D}(x, x'; t) = \mathcal{D}(x - t, x' - t) \tag{10.6}$$

A parametric curve is fitted for the effective volatility [27] using historical data from the prices of Libor before 4 May 2003. More precisely, the forward interest rates that are used for fixing the input volatility and propagator are the daily rates for the Eurodollar futures from 4 May 1998 to 29 April 2003; the length of the data set is 1256 trading days for daily prices of Libor seven years into the future. For $\ell = 90$ days, the following approximation is made

$$L(t, T) = \frac{e^{\int_T^{T+\ell} f(t,x)dx} - 1}{\ell} \tag{10.7}$$

$$\simeq f(t, T) \tag{10.8}$$

Libor is treated as being approximately equal to the bond forward interest rates, and a moving average over the last 63 days – from 29 January 2003 to 29 April 2003 – is taken for evaluating the statistical average. Note that averaging over 63 trading days carries the most relevant information.

The connected correlator, from Eq. (7.10), is given by

$$E[(\delta f(t,\theta))^2]_c \equiv E[(\delta f(t,\theta))^2] - E[\delta f(t,\theta)]E[\delta f(t,\theta)] = \epsilon \sigma_H^2(\theta)\mathcal{D}(\theta,\theta)$$

with $\epsilon = 1/260$, since there are 260 trading days in one year. To be able to compare the volatilities of different Gaussian models, the field $\mathcal{A}(t,\theta)$ is re-scaled, as in Section 7.3, so that $\mathcal{D}(\theta,\theta) = 1/\epsilon$. The re-scaled frame yields the usual definition of volatility of the forward interest rates, given as follows

$$\langle (\delta f(t,\theta))^2 \rangle_c = \sigma_H^2(t,\theta) \tag{10.9}$$

Note ϵ has been canceled by the scale chosen for the correlator.

Table 10.1 *The best fit for parameters* λ, μ, b, *and* η *of a Libor caplet obtained by minimizing the overall root mean square of the fit.*

Parameters from Libor caplet data				
$\tilde{\lambda}$	$\tilde{\mu}$	b	η	rms error
16.578657/year	8.0761/year	1.376644	0.044127	1.09%

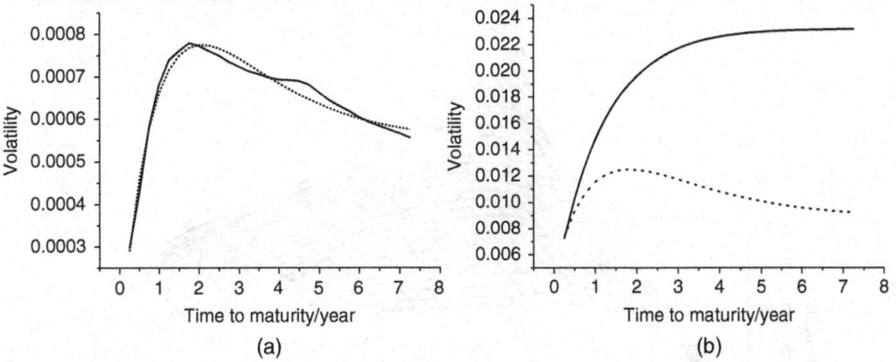

(a)

(b)

Figure 10.2 Note the historical market volatility's graph on the left has a scale two orders of magnitude smaller than the one for implied volatility on the right. (a) Volatility of Libor forward rates $\sigma\,(\text{year}^{-1/2})$ versus remaining time to maturity, both from data (unbroken line) and from formula with fitted parameters (dotted line). The data are from 29 January 2003 to 29 April 2003. The normalized root mean square error is 2.76%. (b) Historical volatility σ_H $(\text{year}^{-1/2})$ (dashed line) and implied volatility σ_I $(\text{year}^{-1/2})$ (unbroken line) fitted from caplet data, for 12 September 2003–4 February 2004, versus time to maturity.

Following Bouchaud and Matacz [27], a parametric formula for volatility is assumed as follows[4]

$$\sigma_H(\theta) = 0.00055 - 0.00026 \exp(-0.71826(\theta - \theta_{\min}))$$
$$+ 0.0006(\theta - \theta_{\min}) \exp(-0.71826(\theta - \theta_{\min})) \quad (10.10)$$

where $\theta_{\min} = 3$ months. The market fit for the volatility of Libor, following the analysis developed in Section 7.3, is given in Figure 10.2(a); The parameters are fixed using the data set used for obtaining the parameters given in Table 10.1. Historical and implied σ_I volatility are shown in Figure 10.2(b).

The empirical values of the three parameters μ, λ, η for the stiff Lagrangian are obtained by fitting the propagator to the forward interest rates taken from Libor

---

[4] σ_H denotes the volatility obtained from historical Libor rates.

Table 10.2 *The best fit for parameters* λ, μ, *and b of a Libor caplet obtained by minimizing the overall root mean square of the fit. The value of* η = 0.34 *is fixed from earlier data on the forward interest rates' correlator.*

Parameters for Libor caplet with fixed $\eta = 0.34$			
$\tilde{\lambda}$	$\tilde{\mu}$	b	rms error for the entire fit
1.354/year	0.847/year	0.727	2.83%

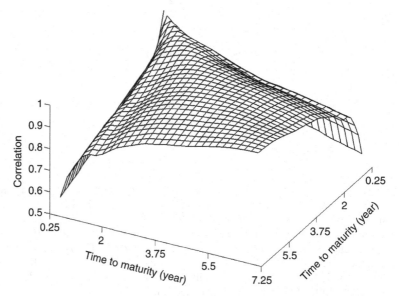

Figure 10.3 Correlation of Libor forward interest rates versus time to maturity, with daily data selected from 29 January 2003 to 29 April 2003.

data, which run from 29 January 2003 to 29 April 2003. The results are given in Table 10.1. The empirical correlation given in Figure 10.3 shows that the underlying Libor interest rates are not perfectly correlated.

The parameters are given in Table 10.1 and are different from the earlier estimates for Libor obtained in Section 7.6. This is because an earlier Libor data set for the period 1990–1996 was used for estimating the parameters. Another fit can be done for the stiff propagator in which the market time index η is taken to be equal to the one obtained from the older data from the period 1990–1996, for which $\eta = 0.34$. One then obtains the 'best' fit given in Table 10.2, but with larger errors than the fit with $\eta = 0.04$.

Data for a relatively short time period of only three months, that is from 29 January 2003 to 29 April 2003, were used for finding the stiff parameters. This

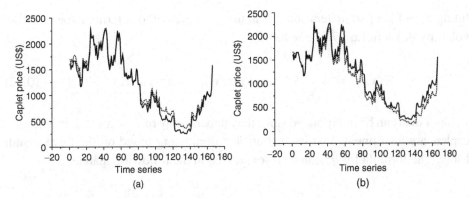

Figure 10.4 Caplet prices that mature at 12 December 2004 versus time t_0 (12 September 2003–7 May 2004): market (unbroken line) and model (dashed line). (a) Prices computed using historical volatility and correlation; the normalized root mean square error is 17.39%. (b) The effective volatility q is computed directly from Libor rates. The normalized root mean square error is 17.89%.

partly explains the significant changes in the parameters as one varies the market time index η – as can be seen by comparing the results obtained in Tables 10.1 and 10.2.

When the covariance is used as an input in a calculation, such that the propagator $\tilde{D}(x, x'; t)$ is normalized to unity, that is $\mathcal{D}(x, x'; t) = \tilde{\mathcal{D}}(x, x'; t)/\epsilon$, the effective volatility q is then given by

$$q^2 = q^2(t_0, t_*, T_n)$$

$$= \frac{1}{\epsilon} \int_{t_0}^{t_*} dt \int_{T_n}^{T_n+\ell} dx dx' \sigma_H(t, x) \tilde{\mathcal{D}}(x, x'; t) \sigma_H(t, x') \quad (10.11)$$

Using the initial forward rates curve and volatility, as well as correlation as an input and from the pricing formula, one obtains the empirical linear caplet price. It can be seen from Figure 10.4 (a) that the computed caplet price does not match the market value very well, with the normalized root mean square error being 17.39%.

10.4 Linear caplet price: market correlator

Note the parametric fit for σ_H, as given in Eq. (10.10), and the propagator combine to yield the covariance given by

$$M(x, x'; t) = \sigma(t, x)\mathcal{D}(x, x'; t)\sigma(t, x')$$

Although the parameters for the propagator $\mathcal{D}(x, x'; t)$ provide insights on the linear pricing model, one can also obtain $M(t, x, x')$ directly from data without fitting any of the parameters and this in turn is sufficient to determine the effective volatility q. From Eq. (7.6) one has

$$M(x, x'; t) = \frac{1}{\epsilon} \langle \delta f(t, x) \delta f(t, x') \rangle_c = M(x - t, x' - t) \tag{10.12}$$

Libor data can be interpolated since they depend only on $\theta = x - t$. Furthermore, caplets are instruments that have short duration, being based on the three-month Libor. The formula for q^2 can be re-expressed in the following manner

$$q^2 = \int_{t_0}^{t_*} dt \int_{T_n - t}^{T_n + \ell - t} d\theta d\theta' M(\theta, \theta') \tag{10.13}$$

The integration on time requires the future numerical values of M; since M is a function of remaining future time θ and θ', the average block of $M(\theta, \theta')$ is shifted back to its historical values. For calculating q^2, one needs to do one integration on t, which is reduced to a summation of the average value on different blocks of historical Libor data; the difference among the parallelogram blocks is only a horizontal shift and all of them end at time t_0. A detailed discussion is given in Section 12.7 on how to evaluate Eq. (10.13).

Libor data are expressed in Eq. (10.13) as an integral over θ and θ'; two integrations can be saved by directly using the Libor ZCYC without approximating Libor by the forward interest rates as in Eq. (10.8). The caplet price can hence be evaluated by directly obtaining q^2 from the ZCYC data; this is more efficient and more accurate than first finding the forward interest rates. Market data yield q^2 via the following correlator

$$q^2 = \int_{t_0}^{t^*} dt \langle \delta Y(t, T_n) \delta Y(t, T_n) \rangle_c \tag{10.14}$$

where the ZCYC data yield

$$Y(t, T_n) = \int_{T_n}^{T_n + \ell} dx f(t, x)$$
$$= \ln(1 + \ell L(t, T_n)) \tag{10.15}$$

The computed caplet prices are given in Figure 10.4(b). The linear caplet price does not match the market price very well, with the normalized root mean square error still being 17.89%, which is approximately as in the previous case given in Figure 10.4(a).

10.5 Effective volatility: parametric fit

Note that q was computed by both fitting the parameters of the linear caplet pricing model and by directly using the market correlator – both of which use historical Libor data.

Another alternative is to directly fit q from the market caplet prices, thus yielding the implied volatility σ_I. Recall that, in contrast, σ_H is obtained by empirically evaluating $E[(\delta f(t,\theta))^2]_c$.

The first approximate fit for the effective volatility, which is both accurate and simple, is to fit q as a linear function $q = b\theta$ and implied volatility is then a square root function of future time.[5]

The linear fit $q = b\theta$, in remaining future time $\theta = x - t$, obviously cannot explain the behavior of implied volatility since it diverges linearly as remaining future time increases indefinitely. However, for the market price of a caplet over only a short duration, the square root of volatility provides a very good fit.

For time far into the future, the implied volatility is directly fitted with an exponential formula, as in Eq. (10.10). The fitting is for the first 100 days in the same data set, from 12 September 2003 to 4 February 2004. The best fit for σ_I is given in Figure 10.2, and is the following

$$\sigma_I(\theta) = 0.00144 - 0.00122 \exp(-0.71826(\theta - \theta_{\min}))$$

$$+ 0.00014(\theta - \theta_{\min}) \exp(-0.71826(\theta - \theta_{\min})) \qquad (10.16)$$

The effective volatility is fitted using the first 100 days to price the remaining 168 days caplet prices using the linear caplet pricing formula Eq. (10.1). Caplet prices, both market and model, are shown in Figure 10.5(a), with their difference yielding a normalized root mean square error of 6.67%; similarly floorlet prices are shown in Figure 10.5(b), with a normalized root mean square error of 7.9%.

Given in Table 10.3 is the normalized root mean square error of the above three approaches for fitting the linear caplet price. The best fit is given by implicit volatility $\sigma_I(\theta)$, but the errors are still large, almost 7%. Figure 10.6(a) shows the linear fit for the linear caplet pricing formula.

10.5.1 Parametric fit and Black's formula

The price of a caplet is equivalent to an effective value for Black's implied volatility σ_B, and one obtains a daily implied volatility from the caplet price; σ_B is shown in

---

[5] Fitting effective volatility is much easier than fitting a correlator from caplet price data. Furthermore, the impact of changing correlation is insignificant compared with changing effective volatility; the reason being that a caplet only involves the correlation between two neighboring forward interest rates within the range of a single caplet, and hence over a maximum future time difference of 90 days; to a good approximation, for x, x' differing by $\ell = 90$ days, $\mathcal{D}(x, x') \simeq 1$.

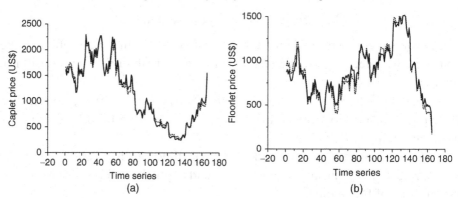

Figure 10.5 Both the market (unbroken line) and model (dashed line) are fitted with implied volatility directly from first 100 days caplet prices. (a) Caplet prices mature at 12 December 2004 versus time t_0 (12 September 2003–7 May 2004). The normalized root mean square error is 6.67%. (b) Floorlet prices mature at 12.12.2004 versus time t_0 (12 September 2003–7 May 2004). The normalized root mean square error is 7.9%.

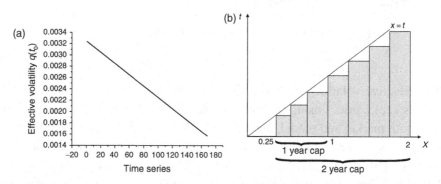

Figure 10.6 (a) Effective volatility q for caplet which matures at 12 December 2004 versus time t_0 (12 September 2003–7 May 2004). (b) Domain for one-year and two-year caps. For a one-year cap, three caplets are involved. Seven caplets are required for a two-year cap.

Table 10.3 *The normalized root mean square (rms) error is given by the rms error divided by the price of the caplet.*

	σ_H from Libor	Market correlator	σ_I from caplet
Normalized rms error in caplet price	17.39%	17.89%	6.67%

Table 10.4

	cap(t_0,2) from σ_I	cap(t_0,3) from σ_I	cap(t_0,3) from market correlator
Cap prices from implied volatility σ_I and market correlator			
Normalized rms error in cap price	6.7%	5.54%	5.59%

Figure 8.3. The shape of Black's implied volatility is very irregular and cannot be fitted well by any smooth formula. No prediction can be made for the future value of Black's volatility and hence one cannot make a prediction for the future price of a caplet.

One can sacrifice expressing the caplet's market price in terms of an implied σ_B, and instead do a best fit for σ_B similar to the fit done to obtain an implicit volatility of σ_I. The error for a parametric fit for σ_B is about 6% and is comparable to the error obtained for σ_I.

10.6 Pricing an interest rate cap

Caps and floors are important financial instruments for managing interest rate risk. However, the multiple payoffs underlying these contracts complicate their pricing as the Libor term structure dynamics are not perfectly correlated.

The linear caplet pricing formula is applied to the pricing of an interest rate cap and, in particular, a cap with a fixed maturity is analyzed. The market price of a cap for 494 trading days, which matures one, two, and three years in the future, is compared to the linear cap price.

Fixed maturity interest rate cap is a sum of interest rate caplets and is given as follows

$$cap(t_0, T_N) = \sum_{n=1}^{N-1} caplet(t_0, T_n) \qquad (10.17)$$

The caplet price is based on a three-month Libor, and the first caplet matures in three months. A one-year cap can be expressed as a sum of three caplets, a two-year cap is a sum of seven caplets. The domain of the forward interest rates required for pricing a cap is given in Figure 10.6(b).

The linear caplet implied volatility σ_I is evaluated from the fixed maturity date cap for the same period and is then used for pricing the two-year and three-year caps. The computed and market prices are shown in Figure 10.7(a). Since there is

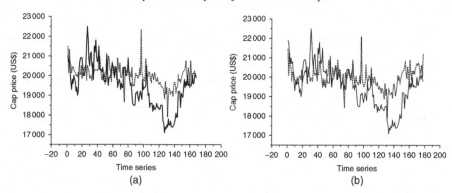

Figure 10.7 Cap price that matures three years in the future versus time t_0 (12 September 2003–7 May 2004) with market cap prices (unbroken line) and linear model prices (dashed line). (a) Diagram shows the cap price using effective volatility computed directly from historical Libor rates. The normalized root mean square error is 5.54%. (b) Diagram shows the cap price computed directly from historical Libor rates. The normalized root mean square error is 5.59%.

a new instrument everyday, one can always improve accuracy by fitting a moving effective volatility directly from the one-, two-, and three-year caps.[6]

The price of a three-year cap, from Eq. (10.14) and using historical Libor data, is shown in Figure 10.7(b). The normalized root mean square error in cap price is the value of the root mean square normalized by the price of the cap. The errors for interest rate caps are smaller than those for caplets in Figures 10.4(a), 10.4(b), and 10.5; this is due to the price of the cap being much larger than the caplet price.

10.7 Summary

Interest rate caplets are one of the simplest interest rate options and provide a useful theoretical 'laboratory' for studying the properties of interest rate options. The bond forward interest rates have deterministic drift and volatility; for this reason, exact calculations can be carried out for the linear price of many instruments, including the caplet.

The effective volatility q for the linear caplet pricing formula can be derived from the underlying historical Libor rate, and hence the linear caplet pricing formula yields a *prediction* for the caplet price: given the input of Libor data, the linear

---

[6] q needs to be fitted at Libor times only.

caplet model generates the daily caplet prices. The effective volatility q can also be used to price other Libor-based options.

The current practice of the financial markets is that Black's model fixes an implied σ_B based on existing market caplet prices. Black's formula, in effect, is a (nonlinear) representation of the market price, with a one-to-one relation between market price and implied Black volatility σ_B.

Historical caplet prices served to obtain a best fit for the effective volatility q, which was then used for predicting the linear caplet's price – including the prices of caplets in the distant future. The empirical study demonstrated that the accuracy of the linear model for pricing interest rate caps is not very accurate, with errors larger than 6%.

The empirical study of the linear caplet pricing formula used three alternative approaches; for all of the three approaches, the empirical errors remain substantial. The fit for q^2 was done only once and used for pricing the whole time series. One can always do a daily moving fit for q^2 to improve the accuracy of the calculation, but this procedure is not expected to lead to any significant improvement since the main reason for the inaccuracy of the predicted prices is the linear caplet pricing formula, which neglects the problem's inherent nonlinearities.

The empirical studies of Libor in Eq. (7.32) showed that $v(t, x)$, the stochastic volatility of Libor forward interest rates $f_L(t, x)$, has the largest fluctuations of about 1.5 years in remaining future time $x - t$. Predictions of linear caplet model prices based on the bond forward interest rates yield caplet prices with errors larger than 17%. The large errors in caplet prices are most likely due to replacing stochastic volatility $v(t, x)$ by the deterministic volatility $\sigma(t, x)$ of bond forward interest rates.

In the Libor Market Model, caplet prices are a function of the log Libor rate $\phi(t, x)$'s volatility function $\gamma(t, x)$. The order of magnitude for implied volatility σ_I given in Figure 10.2(b) is about 0.02, which is the same as that of $\ell\gamma$, as given in Figure 7.7(b). This is an empirical indication that the caplet market prices are better described by the quantum finance Black's formula than by the linear caplet price.

There should be a significant improvement in the accuracy of caplet prices based on the dynamics of $\phi(t, x)$ as these rates are a representation of the nonlinear Libor forward interest rates $f_L(t, x)$ and, hence, are expected to describe Libor options more accurately.

The main conclusion of the study of caplet prices is that the nonlinear effects of the Libor forward interest rates are important for caplet pricing – with nonlinearities accounting for at least 15% of the caplet's price. The methods used for finding effective q for the linear caplet pricing formula can also be applied to Black's case.

σ_B can be determined from Eq. (8.13) as follows

$$\sigma_B^2 = \frac{1}{t_* - t_0} \int_{t_0}^{t_*} dt \int_{T_n}^{T_{n+1}} dx \int_{T_n}^{T_{n+1}} dx' M_\gamma(x, x'; t)$$

Once $M_\gamma(x, x'; t)$ is determined from historical Libor data, the quantum finance generalization of Black's formula can be used for predicting caplet prices and should yield errors less than a few percent.

11

Coupon bond European and Asian options

European options on coupon bonds are studied in the framework of bond forward interest rates $f(t, x)$ studied in Chapter 5 as a *linear* (Gaussian) quantum field [9, 40].[1] One of the advantages of the Gaussian formulation is that the coupon bond option has a representation that is tractable and allows for various analytical approximation schemes. More precisely, including the payoff function for the coupon bond option into the path integral for the option price makes it nonlocal and nonlinear. A perturbation expansion using Feynman diagrams gives a closed-form approximation for the price of European and Asian coupon bond options. The approximate bond option is studied for two limiting cases, namely (a) the industry standard one-factor exponential volatility HJM formula and (b) the BGM–Jamshidian model's swaption price.

11.1 Introduction

Coupon bonds and interest rate *swaps* are the most important derivatives of the debt markets and options on these instruments are widely traded. The pricing of European options on coupon bonds is studied in some detail using the quantum finance approach. The volatility of the forward interest rates is a small quantity, of the order of 10^{-2}/year, and hence provides a small parameter for obtaining a volatility expansion for the option price. A perturbation expansion for the option price is developed, using Feynman diagrams, in a power series to fourth power in the forward interest rates' volatility.

A perturbative study of the coupon bond option price has been discussed in Section 8.5 for the Libor forward interest rates similar to the one that is the main focus of this chapter. The objective of repeating the Libor calculation for the bond forward interest rates is two fold.

---

[1] In this chapter, only the bond forward interest rates are discussed and the term 'bond' is used only if necessary so as to avoid repetition.

- The volatility expansion in Section 8.5 was carried out for the nonlinear Libor Market Model to only second order in the volatility function γ. One would like to develop the volatility expansion to higher order so as to more clearly understand the nature of the perturbation expansion. The bond forward interest rates are Gaussian and hence a high-order calculation is far simpler than a similar high-order computation for the Libor Market Model.
- A comparison of the linear and nonlinear models of the forward interest rates, as encoded in the bond and Libor forward interest rates is of intrinsic interest. In particular, one would like to be able to have a better understanding of the role that nonlinearities play in interest rate option pricing.

The approximate coupon bond price is verified to yield the correct limiting price for the zero coupon bond. An approximate price for the Asian coupon bond option is obtained in the same framework as the European option, and, as expected, yields a price less than the European case. The HJM and BGM–Jamshidian limits are taken of the approximate European coupon bond option price and it is seen that there is a loss of crucial nontrivial correlations for these two limiting cases.

11.2 Payoff function's volatility expansion

The coupon bond option payoff is a nonlinear function of the forward interest rates and is fairly intractable. To leverage on the smallness of $\sigma(t, x)$, the coupon bond option payoff is re-expressed in a manner that can generate a volatility expansion, as in Section 3.14, for the coupon bond option prices.

The forward bond numeraire has been discussed in Section 9.3 and is the most suitable for calculations in this chapter. Call and put options for the coupon bonds, using the forward bond numeraire $B(t, t_*)$, are given by Eq. (3.8) as follows

$$C(t_0, t_*, K) = B(t_0, t_*)E[\mathcal{P}_*]$$

$$P(t_0, t_*, K) = B(t_0, t_*)E\left[\left(K - \sum_{i=1}^{N} c_i B(t_*, T_i)\right)_+\right]$$

$$\mathcal{P}_* = \left(\sum_{i=1}^{N} c_i B(t_*, T_i) - K\right)_+ \tag{11.1}$$

where $E[\ldots]$ refers to the expectation being evaluated using the forward bond measure. The payoff function $\mathcal{P}_*$ includes, as a special case, the swaption payoff given in Eq. (4.32), with the coefficients c_n and strike price K taking specific values.

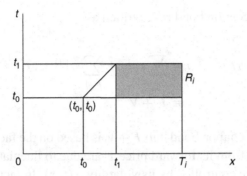

Figure 11.1 The shaded area is the domain of integration R_i.

Recall from Eq. (5.2) that the bond forward interest rates can be written as

$$f(t, x) = f(t_0, x) + \int_{t_0}^t dt' \alpha_*(t', x) + \int_{t_0}^t dt' \sigma(t', x) \mathcal{A}(t', x)$$

The drift for the forward numeraire, from Eq. (9.3), is given by

$$\alpha_*(t, x) = \int_{t_*}^x dx' M(x, x'; t): \text{ forward drift} \tag{11.2}$$

and yields the following for the zero coupon bond[2]

$$B(t_*, T_i) = \exp\left\{-\int_{t_*}^{T_i} dx f(t_*, x)\right\} = e^{-\alpha_i - Q_i} F_i(t_0, t_*, T_i) \tag{11.3}$$

where

$$F_i = F(t_0, t_*, T_i) = \exp\left\{-\int_{t_*}^{T_i} dx f(t_0, x)\right\}; \quad \alpha_i = \int_{R_i} \alpha_*(t, x) \tag{11.4}$$

$$Q_i = \int_{R_i} \sigma(t, x) \mathcal{A}(t, x) \equiv \int_{t_0}^{t_*} dt \int_{t_*}^{T_i} dx \sigma(t, x) \mathcal{A}(t, x) \tag{11.5}$$

The domain of integration R_i is given in Figure 11.1.

The coefficient α_i, the integrated form of the forward measure drift, is

$$\alpha_i = \int_{R_i} \alpha_*(t, x) \equiv \frac{1}{2} \int_{t_0}^{t_*} dt \int_{t_*}^{T_i} dx dx' M(x, x'; t) \tag{11.6}$$

---

[2] Recall from Eq. (2.14) that $F_i \equiv F(t_0, t_*, T_i)$ is the forward price, at time t_0, of a zero coupon bond $B(t_*, T)$ that is to be issued at time $t_* > t_0$ in the future and matures at future calendar time T_i.

The price of the coupon bond is re-written as

$$\sum_{i=1}^{N} c_i B(t_*, T_i) = \sum_{i=1}^{N} c_i F_i + \sum_{i=1}^{N} c_i [B(t_*, T_i) - F_i]$$

$$\equiv F + V \tag{11.7}$$

The break-up of the coupon bond into $F + V$ is based on the fact that all deviations of the coupon bond from its forward price F are due to fluctuations in the forward interest rates that are controlled by its volatility $\sigma(t, x)$. In fact, V has an order of magnitude equal to $O(\sigma)$, and hence an expansion in power of V would, in effect, result in the volatility expansion that one is aiming for.

From Eq. (11.3), the following are a few definitions.

$$J_i \equiv c_i F_i; \quad F_i = \exp\left\{ -\int_{t_*}^{T_i} dx f(t_0, x) \right\} \tag{11.8}$$

$$F \equiv \sum_{i=1}^{N} c_i F_i = \sum_{i=1}^{N} J_i \tag{11.9}$$

$$V \equiv \sum_{i=1}^{N} c_i [B(t_*, T_i) - F_i] = \sum_{i=1}^{N} J_i \left[e^{-\alpha_i - \mathcal{Q}_i} - 1 \right] \tag{11.10}$$

The payoff function is given by

$$\left[\sum_{i=1}^{N} c_i B(t_*, T_i) - K \right]_+ = \left[F + \sum_{i=1}^{N} J_i (e^{-\alpha_i - \mathcal{Q}_i} - 1) - K \right]_+ \tag{11.11}$$

$$= [F + V - K]_+$$

and is re-written using Eq. (A.7) as follows[3]

$$\left(\sum_{i=1}^{N} c_i B(t_*, T_i) - K \right)_+ = \frac{1}{2\pi} \int_{-\infty}^{+\infty} d\mathcal{Q} d\eta \, e^{i\eta(V - \mathcal{Q})} (F + \mathcal{Q} - K)_+$$

The price of the call option, from Eq. (11.1), can be written as

$$C(t_0, t_*, K) = B(t_0, t_*) \frac{1}{2\pi} \int_{-\infty}^{+\infty} d\mathcal{Q} d\eta (F + \mathcal{Q} - K)_+ e^{-i\eta \mathcal{Q}} Z(\eta) \tag{11.12}$$

---

[3] The integration variable $\mathcal{Q}$ should not be confused with $\mathcal{Q}_i$.

with the partition function given by

$$Z(\eta) = \frac{1}{Z} \int D\mathcal{A} e^S e^{i\eta V}; \quad Z = \int D\mathcal{A} e^S \quad (11.13)$$

All the random terms are contained in $Z(\eta)$. A perturbation expansion is developed that evaluates the partition function $Z(\eta)$ as a series in the volatility function $\sigma(t, x)$.

11.2.1 Put–call parity

To see how put–call parity is expressed in terms of the partition function $Z(\eta)$, note from Eqs. (4.15), (11.1), and (11.12)

$$C(t_0, t_*, K) - P(t_0, t_*, K) = \frac{B(t_0, t_*)}{2\pi} \int_{-\infty}^{+\infty} dQ d\eta (F + Q - K) e^{-i\eta Q} Z(\eta) \quad (11.14)$$

The integration over the variable Q in the equation above can be performed exactly and, from Eq. (A.7), yields a Dirac-delta function and its derivative in the η variable; hence, one obtains from Eq. (11.14) that

$$
\begin{aligned}
C(t_0, t_*, K) &- P(t_0, t_*, K) \\
&= B(t_0, t_*) \int_{-\infty}^{+\infty} d\eta \left[(F - K)\delta(\eta) + i\frac{\partial}{\partial\eta}\delta(\eta) \right] Z(\eta) \\
&= B(t_0, t_*) \left[(F - K)Z(0) - i\frac{\partial}{\partial\eta}Z(\eta)|_{\eta=0} \right] \quad (11.15)
\end{aligned}
$$

Comparing Eqs. (4.23) and (11.15) yields the following two conditions

$$Z(0) = 1; \quad \frac{\partial}{\partial\eta}Z(\eta)|_{\eta=0} = 0 \quad (11.16)$$

$$\Rightarrow Z(\eta) = 1 + O(\eta^2)$$

Any approximation scheme for evaluating the partition function $Z(\eta)$ must satisfy the put–call parity given above in Eq. (11.16).

11.3 Coupon bond option: Feynman expansion

In general, computing the coupon bond European option price is a nonlinear problem that needs to be studied numerically or perturbatively. In this section, an analytic expansion for the approximate price of the coupon bond option is derived [9].

From Figure 7.1 the volatility of the forward interest rates is a small quantity, of the order of $\sigma(t, x) \simeq 10^{-2}$/year. The volatility expansion yields the price of the coupon bond option as a power series in $\sigma(t, x)$ and, in effect, provides a rapidly convergent series for the partition function $Z[\eta]$.

The partition function of the coupon bond option price can be written more explicitly. Recall from Eq. (11.13)

$$Z(\eta) = \frac{1}{Z} \int \mathcal{D}\mathcal{A} e^{i\eta V} e^{S[\mathcal{A}]} \tag{11.17}$$

The effective action for the pricing of the coupon bond option, from Eqs. (11.10) and (11.17), is given by

$$S_{\text{Effective}} \equiv S[\mathcal{A}] + i\eta V \tag{11.18}$$

$$= S[\mathcal{A}] + i\eta \sum_{i=1}^{N} J_i \left[e^{-\alpha_i - Q_i} - 1 \right]$$

$$= S[\mathcal{A}] + i\eta \sum_{i=1}^{N} J_i \left[e^{-\alpha_i} e^{-\int_{R_i} \sigma \mathcal{A}} - 1 \right] \tag{11.19}$$

Eq. (11.19) yields a highly nonlinear and nonlocal two-dimensional quantum field theory, with the coupon bond option payoff function providing an effective nonlocal exponential potential for the quantum field $\mathcal{A}(t, x)$.

Nonlinear quantum field theories are usually intractable, and the best that one can do is to develop a consistent perturbation expansion for the partition function $Z(\eta)$. Feynman diagrams provide the standard technique in quantum field theory for perturbatively studying nonlinear systems [95]. The quantum finance formulation of the forward interest rates provides a Feynman perturbation expansion of the partition function.

A cumulant expansion [95] of the partition function in a power series in η yields

$$Z(\eta) = e^{i\eta C_1 - \frac{1}{2}\eta^2 C_2 - i\frac{1}{3!}\eta^3 C_3 + \frac{1}{4!}\eta^4 C_4 + \dots} \tag{11.20}$$

The coefficients $C_1, C_2, C_3, C_4, \dots$ are evaluated using Feynman diagrams.

From the put–call parity constraint given in Eq. (11.16), the first condition $Z(0) = 1$ is satisfied automatically, and the second condition implies that $C_1 = 0$. Hence any approximate scheme for $Z(\eta)$ that fulfills the put–call parity relation must yield

$$Z(\eta) = e^{-\frac{1}{2}\eta^2 C_2 - i\frac{1}{3!}\eta^3 C_3 + \frac{1}{4!}\eta^4 C_4 \dots} \tag{11.21}$$

Expanding the right-hand side of Eq. (11.17) in the power series to fourth order in η yields

$$Z(\eta) = \frac{1}{Z} \int \mathcal{D}\mathcal{A} e^{i\eta V} e^{S[\mathcal{A}]}$$

$$= \frac{1}{Z} \int \mathcal{D}\mathcal{A} e^{S[\mathcal{A}]} \Big[1 + i\eta V + \frac{1}{2!} (i\eta)^2 V^2$$

$$+ \frac{1}{3!} (i\eta)^3 V^3 + \frac{1}{4!} (i\eta)^4 V^4 + \dots \Big] \tag{11.22}$$

Comparing Eqs. (11.20) and (11.22) yields, in the notation of Eq. (3.66) and to fourth order in η, the following

$$C_1 = E[V] \tag{11.23}$$

$$C_2 = E[V^2] - C_1^2 \tag{11.24}$$

$$C_3 = E[V^3] - C_1^3 \tag{11.25}$$

$$C_4 = E[V^4] - 3C_2^2 - C_1^4 \tag{11.26}$$

The coefficient C_1 is exactly zero since the martingale condition for the forward measure yields

$$C_1 = E[V] = \sum_{i=1}^{N} J_i \big[E_F \big(e^{-\alpha_i - Q_i} \big) - 1 \big] = 0 \tag{11.27}$$

Put–call parity is satisfied in the approximation scheme since $C_1 = 0$; one can see that the martingale condition is essential in the realization of put–call parity.

Define the dimensionless forward bond price correlator by

$$G_{ij} \equiv G_{ij}(t_0, t_*, T_i, T_j; \sigma)$$

$$= \int_{t_0}^{t_*} dt \int_{t_*}^{T_i} dx \int_{t_*}^{T_j} dx' M(x, x'; t) \tag{11.28}$$

$$= G_{ji} \quad : \quad \text{real and symmetric}$$

where $M(x, x'; t) = \sigma(t, x)\mathcal{D}(x, x'; t)\sigma(t, x')$. The integration domain for G_{ij} is illustrated in Figure 11.2(a), and Figure 11.2(b) shows its dependence on T_i and T_j.

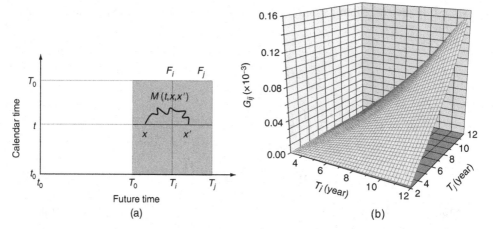

Figure 11.2 (a) The shaded domain of the forward interest rates contributes to the correlator $G_{ij} = \int_{t_0}^{t_*} dt \int_{t_*}^{T_i} dx \int_{t_*}^{T_j} dx' M(x, x'; t)$. For a typical point t in the time integration the figure shows the correlation function $M(x, x'; t)$ connecting two different values of the forward interest rates at future time x and x'. (b) The forward bond price correlator $G_{ij} = G_{ij}(t_0, t_*, T_i, T_j)$, is plotted against T_i and T_j with $t_* - t_0 = 2$ years. $M(x, x'; t)$ is taken from swaption data.

G_{ij} is the forward bond propagator that expresses the correlation in the fluctuations of the forward bond prices $F_i = F(t_0, t_*, T_i)$ and $F_j = F(t_0, t_*, T_j)$.

For any application of the coupon bond option price to the financial markets, one has to take into account market future time defined by Eq. (5.25) and given by

$$\frac{\partial f(t, t + \theta)}{\partial t} = \alpha(t, z(\theta)) + \sigma(t, z(\theta))\mathcal{A}(t, z(\theta))$$

$$\theta = x - t; \quad z = \theta^\nu$$

The forward interest rates' correlator has the property that, for a given period of time, the market's correlator depends only on remaining future time; that is, $M(x, x'; t) = M(x - t, x' - t)$. The market correlator for the forward bond prices is then given by

$$G_{ij}^{\text{Market}} \equiv G_{ij}^{\text{Market}}(t_0, t_*; \sigma; \nu)$$

$$= \int_{t_0}^{t_*} dt \int_{t_*-t}^{T_i-t} d\theta \int_{t_*-t}^{T_j-t} d\theta' M(z(\theta), z(\theta'))$$

11.4 Cumulant coefficients

A computation for the cumulant's coefficients is carried out in Section 11.12 and yields the following result

$$C_2 = \sum_{ij=1}^{N} J_i J_j [e^{G_{ij}} - 1] \tag{11.29}$$

$$C_3 = \sum_{ijk=1}^{N} J_i J_j J_k [e^{G_{ij}+G_{jk}+G_{ki}} - e^{G_{ij}} - e^{G_{jk}} - e^{G_{ki}} + 2] \tag{11.30}$$

and

$$C_4 = \sum_{ijkl=1}^{N} J_i J_j J_k J_l \Big[e^{G_{ij}+G_{ik}+G_{il}+G_{jk}+G_{jl}+G_{kl}}$$
$$- e^{G_{ij}+G_{jk}+G_{ki}} - e^{G_{ij}+G_{jl}+G_{li}} - e^{G_{ik}+G_{kl}+G_{li}} - e^{G_{jk}+G_{kl}+G_{lj}}$$
$$- e^{G_{ij}+G_{kl}} - e^{G_{jk}+G_{il}} - e^{G_{ik}+G_{jl}}$$
$$+ 2(e^{G_{ij}} + e^{G_{ik}} + e^{G_{il}} + e^{G_{jk}} + e^{G_{jl}} + e^{G_{kl}}) - 6 \Big] \tag{11.31}$$

The terms required to determine the coefficients rapidly proliferate.

As things stand, all coefficients $C_2, C_3, C_4, \ldots$ seem to be of equal magnitude. A consistent expansion is obtained if one assumes that $\sigma(t, x)$ is small for all values of its argument. For Libor, data indicate that $\sigma(t, x) \simeq 10^{-2}$/year; furthermore, normalizing the propagator, as in Section 7.3, yields $M(x, x; t) = \sigma(t, x)^2$ and $\mathcal{D}(x, x'; t) \leq 1$ for all x, x'.

G_{ij} is dimensionless and is of order of magnitude of σ^2, which yields that $G_{ij} \simeq 10^{-4}$. Hence G_{ij} can be taken to be a small expansion parameter, with all the coefficients $C_2, C_3, C_4, \ldots$ being expressed as power series in G_{ij}. Expanding the exponential functions in Eqs. (11.29), (11.30), and (11.31) yields the following result

$$C_2 = \sum_{ij=1}^{N} J_i J_j \left[G_{ij} + \frac{1}{2} G_{ij}^2 \right] + O(G_{ij}^3) \tag{11.32}$$

$$C_3 = 3 \sum_{ijk=1}^{N} J_i J_j J_k G_{ij} G_{jk} + O(G_{ij}^3) \tag{11.33}$$

$$C_4 = 16 \sum_{ijkl=1}^{N} J_i J_j J_k J_l G_{ij} G_{jk} G_{kl} + O(G_{ij}^4) \tag{11.33}$$

Denote the magnitude of the matrix elements G_{ij} by G; using the fact that $G \simeq \sigma^2$, the partition function Z has an order of magnitude expansion given by

$$Z(\eta) \simeq e^{-\sum_{l=2}^{\infty} a_l \eta^l G^{l-1}}$$
$$\simeq e^{-\sum_{l=2}^{\infty} a_l \eta^l \sigma^{2l-2}}$$
$$= e^{-a_2 \zeta^2 - a_3 \zeta^3 \sigma - a_4 \zeta^4 \sigma^2 + \cdots}; \quad \zeta = \sigma \eta \qquad (11.34)$$

where all the coefficients $a_l \simeq O(1)$.

The quadratic term in the exponential for $Z(\eta)$ fixes the magnitude of the fluctuations of the $\zeta = \sigma \eta$ variable to be of $O(1)$; hence, the remaining terms are of order σ, σ^2, and so on to higher and higher order. The perturbation expansion for the partition function $Z(\eta)$ is consistent, with the higher-order terms in η being smaller than the lower-order ones. One can obtain any degree of accuracy in the expansion parameter G, or equivalently in σ, by going to high enough order and self-consistently terminating the expansion at that order.

The perturbation expansion for the partition function $Z(\eta)$ has an intuitive representation using Feynman diagrams. The forward bond propagator G_{ij} that

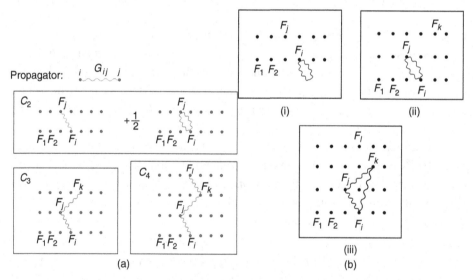

(a)   (b)

Figure 11.3 (a) Feynman diagrams for the partition function $Z(\eta)$. In the diagrams the dots represent the forward bond prices F_i and F_j and the wavy lines are their correlators G_{ij}. Coefficient C_2 is evaluated by the sum of two diagrams; coefficients C_3 and C_4 are each evaluated by a single diagram. (b) *Disconnected* diagrams are ones in which some of the forward bond prices, indicated by the top line, do not couple to the other forward bond prices. Figures (i), (ii), and (iii), are second-, third-, and fourth-order disconnected diagrams that do not contribute to $\ln(Z(\eta))$ and, hence, to any of the coefficients.

represents the correlation between forward bond price F_i and F_j is indicated with wavy lines in Figure 11.3; note that all the diagrams for C_2, C_3, C_4 that yield the partition function $Z(\eta)$ are *connected*, in that none of the forward bond prices is decoupled from the forward bond propagator G_{ij}. In contrast Figure 11.3(b) has examples of disconnected Feynman diagrams; for example in Figure 11.3(b), (i) none of the forward bond prices in the second line, denoted by F_j, couples to the propagator. It can be shown that, in general, only the connected diagrams contribute to $\ln(Z(\eta))$ [95].

11.5 Coupon bond option: approximate price

From Eqs. (11.12) and (11.25) expand the partition function up to the quartic terms in η, and then perform the Gaussian integrations over η [9]. This yields

$$C(t_0, t_*, K)$$

$$= B(t_0, t_*) \frac{1}{2\pi} \int_{-\infty}^{+\infty} dQ d\eta (F + Q - K)_+ e^{-i\eta Q} e^{-\frac{1}{2}\eta^2 C_2 - i\frac{1}{3!}\eta^3 C_3 + \frac{1}{4!}\eta^4 C_4 + \dots}$$

$$= B(t_0, t_*) \frac{1}{\sqrt{2\pi}} \int_{-\infty}^{+\infty} dQ (F + Q\sqrt{C_2} - K)_+ f(\partial_Q) e^{-\frac{1}{2}Q^2} + O(\sigma^4) \quad (11.35)$$

where, for $\partial_Q \equiv \partial/\partial Q$, one has the following

$$f(\partial_Q) \equiv 1 - \left(\frac{C_3}{6C_2^{3/2}}\right) \partial_Q^3 + \left(\frac{C_4}{24C_2^2}\right) \partial_Q^4 + \frac{1}{2}\left(\frac{C_3}{6C_2^{3/2}}\right)^2 \partial_Q^6 + O(\sigma^4)$$

$$C_2 \simeq O(\sigma^2); \quad \frac{C_3}{C_2^{3/2}} \simeq O(\sigma); \quad \frac{C_4}{C_2^2} \simeq O(\sigma^2) \quad (11.36)$$

Due to the properties of $\Theta(x)$, the Heaviside theta function given in Eq. (A.3), the second derivative of the payoff is equal to the Dirac-delta function, namely

$$\partial_Q^2 (F + Q\sqrt{C_2} - K)_+ = \sqrt{C_2}\delta(Q - X) \quad (11.37)$$

$$X = \frac{K - F}{\sqrt{C_2}} : \text{ Dimensionless} \quad (11.38)$$

Using the equation above and Eqs. (11.35) and (11.36) yields to $O(\sigma^4)$, after an integration by parts, the following

$$C(t_0, t_*, K) = B(t_0, t_*)\frac{1}{\sqrt{2\pi}}\int_{-\infty}^{+\infty}dQ\left[(F + Q\sqrt{C_2} - K)_+\right.$$

$$\left. +\sqrt{C_2}\delta(Q - X)\left\{-\frac{C_3}{6C_2^{3/2}}\partial_Q + \frac{C_4}{24C_2^2}\partial_Q^2 + \frac{1}{2}\left(\frac{C_3}{6C_2^{3/2}}\right)^2\partial_Q^4\right\}\right]e^{-\frac{1}{2}Q^2}$$

$$(11.39)$$

where, as in Eq. (3.70)

$$I(X) = \int_{-\infty}^{+\infty}dQ(Q - X)_+e^{-\frac{1}{2}Q^2} = e^{-\frac{1}{2}X^2} + \sqrt{2\pi}X[N(X) - 1] \quad (11.40)$$

Hence the price of the coupon bond is given by

$$C(t_0, t_*, K) = B(t_0, t_*)\sqrt{\frac{C_2}{2\pi}}I(X) + O(\sigma^4) \quad (11.41)$$

$$+B(t_0, t_*)\sqrt{\frac{C_2}{2\pi}}\left[\frac{C_3}{6C_2^{3/2}}X + \frac{C_4}{24C_2^2}(X^2 - 1) + \frac{1}{72}\frac{C_3^2}{C_2^3}(X^4 - 6X^2 + 3)\right]e^{-\frac{1}{2}X^2}$$

The leading behavior of option price $C(t_0, t_*, K)$ is graphed in Figure 11.4 (a); the surface is smooth because variables X and A are varied continuously.

For the coupon bond and swaption, at the money is given by $F = K$; hence, the option's price close to at the money has $X \approx 0$ and to leading order, from Eq. (3.72), yields the price to be

$$C(t_0, t_*, K) \approx B(t_0, t_*)\sqrt{\frac{C_2}{2\pi}} - \frac{1}{2}B(t_0, t_*)(K - F) + O(X^2) \quad (11.42)$$

11.5.1 Put–call parity for approximate option price

The approximate price of the coupon bond call option in Eq. (11.41), expressed in terms of the expansion coefficients, can be written symbolically as $C(t_0, t_*, K) \equiv C_{[t_0, t_*, K]}(C_2, C_3, C_4, X)$. The put option is given by an expression similar to Eq. (11.35) for the call option, namely

$$P(t_0, t_*, K)$$

$$= B(t_0, t_*)\frac{1}{2\pi}\int_{-\infty}^{+\infty}dQd\eta(K - F - Q)_+e^{-i\eta Q}e^{-\frac{1}{2}\eta^2C_2 - i\frac{1}{3!}\eta^3C_3 + \frac{1}{4!}\eta^4C_4 + \cdots}$$

$$= C_{[t_0, t_*, K]}(C_2, -C_3, C_4, -X)$$

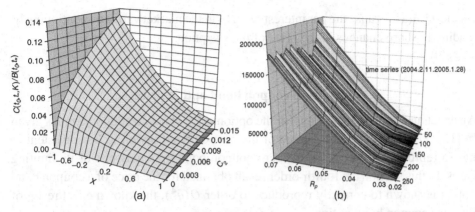

Figure 11.4 (a) The value of the swaption $C(t_0, t_*, K)/B(t_0, t_*) = \sqrt{C_2/2\pi}\, I(X)$ $+ O(\sigma^2)$, plotted as a function of C_2 and X. (b) The price of a 1x3 swaption, which has three years' duration that matures one year in the future. The floating payments are made at 90-day intervals and fixed payments are made at 180-day intervals. The x-axis is the par value R_P and time is plotted along the y-axis.

For the put and call options' approximate price, since $X = (K - F)/\sqrt{C_2}$, put–call parity yields

$$C(t_0, t_*, K) - P(t_0, t_*, K)$$

$$= C_{[t_0, t_*, K]}(C_2, C_3, C_4, X) - C_{[t_0, t_*, K]}(C_2, -C_3, C_4, -X)$$

$$= B(t_0, t_*)\sqrt{\frac{C_2}{2\pi}} \times \left(-2\sqrt{\frac{\pi}{2}} X\right) + O(\sigma^4) = B(t_0, t_*)[F - K]$$

$$= \sum_i c_i B(t_0, T_i) - K B(t_0, t_*)$$

Hence, the volatility expansion in σ yields an approximate price for the call and put option that obeys put–call parity order by order.

11.5.2 Numerical price of swaptions

The swaption price is given by Eq. (11.41), with the factor of $B(t_0, t_*)$ for the case of coupon bonds replaced by $V B(t_0, T_0)$. Putting c_n to be equal to its values for the swaption coefficients, as given in Eq. (4.32), and setting the strike price of $K = 1$ in Eq. (11.41) yield the price of the swaption $C_L(t_0, T_0, R_S)$ at time t_0.

Figure 11.4(b) shows the time series for the price of a swaption on Libor with values for the fixed interest rate being equal to the daily par value R_P given by Eq. (4.36). The surface shown in Figure 11.4 (b) is rough compared to the surface in

Figure 11.4(a) because the par interest rate R_P in the market varies discontinuously, leading to sharp changes in the price of the swaption.

11.6 Zero coupon bond option price

An exact quantum finance result for the option price of a zero coupon bond is given in [12]. For the approximate price of the coupon bond option to be consistent, it has to reproduce the price of the zero coupon bond option as one of its limiting cases. In this section the fourth-order result obtained for the price of a coupon bond option is shown to correctly reproduce to order $O(q^4)$, the known exact result of the zero coupon bond option.

The zero coupon bond is a special case of the coupon bond when only one of the coefficient functions c_i is nonzero. Let $c_1 = 1$ and $T_1 = T$, with the rest of the coupon payments being zero, that is $c_i = 0$, $i = 2, 3, \ldots, N$. From Eq. (11.1) the payoff function, the forward price for the zero coupon bond, and the propagator are as follows

$$\mathcal{P}(t_*) = \big(B(t_*, T) - K\big)_+$$

$$F = c_1 F_1 = \exp\left\{-\int_{t_*}^{T} dx f(t_0, x)\right\}$$

$$G_{11} \equiv q^2 = \int_{t_0}^{t_*} dt \int_{t_*}^{T} dx \int_{t_*}^{T} dx' M(x, x'; t) \qquad (11.43)$$

The coefficients in the expansion for the price of the coupon bond option yield

$$C_2 = F_1^2 \left[G_{11} + \frac{1}{2} G_{11}^2 \right] = F^2 \left[q^2 + \frac{1}{2} q^4 \right] + O(q^6) \qquad (11.44)$$

$$\frac{C_3}{C_2^{3/2}} = \frac{3 F_1^3 G_{11}^2}{A^{3/2}} = 3q + O(q^3); \quad \frac{C_4}{C_2^2} = \frac{16 F_1^4 G_{11}^3}{A^2} = 16q^2 + O(q^4)$$

$$X = \frac{K - F}{qF} + O(q)$$

Note that the expansion for coefficient C_2 has to be kept to $O(q^4)$; since the square root of C_2 appears in the payoff function, the $O(q^4)$ term yields the next leading order term for the payoff function, which is a term of $O(q^3)$.

The price of the coupon bond call option, from Eqs. (11.39) and (11.44), simplifies in the case of the zero coupon option to

$$C_{zcb}(t_0, t_*, K) \tag{11.45}$$

$$= B(t_0, t_*) \frac{qF}{\sqrt{2\pi}} \left[\frac{1}{2}qX + \frac{2}{3}q^2(X^2 - 1) + \frac{1}{8}q^2(X^4 - 6X^2 + 3) \right] e^{-\frac{1}{2}X^2}$$

$$+ B(t_0, t_*) \frac{1}{\sqrt{2\pi}} \int_{-\infty}^{+\infty} dQ \left(F + qF(1 + \frac{1}{4}q^2)Q - K \right)_+ e^{-\frac{1}{2}Q^2} + O(q^4)$$

Consider the following Taylor's expansion of the payoff function to $O(q^3)$

$$\left\{ F + qF \left(1 + \frac{1}{4}q^2 \right) Q - K \right\}_+ \simeq \left(1 + \frac{1}{4}q^2 Q \partial_Q \right) (F + qFQ - K)_+$$

$$\tag{11.46}$$

Inserting Eq. (11.46) into the last term in Eq. (11.45), doing an integration by parts using $Qe^{-Q^2/2} = -\partial_Q e^{-Q^2/2}$ and from Eq. (11.37)

$$B(t_0, t_*) \frac{1}{\sqrt{2\pi}} \int_{-\infty}^{+\infty} dQ e^{-\frac{1}{2}Q^2} \left(1 + \frac{1}{4}q^2 Q \partial_Q \right) (F + qFQ - K)_+$$

$$= B(t_0, t_*) \frac{1}{\sqrt{2\pi}} \int_{-\infty}^{+\infty} dQ \left[(F + qFQ - K)_+ + \frac{1}{4}q^3 F\delta(Q - X) \right] e^{-\frac{1}{2}Q^2}$$

$$= B(t_0, t_*) \frac{qF}{\sqrt{2\pi}} \left[I(X) + \frac{1}{4}q^2 e^{-\frac{1}{2}X^2} \right] \tag{11.47}$$

$I(X)$ is given in Eq. (3.70). Hence from Eqs. (11.45) and (11.47), the price of the zero coupon bond option, after some simplifications, is given by

$$C_{zcb}(t_0, t_*, K) = B(t_0, t_*) \frac{qF}{\sqrt{2\pi}} \left[\frac{1}{2}qX + \frac{1}{6}q^2(X^2 - 1) + \frac{1}{8}q^2(X^2 - 1)^2 \right] e^{-\frac{1}{2}X^2}$$

$$+ B(t_0, t_*) \frac{qF}{\sqrt{2\pi}} I(X) + O(q^4) \tag{11.48}$$

It is shown in Section 11.13 that the exact result for the zero coupon bond option price, when expanded in a power series in q^2, yields the same result as the one obtained in Eq. (11.48).

11.6.1 Zero coupon bond option numerical estimate

The accuracy of the volatility expansion of the zero coupon bond option price is studied by comparing the approximate expression for the call option price given in Eq. (11.48) with the exact expression for the zero coupon bond option given in Eq. (11.66). The zero coupon bond option has been derived in [12] based on the bond forward interest rates and is given by

$$C(t_0, t_*, T, K) = B(t_0, t_*)[F(t_0, t_*, T)N(d_+) - KN(d_-)] \qquad (11.49)$$

$$F \equiv F(t_0, t_*, T) = \exp\left\{-\int_{t_*}^{T} dx f(t_0, x)\right\}; \quad d_\pm = \frac{1}{q}\left[\ln\frac{F}{K} \pm \frac{q^2}{2}\right]$$

The normalized difference of the exact and approximate option prices, namely $(C_{zcb}^{Exact} - C_{zcb}^{Approx})/C_{zcb}^{Exact}$, is plotted in Figure 11.5(a) for various values of $X = (K - F)/qF$ and for different values of q^2; the result shows that the approximate value of the option price has a negligible normalized error of about 10^{-5} for $0 \le q^2 \le 0.01$. The (normalized) root mean square is computed for the entire fit, and is given in Figure 11.5(b); the normalized error is again about 10^{-5} over the same range of q^2.

For the coupon bond with N-coupons, the effective expansion is approximately Nq^2; from the results of the zero coupon bond, one can estimate that as long as $Nq^2 \le 0.01$ the approximation has an accuracy of about 10^{-3}; for a typical value of $q^2 \simeq 10^{-3}$, one can conservatively conclude that, for the coupon bond option, the perturbation expansion is valid for $N \simeq 100$.

11.7 Coupon bond Asian option price

The path integral formulation of option pricing is ideally suited for studying the Asian option since its payoff is a function of all the values ('paths') that the underlying security takes, from its initial to its final value. The coupon bond option is a particularly complex instrument since the payoff is an exponential function of the underlying forward interest rates.

Suppose a coupon bond is issued at time t_*

$$\mathcal{B}(t_*) = \sum_{i=1}^{N} c_i B(t_*, T_i)$$

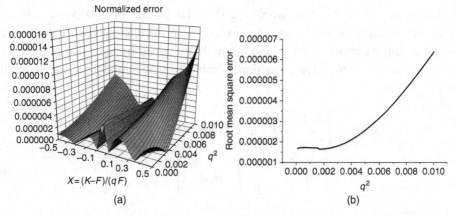

Figure 11.5 (a) The normalized difference of the exact and approximate option prices for the zero coupon bond, namely $(C_{zcb}^{Exact} - C_{zcb}^{Approx})/C_{zcb}^{Exact}$, for various values of q^2 and strike price $X = (K - F)/qF$. (b) The normalized root mean square error $\sqrt{\text{var}((C_{zcb}^{Exact} - C_{zcb}^{Approx})/C_{zcb}^{Exact})}$ of the fit for the entire zero coupon bond option price, as a function of effective volatility q^2.

with $T_i > t_*$ for all i. The forward coupon bond price, at time $t < t_*$, is given by

$$\mathcal{F}(t) = \sum_{i=1}^{N} c_i F(t, t_*, T_i); \quad F(t, t_*, T_i) = \exp\left\{ -\int_{t_*}^{T_i} dx f(t, x) \right\}; \ t_0 < t < t_*$$

The payoff of the Asian call option, issued at time t_0, maturing at t_* and with strike price K, is given by

$$\mathcal{P}_{Asn} = \left[\frac{1}{t_* - t_0} \int_{t_0}^{t_*} dt \rho(t) \mathcal{F}(t) - K \right]_+ \tag{11.50}$$

where $\rho(t)$ is a weighting function, equal to 1 for the arithmetic average; if one wants to place greater emphasis on the price of a coupon bond near the maturity of the option one can take $\rho(t) = \exp\{-\lambda(t - t_*)\}$. The forward interest rates domain of the Asian coupon bond option is shown in Figure 5.2(b).

The Asian option price, using the forward bond measure, is given by Eq. (3.8) as

$$C(t_0, t_*, K) = B(t_0, t_*) E[\mathcal{P}_{Asn}]$$

$$E[\mathcal{P}_{Asn}] = \frac{1}{Z} \int \mathcal{D} A e^{S[A]} \mathcal{P}_{Asn}$$

The Asian coupon bond option is studied using the bond forward interest rates defined by Eq. (5.1). A Feynman perturbation expansion for the Asian option price, similar to the European option case, is generated in powers of $\sigma^2(t, x)$, the volatility of the forward interest rates. One expects that all fluctuations of the payoff function from its initial value should be small, governed by the value of $\sigma^2(t, x)$; hence, one re-writes the payoff function by subtracting its initial forward value. Eq. (11.50) yields

$$
\mathcal{P}_{Asn} = \left[\frac{1}{t_* - t_0} \int_{t_0}^{t_*} dt \rho(t) \mathcal{F}(t) - F_\rho + F_\rho - K \right]_+
$$

$$
F_\rho = \mathcal{F}(t_0) \left[\frac{1}{t_* - t_0} \int_{t_0}^{t_*} dt \rho(t) \right]
\tag{11.51}
$$

From Eq. (5.2)

$$
f(t, x) = f(t_0, x) + \int_{t_0}^{t} dt' \alpha_*(t', x) + \int_{t_0}^{t} dt' \sigma(t', x) \mathcal{A}(t', x)
$$

and yields, as in Eqs. (11.4) and (11.5), the following forward prices of the zero coupon bonds

$$
F(t, t_*, T_i) = \exp\left\{ -\int_{t_*}^{T_i} dx f(t, x) \right\} = e^{-Q_i(t) - \beta_i(t)} F(t_0, t_*, T_i)
$$

$$
Q_i(t) = \int_{t_0}^{t} dt' \int_{t_*}^{T_i} dx \sigma(t', x) \mathcal{A}(t', x); \quad \beta_i(t) = \int_{t_0}^{t} dt' \int_{t_*}^{T_i} \alpha_*(t', x)
$$

Similar to the European case, the random term in the payoff function is V_{Asn} and is re-written as follows

$$
V_{Asn} \equiv \frac{1}{t_* - t_0} \int_{t_0}^{t_*} dt \rho(t) \mathcal{F}(t) - F_\rho
$$

$$
= \frac{1}{t_* - t_0} \int_{t_0}^{t_*} dt \sum_i J_i(t) [e^{-Q_i(t) - \beta_i(t)} - 1]; \quad J_i(t) = c_i \rho(t) F(t_0, t_*, T_i)
$$

One needs to evaluate, similar to Eq. (11.17), the following functional integral

$$
Z_{Asn}(\eta) = \frac{1}{Z} \int D\mathcal{A} e^{i\eta V_{Asn}} e^{S[\mathcal{A}]}
$$

To illustrate the new features of the Asian option *vis-à-vis* the European option it is sufficient to carry out the calculation to only $O(\sigma^2)$; a calculation to fourth or higher order can be carried out, using the technology of Feynman diagrams, in a manner very similar to the European case.

The martingale property, given in Eq. (11.6), namely that

$$\beta_i(t) = \frac{1}{2} \int_{t_0}^{t_*} dt \int_{t_*}^{T_i} dx dx' M(x,x';t)$$

yields the following

$$C_0 = E[1] = 1; \quad C_1 = E[V_{Asn}] = 0$$

The nontrivial coefficient is given by

$$C_2 = E[V_{Asn}^2]$$

$$= \frac{1}{(t_* - t_0)^2} \int_{t_0}^{t_*} dt dt' \sum_{ij} J_i(t) J_j(t') E\left[[e^{-Q_i(t) - \beta_i(t)} - 1][e^{-Q_j(t') - \beta_j(t')} - 1] \right]$$

$$\equiv \frac{1}{(t_* - t_0)^2} \int_{t_0}^{t_*} dt dt' \sum_{ij} J_i(t) J_j(t') [e^{\mathcal{H}_{ij}(t,t')} - 1] \tag{11.52}$$

The martingale property for $\beta_i(t)$, given in Eq. (11.6), yields

$$\mathcal{H}_{ij}(t,t') = \int_{t_*}^{T_i} dx \int_{t_*}^{T_j} dx' \int_{t_0}^{t} d\tau \int_{t_0}^{t'} d\tau' M(x,x';\tau)\delta(\tau - \tau')$$

$$= \int_{t_0}^{t} d\tau G_{ij}(\tau)\theta(t' - \tau) \tag{11.53}$$

$$G_{ij}(\tau) = \int_{t_*}^{T_i} dx \int_{t_*}^{T_j} dx' M(x,x';\tau) \tag{11.54}$$

Hence, to leading order in σ^2, one has

$$C_2 = \frac{1}{(t_* - t_0)^2} \int_{t_0}^{t_*} dt dt' \sum_{ij} J_i(t) J_j(t') \mathcal{H}_{ij}(t,t') + O(\sigma^3)$$

$$= \frac{1}{(t_* - t_0)^2} \int_{t_0}^{t_*} dt dt' \sum_{ij} J_i(t) J_j(t') \int_{t_0}^{t} d\tau G_{ij}(\tau)\theta(t' - \tau)$$

$$= \frac{1}{(t_* - t_0)^2} \int_{t_0}^{t_*} dt' \int_{t_0}^{t'} dt \sum_{ij} J_i(t) J_j(t') \int_{t_0}^{t} d\tau G_{ij}(\tau) \tag{11.55}$$

The $\rho(t) = 1$ limit yields, similar to Eq. (8.43), the following

$$C_2 = \frac{1}{(t_* - t_0)^2} \sum_{ij} J_i J_j \int_{t_0}^{t_*} dt(t - t_0)(t_* - t) G_{ij}(t); \quad J_i = c_i F(t_0, t_*, T_i)$$

Once one has evaluated the coefficients C_0, C_1, and C_2, the price of the Asian option can be obtained from Eq. (3.69).

The HJM limit for $\rho(t) = 1$ is taken in Section 11.10 in order to gain some insight into the approximate price of the Asian coupon bond option.

11.8 Coupon bond European option: HJM limit

In Section 4.11, the limiting cases for the quantum finance formulation of coupon bonds and interest rates were discussed. The HJM and BGM–Jamshidian models are special cases of forward interest rates being *exactly* correlated and correspond to the propagator being a constant. The limit of $\mathcal{D}(x, x'; t) \to 1$ for all x, x' in turn yields $M(x, x'; t) = \sigma(x - t)\mathcal{D}(x, x'; t)\sigma(x' - t) \to \sigma(x - t)\sigma(x' - t)$.

The limit of $\mathcal{D}(x, x'; t) \to 1$ is studied in order to determine the importance of having a nontrivial correlation for the forward interest rates.

Taking $\sigma(x - t)$ equal to the market volatility of the forward interest rates, the percentage *difference* between the daily price of a 2 by 10 swaption at the money and its HJM limit of $\mathcal{D}(x, x'; t) \to 1$ is plotted in Figure 11.6(a); the daily HJM option price C_{HJM} is seen to be overpriced by 4–9% in comparison with the correlated quantum finance option price C_{QF}. This result shows the important role of the nontrivial correlations in pricing coupon bond options.

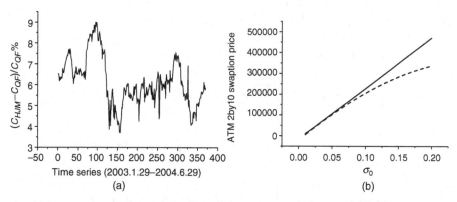

Figure 11.6 (a) The percentage difference of a 2 by 10 swaption daily price, at the money, for the quantum finance model's price C_{QF} and of its HJM limit C_{HJM}. The volatility function $\sigma(t,x)$ is taken from the Libor market as are the daily forward bond prices F_i. Note the HJM model systematically overprices the swaption by 4–9%. (b) The difference of the swaption price, at the money, as a function of σ_0, of the one-factor HJM model and the quantum finance model with $\mathcal{D}(x, x'; t) \to 1$; the volatility function is $\sigma_0 e^{-\lambda(x-t)}$ with $\lambda = 0.1$/year.

The quantum finance option price has been derived as a perturbation expansion in the volatility. From Eq. (3.72), to leading order, the option price is given by

$$C(t_0, t_*, K) = B(t_0, t_*)\sqrt{\frac{C_2}{2\pi}} - \frac{1}{2}B(t_0, t_*)(K - F) + O(X^2, \sigma^2)$$

To lowest order, from Eqs. (11.8), (11.28), and (11.32)

$$C_2 = \sum_{ij=1}^{N} J_i J_j G_{ij}$$

$$G_{ij} = \int_{t_0}^{t_*} dt \int_{t_*}^{T_i} dx \int_{t_*}^{T_j} dx' \sigma(t, x) \mathcal{D}(x, x'; t) \sigma(t, x')$$

Taking the HJM limit of $\mathcal{D}(x, x'; t) \to 1$ and for exponential volatility $\sigma(t, x) = \sigma_0 e^{-\lambda(x-t)}$ yields, from Eq. (4.43), the following

$$G_{ij} \to \sigma_0^2 \int_{t_0}^{t_*} dt \int_{t_*}^{T_i} dx e^{-\lambda(x-t)} \int_{t_*}^{T_j} dx' e^{-\lambda(x'-t)} = \sigma_E^2 Y(t_*, T_i) Y(t_*, T_j)$$

$$\Rightarrow \sqrt{C_2} = \sqrt{\sigma_E^2 \sum_{ij=1}^{N} J_i J_j Y(t_*, T_i) Y(t_*, T_i)} = \sigma_E \sum_{i=1}^{N} J_i Y(t_*, T_i)$$

Hence, the HJM limit of the quantum finance option price is given by

$$C(t_0, t_*, K) \simeq B(t_0, t_*) \left[\sqrt{\frac{1}{2\pi}} \sigma_E \sum_{i=1}^{N} J_i Y(t_*, T_i) - \frac{1}{2}(K - F) \right]$$

$$+ O((K - F)^2, \sigma_0^2) \qquad (11.56)$$

From Eq. (4.48), to leading order in σ_0, Eq. (11.56) is seen to be equal to $C_{HJM}(t_0, t_*, K)$.

Figure 11.6(b) shows the result of a numerical evaluation of both the HJM and the HJM limit of the quantum finance option price with $K = F$ (at the money) – with exponential volatility $\sigma_0 e^{-\lambda(x-t)}$ and for fixed forward bond prices F_i; only σ_0 is allowed to vary. Figure 11.6(b) shows that the HJM limit of the approximate quantum coupon bond price starts to deviate from the HJM price for $\sigma_0 \geq 0.1$; this is to be expected since the quantum finance approximation, in the first place, is expected to hold for only small values of $\sigma_0^2 \simeq 10^{-2}$.

The result of Figure 11.6(b) does not mean that the HJM model price is as accurate as the correlated quantum finance model for small σ_0; rather the result shows that the HJM *approximation* of the quantum finance model agrees, for small σ_0, with the result of the one factor HJM model. In fact, as shown in Figure 11.6(a), the best version of the HJM model, which uses the market volatility and forward interest rates that are taken to be exactly correlated, systematically overprices the actual market price of the swaption, with the quantum finance approximation being more accurate.

The HJM limit shows a number of special features of the HJM option price.

- The one-factor HJM coupon bond option price for exponential volatility is given by a single sum, each term of which is similar to the zero coupon bond option price given in Eq. (11.49); the single sum is due to the specific exponential form chosen for the volatility.
- The quantum finance and one-factor HJM option prices are seen to be equal in the limit of $\mathcal{D}(x, x'; t) \to 1$, at the money, and to lowest order in the exponential volatility.
- For a general volatility, which in particular is not an exponential, the limit of $\mathcal{D}(x, x'; t) \to 1$ does not remove the square root on C_2.

11.9 Coupon bond option: BGM–Jamshidian limit

The correlation of the Libor rates $L_n(t)$ in the BGM–Jamshidian model, as a function of future time T_n, is given by Eq. (6.9) as

$$E\left[\frac{dL(t, T_n)}{dt}\frac{dL(t, T_{n'})}{dt}\right]_c = \gamma_n(t)\gamma_{n'}(t)E[R(t)R(t')]$$
$$= \gamma_n(t)\gamma_{n'}(t)\delta(t - t') \qquad (11.57)$$

As in the HJM case, in the BGM–Jamshidian formulation of the Libor Market Model, changes in the Libor rates are *exactly* correlated; that is, $\mathcal{D}_{BGM}(t, T, T') = 1$. Future time runs over Libor time $T_n = T_0 + n\ell$; hence, the quantum finance formula for the coupon bond option has the following BGM–Jamshidian limit

$$\sum_{ij=1}^{N} J_i J_j G_{ij} \to \sum_{ij=1}^{N} J_i J_j \int_{t_0}^{T_0} dt \int_{T_0}^{T_i} dx \int_{T_0}^{T_j} dx' \gamma(t, x)\gamma(t, x')\mathcal{D}_{BGM}(t, x, x')$$

$$\simeq \sum_{ij=1}^{N} J_i J_j \int_{t_0}^{T_0} dt \sum_{n=n_0}^{i-1}\sum_{m=n_0}^{j-1} \gamma_n(t)\gamma_{n'}(t) = \sum_{ij=1}^{N} J_i J_j \int_{t_0}^{T_0} dt \sum_{n=n_0}^{i}\sum_{m=n_0}^{j} \Delta_{mn}$$

where

$$\gamma_n(t) = \int_{T_n}^{T_{n+1}} dx\gamma(t, x); \quad \int_{t_0}^{T_0} dt\gamma_n(t)\gamma_m(t) = \Delta_{mn}$$

The BGM–Jamshidian approximation, as given in Eq. (8.50), yields the following

$$\Delta_{mn} \simeq \Gamma_m \Gamma_n$$

and this leads to the following factorization of G_{ij} [34]

$$\sum_{ij=1}^{N} J_i J_j G_{ij} \rightarrow \left[\sum_{k=1}^{N} \sum_{n=n_0}^{k-1} J_i \Gamma_n \right]^2$$

Collecting all the above results, one obtains that, for small volatility, the BGM–Jamshidian model limit of the quantum finance pricing formula is given by

$$\frac{B(t_0, T_0)}{\sqrt{2\pi}} \sqrt{\sum_{i,k=1}^{N} J_i J_j G_{ij}} \rightarrow \frac{B(t_0, T_0)}{\sqrt{2\pi}} \sum_{k=1}^{N} c_i F(t_0, T_0, T_k) \sum_{n=n_0}^{k-1} \Gamma_n(t_0, T_0, T_n)$$

$$\simeq \mathcal{C}_{\text{BGM}}(t_0; T_0, K)$$

The BGM–Jamshidian approximation yields a factorization of $\sqrt{C_2}$ quite distinct from the factorization of the HJM model with exponential volatility.

11.9.1 BGM–Jamshidian approximation for exponential volatility

Similar to the HJM model, the BGM–Jamshidian approximation can be explicitly evaluated for the case of exponential volatility.

$$\gamma(t, x) = \gamma_0 \exp\{-\lambda(x - t)\}; \quad \gamma_n(t) = \frac{\gamma_0}{\lambda} e^{\lambda(t-n\ell)}[1 - e^{-\lambda\ell}]$$

The matrix Δ_{mn} factorizes exactly; for $T_0 = n_0 \ell$, one has

$$\Delta_{mn} = \Gamma_n \Gamma_m$$

$$\Gamma_n = \gamma_L e^{-(n-n_0)\lambda\ell}[1 - e^{-\lambda\ell}]; \quad \gamma_L^2 = \frac{\gamma_0^2}{2\lambda^3}[1 - e^{-2\lambda(t_0 - T_0)}]$$

Hence, for exponential volatility the BGM–Jamshidian approximate price is

$$\mathcal{C}_{\text{BGM}}(t_0; T_0, K) \simeq \frac{\gamma_L}{\sqrt{2\pi}} B(t_0, T_0) \sum_{k=1}^{N} c_k F(t_0, T_0, T_k) \sum_{n=n_0}^{k-1} \Gamma_n$$

$$= \frac{\gamma_L}{\sqrt{2\pi}} B(t_0, T_0) \sum_{k=1}^{N} c_k F(t_0, T_0, T_k)[1 - \exp\{-(k - n_0)\lambda\ell\}]$$

In summary:

- For the HJM and BGM–Jamshidian models, the bond forward interest rates and Libors, in the future time direction, are exactly correlated and result in $D_{HJM}(t, x, x') = 1 = D_{BGM}(t, x, x')$. The forward bond correlator G_{ij} is, consequently, factorized, that is $G_{ij} \rightarrow \Gamma_i \Gamma_j$.
- The factorization of G_{ij} results in the following

$$\sqrt{\sum_{ij=1}^{N} J_i J_j G_{ij}} \rightarrow \sum_{i=1}^{N} \Gamma_i J_i$$

- Since $J_i = c_i F(t_0, T_0, T_i)$, the factorization leads to a crucial loss of correlations between the F_is, leading to systematic inaccuracies in option pricing.

11.10 Coupon bond Asian option: HJM limit

The coefficient C_2 yields the Asian coupon bond option price from Eq. (3.69) and is given in full generality by Eq. (11.52) as follows

$$C_2 = \frac{1}{(t_* - t_0)^2} \int_{t_0}^{t_*} dt' \int_{t_0}^{t'} dt \sum_{ij} J_i(t) J_j(t') \int_{t_0}^{t} d\tau \, G_{ij}(\tau)$$

$$G_{ij}(\tau) = \int_{t_*}^{T_i} dx \int_{t_*}^{T_j} dx' M(x, x'; \tau); \quad J_i(t) = c_i F(t_0, t_*, T_i) \rho(t)$$

For definiteness, let $\rho(t) = 1$. The HJM limit for the one-factor model, with $\sigma(t, x) = \sigma_0 \exp\{-\lambda(x - t)\}$, yields

$$M(x, x'; \tau) = \sigma_0^2 e^{2\lambda \tau} e^{-\lambda(x+x')}; \quad Y(t_*, T_i) \equiv \frac{1}{\lambda}[1 - e^{-\lambda(T_i - t_*)}]$$

$$\Rightarrow G_{ij}(\tau) = \sigma_0^2 e^{-2\lambda(t_* - \tau)} Y(t_*, T_i) Y(t_*, T_j)$$

and hence

$$C_2 = \sigma_{Asn}^2 \left[\sum_{i=1}^{N} J_i Y(t_*, T_i) \right]^2; \quad J_i = c_i F(t_0, t_*, T_i)$$

$$\Rightarrow \sigma_{Asn}^2 = \frac{\sigma_0^2}{(t_* - t_0)^2} e^{-2\lambda t_*} \int_{t_0}^{t_*} dt' \int_{t_0}^{t'} dt \int_{t_0}^{t} d\tau e^{2\lambda \tau}$$

$$= \frac{\sigma_0^2}{8\lambda^3 (t_* - t_0)^2} \left[1 - 2\left\{ \lambda^2 (t_* - t_0)^2 - \lambda(t_* - t_0) + \frac{1}{2} \right\} e^{-2\lambda(t_* - t_0)} \right]$$

$$< \sigma_E^2; \quad \sigma_E^2 = \frac{\sigma_0^2}{2\lambda}[1 - \exp\{-\lambda(t_* - t_0)\}] \qquad (11.58)$$

Note σ_E^2 is the corresponding volatility for the European option given in Eq. (4.47).[4] The only difference between the HJM limit of the European and Asian coupon bond options lies in the functions σ_A^2 and σ_E^2 and, as expected, the Asian option price has a volatility lower than the European case. In particular, for the case of $\lambda = 0$ it is the extra integrations used in defining the average of the coupon bond that lowers the value of σ_A^2 below σ_E^2.

The BGM–Jamshidian approximation of the coupon bond Asian option can be taken in a straightforward manner and yields results that are similar to the HJM case.

11.11 Summary

Coupon bond options and swaptions are amongst the most complex of financial instruments, and the pricing and hedging of these is of great interest for the debt market.

The quantum finance formalism of bond forward interest rates provides a scheme for approximately evaluating the price of coupon bond options and swaptions. The empirical value of the forward interest rates' volatility is a small parameter and was used for developing a power series expansion for the price of the coupon bond options.

In the earlier second-order volatility expansion for the coupon bond option carried out in Section 8.5 for the Libor Market Model, there were only a few terms in the expansion and they did not need to be organized in any particular manner. One of the main lessons of this chapter is that in a high-order perturbation expansion the terms proliferate and need to be organized in a systematic manner. Feynman diagrams provided a graphical representation of the increasingly complex terms generated by higher and higher orders of the expansion.

The perturbation expansion using Feynman diagrams was realized by expanding the nonlinear terms in the partition function and performing the path integral, order by order, using Gaussian path integrations. The approximate coupon bond option price shows that the nontrivial *correlation* between forward prices of the zero coupon bonds of different maturities plays a crucial role in yielding an accurate price for the swaptions. This result agrees with our intuition since it is the interaction between the various forward bond prices that should determine the price of a coupon bond option.

The Feynman expansion was carried out to fourth order so as to demonstrate a very general advantage of the quantum finance formulation: the price of many debt instruments can be evaluated to a high order of accuracy by systematically computing higher- and higher-order Feynman diagrams. The high-order coupon

---

[4] In the limit of $\lambda \to 0$, $\sigma_{Asn}^2 \to \sigma_0^2 (t_* - t_0)/6$ and $\sigma_E^2 \to \sigma_0^2 (t_* - t_0)/2 > \sigma_{Asn}^2$.

bond option price computation is an exemplar for the nature of the expansion that holds true for many of the instruments. In most cases, however, the second-order result suffices, as the expansion parameter tends to be small.

There are other instruments, such as the coupon bond American and barrier options, that do not lend themselves to a Feynman expansion. Numerical and analytical techniques are discussed in later chapters for addressing these options.

The coupon bond option price obtained for the Libor Market Model in Section 8.5 is very different from the results obtained in this chapter using the bond forward interest rates. Similar to the study of caplet pricing, an empirical study needs to be carried out to compare the predictions of the two models of the forward interest rates. In particular, the empirical study would be able to determine the domain of applicability, if at all, of the two approaches to the modeling of forward interest rates.

The case of the one-factor HJM swaption price with exponential volatility was seen to be a particular limit of the quantum finance formula. The HJM limit of an exactly correlated bond forward interest rate was seen to be inaccurate, leading to a systematic overpricing of the swaptions.

It is seen that the formalism of quantum finance can solve problems that otherwise would be analytically intractable and only amenable to numerical analysis. In particular, in the conventional formulation of finance, relying heavily as it does on stochastic calculus, the techniques developed in quantum finance are far from obvious. In conclusion, the formalism of quantum finance is a useful, flexible, and transparent theoretical tool that yields accurate results for coupon bond options.

11.12 Appendix: Coupon bond option price

A detailed field theory derivation is given of the coefficients C_2, C_3, C_4. Since $C_1 = 0$, Eqs. (11.24), (11.25), and (11.26) yield the following

$$C_2 = E[V^2] \tag{11.59}$$
$$C_3 = E[V^3] \tag{11.60}$$
$$C_4 = E[V^4] - 3C_2^2 \tag{11.61}$$

The computation for the coefficient C_2 is carried in complete detail. Writing out the expression for C_2 yields

$$C_2 = \sum_{i,j=1}^{N} J_i J_j \int D\mathcal{A}[e^{-\alpha_i - Q_i - \alpha_j - Q_j} - e^{-\alpha_i - Q_i} - e^{-\alpha_j - Q_j} + 1]e^{S[\mathcal{A}]}/Z$$

$$= \sum_{i,j=1}^{N} J_i J_j \int D\mathcal{A}[e^{-\alpha_i - Q_i - \alpha_j - Q_j} - 1]e^{S[\mathcal{A}]}/Z$$

since $E[e^{-\alpha_i - Q_i}] = 1$ due to the martingale condition for the forward bond measure.[5] Note from the definition of Q_i given in Eq. (11.5)

$$Q_i + Q_j = \int_{R_i} \sigma \mathcal{A} + \int_{R_j} \sigma \mathcal{A} = \int_{t_0}^{t_*} dt \int_{t_*}^{\infty} (h_i + h_j)(t,x)\mathcal{A}(t,x)$$

$$\equiv \int_{R_{ij}} (h_i + h_j)\mathcal{A} \tag{11.62}$$

On performing the Gaussian integration over the quantum field $\mathcal{A}(t,x)$ using Eq. (5.21) one obtains, in abbreviated notation

$$C_2 = \sum_{i,j=1}^{N} J_i J_j [e^{-\alpha_i - \alpha_j + \frac{1}{2} \int_{R_{ij}} (h_i + h_j)(t,x) D(x,x';t)(h_i + h_j)(t,x')} - 1]$$

$$= \sum_{i,j=1}^{N} J_i J_j [e^{G_{ij}} - 1] \tag{11.63}$$

$$\Rightarrow G_{ij} = \int_{R_{ij}} h_i(t,x) D(x,x';t) h_j(t,x')$$

$$= \int_{t_0}^{t_*} dt \int_{t_*}^{T_i} dx \int_{t_*}^{T_j} dx' M(x,x';t) \tag{11.64}$$

Note $\int_{R_i} h_i(t,x) D(x,x';t) h_i(t,x')$ and $\int_{R_j} h_j(t,x) D(x,x';t) h_j(t,x')$ in Eq. (11.63) are the diagonal terms that cancel against the drift terms α_i, α_j terms respectively. The cross term $G_{ij} = \int_{R_{ij}} h_i(t,x) D(x,x';t) h_j(t,x')$ yields the final result for the coefficient C_2.

A similar calculation yields the coefficient C_3 given in Eq. (11.25). To evaluate coefficient C_4 note, from Eq. (11.61) and writing out the coefficient C_2^2 using Eq. (11.59) in a symmetric form, one obtains

$$C_4 = \langle V^4 \rangle - 3C_2^2$$

$$= \left\langle \sum_{ijkl=1}^{N} J_i J_j J_k J_l [e^{-\alpha_i - Q_i} - 1][e^{-\alpha_j - Q_j} - 1][e^{-\alpha_k - Q_k} - 1][e^{-\alpha_l - Q_l} - 1] \right\rangle$$

$$- \sum_{ijkl=1}^{N} J_i J_j J_k J_l [e^{G_{ij} + G_{kl}} + e^{G_{jk} + G_{li}} + e^{G_{ik} + G_{jl}} - 3]$$

---

[5] Field theorists will recognize that $e^{-\alpha_i - Q_i}$ is equal to the normal ordered expression e^{-Q_i} :.

Doing a calculation similar to the one carried out for the C_2 coefficient, and using the martingale condition for the forward bond measure, yields

$$
\begin{aligned}
C_4 = \sum_{ijkl=1}^{N} J_i J_j J_k J_l \Big[& e^{G_{ij}+G_{ik}+G_{il}+G_{jk}+G_{jl}+G_{kl}} \\
& - e^{G_{ij}+G_{jk}+G_{ki}} - e^{G_{ij}+G_{jl}+G_{li}} - e^{G_{ik}+G_{kl}+G_{li}} - e^{G_{jk}+G_{kl}+G_{lj}} \\
& - e^{G_{ij}+G_{kl}} - e^{G_{jk}+G_{il}} - e^{G_{ik}+G_{jl}} \\
& + 2(e^{G_{ij}} + e^{G_{ik}} + e^{G_{il}} + e^{G_{jk}} + e^{G_{jl}} + e^{G_{kl}}) - 6 \Big]
\end{aligned}
\tag{11.65}
$$

To understand the significance of the various terms for coefficient C_4 in Eq. (11.65), consider the case of the forward bond propagator G_{ij} being a small parameter; an expansion of the coefficient C_4 as a power series in G_{ij} yields

$$
C_4 = 16 \sum_{ijkl=1}^{N} J_i J_j J_k J_l G_{ij} G_{jk} G_{kl} + O(G_{ij}^4)
$$

One sees that the terms in Eq. (11.65) combine to cancel terms that are of lower order than the cubic term in the propagator, yielding the leading term to be of $O(G_{ij}^3)$. Furthermore, all the disconnected Feynman diagrams, generically represented in Figure 11.3(b), are canceled out by the terms appearing after the leading term. The final result is the leading-order quartic term that consists of only the *connected* Feynman diagram given in Figure 11.3(a).

In general, for all quantum field theories, the partition function $Z(\eta)$ is given by the sum of all Feynman diagrams, both connected and disconnected, whereas the log of the partition function $\ln(Z(\eta))$ is given by the sum of only the connected Feynman diagrams [95]. For this reason, all the coefficients $C_2, C_3, C_4, \ldots$ are given by only the connected Feynman diagrams given in Figure 11.3(a).

11.13 Appendix: Zero coupon bond option price

The exact zero coupon bond option price is given by [12]

$$
C_{zcb}(t_0, t_*, K) = \frac{B(t_0, t_*)}{\sqrt{2\pi q^2}} \int_{-\infty}^{+\infty} dQ \, e^{-\frac{1}{2q^2}(Q + \int_{t_*}^{T} dx f(t_0, x) + \frac{q^2}{2})^2} (e^Q - K)_+
\tag{11.66}
$$

The explicit expression for $C_{zcb}(t_0, t_*, K)$ given in Eq. (11.49) is obtained by doing the integration over Q. Eq. (11.66) is a convenient form of the zero coupon bond price for deriving the volatility expansion.

Making a change of variable yields

$$C_{zcb}(t_0, t_*, K) = B(t_0, t_*) \frac{1}{\sqrt{2\pi}} \int_{-\infty}^{+\infty} dQ e^{-\frac{1}{2}Q^2} (Fe^{qQ-\frac{q^2}{2}} - K)_+$$

(11.67)

where $F \equiv \exp(-\int_{t_*}^{T} dx f(t_0, x))$. A Taylor's expansion in the Q-variable for the payoff function to $O(q^4)$, for $X = (K - F)/qF$, yields

$$(Fe^{qQ-\frac{q^2}{2}} - K)_+ = (F + qFQ - K)_+ + qF \left[\left\{ \frac{q}{2}(Q^2 - 1) \right. \right.$$

$$\left. \left. + \frac{q^2}{6}(Q^3 - 3Q) \right\} \partial_Q(Q - X)_+ + \frac{q^2}{8}(Q^2 - 1)^2 \partial_Q^2(Q - X)_+ \right] \quad (11.68)$$

Using $(Q^2 - 1)e^{-\frac{1}{2}Q^2} = \partial_Q^2 e^{-\frac{1}{2}Q^2}$ and $(Q^3 - 3Q)e^{-\frac{1}{2}Q^2} = -\partial_Q^3 e^{-\frac{1}{2}Q^2}$, doing integrations by parts and using Eqs. (11.68) and (11.37), yields

$$C_{zcb}(t_0, t_*, K)$$

$$= \frac{B(t_0, t_*)}{\sqrt{2\pi}} \int_{-\infty}^{+\infty} dQ \left[(F + qFQ - K)_+ e^{-\frac{1}{2}Q^2} \right]$$

$$+ qF\delta(Q - X) \left\{ -\frac{q}{2}\partial_Q e^{-\frac{1}{2}Q^2} + \frac{q^2}{6}\partial_Q^2 e^{-\frac{1}{2}Q^2} + \frac{q^2}{8}(Q^2 - 1)^2 e^{-\frac{1}{2}Q^2} \right\} + O(q^4)$$

$$= B(t_0, t_*) \frac{qF}{\sqrt{2\pi}} I(X) \quad (11.69)$$

$$+ B(t_0, t_*) \frac{qF}{\sqrt{2\pi}} \left[\frac{1}{2}qX + \frac{1}{6}q^2(X^2 - 1) + \frac{1}{8}q^2(X^2 - 1)^2 \right] e^{-\frac{1}{2}X^2} + O(q^4)$$

where $I(X)$ is given in Eq. (11.40). The result obtained above agrees with the zero coupon bond limit of the coupon bond option price given in Eq. (11.48).

12

Empirical analysis of interest rate swaptions

The pricing formulas for coupon bond options derived in Chapter 11 are employed for an empirical study. This chapter studies the realization of swaptions as a special case of coupon bond options. Similar to the analysis of interest rate caplets in Chapter 10, by considering swaptions as a special case of coupon bonds, one in effect is using bond forward interest rates to model the swaptions. An empirical study of the swaption market is carried out in some detail and an efficient computational procedure is developed for analyzing swaption data [18]. Empirical results of the swaption price, swaption volatility, and swaption correlation are compared with the predictions of the quantum finance model that generates, up to a scaling factor, the market swaption prices to an accuracy of over 90% [19, 40].

12.1 Introduction

Interest rate swaptions have a deep and liquid market and arguably are today the most liquid option on interest rates; one of their major components is the highly liquid European swaptions. Swaption pricing is a nonlinear problem that has been widely studied using numerical techniques [38, 42, 44, 51, 84].

The swaption market is studied within the quantum finance framework. The theoretical price of a swaption can be modeled in terms of two distinct ways.

- In Section 4.9 the swaption price was shown to be equivalent to a specific case of a coupon bond option. As discussed in Chapters 5 and 11, one can consistently model (zero) coupon bonds using a Gaussian model of the bond forward interest rates, given in Eqs. (2.12) and (5.1), as follows

$$
B(t, T) = \exp\left\{-\int_t^T dx f(t, x)\right\}
$$

$$
\frac{\partial f(t, x)}{\partial t} = \alpha(t, x) + \sigma(t, x)\mathcal{A}(t, x)
$$

268

- In Section 8.6 the swaption price was derived entirely in terms of observed Libor rates $L(t, T_n)$ that, from Eqs. (6.1) and (6.10), can be expressed as follows

$$B(t, T) = \exp\left\{-\int_t^T dx f_L(t, x)\right\}$$

$$\frac{\partial f_L(t, x)}{\partial t} = \mu(t, x) + v(t, x)\mathcal{A}_L(t, x)$$

- Due to their complexity, the prices of Libor and coupon bond options can only be evaluated perturbatively and one needs to decide on a consistent scheme for working out these approximations. If one is evaluating coupon bond options, a consistent approximation is to use the bond forward interest rates for $f(t, x)$. For studying Libor options one needs to study the nonlinear Libor forward interest rates given by $f_L(t, x)$.
- In this chapter, swaption pricing is analyzed using its equivalence to a coupon bond option. All prices are derived by considering $f(t, x)$ to be the bond forward interest rates discussed in Chapter 5. The empirical techniques used for obtaining the swaption price are also valid for coupon bond options.
- Swaptions have been studied in Section 8.6 in the framework of the Libor Market Model. Swaption pricing, based of the LMM representation of Libor, is quite distinct from pricing obtained using the equivalence of a swaption to a coupon bond option. In this chapter, the LMM swaption price is not empirically studied.

12.2 Swaption price

Recall from Section 4.2 that interest rate swaps are derivatives in which one party pays the floating interest rate, determined by the prevailing three month Libor at the time of the payment, with the other party paying at a pre-fixed interest rate R_S. The swaps that are being considered have floating interest rate payments that are paid at $\ell = 3$-month intervals and fixed rate payments that are paid at intervals of $2\ell = 6$ months. There are N floating rate payments, at times $T_0 + n\ell$ for $n = 1, 2, \ldots, N$, made at three-monthly intervals. For six-monthly fixed rate payments there are only $N/2$ payments of amount $2R_S$, made at times $T_0 + 2n\ell$, $n = 1, 2, \ldots, N/2$.[1]

Recall from Section 4.9 that the payoff function for a fixed rate receiver *swaption* – in which the holder of the option receives at the fixed rate and pays at the floating rate – is given by

$$C_F(T_0; R_S) = V\left[B(T_0, T_0 + N\ell) + 2\ell R_S \sum_{n=1}^{N/2} B(T_0, T_0 + 2n\ell) - 1\right]_+$$

---

[1] Suppose the swap has a duration such that N is even. Note $N = 4$ for a year-long swap.

$$= V \left[\sum_{n=1}^{N/2} c_n B(T_0, T_0 + 2n\ell) - 1 \right]_+ \tag{12.1}$$

where $B(t, T)$ is the price of a zero coupon bond at time t that matures at time $T > t$. The coefficients and strike price for a swaption are hence, from Eq. (4.35), given by

$$c_n = 2\ell R_S; \quad n = 1, 2, \ldots, (N-2)/2; \quad \text{payment at time } T_0 + 2n\ell$$

$$c_{N/2} = 1 + 2\ell R_S; \quad \text{payment at time } T_0 + N\ell$$

$$K = 1 \tag{12.2}$$

The fixed interest rate par value R_P, at time t_0, is such that the interest rate swap has zero value. From Eq. (4.37)

$$2\ell R_P(t_0) = \frac{B(t_0, T_0) - B(t_0, T_0 + N\ell)}{\sum_{n=1}^{N/2} B(t_0, T_0 + 2n\ell)} \tag{12.3}$$

The price of a coupon bond option $C(t_0, t_*, R_S)$ at time $t_0 < t_*$, using the money market measure and discounting the value of the payoff function using the spot interest rate $r(t) = f(t, t)$, is given from Eq. (12.1) by

$$C(t_0, t_*, R_S) = V E \left[e^{-\int_{t_0}^{t_*} dt r(t)} \left(\sum_{n=1}^{N/2} c_n B(t_*, t_* + 2n\ell) - 1 \right)_+ \right] \tag{12.4}$$

where V is the notional deposit on which the interest is calculated; the swaption prices quoted by the market are for $V = \text{US\$1 million}$.

The coupon bond European option price has been derived in Section 11.5. Since $\sigma(t, x)$, the volatility of the forward interest rates $f(t, x)$ is $O(10^{-2})$, only the lowest order result in the perturbation expansion needs to be retained for the empirical study; hence, from Eq. (11.41)

$$C(t_0, t_*, R_S) = B(t_0, t_*) \sqrt{\frac{C_2}{2\pi}} I(X) + O(\sigma^2) \tag{12.5}$$

$$I(X) = e^{-\frac{1}{2}X^2} + \sqrt{2\pi} X N(X); \quad X = \frac{K - F}{\sqrt{C_2}}$$

$$F_i \equiv F_i(t_0, t_*, T_i) = \exp \left\{ -\int_{t_*}^{T_i} dx f(t_0, x) \right\}$$

$$F = \sum_{i=1}^{N} J_i; \quad J_i \equiv c_i F_i \tag{12.6}$$

F_i are the forward bond prices; coefficients c_i and strike price K are given in Eq. (12.2) in terms of the fixed interest rate R_s. For a swaption initialized at time t_0 to be at the money, the fixed interest rate R_s is equal to the par value $R_P(t_0)$.

The coefficients in the option price are given in Eq. (11.32) as follows

$$C_2 = \sum_{ij=1}^{N} J_i J_j \left[G_{ij} + \frac{1}{2} G_{ij}^2 \right] + O(G_{ij}^3)$$

The market correlator G_{ij} of the forward bond prices is given in Eq. (11.28). G_{ij} for different quantities is defined over different domains of the forward interest rates and this results in the integration of the forward interest rates' correlation function over different integration limits. The exact form of the various integrations will be discussed later, together with the correlators that are required for the computation of swaption volatility.

12.3 Swaption price 'at the money'

Recall that for the par value $R_P(t_0)$ of the fixed interest payments, the value of the swap at time t_0 is zero. From Eqs. (12.6) and (12.3), the fixed interest rate par value, namely $R_S = R_P$, implies the following

$$F \equiv F(t_0) = \sum_{i=1}^{N/2} c_i F(t_0, T_0, T_0 + 2i\ell)$$

$$= \sum_{i=1}^{N/2} 2\ell R_P F(t_0, T_0, T_0 + 2i\ell) + F(t_0, T_0, T_0 + N\ell)$$

Hence, from Eq. (12.3)

$$F = \frac{B(t_0, T_0) - B(t_0, T_0 + N\ell)}{\sum_{n=1}^{N/2} B(t_0, T_0 + 2n\ell)} \sum_{i=1}^{N/2} F(t_0, T_0, T_0 + 2i\ell) + F(t_0, T_0, T_0 + N\ell)$$

$$\Rightarrow F = 1: \text{ at the money}$$

In the coupon bond option pricing formula, $X = (F - K)/\sqrt{C_2}$ and for swaptions, $K = 1$. Hence when the fixed interest rate R_S for the swaption is at the money $F = 1$ and this leads to $X = (F - K)/\sqrt{C_2} = 0$. To leading order, from Eq. (3.72), the swaption price close to 'at the money', is given by

$$C(t_0, t_*, R_P) \simeq B(t_0, t_*)\sqrt{\frac{C_2}{2\pi}} + \frac{1}{2} B(t_0, t_*)(F - K) + 0(X^2) \qquad (12.7)$$

12.4 Volatility and correlation of swaptions

The volatility and correlation of swaption prices are important quantities since they are indicators of the market's direction and provide insights into portfolio behavior.

Consider the volatility and correlation of the change of swaption price for infinitesimal time steps. Let $C_I \equiv C(t_0, t_1, R_1)$ and $C_{II} \equiv C(t_0, t_2, R_2)$ denote two swaptions. Introduce the notation

$$\sqrt{2\pi}C(t_0, t_1, R_I) \simeq B(t_0, t_1)\sqrt{C_{2,I}} + \sqrt{\frac{\pi}{2}}B(t_0, t_1)(F_I - K_I) + 0(X_I^2) \quad (12.8)$$

and a similar expression for C_{II}.

Denote the time derivative by an upper dot; for infinitesimal time step ϵ

$$\langle \dot{C}_I \dot{C}_{II} \rangle_c = \frac{1}{\epsilon^2}\langle((C_I(t_0 + \epsilon) - C_I(t_0))(C_{II}(t_0 + \epsilon) - C_{II}(t_0)))\rangle_c$$

$$= \frac{1}{\epsilon^2}\langle \delta C_I(t_0)\delta C_{II}(t_0)\rangle_c \quad (12.9)$$

where recall that the connected correlator is defined by $\langle AB \rangle_c \equiv \langle AB \rangle - \langle A \rangle \langle B \rangle$.[2]

The swaption prices C_I, C_{II} depend on the forward bond prices F_i, which take random values every day. Random changes in the price of the forward bond prices lead to random changes in the price of a swaption. The correlation function $\langle \delta C_I(t_0)\delta C_{II}(t_0)\rangle_c$ can be evaluated by a historical average over the daily swaption prices, considered as the random outcomes of the swaption price. Hence, a historical average of the correlator of changes in the swaption price, taken over the random fluctuations of the forward bond prices, can be equated to the ensemble average of the correlators.

Eqs. (5.1) and (5.22) yield the following bond forward interest rate covariance

$$\langle \dot{f}(t, x)\dot{f}(t, x') \rangle_c = \frac{1}{\epsilon}M(x, x'; t) \quad (12.10)$$

From the pricing formula given in Eq. (12.8), the swaption's rate of change at the money – namely for $X_I = 0$ – is given by the following

$$\sqrt{2\pi}\frac{dC(t_0, t_1, R_I)}{dt_0} = \frac{dB(t_0, t_1)}{dt_0}\sqrt{C_{2,I}} + \frac{B(t_0, t_1)}{2\sqrt{C_{2,I}}}\frac{dC_{2,I}}{dt_0} + \sqrt{\frac{\pi}{2}}B(t_0, t_1)\frac{dF_I}{dt_0}$$

$$= D_I - C(t_0, t_1, R_I)\int_{t_0}^{t_1} dx \dot{f}(t_0, x) - \frac{B(t_0, t_1)}{\sqrt{C_{2,I}}}\sum_{ij=1}^{NI} J_i J_j G_{ij}\int_{t_1}^{T_j} dx \dot{f}(t_0, x)$$

---

[2] To simpify the notation, in the chapter this notation for expectation value, namely $E[A]$ will be denoted by $\langle A \rangle$.

$$-\sqrt{\frac{\pi}{2}} B(t_0, t_I) \sum_{i=1}^{NI} J_i \int_{t_I}^{T_i} dx \, \dot{f}(t_0, x) \tag{12.11}$$

where t_I denotes t_1 and NI denotes $N1$. D_I contains all the deterministic (non-stochastic) factors that are subtracted out in forming the connected correlation functions. One can obtain the expression for the other swaption by replacing I by II; t_{II} denotes t_2 and NII denotes $N2$.

To determine $\dot{C}_I$, as in the equation above, one needs $\dot{f}(t_0, x)$, namely the evolution equation of the quantum field $f(t, x)$ given in Eq. (5.1); hence, together with Eqs. (12.10) and (12.11), the correlator $\langle \dot{C}_I \dot{C}_{II} \rangle$ is given by

$$2\pi \epsilon \langle \delta C_I(t_0) \delta C_{II}(t_0) \rangle_c = C_I C_{II} \int_{t_0}^{t_1} dx \int_{t_0}^{t_2} dx' \, M(t_0, x, x')$$

$$+ \frac{B(t_0, t_2)}{\sqrt{C_{2,II}}} C_I \sum_{jj'=1}^{N2} G_{jj'} J_j J_{j'} \int_{t_0}^{t_1} dx \int_{t_2}^{T_j} dx' M(t_0, x, x')$$

$$+ \frac{B(t_0, t_1)}{\sqrt{C_{2,I}}} C_{II} \sum_{ii'=1}^{N1} G_{ii'} J_i J_{i'} \int_{t_0}^{t_2} dx \int_{t_1}^{T_i} dx' M(t_0, x, x')$$

$$+ \frac{B(t_0, t_1) B(t_0, t_2)}{\sqrt{C_{2,I} C_{2,II}}} C_I C_{II} \sum_{ii'=1}^{N1} \sum_{jj'=1}^{N2} G_{ii'} J_i J_{i'} G_{jj'} J_j J_{j'} \int_{t_1}^{T_i} dx \int_{t_2}^{T_j} dx' M(t_0, x, x')$$

$$+ \sqrt{\frac{\pi}{2}} B(t_0, t_2) C_I \sum_{j=1}^{N2} J_j \int_{t_0}^{t_1} dx \int_{t_2}^{T_j} dx' M(t_0, x, x')$$

$$+ \sqrt{\frac{\pi}{2}} B(t_0, t_1) C_{II} \sum_{i=1}^{N1} J_i \int_{t_0}^{t_2} dx \int_{t_1}^{T_i} dx' M(t_0, x, x')$$

$$+ \sqrt{\frac{\pi}{2}} \frac{B(t_0, t_1) B(t_0, t_2)}{\sqrt{C_{2,I}}} \sum_{ii'=1}^{N1} \sum_{j=1}^{N2} G_{ii'} J_i J_{i'} J_j \int_{t_1}^{T_i} dx \int_{t_2}^{T_j} dx' M(t_0, x, x')$$

$$+ \sqrt{\frac{\pi}{2}} \frac{B(t_0, t_1) B(t_0, t_2)}{\sqrt{C_{2,II}}} \sum_{i=1}^{N1} \sum_{jj'=1}^{N2} J_i G_{jj'} J_j J_{j'} \int_{t_1}^{T_i} dx \int_{t_2}^{T_j} dx' M(t_0, x, x')$$

$$+ \frac{\pi}{2} B(t_0, t_1) B(t_0, t_2) \sum_{i=1}^{N1} \sum_{j=1}^{N2} J_i J_j \int_{t_1}^{T_i} dx \int_{t_2}^{T_j} dx' M(t_0, x, x') \tag{12.12}$$

Table 12.1 *The various domains of integration for evaluating the integral* $\mathcal{I} = \int_{t_0}^{m_1} dt \int_{m_2}^{d_1} dx \int_{m_3}^{d_2} dx' M(x,x';t)$ *that are required for computing the coefficients in the swaption price and correlators.*

$\mathcal{I}$	m_1	m_2	m_3	d_1	d_2
G_{ij}	t_*	t_*	t_*	T_i	T_j
$G_{ii'}$	t_1	t_1	t_1	T_i	$T_{i'}$
$G_{jj'}$	t_2	t_2	t_2	T_j	$T_{j'}$

where $C_{2,I}$ and $C_{2,II}$ denote the coefficient C_2 for the two swaptions C_I and C_{II} respectively.

In Eq. (12.12) the indices i, i' refer to C_I and j, j' refer to C_{II}. For the swaption correlation, options mature at two different times $t_2 \geq t_1$, and hence two indices i, j have the range $i = 1, 2, \ldots, N1$ and $j = a, a+1, \ldots, N2$ where the last payments are made at T_{N1} and T_{N2} respectively. In the next section, data is examined in detail in order to compute the swaption price.

12.4.1 Market correlator

The forward bond price correlator G_{ij}, the swaption correlator, and volatility are all computed, with various integration limits, from a set of three-dimensional integrations on $M(x,x';t)$. For a single swaption, the swaption maturity is at t_* and the two indices i and j run from 1 to N, with the last payment being made at T_N.

A general form of all the integration is given as follows

$$\mathcal{I} = \int_{t_0}^{m_1} dt \int_{m_2}^{d_1} dx \int_{m_3}^{d_2} dx' M(x,x';t) \tag{12.13}$$

and the limits of integration are listed in Table 12.1.

12.5 Data from swaption market

The input data that are required for computing the swaption price consist of the underlying empirical forward interest rates, the coupon bond price, the forward bond price, and the fixed rate par value R_P. The swaption market provides daily data for X by Y swaptions. These swaptions mature X years from today, with the underlying swap starting at time X and the last payment being paid $X + Y$ years in the future. The domain for the swaption instrument is given in the time and future time $x - t$ plane in Figure 12.1(a).

All the prices are presented with interest rates in basis points (100 basis points = 1% annual interest rate) and have to be multiplied by the notional value of one

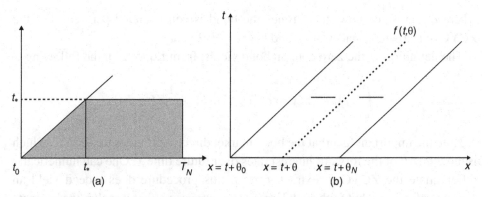

Figure 12.1 (a) The shaded area is the domain of the forward interest rates for evaluating the price of a swaption. For 2by10 swaption $t_* = t_0 + 2$ years and $T_N = t_* + 10$ years. (b) Zero coupon yield curve data on lines of constant θ; the θ interval is three months with $\theta_N = 30$ years.

million Dollars. The swaption analysis uses Bloomberg data for the ZCYC, denoted by $Z(t_0, T)$, from 29 January 2003 to 28 January 2005, and yields, in total, 523 daily ZCYC data. Daily swaption prices are quoted only for 'at the money'. In order to get accurate results, actual days in the 12 calendar months are divided by 360, since the convention for total number of days in a calendar year is 360.

12.6 Zero coupon yield curve

In order to generate swaption prices and swaption correlation from the model, both the historical and current forward interest rates are required. The value of the coupon bonds and forward bond prices as well as the par fixed rate R_P are computed from the current forward interest rates, as encoded in the ZCYC. The integrand of the forward bond correlator G_{ij}, namely $M(x, x'; t)$, is derived from historical forward interest rates' data.

The ZCYC is necessary for evaluating long duration swaptions since ZCYC data with maturity of up to 30 years is available. The ZCYC is given in the $\theta = x - t =$ constant direction, as shown in Fig. 12.1(b), with the interval of θ between two data points not being a constant. Cubic spline is used for interpolating the data to a three-month interval.

From the discussion in Section 2.11, the zero coupon bond is given by[3]

$$B(t_0, T) = \frac{1}{(1 + \frac{1}{c}Z(t_0, T))^{c[T-t_0]}} \tag{12.14}$$

---

[3] The number of years in the time interval $T - t$ is given by integer $[T - t] = (T - t)/1$ year.

where c represents how many times the bond is compounded per year. For the ZCYC, c is given as half yearly, and hence $c = 2$/year.

The definition of the zero coupon bond yields, from Eq. (2.28), the following

$$\int_{t_0}^{T} f(t_0, x) dx = c[T - t] \ln \left(1 + \frac{1}{c} Z(t_0, T)\right)$$

Note the important fact that the bond market directly provides the ZCYC, which is the *integral* of the forward interest rates over future time x. One can numerically differentiate the ZCYC to extract $f(t, x)$; this procedure does indeed yield an estimate of $f(t, x)$, but with such large systematic errors that it makes the estimate quite useless for empirically analyzing swaption pricing. Hence all the numerical procedures are directly based on the ZCYC [18].

All the data required for calculating a swaption's price can be obtained directly from the ZCYC data. The interpolation of ZCYC data and the convention used by Bloomberg have been empirically tested by comparing the computed value of R_P, using Eq. (12.3), with the value given by the market; the result confirms the correctness of the computation.

12.7 Evaluating $\mathcal{I}$: the forward bond correlator

The market value of the forward bond price correlator $\mathcal{I}$, given in Eq. (12.10), can be derived from ZCYC data. From Eq. (12.10) and for discrete time $\dot{f} \simeq \delta f / \epsilon$ the correlation for changes in the forward interest rates is given by [12]

$$M(x, x'; t) = \frac{1}{\epsilon} \langle \delta f(t, x) \delta f(t, x') \rangle_c \; ; \; \delta f(t, x) = f(t + \epsilon, x) - f(t, x) \quad (12.15)$$

Thus, the forward bond correlator is given by the following

$$\mathcal{I} = \frac{1}{\epsilon} \int_{t_0}^{m1} dt \int_{m2}^{d1} dx \int_{m3}^{d2} dx' \langle \delta f(t, x) \delta f(t, x') \rangle_c \quad (12.16)$$

From Table 12.1, it can be seen that none of the limits on the integrations over x, x' depends on the time variable t; hence, the calendar time finite difference operator δ can be moved out of the x, x' integrations and yields

$$\mathcal{I} = \frac{1}{\epsilon} \int_{t_0}^{m1} dt \left\langle \left[\delta \int_{m2}^{d1} dx f(t, x)\right] \left[\delta \int_{m3}^{d2} dx' f(t, x')\right] \right\rangle_c \quad (12.17)$$

The x and x' integration variables are directly used instead of changing them to θ and θ' since, as discussed earlier, ZCYC data directly yield the integrals of forward interest rates over future time x. The numerical values of $\int_{m2}^{d1} dx f(t, x)$ and $\int_{m3}^{d2} dx' f(t, x')$ are obtained from the market values of the ZCYC.

To evaluate the market correlator $\mathcal{I}$ one needs to know the value of the correlator $M(x, x'; t)$ for *future calendar time*; the reason being that the time integration t in $\mathcal{I}$ runs from present calendar time t_0 to future calendar time $m1 > t_0$. The problem of obtaining the future values of $M(x, x'; t)$ can be solved by assuming that the correlation function, for changes in the forward interest rates, is invariant under time translations; that is

$$M(x, x'; t) = M(x - a, x' - a; t - a) \qquad (12.18)$$

The assumption of time translation invariance of the forward rates correlation function has been empirically tested in [28]; although this assumption cannot be indefinitely extended, a two-year shift is considered to be reasonable [28].

12.7.1 Shifting integration over calendar time

The integration on the t axis can be converted into a summation by discretizing time into a lattice with spacing ϵ'; Eq. (12.17) yields

$$t = t_0 + t_k; \quad t_k = k\epsilon'$$

$$\Rightarrow \mathcal{I} = \epsilon' \sum_{t_k=0}^{m1-t_0} \int_{m2}^{d1} dx \int_{m3}^{d2} dx' M(t_0 + t_k, x, x') \qquad (12.19)$$

Consider the following change of variables for Eq. (12.19)

$$x = y + t_k \ ; \ x' = y' + t_k$$

Hence, from Eq. (12.19)

$$\mathcal{I} = \epsilon' \sum_{t_k} \int_{m2-t_k}^{d1-t_k} dy \int_{m3-t_k}^{d2-t_k} dy' M(t_0 + t_k, y + t_k, y' + t_k)$$

$$= \epsilon' \sum_{t_k} \int_{m2-t_k}^{d1-t_k} dy \int_{m3-t_k}^{d2-t_k} dy' M(t_0, y, y') \qquad (12.20)$$

where the condition given in Eq. (12.18) has been used to obtain Eq. (12.20). The shift of the future time integration to the present and the domain used for doing the averages for the correlator are illustrated in Figure 12.2.

The integration on future calendar time has been replaced by a summation on the *current* value of $M(t_0, x, x')$, with x, x' taking values on various intervals. The

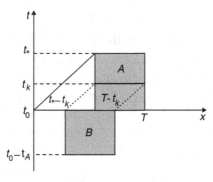

Figure 12.2 The expectation value $\langle\delta Y(t_0 + t_k, t_*, T)\delta Y(t_0 + t_k, t_*, T)\rangle$ for *future calendar time* is shifted back to t_0 using the invariance under time translations and yields an equivalent expression that is evaluated at *present calendar time* t_0, namely $\langle\delta Y(t_0, t_* - t_k, T - t_k)\delta Y(t_0, t_* - t_k, T - t_k)\rangle$. A historical average is done over the rectangular area B, which is in the *past calendar time* of t_0; the optimum days for evaluating the historical averages is $t_A = 180$ days.

current value of $M(t_0, x, x')$ in turn is evaluated by taking averages of the correlator over its *past* values.

From above and Eqs. (12.16), (12.17), and (12.20) one has

$$\mathcal{I} = \frac{\epsilon'}{\epsilon}\sum_{t_k}\left\langle\int_{m2-t_k}^{d1-t_k}\delta f(t_0, y)dy \int_{m3-t_k}^{d2-t_k}\delta f(t_0, y')dy'\right\rangle_c$$

$$= \frac{\epsilon'}{\epsilon}\sum_{t_k}\left\langle\left[\delta\int_{m2-t_k}^{d1-t_k} f(t_0, y)dy\right]\left[\delta\int_{m3-t_k}^{d2-t_k} f(t_0, y')dy'\right]\right\rangle_c$$

As discussed earlier, in order to directly use the ZCYC data the finite time difference operator δ is taken outside the future time integrations. Note ϵ' is the time integration interval and is equal to ϵ; for the time summation with daily intervals $\epsilon = \epsilon' = 1/260$ (since 260 is the actual number of trading days in one year).

Re-expressing $\mathcal{I}$ in terms of the ZCYC data yields

$$\mathcal{I} = \frac{\epsilon'}{\epsilon}\sum_{t_k}\langle\delta Y(t_0, m2 - t_k, d1 - t_k)\delta Y(t_0, m3 - t_k, d2 - t_k)\rangle_c$$

where, from Eq. (2.28)

$$Y(t_0, t_*, T) = \int_{t_*}^{T} f(t_0, x)dx = \int_{t_0}^{T} f(t_0, x)dx - \int_{t_0}^{t_*} f(t_0, x)dx$$

$$= \log[(1 + Z(t_0, T)/c)^{(T-t_0)c}] - \log[(1 + Z(t_0, t_*)/c)^{(t_*-t_0)c}]$$

$$(12.21)$$

The forward bond price correlator's present value (at time t_0) is obtained by averaging the correlator $\langle \delta Y(t_0, m2 - t_k, d1 - t_k) \delta Y(t_0, m3 - t_k, d2 - t_k) \rangle$ over the last $t_0 - t_A$ days with $t_A = 180$ days. The program was run by adding 30 days to the time averaging for evaluating the expectation values of the correlators; the best fit is given when the averaging is done over the past 180 days; see Section 12.8. Since the computation requires the value of δY for different future time intervals x, x', one has to use cubic splines to interpolate ZCYC for obtaining daily values of the ZCYC.

12.8 Empirical results

The 2by10 and 5by10 swaptions are priced for time series 6 April 2004–28 January 2005 using the pricing formula from Section 12.3. When computing the forward interest rates' correlator $M(x, x'; t)$ the daily swaption prices are stable when more than 270 days of historical data for ZCYC were used; but a 270-day average does not give the best fit of the predictions of model swaption price with the swaption's market value. This may be due to too much old information creating large errors in the predictions for the present-day swaption prices. However, averaging on less historical data causes the swaption price curve to fluctuate strongly since it is likely that new information dominates swaption pricing and makes the price too sensitive to small changes.

The empirical study showed that a moving averaging of 180 days of historical data gives the best result for this period. One can most likely improve the accuracy by higher frequency sampling of 180 days of historical data.

The results obtained from the quantum finance model are compared with daily market data and are shown in Figures 12.3(a) and 12.3(b); the normalized root mean square of errors are 3.31% and 6.31% respectively. The perturbative model result given in Eq. (12.5) had to be rescaled by an overall factor of $1/\sqrt{\pi}$ to match it with the market swaption values [18]; the explanation of this single overall factor needs further analysis.

The results for the swaption volatility and correlation discussed in Section 12.4 are derived for the change on the same instruments. From Eq. (12.9)

$$\delta C_I \equiv C_I(t_0 + \epsilon) - C_I(t_0) \equiv C_I(t_0 + \epsilon, R_s) - C_I(t_0, R_s) \qquad (12.22)$$

where $C_I(t_0 + \epsilon)$ and $C_I(t_0)$ are the same contract being traded on successive days. Par fixed interest rate R_P is determined when the contract is initiated at time t_0, and the swaption $C_I(t_0)$ is at the money. However, in general, $C_I(t_0 + \epsilon)$ is away from the money; the reason being that the swaption depends on the forward bond prices F_i, and these change every day and hence there is a daily change in the par fixed rate R_P.

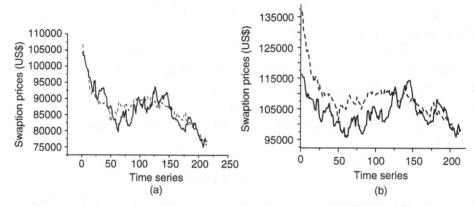

Figure 12.3 (a) 2by10 swaption price versus time t_0 (6 April 2004–28 January 2005), for both market (unbroken line) and the quantum finance model (dashed line). Normalized root mean square error = 3.31%. (b) 5by10 swaption price versus time t_0 (6 April 2004–28 January 2005), both market (unbroken line) and model (dashed line). The normalized root mean square error = 6.31%.

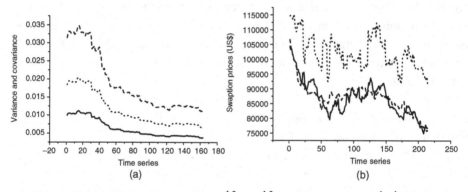

Figure 12.4 (a) Swaption variance $\langle \dot{C}_1^2 \rangle_c$, $\langle \dot{C}_2^2 \rangle_c$ and covariance $\langle \dot{C}_1 \dot{C}_2 \rangle_c$ versus time t_0 (15 June 2004–27 January 2005) computed from the quantum finance model, with the value of the forward bond prices taken from market data. The unbroken line is the variance of a 2by10 swaption, the dashed line is the variance of a 5by10 swaption, and the dotted line is the covariance of the two swaptions. (b) 2by10 swaption price, at the money, from the market (unbroken line), from the quantum finance model (large dashes), and from the HJM model (dotted line). Time t_0 is in the interval (6 April 2004–28 January 2005). The normalized root mean square error for HJM = 18.87% compared with the far more accurate quantum finance swaption formula with error = 3.31%.

Bloomberg provides historical daily data only for the prices of the swaptions *at the money*; swaption prices 'in the money' and 'out of the money' are not quoted. Hence, only the swaption volatility and correlation computed from the model are shown in Figure 12.4(a), without any comparison made with the market value for these quantities.

12.9 Swaption pricing and HJM model

In order to see how the quantum finance model compares with the industry standard one-factor HJM model, the HJM model swaption price was empirically studied. By considering the volatility function to have the special form of $\sigma(t, x) = \sigma_0 e^{-\lambda(x-t)}$, the one-factor HJM model yields the following explicit expression, given in Eq. (4.47), for the coupon bond option

$$C_{HJM}(t_0, t_*, K) = \sum_{i=1}^{N} c_i B(t_0, T_i) N(d_i) - K B(t_0, t_*) N(d)$$

$$d_i \equiv \frac{r'}{\sigma_R} + W(t_*, T_i)\sigma_R; \quad d = \frac{r'}{\sigma_R}$$

$$W(t_*, T_i) \equiv \frac{1}{\lambda}\left[1 - e^{-\lambda(T_i - t_*)}\right]; \quad \sigma_E^2 = \frac{\sigma_0^2}{2\lambda}[1 - e^{-2\lambda(t_* - t_0)}]$$

As shown in Section 11.8, to leading order in σ_0 the HJM limit of the quantum finance pricing formula with exponential volatility yields the HJM pricing formula.

The HJM swaption price is evaluated for exponential volatility $\sigma(t, x) = \sigma_0$ $\exp\{-\lambda(x - t)\}$ and using the daily forward bond prices obtained from ZCYC; σ_0 and λ are estimated from historical ZCYC data. The HJM pricing formula for the swaption price is shown in Figure 12.4(b) together with the market price and the quantum finance swaption price.

The results show that the HJM model is inadequate for pricing swaptions since it systematically over prices the swaption by a large amount. The highly jagged (nondifferentiable) shape of the one-factor HJM swaption price will give incorrect results if one tries to take derivatives that are required for hedging the swaption.

Instead of using the HJM formula for pricing the coupon bond options, practitioners may consider representing the price of the swaption by an implied volatility using the HJM pricing formula. However, unlike the case for the price of caplets where this procedure is possible, the entire swaption curve cannot be fitted by adjusting only one quantity σ_0. Furthermore, the implied volatility $\sigma(t, x)$ in the first place may not be able to fit the price of all swaptions, and, secondly, it will depend on time; it is quite impractical to numerically evaluate daily implied volatility from daily swaption prices.

12.10 Summary

The quantum finance swaption pricing formula was empirically tested by comparing its predictions with the market values. There is over 90% agreement of the theoretical predictions for the swaption's price with its market value, with errors around

6% for most swaptions and with an accuracy of about 3% for the shorter maturity swaptions. The quantum finance formulation directly uses the market correlator $M(x, x'; t)$ and all the market information is fully accounted for in the swaption price.

If one needs to price swaptions that are far in the future, such as a 2 by 10 swaption, one cannot evaluate $M(x, x'; t)$ from the market because data on swaption historical prices are not available over a sufficiently long time period. For such long-dated swaptions, the only way to price them is to first obtain the best fit for volatility $\sigma(t, x)$ from market data and then use the quantum finance model for the propagator $\mathcal{D}(t, x, x')$ to construct the correlator $M(x, x'; t) = \sigma(t, x)\mathcal{D}(t, x, x')\sigma(t, x')$.

The necessity of using an overall scale factor equal to $1/\sqrt{\pi}$ to match the theoretical swaption price with that of the market remains inexplicable. Further analysis is required to fully understand the coupon bond option approach to swaption pricing.

The HJM model is not suited for pricing swaptions because the volatility parameter that goes into the pricing formula cannot be extracted from the swaption data. A comparison of quantum finance and the HJM model for the swaption price shows that the quantum finance model gives a more accurate and stable result than the HJM model.

The quantum finance swaption pricing formula provides an approximate analytical result that can in turn be used to analytically compute the correlation and volatility of swaptions; based on these analytical results one can construct and hedge interest rate portfolios.

The swaption price derived, in Section 8.6, from the Libor Market Model, is not considered. Further empirical studies need to be carried out to decide whether, within the framework of quantum finance, the coupon bond option approach or the Libor Market Model approach is more accurate for pricing swaptions.

13

Correlation of coupon bond options

The correlation of two different coupon bond options is studied in the framework of bond forward interest rates discussed in Chapter 5. Coupon bond options are discounted using the money market numeraire. The correlation is studied for illustrating the mathematics required for pricing more complex instruments, including a more general version of the volatility expansion. The correlation of coupon bonds can lead to the definition of new derivative instruments. This chapter is based on the results of [21].

13.1 Introduction

Exotic equity options often combine a basket of equities that are correlated; the price of the options reflect the effects of equity correlations, which are also required for hedging a portfolio of equities.

The correlation of coupon bond options has many new features not present in the pricing of a single coupon bond option. The calculation for the coupon bond option correlation generalizes the pricing formula obtained for the coupon bond option. The correlation results extend in a straightforward manner to the correlation of swaptions.

A major new feature of the coupon bond option correlation is that – not being traded in the financial markets – it does not have a martingale evolution; in particular, the drift is not fixed by the martingale condition. Instead, the drift for the individual coupon bond option has to be evaluated from market data.

The forward bond numeraire can no longer be used to simplify the option price calculations since the two coupon bond options, in principle, have different maturities. It turns out that the most efficient approach for evaluating the correlation function is to use the money market numeraire for discounting the value of the individual coupon bond options.

13.2 Correlation function of coupon bond options

The volatility expansion developed in Section 3.14 is used for calculating the correlation of two different coupon bond options. The coupon bond options are correlated since they are driven by the same underlying bond forward interest rates. Let the options expire at times $t_2 \geq t_1$, respectively; in a notation that generalizes the payoff function given in Eq. (4.20) let the payoff functions for the coupon bond options be given as below

$$\mathcal{P}_1 = \left(\sum_{i=1}^{N1} c_i B(t_1, T_i) - K_1 \right)_+$$

$$\mathcal{P}_2 = \left(\sum_{j=a}^{N2} c_i B(t_2, T_j) - K_2 \right)_+ \tag{13.1}$$

where $T_i = T_0 + i\ell$ are the fixed times for the coupon payments.

The first payoff function $\mathcal{P}_1$ matures at time T_1, with the zero coupon bond maturing earliest at time $T_1 > t_0$ and the last zero coupon bond maturing at time T_{N1}. Similarly, the second payoff function $\mathcal{P}_2$ matures at time t_2, with the earliest zero coupon bond maturing at $T_a > t_2$ and the last zero coupon bond maturing at time T_{N2}. The domain of each of the two payoff functions is similar to the domain for a single coupon bond option given in Figure 4.10(a); their joint domain is given in Figure 13.1.

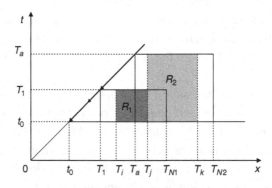

Figure 13.1 The rectangular domains R_1 and R_2 for the bond forward interest rates that determine $\mathcal{M}$: the correlation function of two coupon bond options. Domains R_1 and R_2 overlap, and the dashed line at time T_j indicates that both the options have bonds maturing at T_j and out to T_{N1}.

The connected correlation function of the discounted coupon bond options is given by

$$\mathcal{M} = \mathcal{M}(t_0, t_1, t_2, K_1, K_2) = \langle e^{-\int_{t_0}^{t_1} dt r(t)} \mathcal{P}_1 e^{-\int_{t_0}^{t_2} dt r(t)} \mathcal{P}_2 \rangle_c \qquad (13.2)$$

$$\equiv \langle e^{-\int_{t_0}^{t_1} dt r(t)} \mathcal{P}_1 e^{-\int_{t_0}^{t_2} dt r(t)} \mathcal{P}_2 \rangle - \langle e^{-\int_{t_0}^{t_1} dt r(t)} \mathcal{P}_1 \rangle \langle e^{-\int_{t_0}^{t_2} dt r(t)} \mathcal{P}_2 \rangle$$

$$\equiv \mathcal{M}_{12} - \mathcal{M}_1 \mathcal{M}_2 \qquad (13.3)$$

The correlator of two swaptions is not a traded financial instrument and hence one does not expect its numerical value to be the price of a financial instrument. The expectation value of the correlator consequently need not be evaluated using the martingale measure, and could equally consistently be evaluated using the market evolution of the underlying forward interest rates with the market drift of the forward interest rates not being equal to the martingale drift. The precise probability measure used for performing the averaging $\langle \ldots \rangle$ need not, for now, be completely specified.

The quantities $\mathcal{M}_1$, $\mathcal{M}_2$ are similar to the price of a coupon bond option, except that, unlike the coupon bond option, they need to be evaluated using the market drift. The new piece for the connected correlator is given by $\mathcal{M}_{12}$; to evaluate it, a natural generalization of the notation of Eq. (11.12) yields

$$\mathcal{M}_{12} = B(t_0, t_2) B(t_0, t_1) \left(\frac{1}{2\pi}\right)^2 \int_{-\infty}^{+\infty} dW_1 d\eta_1 dW_2 d\eta_2 (\mathcal{F}_1 + W_1 - K_1)_+$$

$$\times (\mathcal{F}_2 + W_2 - K_2)_+ e^{-i(\eta_1 W_1 + \eta_2 W_2)} Z(\eta_1, \eta_2) \qquad (13.4)$$

The correlator partition function for the two bond options, namely $Z(\eta_1, \eta_2)$, is given by the appropriate generalization of Eq. (11.13), as follows

$$Z(\eta_1, \eta_2) = \langle M_1 e^{i\eta_1 V_1} M_2 e^{i\eta_2 V_2} \rangle \qquad (13.5)$$

The terms $\mathcal{F}_1$, $\mathcal{F}_2$, M_1, M_2, V_1, V_2 are defined in Section 13.4.

13.3 Perturbation expansion for correlator

To leading order, the perturbation expansion for the partition function, from Eq. (13.5), yields

$$Z(\eta_1, \eta_2) = \langle M_1 M_2 \rangle \, e^{ia_1\eta_1 + ia_2\eta_2 - \frac{1}{2}\sum_{ij=1}^{2} \eta_i A_{ij}\eta_j} + O(\eta_1^3, \eta_2^3) \qquad (13.6)$$

The explicit expressions for the coefficients a_1, a_2, and matrix A_{ij} are given in Eqs. (13.17)–(13.20). The coefficients are evaluated for both the martingale and the market drift.

From Eqs. (13.4), (13.5), and (13.6), the correlation function, after performing the η_1, η_2 integrations and some simplifications, is given by

$$\mathcal{M}_{12} = \mathcal{M}_0 \frac{1}{\sqrt{1-\rho^2}} \int_{-\infty}^{+\infty} dW_1 dW_2 (W_1 - X_1)_+ (W_2 - X_2)_+$$

$$\times e^{-\frac{1}{2(1-\rho^2)}\left(W_1^2 + W_2^2 - 2\rho W_1 W_2\right)} \tag{13.7}$$

where

$$\mathcal{M}_0 = \frac{1}{2\pi} B(t_0, t_1) B(t_0, t_2) A$$

$$A = \langle M_1 M_2 \rangle \sqrt{A_{11} A_{22}}$$

$$\rho = \frac{A_{12}}{\sqrt{A_{11} A_{22}}}$$

$$X_1 = \frac{K_1 - \mathcal{F}_1 - a_1}{\sqrt{A_{11}}};$$

$$X_2 = \frac{K_2 - \mathcal{F}_2 - a_2}{\sqrt{A_{22}}} \tag{13.8}$$

The correlator $\mathcal{M}_{12}$ has two possible expansions, namely the case where:

- X_1, X_2 are small and ρ is arbitrary
- ρ is small and X_1, X_2 are arbitrary

13.3.1 Expansion in X_1, X_2

Expanding the payoff function about $X \simeq 0$ yields

$$(W - X)_+ \simeq (W - X)\theta(W) + \frac{X^2}{2}\delta(W) + O(X^3)$$

Hence, performing the integrations in Eq. (13.7) using the properties of the error function yields

$$\mathcal{M}_{12} = \mathcal{M}_0 \left[m_0 + m_1(X_1 + X_2) + m_2 X_1 X_2 + m_3(X_1^2 + X_2^2) \right]$$

$$+ O(X_1^3, X_2^3)$$

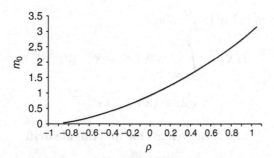

Figure 13.2 Graph of m_0, with ρ plotted along the x-axis and value of m_0 along the y-axis.

with the coefficients being given by

$$
m_0 = \begin{cases} \rho\left[\pi + \sqrt{\frac{1-\rho^2}{\rho^2}} - \tan^{-1}\left(\sqrt{\frac{1-\rho^2}{\rho^2}}\right)\right]; & \rho \geq 0 \\ |\rho|\left[\sqrt{\frac{1-\rho^2}{\rho^2}} - \tan^{-1}\left(\sqrt{\frac{1-\rho^2}{\rho^2}}\right)\right]; & \rho \leq 0 \end{cases}
$$

$$
m_1 = \sqrt{\frac{\pi}{2}}(1+\rho)
$$

$$
m_2 = \begin{cases} \pi - \tan^{-1}\left(\sqrt{\frac{1-\rho^2}{\rho^2}}\right); & \rho \geq 0 \\ \tan^{-1}\left(\sqrt{\frac{1-\rho^2}{\rho^2}}\right); & \rho \leq 0 \end{cases}
$$

$$
m_3 = \frac{1}{2}\sqrt{1-\rho^2}
$$

Note that $\mathcal{M}_{12}$ is a continuous function of ρ, with the graph of m_0 given in Figure 13.2.

13.3.2 Expansion in ρ

To $O(\rho^2)$, the W_1, W_2 integrations in Eq. (13.7) completely factorize. Expanding $\mathcal{M}_{12}$ in a power series in ρ yields

$$
\mathcal{M}_{12} = \mathcal{M}_0 \int_{-\infty}^{+\infty} dW_1 dW_2 \left[1 + \rho W_1 W_2 + O(\rho^2)\right]
$$

$$
\times \left(W_1 - X_1\right)_+ \left(W_2 - X_2\right)_+ e^{-\frac{1}{2}(W_1^2 + W_2^2)}
$$

$$
= \mathcal{M}_0\left[I(X_1)I(X_2) + \rho J(X_1)J(X_2)\right] + O(\rho^2) \tag{13.9}
$$

where $I(X)$ is given in Eq. (3.70) and

$$J(X) = \int_{-\infty}^{+\infty} dW(W - X)_{+} W e^{-\frac{1}{2}W^{2}}$$

$$= \sqrt{2\pi} N(X) - 2X e^{-\frac{X^{2}}{2}} \tag{13.10}$$

13.4 Coefficients for martingale drift

The explicit expressions for the coefficients a_1, a_2, and matrix A_{ij} required for obtaining $\mathcal{M}_{12}$ are computed.

For the purpose of illustrating the computation required, the calculation for the correlated coupon bond option is analytically carried out using the money market numeraire [11] given by $\exp(\int_{t_0}^{t} dt' r(t'))$, which yields a martingale measure for all zero coupon bonds. All expectation values in this section are defined using the money market numeraire, and the martingale condition states, from Eq. (3.4), that

$$\langle e^{-\int_{t_0}^{t} dt' r(t')} B(t, T)\rangle \equiv E[e^{-\int_{t_0}^{t} dt' r(t')} B(t, T)] = B(t_0, T) \tag{13.11}$$

The money market numeraire drift velocity, as derived in Section 9.2, is given by

$$\alpha(t, x) = \int_{t}^{x} dx' M(x, x'; t) \tag{13.12}$$

The money market numeraire is the most suitable numeraire for finding the correlation function as it treats both the payoff functions on an equal basis.

The correlator partition function from Eq. (13.5) is given by

$$Z(\eta_1, \eta_2) = \langle M_1 e^{i\eta_1 V_1} M_2 e^{i\eta_2 V_2}\rangle \tag{13.13}$$

with the following definitions[1]

$$M_1 = e^{-\int_{t_0}^{t_1} dtr(t)}/B(t_0, t_1); \quad M_2 = e^{-\int_{t_0}^{t_2} dtr(t)}/B(t_0, t_2)$$

$$\mathcal{F}_1 = \sum_{i=1}^{N1} c_i F_{1i}; \quad \mathcal{F}_2 = \sum_{j=a}^{N2} c_i F_{2i}; \quad J_{1i} = c_{1i} F_{1i}; \quad J_{2i} = c_{2i} F_{2i}$$

$$V_1 = \sum_{i=1}^{N1} J_{1i}[e^{-\alpha_{1i} - \mathcal{Q}_{1i}} - 1]; \quad V_2 = \sum_{j=a}^{N2} J_{2j}[e^{-\alpha_{2j} - \mathcal{Q}_{2j}} - 1]$$

---

[1] Generalizing the notation of Eq. (11.3), namely that

$$B(t_*, T_i) = \exp\left\{-\int_{t_*}^{T_i} dx f(t_*, x)\right\} = e^{-\alpha_i - \mathcal{Q}_i} F(t_0, t_*, T_i) \tag{13.14}$$

The two payoff functions are shown in Figure 13.3.
Using the money market drift velocity given in Eq. (13.12) yields

$$\alpha_{1i} = \int_{t_0}^{t_1} dt \int_{t_1}^{T_i} dx \alpha(t,x) = \int_{t_0}^{t_1} dt \int_{t_1}^{T_i} dx \int_{t}^{x} dx' M(x,x';t)$$

$$Q_{1i} = \int_{R_i} \sigma(t,x) \mathcal{A}(t,x) = \int_{t_0}^{t_1} dt \int_{t_1}^{T_i} dx \sigma(t,x) \mathcal{A}(t,x)$$

$$\alpha_{2j} = \int_{t_0}^{t_2} dt \int_{t_2}^{T_j} dx \int_{t}^{x} dx' M(x,x';t)$$

$$Q_{2j} = \int_{R_j} \sigma(t,x) \mathcal{A}(t,x) = \int_{t_0}^{t_2} dt \int_{t_2}^{T_j} dx \sigma(t,x) \mathcal{A}(t,x) \qquad (13.15)$$

and the discount factors yield, from Eq. (5.1) and for $i = 1, 2$, the following

$$M_i = e^{-\int_{\Delta_i} \alpha - \int_{\Delta_i} \sigma \mathcal{A}}$$

$$\int_{\Delta_i} \alpha = \int_{t_0}^{t_i} dt \int_{t}^{t_i} dx \alpha(t,x)$$

$$\int_{\Delta_i} \sigma \mathcal{A} = \int_{t_0}^{t_i} dt \int_{t}^{t_i} dx \sigma(t,x) \mathcal{A}(t,x)$$

Domains R_i, R_j, Δ_1, Δ_2 are given in Figure 13.3.
The partition function for the correlation function of two swaptions, from Eq. (13.5), is given by

$$Z(\eta_1, \eta_2) = \langle M_1 M_2 \rangle \, e^{ia_1\eta_1 + ia_2\eta_2 - \frac{1}{2}\sum_{ij=1}^{2} \eta_i A_{ij} \eta_j} + O(\eta_1^3, \eta_2^3) \qquad (13.16)$$

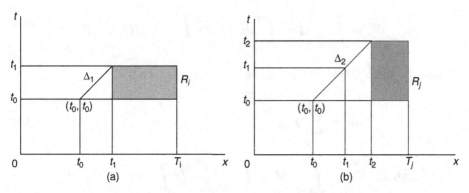

Figure 13.3 (a) Payoff $\mathcal{P}_1$ matures at time t_1, with a typical bond maturing at time T_i. (b) Payoff $\mathcal{P}_2$ matures at time t_2, with a typical bond maturing at time T_j.

Using the fact that for the money market numeraire $\langle M_1 \rangle = \langle M_2 \rangle = 1$ the coefficients up to terms of $O(\eta_1^3, \eta_2^3)$ are given, from Eqs. (13.5) and (13.16), by the following

$$a_1 = \frac{1}{\langle M_1 M_2 \rangle} \langle M_1 M_2 V_1 \rangle \qquad (13.17)$$

$$a_2 = \frac{1}{\langle M_1 M_2 \rangle} \langle M_1 M_2 V_2 \rangle \qquad (13.18)$$

$$A_{ii} = \frac{1}{\langle M_1 M_2 \rangle} \langle M_1 M_2 V_i^2 \rangle - a_i^2; \quad i = 1, 2 \qquad (13.19)$$

$$A_{12} = \frac{1}{\langle M_1 M_2 \rangle} \langle M_1 M_2 V_1 V_2 \rangle - a_1 a_2 = A_{21} \qquad (13.20)$$

All the calculations for the coefficients of η_1, η_2 are given by Gaussian path integrations, as was the case for evaluating the price of the coupon bond option. The path integrals for evaluating the coefficients a_1, a_2, and A_{ij} are carried out on the various sub-domains of R_1, R_2 shown in Figure 13.1.

The definition of M_1 and M_2 yields the following

$$\langle M_1 M_2 \rangle = e^{\Omega_{12}} \qquad (13.21)$$

$$\Omega_{12} = \int_{\mathcal{T}_{12}} M(x, x'; t) \equiv \int_{t_0}^{t_1} dt \int_{t}^{t_1} dx \int_{t}^{t_2} dx' M(x, x'; t)$$

The domain $\mathcal{T}_{12}$ is given in Figure 13.4(a). Furthermore, Gaussian integrations yield

$$a_1 = \frac{1}{\langle M_1 M_2 \rangle} \langle M_1 M_2 V_1 \rangle = e^{-\Omega_{12}} \langle M_1 M_2 V_1 \rangle = \sum_{i=1}^{N1} J_i \left(e^{\Gamma_{2i}} - 1 \right)$$

where

$$\Gamma_{2i} = \int_{\Delta_2 R_i} M = \int_{t_0}^{t_1} dt \int_{t}^{t_2} dx \int_{t_1}^{T_i} dx' M(x, x'; t)$$

Similarly, since $\langle M_2 V_2 \rangle = 0$, the coefficient a_2 is given by

$$a_2 = \sum_{j=a}^{N2} J_j \left(e^{\Gamma_{1j}} - 1 \right) \qquad (13.22)$$

$$\text{where} \quad \Gamma_{1j} = \int_{\Delta_1 R_j} M = \int_{t_0}^{t_1} dt \int_{t}^{t_1} dx \int_{t_2}^{T_j} dx' M(x, x'; t)$$

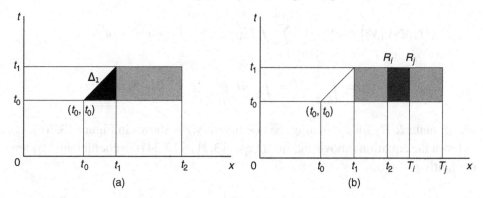

Figure 13.4 (a) Domain $\mathcal{T}_{12}$ for evaluating $\Omega_{12} = \int_{\mathcal{T}_{12}} M(x,x';t)$. (b) $\gamma_{ij} = \int_{R_i R_j} M = \int_{t_0}^{t_1} dt \int_{t_1}^{T_i} dx \int_{t_2}^{T_j} dx' M(x,x';t)$ is evaluated on domains R_i and R_j.

The coefficients A_{ij} are now evaluated in a manner similar to the derivation of Eqs. (11.24) and (11.29). Since Eq. (13.12) yields $\langle M_1 V_1 \rangle = 0$, one obtains the following

$$\langle M_1 V_1^2 \rangle = \sum_{ii'=1}^{N1} J_i J_{i'} \left(e^{G_{ii'}} - 1 \right) \tag{13.23}$$

$$\langle M_2 V_2^2 \rangle = \sum_{jj'=a}^{N2} J_j J_{j'} \left(e^{G_{jj'}} - 1 \right) \tag{13.24}$$

with $G_{ii'} = \int_{R_i R_{i'}} M = \int_{t_0}^{t_1} dt \int_{t_1}^{T_i} dx \int_{t_1}^{T_{i'}} dx' M(x,x';t)$

$G_{jj'} = \int_{R_j R_{j'}} M = \int_{t_0}^{t_2} dt \int_{t_2}^{T_j} dx \int_{t_2}^{T_{j'}} dx' M(x,x';t)$

The indices i, i' refer to payoff $\mathcal{P}_1$ and indices j, j' refer to $\mathcal{P}_2$.

Consider the expectation values

$$\langle M_1 M_2 V_1^2 \rangle = e^{\Omega_{12}} \sum_{ii'=1}^{N1} J_i J_{i'} \left(e^{\Gamma_{2i} + \Gamma_{2i'} + G_{ii'}} - e^{\Gamma_{2i}} - e^{\Gamma_{2i'}} + 1 \right)$$

$$\langle M_1 M_2 V_2^2 \rangle = e^{\Omega_{12}} \sum_{jj'=a}^{N2} J_j J_{j'} \left(e^{\Gamma_{1j} + \Gamma_{1j'} + G_{jj'}} - e^{\Gamma_{1j}} - e^{\Gamma_{1j'}} + 1 \right)$$

$$\langle M_1 M_2 V_1 V_2 \rangle = e^{\Omega_{12}} \sum_{i=1}^{N1} \sum_{j=a}^{N2} \mathcal{J}_i \mathcal{J}_j \left(e^{\Gamma_{1j} + \Gamma_{2i} + \gamma_{ij}} - e^{\Gamma_{2i}} - e^{\Gamma_{1j}} + 1 \right)$$

$$\text{with } \gamma_{ij} = \int_{R_i R_j} M = \int_{t_0}^{t_1} dt \int_{t_1}^{T_i} dx \int_{t_2}^{T_j} dx' M(x, x'; t)$$

The domain $R_i R_j$ for evaluating the coefficient γ_{ij} is shown in Figure 13.4(b).

From the equations above and from Eqs. (13.21)–(13.24) the coefficients A_{ij} are explicitly given by

$$A_{11} = \sum_{ii'=1}^{N1} \mathcal{J}_i \mathcal{J}_{i'} \left(e^{G_{ii'}} - 1 \right)$$

$$A_{22} = \sum_{jj'=a}^{N2} \mathcal{J}_j \mathcal{J}_{j'} \left(e^{G_{jj'}} - 1 \right)$$

$$A_{12} = A_{21} = \sum_{i=1}^{N1} \sum_{j=a}^{N2} \mathcal{J}_i \mathcal{J}_j \left(e^{\gamma_{ij}} - 1 \right)$$

where

$$\mathcal{J}_i = J_i e^{\Gamma_{1i}}; \quad i = 1, 2, \ldots, N1$$

$$\mathcal{J}_j = J_j e^{\Gamma_{2j}}; \quad j = a, a+1, \ldots, N2$$

To compute $\mathcal{M} = \mathcal{M}_{12} - \mathcal{M}_1 \mathcal{M}_2$, one needs to determine $\mathcal{M}_1$ and $\mathcal{M}_2$. For the martingale measure, $\mathcal{M}_1$ and $\mathcal{M}_2$ are simply the coupon bond option prices, and hence

$$\mathcal{M}_1 = C_1(t_0, t_1; K_1)$$

$$\mathcal{M}_2 = C_2(t_0, t_2; K_2)$$

The result obtained for the correlator of two coupon bond options $\mathcal{M}$ in Eq. (13.2) contains the special case of the variance of the swaptions. For $t_1 = t_2, K_1 = K_2, N = N1 = N2$, and $a = 1$, one has $\mathcal{P}_1 = \mathcal{P}_2$. The *auto-correlation* of $\tilde{\mathcal{P}}_1$ is given by

$$\sigma^2(\tilde{\mathcal{P}}_1) = \langle \tilde{\mathcal{P}}_1^2 \rangle - \left(\langle \tilde{\mathcal{P}}_1 \rangle \right)^2$$

$$= \mathcal{M}(t_0, t_1, t_1, K_1, K_1)$$

$$\text{with } \tilde{\mathcal{P}}_1 = e^{-\int_{t_0}^{t_1} dt \, r(t)} \mathcal{P}_1 \qquad (13.25)$$

Note, $\tilde{\mathcal{P}}_2 = e^{-\int_{t_0}^{t_2} dt\, r(t)} \tilde{\mathcal{P}}_2$ yields a similar expression for $\sigma^2(\tilde{\mathcal{P}}_2)$.

The variance of the coupon bond option prices is derived in Section 13.8 and provides a check for the correlation function $\mathcal{M}$.

13.5 Coefficients for market drift

Recall, as discussed earlier, the explicit expressions for the coefficients a_1, a_2, and matrix A_{ij} required for obtaining $\mathcal{M}_{12}$ can also be evaluated from the market evolution, in particular using the market drift, of the underlying empirical forward interest rates.

The drift terms need to be redefined in order to account for the market drift

$$\alpha_{1i} = \int_{t_0}^{t_1} dt \int_{t_1}^{T_i} dx\, \alpha(t, x)$$

$$\alpha_{2j} = \int_{t_0}^{t_2} dt \int_{t_2}^{T_j} dx\, \alpha(t, x) \tag{13.26}$$

where $\alpha(t, x)$ is the market drift which has to be computed from the empirical forward interest rates.

The coefficients in Eq. (13.16) are now given by the following

$$a_1 = \frac{1}{\langle M_1 M_2 \rangle_c} \left[\langle M_1 M_2 V_1 \rangle - \langle M_1 V_1 \rangle \langle M_2 \rangle \right] \tag{13.27}$$

$$a_2 = \frac{1}{\langle M_1 M_2 \rangle_c} \left[\langle M_1 M_2 V_2 \rangle - \langle M_2 V_2 \rangle \langle M_1 \rangle \right] \tag{13.28}$$

$$A_{ii} = \frac{1}{\langle M_1 M_2 \rangle_c} \left[\langle M_1 M_2 V_i^2 \rangle - \langle M_i V_i^2 \rangle \langle M_j \rangle \right] - a_i^2; \quad j \neq i \tag{13.29}$$

$$A_{12} = \frac{2}{\langle M_1 M_2 \rangle_c} \left[\langle M_1 M_2 V_1 V_2 \rangle - \langle M_1 V_1 \rangle \langle M_2 V_2 \rangle \right] - 2 a_1 a_2 \tag{13.30}$$

$$= A_{21}$$

The definition of M_i for nonmartingale drift yields the following

$$\langle M_i \rangle = e^{-\int_{\Delta_i} \alpha + \frac{1}{2} \int_{\Delta_i} M(x, x'; t)} \quad\quad i = 1, 2 \tag{13.31}$$

$$\langle M_1 M_2 \rangle = e^{\Omega_{12} - \int_{\Delta_1} \alpha - \int_{\Delta_2} \alpha + \frac{1}{2} \int_{\Delta_1} M(x, x'; t) + \frac{1}{2} \int_{\Delta_2} M(x, x'; t)} \tag{13.32}$$

$$\langle M_1 M_2 \rangle_c = \langle M_1 M_2 \rangle - \langle M_1 \rangle \langle M_2 \rangle \tag{13.33}$$

where

$$\int_{\Delta_i} \alpha = \int_{t_0}^{t_i} dt \int_t^{t_i} dx\, \alpha(t, x)$$

$$\int_{\Delta_i} M(x, x'; t) = \int_{t_0}^{t_i} dt \int_t^{t_i} dx \int_t^{t_i} dx'\, M(x, x'; t)$$

$$\Omega_{12} = \int_{T_{12}} M(x, x'; t) \equiv \int_{t_0}^{t_1} dt \int_t^{t_1} dx \int_t^{t_2} dx'\, M(x, x'; t)$$

Gaussian path integrations yield

$$\langle M_1 M_2 V_1 \rangle = \langle M_1 M_2 \rangle \sum_{i=1}^{N1} J_i \left(e^{-\alpha_{1i} + \frac{1}{2} G_{ii'} + \Gamma_{1i} + \Gamma_{2i}} - 1 \right) \tag{13.34}$$

where $\Gamma_{1i} = \int_{\Delta_1 R_i} M = \int_{t_0}^{t_1} dt \int_t^{t_1} dx \int_{t_1}^{T_i} dx'\, M(x, x'; t)$

$$\Gamma_{2i} = \int_{\Delta_2 R_i} M = \int_{t_0}^{t_1} dt \int_t^{t_2} dx \int_{t_1}^{T_i} dx'\, M(x, x'; t)$$

$$G_{ii'} = \int_{R_i R_{i'}} M = \int_{t_0}^{t_1} dt \int_{t_1}^{T_i} dx \int_{t_1}^{T_{i'}} dx'\, M(x, x'; t)$$

Similarly

$$\langle M_1 M_2 V_2 \rangle = \langle M_1 M_2 \rangle \sum_{j=a}^{N2} J_j \left(e^{-\alpha_{2j} + \frac{1}{2} G_{jj'} + \Gamma_{1j} + \Gamma_{2j}} - 1 \right) \tag{13.35}$$

where $\Gamma_{1j} = \int_{\Delta_1 R_j} M = \int_{t_0}^{t_1} dt \int_t^{t_1} dx \int_{t_2}^{T_j} dx'\, M(x, x'; t)$

$$\Gamma_{2j} = \int_{\Delta_2 R_j} M = \int_{t_0}^{t_1} dt \int_t^{t_2} dx \int_{t_2}^{T_j} dx'\, M(x, x'; t)$$

$$G_{jj'} = \int_{R_j R_{j'}} M = \int_{t_0}^{t_2} dt \int_{t_2}^{T_j} dx \int_{t_2}^{T_{j'}} dx'\, M(x, x'; t)$$

and

$$\langle M_1 V_1 \rangle = \langle M_1 \rangle \sum_{i=1}^{N1} J_i \left(e^{-\alpha_{1i} + \frac{1}{2} G_{ii'} + \Gamma_{1i}} - 1 \right) \tag{13.36}$$

$$\langle M_2 V_2 \rangle = \langle M_2 \rangle \sum_{j=a}^{N2} J_j \left(e^{-\alpha_{2j} + \frac{1}{2} G_{jj'} + \Gamma_{2j}} - 1 \right) \tag{13.37}$$

Coefficients a_1 and a_2 can now be evaluated by Eqs. (13.27) and (13.28), given the above explicit expressions. The coefficients A_{ij} are now evaluated in a similar manner.

$$\langle M_1 V_1^2 \rangle = \langle M_1 \rangle \sum_{i,i'=1}^{N1} J_i J_{i'} \left(e^{-\alpha_{1i}-\alpha_{1i'}+\frac{1}{2}G_{ii}+\frac{1}{2}G_{i'i'}+G_{ii'}+\Gamma_{1i}+\Gamma_{1i'}} \right.$$
$$\left. - e^{-\alpha_{1i}+\frac{1}{2}G_{ii}+\Gamma_{1i}} - e^{-\alpha_{1i'}+\frac{1}{2}G_{i'i'}+\Gamma_{1i'}} + 1 \right) \tag{13.38}$$

$$\langle M_2 V_2^2 \rangle = \langle M_2 \rangle \sum_{j,j'=a}^{N2} J_j J_{j'} \left(e^{-\alpha_{2j}-\alpha_{2j'}+\frac{1}{2}G_{jj}+\frac{1}{2}G_{j'j'}+G_{jj'}+\Gamma_{2j}+\Gamma_{2j'}} \right.$$
$$\left. - e^{-\alpha_{2j}+\frac{1}{2}G_{jj}+\Gamma_{2j}} - e^{-\alpha_{2j'}+\frac{1}{2}G_{j'j'}+\Gamma_{2j'}} + 1 \right) \tag{13.39}$$

and

$$\langle M_1 M_2 V_1^2 \rangle = \langle M_1 M_2 \rangle \sum_{i,i'=1}^{N1} J_i J_{i'} \left(e^{-\alpha_{1i}-\alpha_{1i'}+\Omega+\frac{1}{2}G_{ii}+\frac{1}{2}G_{i'i'}+G_{ii'}+\Gamma_{1i}+\Gamma_{1i'}+\Gamma_{2i}+\Gamma_{2i'}} \right.$$
$$\left. - e^{-\alpha_{1i}+\Omega+\frac{1}{2}G_{ii}+\Gamma_{1i}+\Gamma_{2i}} - e^{-\alpha_{1i'}+\Omega+\frac{1}{2}G_{i'i'}+\Gamma_{1i'}+\Gamma_{2i'}} + 1 \right) \tag{13.40}$$

$$\langle M_1 M_2 V_2^2 \rangle = \langle M_1 M_2 \rangle \sum_{j,j'=a}^{N2} J_j J_{j'} \left(e^{-\alpha_{2j}-\alpha_{2j'}+\Omega+\frac{1}{2}G_{jj}+\frac{1}{2}G_{j'j'}+G_{jj'}+\Gamma_{1j}+\Gamma_{1j'}+\Gamma_{2j}+\Gamma_{2j'}} \right.$$
$$\left. - e^{-\alpha_{2j}+\Omega+\frac{1}{2}G_{jj}+\Gamma_{1j}+\Gamma_{2j}} - e^{-\alpha_{2j'}+\Omega+\frac{1}{2}G_{j'j'}+\Gamma_{1j'}+\Gamma_{2j'}} + 1 \right) \tag{13.41}$$

$$\langle M_1 M_2 V_1 V_2 \rangle = \langle M_1 M_2 \rangle \sum_{i=1}^{N1} \sum_{j=a}^{N2} J_i J_j \left(e^{-\alpha_{1i}-\alpha_{2j}+\Omega+\frac{1}{2}G_{ii}+\frac{1}{2}G_{jj}+G_{ij}+\Gamma_{1i}+\Gamma_{1j}+\Gamma_{2i}+\Gamma_{2j}} \right.$$
$$\left. - e^{-\alpha_{1i}+\Omega+\frac{1}{2}G_{ii}+\Gamma_{1i}+\Gamma_{2i}} - e^{-\alpha_{2j}+\Omega+\frac{1}{2}G_{jj}+\Gamma_{1j}+\Gamma_{2j}} + 1 \right) \tag{13.42}$$

Furthermore, $\mathcal{M}_1$ and $\mathcal{M}_2$ are not the coupon bond option prices if one consistently uses the market drift. Different from the one derived in [9], one has

$$D_i = \frac{1}{\langle M_i \rangle} \langle M_i V_i \rangle \tag{13.43}$$

$$A_i = \frac{1}{\langle M_i \rangle} \langle M_i V_i^2 \rangle - D_i^2; \quad i = 1,2 \tag{13.44}$$

13.6 Empirical study

The correlation and auto-correlation of coupon bond options have been evaluated for both martingale and nonmartingale evolution of the bond forward interest rates

Table 13.1 *Evaluating* $\mathcal{I} = \int_{t_0}^{m1} dt \int_{m2}^{d1} dx \int_{m3}^{d2} dx' M(t, x, x')$ *for different limits of integration. When an entry has the value of* t, *it means it is not a fixed value and depends on the* t *integration.*

$\mathcal{I}$	Γ_{*i}	G_{ij}	Γ_{1i}	Γ_{2i}	Γ_{2j}	Γ_{2j}	$G_{ii'}$	$G_{jj'}$	γ_{ij}
$m1$	t_*	t_*	t_1	t_1	t_1	t_1	t_1	t_2	t_1
$m2$	t	t_*	t	t	t	t	t_1	t_2	t_1
$m3$	t_*	t_*	t_1	t_1	t_2	t_2	t_1	t_2	t_2
$d1$	t_*	T_i	t_1	t_2	t_1	t_2	T_i	T_j	T_i
$d2$	T_i	T_j	T_i	T_i	T_j	T_j	$T_{i'}$	$T_{j'}$	T_j

by using analytical techniques; these results are now studied empirically. The structure of the ZCYC data and computational procedures discussed in Chapter 12 carry over to the empirical study of this section.

The analytical results show that all the computations finally boil down to a set of three-dimensional integrations on $M(x, x'; t)$

$$M(x, x'; t) = \sigma(t, x)\mathcal{D}(x, x'; t)\sigma(t, x')$$

with various integration limits. A general form for all the integrations, similar to Eq. (12.13), is given by

$$\mathcal{I} = \int_{t_0}^{m1} dt \int_{m2}^{d1} dx \int_{m3}^{d2} dx' M(x, x'; t) \tag{13.45}$$

with the limits of integrations being listed in the Table 13.1.

The evaluation $\mathcal{I}$, given in Eq. (13.45), is similar to the computation carried out in Section 12.7, except now, as can be seen from Table 13.1, the limits on $\mathcal{I}$ are more complicated.

A set of two-dimensional integrations on $\alpha(t, x)$ is required for evaluating market drift and which has the general form

$$\mathcal{D} = \int_{t_0}^{m1} dt \int_{m2}^{d1} dx\, \alpha(t, x) \tag{13.46}$$

with the limits of integration being listed in Table 13.2.

As discussed in Section 4.9 swaptions are equivalent to a special class of coupon bond options. Swaption data will be used to empirically study the formulae obtained for the correlation of coupon bond options. A swaption's price is equivalent to a coupon bond option, which will be taken to mature at t_*; the indices i and j run from 1 to N, with the last payment being made at T_N. For the correlation, swaptions will

Table 13.2 *Limits of integration for evaluating* $\mathcal{D} = \int_{t_0}^{m1} dt \int_{m2}^{d1} dx \, \alpha(t, x)$. *When m2 has the value of t, it means it is not a fixed value and depends on the t integration.*

$\mathcal{D}$	α_{Δ_1}	α_{Δ_2}	α_{1i}	α_{2j}
$m1$	t_1	t_2	t_1	t_2
$m2$	t	t	t_1	t_2
$d1$	t_1	t_2	T_i	T_j

mature at two different times $t_2 \geq t_1$ and the two indices i and j have the following rages: $i = 1, 2, \ldots, N1$ and $j = a, a+1, \ldots, N2$; the last payments are made at T_{N1} and T_{N2} respectively.

Since

$$M(x, x'; t) = \frac{1}{\epsilon} \langle \delta f(t, x) \delta f(t, x') \rangle_c \qquad (13.47)$$

one has, similar to Eq. (12.16), the following

$$\mathcal{I} = \frac{1}{\epsilon} \int_{t_0}^{m1} dt \int_{m2}^{d1} dx \int_{m3}^{d2} dx' \langle \delta f(t, x) \delta f(t, x') \rangle_c \qquad (13.48)$$

A shift on integration domain, as discussed in Section 12.7, converts integration on future data into a summation on current and past data as follows ($t = t_0 + t_k$; $t_k = k\epsilon'$)

$$\mathcal{I} = \frac{\epsilon'}{\epsilon} \sum_{t_k} \int_{m2-t_k}^{d1-t_k} dy \int_{m3-t_k}^{d2-t_k} dy' M(y + t_k, y' + t_k; t_0 + t_k)$$

$$= \frac{\epsilon'}{\epsilon} \sum_{t_k} \int_{m2-t_k}^{d1-t_k} dy \int_{m3-t_k}^{d2-t_k} dy' M(y, y'; t_0)$$

$$= \frac{\epsilon'}{\epsilon} \sum_{t_k} \left\langle \int_{m2-t_k}^{d1-t_k} \delta f(t_0, y) dy \int_{m3-t_k}^{d2-t_k} \delta f(t_0, y') dy' \right\rangle \qquad (13.49)$$

The shift in the integration over future calendar time is shown in Figure 12.2.

In order to use the ZCYC data directly, the operator δ_{t_0} needs to be taken outside the integration. As can be seen from the list of integration limits given in Table 13.1, one of the lower limits of integration can be the integration variable t; hence, unlike the case analyzed in Section 12.7, two different results are possible and are shown below.

- When both the lower limit of y and y' are fixed

$$\mathcal{I} = \frac{\epsilon'}{\epsilon} \sum_{t_k} \langle \delta Y(t_0, m2 - t_k, d1 - t_k) \delta Y(t_0, m3 - t_k, d2 - t_k) \rangle$$

where[2]

$$Y(t_0, t_*, T) = \int_{t_*}^{T} f(t, x) dx = \int_{t_0}^{T} f(t_0, x) dx - \int_{t_0}^{t_*} f(t_0, x) dx$$

$$= \log\left(\left(1 + \frac{1}{c} Z(t, T)\right)^{c[T-t]}\right) - \log\left(\left(1 + \frac{1}{c} Z(t_0, t_*)\right)^{c[t_*-t_0]}\right)$$

- When one of the lower limits is not fixed, say $m2 = t$

$$\mathcal{I} = \frac{\epsilon'}{\epsilon} \sum_{t_k} \left\langle \left\{ \int_{t_0}^{d1-t_k} \delta f(t_0, y) dy \int_{m3-t_k}^{d2-t_k} \delta f(t_0, y') dy' \right\rangle \right.$$ (13.50)

since

$$\int_{t_0}^{T} \delta_{t_0} f(t, x) dx = \delta_{t_0} \int_{t_0}^{T} f(t_0, x) + \epsilon f(t_0, t_0)$$ (13.51)

this yields

$$\mathcal{I} = \frac{\epsilon'}{\epsilon} \sum_{t_k} \langle \{\delta Y(t_0, t_0, d1 - t_k) + \epsilon f(t_0, t_0)\} \delta Y(t_0, m3 - t_k, d2 - t_k) \rangle$$

Similarly, the market drift can be derived from ZCYC data as

$$\alpha(t, x) = \frac{1}{\epsilon} \langle \delta f(t, x) \rangle$$ (13.52)

From Eq. (13.46), by using the technique of shifting the calendar time integration, integrating over future calendar time is converted to an integration on current and past data and yields

$$\mathcal{D} = \frac{\epsilon'}{\epsilon} \sum_{t_k} \left\langle \int_{m2-t_k}^{d1-t_k} \delta f(t_0, y) dy \right\rangle$$ (13.53)

---

[2] Recall from Eq. (2.28), the future integral of the forward interest rates is given in terms of the ZCYC as follows

$$B(t, T) = \frac{1}{[1 + \frac{1}{c} Z(t, T)]^{c[T-t]}} = \exp\left\{-\int_{t}^{T} dx f(t, x)\right\}$$

$$\Rightarrow \int_{t}^{T} dx f(t, x) = c[T - t] \ln\left[1 + \frac{1}{c} Z(t, T)\right]$$

where integers $[T - t] = (T - t)/1$ year and $[t_* - t] = (t_* - t)/1$ year.

Again, there are two different results as shown below:

- When the lower limit of y is fixed

$$\mathcal{I} = \frac{\epsilon'}{\epsilon} \sum_{t_k} \langle \delta Y(t_0, m2 - t_k, d1 - t_k) \rangle$$

- The lower limit is not fixed, say $m2 = t$

$$\mathcal{I} = \frac{\epsilon'}{\epsilon} \sum_{t_k} \left\langle \int_{t_0}^{d1-t_k} \delta f(t_0, y) dy \right\rangle \tag{13.54}$$

and this yields

$$\mathcal{I} = \frac{\epsilon'}{\epsilon} \sum_{t_k} \langle \delta Y(t_0, t_0, d1 - t_k) + \epsilon f(t_0, t_0) \rangle$$

The average $\langle \ldots \rangle$ for the case of variable t_0 is done by a moving average over the last 180 historical days with fixed future time value x. Cubic spline is used for interpolating ZCYC to a daily interval. The whole shift process is illustrated as Figure 12.2.

The auto-correlation was first evaluated, which is a special case of the coupon bond option correlation for $t_1 = t_2$; the numerical results verified that the correlation reduces to the auto-correlation. The results for correlation and auto-correlation of coupon bond options are plotted in Figure 13.5(a) for martingale drift and in Figure 13.5(b) for market drift.

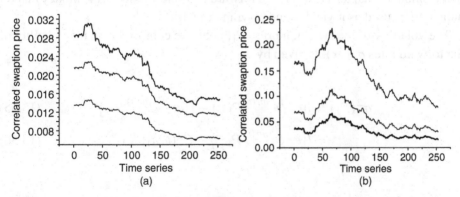

Figure 13.5 The curve on top is the auto-correlation of the 5by10 swaption and the lower curve that of the 2by10 swaption. The curve in the middle is the covariance of the two swaptions. (a) Correlated coupon bond options with martingale drift, computed from the model, plotted versus time t_0 (15 June 2004–27 January 2005). (b) Correlated coupon bond options with market drift, computed from the model, plotted versus time t_0 (15 June 2004–27 January 2005).

13.7 Summary

The quantum finance formulation of bond forward interest rates provides a scheme for approximately evaluating the correlation of coupon bond options and swaptions. The empirical value of the bond forward interest rates' volatility is a small parameter and was used for developing a volatility expansion for the correlation. The perturbation expansion was realized by expanding the nonlinear terms in the partition function, and performing the path integral using Gaussian integrations.

The forward bond measure is an efficient measure for pricing options, whereas the money market numeraire is the appropriate measure for evaluating the correlation of two coupon bond options. The nonmartingale market drift was studied to examine how an instrument with market drift can be analyzed in the framework of quantum finance.

13.8 Appendix: Bond option auto-correlation

The auto-correlation of a coupon bond option is the limit of taking the two coupon bonds in the correlated case to be identical, that for $\mathcal{P}_1 = \mathcal{P}_2 \equiv \mathcal{P}$. Since the computation of auto-correlation is a bit simpler than the case for the correlator, the computation is carried out for its own interest and for providing a check on the correlated case.

With the notation of the correlation function of two coupon bond options (swaptions) in mind, let the option $\mathcal{P}_1$ mature at time t_1. The option's volatility is defined using the *martingale measure*; unlike the computation for the price of a coupon bond option in Chapter 11, where the forward measure was used, the money market numeraire is used as it yields a more symmetric drift.

The volatility of the swaption price $\mathcal{P}_1$, not to be confused with the volatility of the forward rates $\sigma(t, x)$, is given by

$$\sigma^2(\tilde{\mathcal{P}}_1) = \langle [e^{-\int_{t_0}^{t_1} dtr(t)} \mathcal{P}_1]^2 \rangle - [\langle e^{-\int_{t_0}^{t_1} dtr(t)} \mathcal{P}_1 \rangle]^2 \qquad (13.55)$$

$$= \langle [e^{-\int_{t_0}^{t_1} dtr(t)} \mathcal{P}_1]^2 \rangle - C^2(t_0, t_1; K) \qquad (13.56)$$

$$\tilde{\mathcal{P}}_1 = e^{-\int_{t_0}^{t_1} dtr(t)} \mathcal{P}_1$$

since, from Eq. (12.4), the price of the option is given by $C(t_0, t_1; K) = \langle e^{-\int_{t_0}^{t_1} dtr(t)} \mathcal{P}_1 \rangle$.

From Eq. (A.3) the Θ function has the property that $\Theta^2(x) = \Theta(x)$ and this leads to a major simplification for the computation of $\sigma^2(\mathcal{P}_1)$; the payoff function, given in Eq. (13.1), yields the following

$$\mathcal{P}_1 = \left(\sum_{i=0}^{N1} c_i B(t_1, T_i) - K_1 \right)_+ = (F + V - K)\Theta(F + V - K)$$

$$\Rightarrow \mathcal{P}_1^2 = (F + V - K)^2 \Theta(F + V - K)$$

where, recall from Eq. (11.10), that $V = \sum_{i=1}^{N} J_i(e^{-\alpha_i - Q_i} - 1)$.

Since there is only one Θ function in the expectation value needed to evaluate the coupon bond option's volatility, it can be evaluated in a manner similar to the price of the option. Similar to Eqs. (11.12) and (11.13), one has

$$\langle [e^{-\int_{t_0}^{t_1} dt r(t)} \mathcal{P}_1]^2 \rangle = \frac{1}{2\pi} \int_{-\infty}^{+\infty} dW d\eta (F + W - K)^2$$

$$\times \Theta(F + W - K)e^{-i\eta W} Z_{\text{Vol}}(\eta) \qquad (13.57)$$

where the volatility partition function, from Eq. (5.3), is given by

$$Z_{\text{Vol}}(\eta) = \langle e^{-2\int_{t_0}^{t_1} dt r(t)} e^{i\eta V} \rangle \qquad (13.58)$$

Similar to Eq. (11.20), performing the average over the forward interest rates, and factoring out a pre-factor, yields to second order

$$Z_{\text{Vol}}(\eta) = \langle e^{-2\int_{t_0}^{t_1} dt r(t)} \rangle e^{i\eta D_V - \frac{1}{2}\eta^2 A_V + \cdots} \qquad (13.59)$$

where

$$\langle e^{-2\int_{t_0}^{t_1} dt r(t)} \rangle = B^2(t_0, t_1)e^{\Omega}; \quad \Omega = \int_{\Delta_1} M$$

Similar to Eq. (11.20) the coefficients are given by

$$D_V = \frac{1}{B^2(t_0, t_1)e^{\Omega}} \langle e^{-2\int_{t_0}^{t_1} dt r(t)} V \rangle \qquad (13.60)$$

$$A_V = \frac{1}{B^2(t_0, t_1)e^{\Omega}} \langle e^{-2\int_{t_0}^{t_1} dt r(t)} V^2 \rangle - D^2 \qquad (13.61)$$

The coefficient D_V is consequently given by

$$D_V = \frac{1}{B^2(t_0, t_1)e^{\Omega}} \sum_{i=1}^{N} J_i \langle e^{-2\int_{t_0}^{t_1} dt r(t)} (e^{-\alpha_i - Q_i} - 1) \rangle$$

$$= \sum_{i=1}^{N} J_i [e^{2\int_{\Delta_1} R_i M} - 1] \qquad (13.62)$$

since

$$\left\langle e^{-2\int_{t_0}^{t_1} dt\, r(t)} \right\rangle = B^2(t_0, t_1) e^{\Omega}$$

The domain of integration Δ_1 and R_i are shown in Figure 13.3(a). Note $\Delta_1 R_i \equiv \Delta_1 \cup R_i$.

Similarly, the coefficient A_V, from Eqs. (13.61) and (13.62), is given by

$$A_V = \frac{1}{B^2(t_0, t_1)e^{\Omega}} \sum_{i,j=1}^{N} J_i J_j \langle e^{-2\int_{t_0}^{t_1} dt\, r(t)} (e^{-\alpha_i - \mathcal{Q}_i} - 1)(e^{-\alpha_j - \mathcal{Q}_j} - 1) \rangle - D^2$$

$$= \sum_{i,j=1}^{N} \tilde{J}_i \tilde{J}_j [e^{G_{ij}} - 1]; \quad \tilde{J}_i = J_i e^{2\int_{\Delta_1 R_i} M} \tag{13.63}$$

Collecting the results from Eqs. (13.57) and (13.58) yields the following result for the second moment of the coupon bond option price

$$\langle [e^{-\int_{t_0}^{t_1} dt\, r(t)} \mathcal{P}_1]^2 \rangle = \frac{B^2(t_0, t_1)e^{\Omega} A_V}{\sqrt{2\pi}}$$

$$\times \int_{-\infty}^{+\infty} dW (W - X_V)^2 \Theta(W - X_V) e^{-\frac{1}{2}W^2}$$

$$= \frac{B^2(t_0, t_1)e^{\Omega} A_V}{2} H(X_V) \tag{13.64}$$

where

$$H(X_V) = \sqrt{2}(1 + X_V^2)N(X_V) - \sqrt{\frac{2}{\pi}}X_V e^{-\frac{1}{2}X_V^2} \tag{13.65}$$

$$X_V = \frac{K - F - D_V}{\sqrt{A_V}} \tag{13.66}$$

13.8.1 Monte Carlo simulation

Options can be evaluated numerically. Consider a coupon bond option maturing at t_* with payoff function, given from Eq. (13.1) as follows

$$\mathcal{P} = \left(\sum_{i=0}^{N1} c_i B(t_*, T_i) - K_1 \right)_+$$

The price of the option is given by the expectation value of the discounted payoff

$$C(t_0, t_*; K) = E[\tilde{\mathcal{P}}]$$
$$\tilde{\mathcal{P}} = e^{-\int_{t_0}^{t_*} dt r(t)} \mathcal{P}$$

One can generate N random configurations $\mathcal{A}^{(p)}(t, x)$ of the quantum field $\mathcal{A}(t, x)$ with probability distribution given by e^S/Z, where S is the stiff action given in Eq. (5.25) [2]. The price of the option is then given by

$$C(t_0, t_*; K) = \frac{1}{N} \sum_{p=1}^{N} \tilde{\mathcal{P}}[\mathcal{A}^{(p)}] \pm \text{Statistical error}$$

$$\text{Statistical error} = \frac{\sigma(\tilde{\mathcal{P}})}{\sqrt{N}}$$

where $\sigma(\tilde{\mathcal{P}})$ is the auto-correlation of the discounted payoff function given in Eq. (13.55).

For Monte Carlo simulations of the option price, the auto-correlation $\sigma(\tilde{\mathcal{P}})$ provides an estimate of the statistical error. Having an analytical expression for it, as in Eq. (13.55), provides further checks on the accuracy of the simulation.

14

Hedging interest rate options

Hedging is one of the chief tools for managing risk and is a driving force for the invention and development of financial derivatives. Interest rate instruments are much more complicated to hedge compared to equity since there are virtually an unlimited number of interest rates that need to be hedged against.

Hedging Libor derivatives is discussed in the framework of quantum finance [20, 24, 40]. To exemplify the subtleties in hedging interest rate instruments, one of the simplest interest rate options, namely an interest rate caplet is analyzed in some detail. A portfolio is studied in which the fluctuations of a caplet's price are canceled by the negatively correlated movement of Libor futures contract(s).

Interest rates and Libor are modeled by bond forward interest rates $f(t, x)$ that are described by a Gaussian quantum field, as discussed in Chapter 5. Eq. (5.1) yields the following dynamics for $f(t, x)$[1]

$$\frac{\partial f(t, x)}{\partial t} = \alpha(t, x) + \sigma(t, x)\mathcal{A}(t, x); \quad -\infty \le f(t, x) \le +\infty$$

$$1 + \ell L(t, T_n) = \exp\left\{\int_{T_n}^{T_n+\ell} dx f(t, x)\right\}$$

The forward numeraire $B(t, t_*)$ yields a drift $\alpha(t, x)$, given by Eq. (9.3) as follows

$$\alpha(t, x) = \sigma(t, x)\int_{t_*}^{x} dx' \mathcal{D}(x, x'; t)\sigma(t, x')$$

---

[1] In the Libor Market Model, Libor is represented by the Libor forward interest rates $f_L(t, x)$

$$1 + \ell L(t, T_n) = \exp\left\{\int_{T_n}^{T_n+\ell} dx f_L(t, x)\right\}; \quad f_L(t, x) \ge 0$$

Since $f_L(t, x)$ is nonlinear, exact analytical results for it are hard to obtain; hence, the analysis in this chapter is based on the bond forward interest rates.

Gaussian bond forward interest rates are employed for this chapter due to their mathematical tractability. As was the case in the analysis of the caplet's linear pricing formula in Chapter 10 and coupon bond option in Chapter 11, many analytical calculations are possible for the Gaussian model that illustrate important and general features of financial instruments.

New and novel features of hedging interest rate derivatives are obtained in this chapter using Gaussian bond forward interest rates. In particular, delta and gamma hedge parameters are derived for hedging the linear caplet formula, discussed in Chapter 10, against fluctuations in underlying forward interest rates. The results are empirically analyzed to gauge the influence of interest rate correlations in the hedging of caplets.

14.1 Introduction

Quantum finance models provide flexible computational schemes for hedging interest rate instruments. Current research in hedging interest rates instruments are mostly focused on the applications of the HJM and BGM–Jamshidian models [32, 56, 62], for which forward interest rates are exactly correlated. Quantum finance provides a parsimonious alternative to the existing theories of interest rates.

The following two methods are studied for hedging interest rate instruments.

* Stochastic hedging. Interest rate instruments are written, conditioned on the occurrence of a particular value for pre-specified forward interest rates. The interest rate caplet is delta and gamma hedged against the movement of *pre-specified* forward interest rates.
* Residual variance. The portfolio's residual variance is minimized against all fluctuations of the forward interest rates. Minimizing residual variance suppresses *all* changes in the value of a portfolio.

The impact of correlation on a portfolio of interest rate instruments is examined; the delta hedge parameters for a portfolio as well as its residual variance are derived. This chapter extends the concept of stochastic delta hedging developed in [12] to the case of hedging Libor derivatives.

It is shown that the results obtained can be empirically implemented in a straightforward manner; the implications of correlation on the hedge parameters are analyzed. The data used for the empirical study consist of daily closing prices for quarterly Eurodollar futures contracts as described in [27, 29].

14.2 Portfolio for hedging a caplet

Consider a portfolio $\Pi(t)$ composed of a midcurve $caplet(t, t_*, T)$, discussed in Section 4.3 and N Libor futures contracts, which are chosen to ensure fluctuations in the value of the portfolio are minimized. The Libor futures are given by

$$\mathcal{F}_L(t, T_i) = V[1 - \ell L(t, T_i)]$$

The price of a caplet, from Eq. (4.11), is given by

$$caplet(t, t_*, T) = \ell V E\left[e^{-\int_{t_0}^{t_*} r(t)} B(t_*, T + \ell)\left[L(t_*, T) - K\right]_+\right] \qquad (14.1)$$

where V is the principal, $B(t_*, T + \ell)$ is a zero coupon bond, K is the interest rate cap, and $\ell =$ three months.

The bond forward interest rates $f(t, x)$ yield the linear price of the caplet given, from Eq. (9.33), as follows

$$caplet(t, t_*, T) = \int_{-\infty}^{+\infty} dG \Psi(G, T, T + \ell)(X - e^{-G})_+ \qquad (14.2)$$

where

$$\Psi(G, T, T + \ell) = \frac{\tilde{V} B(t, t_*)}{\sqrt{2\pi q^2 (T - t)}}$$

$$\times \exp\left\{\frac{-1}{2q^2(T - t)}\left(G - \int_T^{T+\ell} dx f(t_*, x) - \frac{q^2(T - t)}{2}\right)^2\right\} \qquad (14.3)$$

$$\tilde{V} = V(1 + \ell K); \quad q^2 = \int_t^{t_*} dt \int_T^{T+\ell} dx dx' \sigma(t, x) \mathcal{D}(x, x'; t) \sigma(t, x') \qquad (14.4)$$

The portfolio to be hedged is equal to

$$\Pi(t) = caplet(t, t_*, T) - \sum_{i=1}^{N} \eta_i(t) \mathcal{F}_L(t, T_i), \qquad (14.5)$$

where $\eta_i(t)$ represents the hedge parameter for the ith futures contract included in the portfolio. The parameters $\eta_i(t)$ are chosen so as to ensure that movements in the caplet and futures contracts 'offset' one another so as to minimize the fluctuations in $\Pi(t)$. Re-writing the portfolio yields[2]

$$\Pi(t) = caplet(t, t_*, T) + \ell V \sum_{i=1}^{N} \eta_i(t) L(t, T_i) \qquad (14.6)$$

14.3 Delta-hedging interest rate caplet

Stochastic hedging of interest rate derivatives has been introduced in [12], where the specific case of hedging zero coupon bonds was considered in detail. This technique is applied to the hedging of a Libor caplet against fluctuations in the bond forward interest rates $f(t, x)$.

Consider a portfolio $\Pi(t)$ composed of a $caplet(t, t_*, T)$ and one Libor futures contract. Setting $N = 1$ in Eq. (14.6) yields[3]

$$\Pi(t) = caplet(t, t_*, T) + \ell V \eta_1(t) L(t, T_1) \qquad (14.7)$$

The instantaneous change in the value of this portfolio, at instant time, t is given by a functional Taylor's expansion. Let $\Delta t \equiv \epsilon = 1/360$ year and $\Delta f(t, x) = f(t + \epsilon, x) - f(t, x)$; this yields[4]

$$\Delta\Pi[t, f(t, x)] \equiv \Pi[t + \epsilon, f(t, x) + \Delta f(t, x)] - \Pi[t, f(t, x)]$$

$$= \frac{\partial \Pi}{\partial t} \Delta t + \int dx \frac{\delta \Pi}{\delta f(t, x)} \Delta f(t, x)$$

$$+ \frac{1}{2} \int dx dx' \frac{\delta^2 \Pi}{\delta f(t, x) \delta f(t, x')} \Delta f(t, x) \Delta f(t, x') + O(\delta f^3) \qquad (14.8)$$

Only the lowest order term in ϵ is retained for the purpose of hedging, with higher orders of ϵ being negligible. Taking the limit of $\epsilon \to 0$ yields the following portfolio dynamics; $\Delta\Pi/\Delta t \to d\Pi/dt$ and $\Delta f(t, x)/\Delta t \to \partial f(t, x)/\partial t \equiv \dot{f}(t, x)$; hence, the generalization of Eq. (3.15) yields

$$\frac{d\Pi[t, f(t, x)]}{dt} = \frac{\partial \Pi}{\partial t} + \int dx \frac{\delta \Pi}{\delta f(t, x)} \dot{f}(t, x) \qquad (14.9)$$

$$+ \frac{\epsilon}{2} \int dx dx' \frac{\delta^2 \Pi}{\delta f(t, x) \delta f(t, x')} \dot{f}(t, x) \dot{f}(t, x') + O(\epsilon)$$

---

[2] A constant term equal to $V \sum_{i=1}^{N} \eta_i(t)$ has been dropped in Eq. (14.6) since it is irrelevant to the hedging analysis.

[3] Recall from Eq. (10.7) that $\ell L(t, T_i) = \exp\left\{\int_{T_i}^{T_i + \ell} f(t, x)\right\} - 1$.

[4] Functional differentiation with respect to the function $f(t, x)$, $t = $ constant, denoted by $\delta/\delta f(t, x)$ is discussed in Appendix A.5.

Similar to the simpler case discussed in Eq. (3.15), the instantaneous changes in forward interest rates given by $\dot{f}(t, x)$ are singular for equal time; this is due to the Gaussian quantum field $\mathcal{A}(t, x)$ having a singular quadratic product given in Eq. (5.46), which yields $\dot{f}(t, x) \dot{f}(t, x') \sim 1/\epsilon$; hence the second-order term in $d\Pi/dt$ is as important as the first-order term.[5]

Consider *delta hedging*. The portfolio is required to be invariant to small changes in the forward interest rates that take place due to small changes in time. Thus, from Eq. (14.9), delta hedging the portfolio is given by the following generalization of Eq. (3.27)

$$\text{Delta hedging}: \quad \frac{\delta \Pi(t)}{\delta f(t, x)} = 0 \text{ for each } x; \ t: \text{ fixed} \qquad (14.10)$$

Delta hedging involves a first-order approximation to the change in a portfolio's value as a result of forward interest rates' fluctuations. If the delta hedge parameter has a large variation in time, one needs to gamma hedge the portfolio. In quantum finance, from Eq. (14.9) and similar to the case of equity given in Eq. (3.22), gamma hedging is given by the following

$$\text{Gamma hedging}: \quad \frac{\delta^2 \Pi(t)}{\delta f(t, x) \delta f(t, x')} = 0 \text{ for each } x, x'; \ t: \text{ fixed} \quad (14.11)$$

14.4 Stochastic hedging

In quantum finance, for each time t, there are infinitely many random variables driving the forward interest rates, indexed by x. Therefore, a portfolio with a finite N number of instruments can never be perfectly delta hedged since to fulfill Eq. (14.10) one needs infinitely many instruments. The best alternative is to delta hedge on the average, and this scheme is referred to as stochastic delta hedging [12].

To implement stochastic delta hedging, one considers the *conditional* expectation value of the portfolio $\Pi(t)$, namely, conditioned on the occurrence of some specific value of a forward interest rate $f_h \equiv f(t_*, x_h)$, where t_* is the maturity time for the caplet. The price of a midcurve caplet, given that the value of $f(t_*, x_h)$ is pre-specified, is given by generalizing Eq. (9.32), as follows

$$caplet(t, t_*, T; f_h) = B(t, t_*) E[caplet(t_*, t_*, T) | f_h] \qquad (14.12)$$

$$= \tilde{V} B(t, t_*) E_F \left[\left(X - \exp\left\{ -\int_T^{T+\ell} dx f(t_*, x) \right\} \right)_+ | f(t_*, x_h) \right]$$

---

[5] Normal (nonstochastic) calculus retains only the first-order term since the second-order derivative term is nonsingular and vanishes as $\epsilon \to 0$.

From [12], the conditional probability of a caplet is given by[6]

$$caplet(t, t_*, T; f_h) = \tilde{V} B(t, T) \int_{-\infty}^{\infty} dG(X - e^{-G})_+ \Psi(G|f_h) \qquad (14.13)$$

$$\Psi(G|f_h) = \frac{\int Df \delta\left(\int_T^{T+l} dx f(t_*, x) - G\right)\delta\left(f(t_*, x_h) - f_h\right)e^S}{\int Df \delta\left(f(t_*, x_h) - f_h\right)e^S} \qquad (14.14)$$

A conditional Libor futures is defined by

$$\tilde{L}(t, T_1; f_h) = E[L(t_*, T_1)|f_h] \qquad (14.15)$$

with the conditional probability being given by

$$\tilde{L}(t, T_1; f_h) = \int_{-\infty}^{\infty} dG e^G \Phi(G|f_h; t, T_1) \qquad (14.16)$$

$$\Phi(G|f_h; t, T_1) = \frac{\int Df \delta\left(G - \int_{T_1}^{T_1+\ell} f(t_*, x) dx\right)\delta\left(f(t_*, x_h) - f\right)e^S}{\int Df \delta\left(f(t_*, x_h) - f_h\right)e^S} \qquad (14.17)$$

Generalizing the portfolio given in Eq. (14.7), conditioned on the occurrence of f_h, yields

$$\Pi(t; f_h) = caplet(t, t_*, T; f_h) + \eta_1 L(t, T_1; f_h) \qquad (14.18)$$

Stochastic delta hedging entails that the hedged portfolio be independent of changes in all the forward interest rates. Approximate Eq. (14.10) by the following equation

$$\frac{\partial}{\partial f_h} \Pi(t; f_h) = 0 \qquad (14.19)$$

The hedged portfolio is independent of the small changes of only one forward interest rate f_h. From Eqs. (14.9) and (14.19), stochastic delta hedging yields

$$\eta_1(t) = -\frac{\partial \, caplet(t, t_*, T; f_h)}{\partial f_h} \bigg/ \frac{\partial L(t, T_1; f_h)}{\partial f_h} \qquad (14.20)$$

Thus, changes in the delta hedged portfolio $\Pi(t, x_h)$ are, on average and to lowest order, insensitive to fluctuations in one forward interest rate, namely $f(t, x_h)$.

---

[6] The denominator in Eq. (14.14) is the required normalization for the constrained expectation value in the numerator to be a conditional probability [12].

The conditional probabilities given in Eqs. (14.13) and (14.16), along with the hedge parameter η_1, are evaluated explicitly in Section 14.11. Nontrivial correlations appear in all the terms. The final result, from Eq. (14.39), is[7]

$$\eta_1 = \frac{C}{e^{G_1 + \frac{Q_1^2}{2}} B_1} \left[caplet(t, t_*, T; f_h) - \frac{\eta'}{C} \right]$$

$$\eta' = -B\chi\tilde{V} \left[\frac{X N'(d_+)}{Q} + e^{-G_0 + \frac{Q^2}{2}} N(d_-) - \frac{e^{-G_0 + \frac{Q^2}{2}} N'(d_-)}{Q} \right]$$

The HJM limit of the hedging results is analyzed Section 14.13.

Furthermore, one can gamma hedge the same forward interest rate. The *second-order* gamma hedge recognizes that large movements in the forward interest rates may cause the first-order delta approximation to be inaccurate. In particular, if hedging is not performed frequently, the delta hedge parameter can become outdated. Gamma evaluates changes in the delta hedge parameter as the forward rate term structure evolves over time.

To gamma hedge against the $\partial^2 \Pi(t)/\partial f_h^2$ fluctuations, one needs to form a portfolio with two Libor futures contracts that minimize the change in the value of $E[\Pi(t)|f_h]$ by both delta and gamma hedging. Suppose a caplet needs to be hedged against the fluctuations of two forward interest rates, namely f_h for $h = 1, 2$. The conditional probabilities for the caplet and Libor futures, with two forward rates fixed at f_h, are

$$caplet(t, t_*, T; f_1, f_2) = B(t, t_*) E[caplet(t_*, t_*, T)|f_1, f_2]$$

$$L(t, T_1; f_1, f_2) = E[L(t_*, T_1)|f_1, f_2]$$

A portfolio of two Libor futures contracts with different maturities $T_i \neq T$ and conditioned on two forward interest rates f_1, f_2 is defined as follows

$$\Pi(t; f_1, f_2) = caplet(t, t_*, T; f_1, f_2) + \sum_{i=1}^{2} \eta_i(t) L(t, T_i; f_1, f_2) \quad (14.21)$$

The hedging of this portfolio, at time t, is given by

$$\Delta\Pi(t, f_1, f_2) = \frac{\partial \Pi}{\partial t} \Delta t + \sum_{i=1}^{2} \frac{\partial \Pi}{\partial f_i} \Delta f_i + \frac{1}{2} \sum_{i=1}^{2} \frac{\partial^2 \Pi}{\partial f_i^2} (\Delta f_i)^2$$

$$+ \frac{\partial^2 \Pi}{\partial f_1 \partial f_2} \Delta f_1 \Delta f_2 + O(\epsilon^2) \quad (14.22)$$

---

[7] The notation is defined in Section 14.11.

The stochastic delta and gamma hedging conditions are given by the following *five* constraint equations.

1 Two conditions for stochastic delta hedging

$$\frac{\partial}{\partial f_h} \Pi(t; f_1, f_2) = 0 \quad \text{for } h = 1, 2$$

2 Stochastic gamma hedging requires two conditions

$$\frac{\partial^2}{\partial f_h^2} \Pi(t; f_1, f_2) = 0 \quad \text{for } h = 1, 2$$

3 *Cross gamma* hedging is given by the single condition

$$\frac{\partial^2}{\partial f_1 \partial f_2} \Pi(t; f_1, f_2) = 0$$

Intuitively, one expects that the portfolio that is hedged most effectively should include cross gamma hedging.

Cross gamma hedging makes sense only in the framework of quantum finance, since movements in any number of specific forward interest rates, in particular against f_1, f_2, can be hedged. In contrast, cross gamma hedging in the one-factor HJM model is not possible – one needs at least a two-factor HJM model.

For the portfolio being considered, it can be analytically shown that delta hedge parameters for the two forward rates differ only by a pre-factor A_2/A_{12}, that is

$$\frac{\partial}{\partial f_1} \Pi(t; f_1, f_2) = -\frac{A_2}{A_{12}} \frac{\partial}{\partial f_2} \Pi(t; f_1, f_2) \tag{14.23}$$

where A_2 and A_{12} are defined in Section 14.12. Therefore, delta hedging against two forward interest rates only determines the portfolio's hedge parameters for one Libor future. Gamma hedging two forward rates are also equal except for a pre-factor A_2/A_{12}. Hence, for hedging against two forward interest rates, we are left with only *three* independent constraints from the above five constraints.

In order to study the effect of each set of constraints separately, a portfolio is formed that includes two Libor futures, and the following hedging strategies are adopted that involve only two constraint equations.

• The first strategy implements one delta and one gamma hedge against a single forward rate. Namely

$$\frac{\partial}{\partial f_1} \Pi(t; f_1, f_2) = 0 = \frac{\partial^2}{\partial f_1^2} \Pi(t; f_1, f_2)$$

- The second strategy fixes the two hedge parameters by one delta hedge and an additional cross gamma hedge. Namely

$$\frac{\partial}{\partial f_1} \Pi(t; f_1, f_2) = 0 = \frac{\partial^2}{\partial f_1 \partial f_2} \Pi(t; f_1, f_2)$$

The parameters for each choice of the Libor futures and forward interest rates being hedged are chosen to minimize, by varying the maturity of the Libor futures, the following

$$\sum_{i=1}^{2} |\eta_i| \tag{14.24}$$

This choice of η_is defines the *optimal* portfolio. The additional constraint given by Eq. (14.24) finds the most effective futures contracts measured by requiring the smallest number of contracts and hence minimum transaction costs.

In summary, the optimal portfolio is found by first fixing the two hedge parameters and then minimizing $\sum_{i=1}^{2} |\eta_i|$.

14.5 Residual variance

Hedging a caplet using Libor futures contracts can also be accomplished by minimizing the residual variance of the hedged portfolio [12, 20, 24]. The instantaneous change in the portfolio value is stochastic; the volatility of this change is computed to ascertain the efficacy of the hedging strategy.

To simplify the notation, consider a caplet that matures when it becomes operational; that is, $t_* = T$; denote the price of the caplet by $caplet(t, T)$; let the portfolio given in Eq. (14.6) be denoted by

$$\Pi(t) = caplet(t, T) + \sum_{i=1}^{N} \Delta_i L(t, T_i)$$

The variance of the portfolio fluctuations $Var[d\Pi(t)/dt]$ is given by[8]

$$Var\left[\frac{d\Pi(t)}{dt}\right] = Var\left[\frac{d\ caplet(t, T)}{dt}\right] + Var\left[\sum_{i=1}^{N} \Delta_i \frac{L(t, T_i)}{dt}\right]$$

$$+ \sum_{i=1}^{N} \Delta_i E\left[\frac{d\ caplet(t, T)}{dt} \frac{L(t, T_i)}{dt}\right]_c \tag{14.25}$$

---

[8] Note the variance of $\sum_{i=1}^{N} c_i X_i$, where X_i are random variables, is given by $Var[\sum_{i=1}^{N} c_i X_i] = \sum_{i=1}^{N} c_i^2 Var[X_i] + \sum_{ij=1}^{N} c_i c_j E[X_i X_j]_c$.

A detailed calculation for determining the hedge parameters and portfolio variance is carried out in Section 14.10. The following notation is introduced for simplicity

$$
K_i = \chi \hat{L}(t, T_i) \int_T^{T+\ell} dx \int_{T_i}^{T_i+\ell} dx' \sigma(t, x)\sigma(t, x')\mathcal{D}(x, x'; t) \tag{14.26}
$$

$$
M_{ij} = \hat{L}(t, T_i)\hat{L}(t, T_j) \int_{T_i}^{T_i+\ell} dx \int_{T_j}^{T_j+\ell} dx' \sigma(t, x)\sigma(t, x')\mathcal{D}(x, x'; t)
$$

Equation (14.26) allows the residual variance given in Eq. (14.25) to be succinctly expressed as follows

$$
Var\left[\frac{d\Pi(t)}{dt}\right] = \chi^2 \int_T^{T+\ell} dx \int_T^{T+\ell} dx' \sigma(t, x)\sigma(t, x')\mathcal{D}(x, x'; t)
$$

$$
+ 2\sum_{i=1}^N \Delta_i K_i + \sum_{i=1}^N \sum_{j=1}^N \Delta_i \Delta_j M_{ij} \tag{14.27}
$$

The value of χ, derived in Eq. (14.33), is given below and is useful for the empirical analysis of residual variance.

$$
\chi = V B(t, T)\left[\frac{1}{\sqrt{2\pi q^2}}e^{-d_+^2/2} + \left(\frac{F}{X}\right)\left\{N(d_-) - \frac{1}{\sqrt{2\pi q^2}}e^{-d_-^2/2}\right\}\right]
$$

where $d_\pm = \left(\ln \frac{X}{F} \pm q^2/2\right)/q$ and $F = \exp\{-\int_T^{T+\ell} dx f(t, x)\}$.
The value of χ for at-the-money options has $X = F$, $d_\pm = \pm q/2$ and yields

$$
\chi(t, T)|_{\text{at-the-money}} = V B(t, T)N(d_-)
$$

The residual variance depends on the correlations between forward interest rates that are described by the propagator. Ultimately, the effectiveness of the hedge portfolio is an empirical question since perfect hedging is not possible. This empirical question is addressed in Section 14.6, with the propagator being calibrated from market data.

The hedge parameters Δ_i that minimize the portfolio's residual variance, as in Eq. (14.27), are given by

$$
\Delta_i = -\sum_{j=1}^N K_j M_{ij}^{-1} \tag{14.28}
$$

Eq. (14.28) is derived by differentiating Eq. (14.27) with respect to Δ_i and setting the result to zero. The Δ_i parameters represent the optimal amounts of the futures contracts that need to be included in the hedged portfolio.

The variance of the hedged portfolio is obtained by substituting parameters given in Eq. (14.28) into Eq. (14.27) and yields

$$V_R = \chi^2 \int_T^{T+\ell} dx \int_T^{T+\ell} dx' \sigma(t,x)\sigma(t,x')\mathcal{D}(x,x';t) - \sum_{ij=1}^N K_i M_{ij}^{-1} K_j \quad (14.29)$$

As expected, portfolio variance V_R declines monotonically as N increases.

The residual variance in Eq. (14.29) enables the effectiveness of the hedging strategy to be evaluated and is the basis for studying the impact of including different Libor futures contracts in the hedged portfolio. For $N = 1$, there is only a single Libor maturing at T_1; the residual variance in Eq. (14.29) simplifies

$$\chi^2 \left[\int_T^{T+\ell} dx \int_T^{T+\ell} dx' \sigma(t,x)\sigma(t,x')\mathcal{D}(x,x';t) \right.$$

$$\left. - \frac{\left(\int_T^{T+\ell} dx \int_{T_1}^{T_1+\ell} dx' \sigma(t,x)\sigma(t,x')\mathcal{D}(x,x';t) \right)^2}{\int_{T_1}^{T_1+\ell} dx \int_{T_1}^{T_1+\ell} dx' \sigma(t,x)\sigma(t,x')\mathcal{D}(x,x';t)} \right] \quad (14.30)$$

The second term in Eq. (14.30) represents the reduction in variance attributable to the hedging strategy. To obtain the HJM limit, the propagator is constrained to equal one, namely $\mathcal{D}(x,x';t) \to 1$; as shown below, the residual variance V_R in Eq. (14.30) reduces to zero as follows

$$\chi^2 \left[\left(\int_T^{T+\ell} dx \sigma(t,x) \right)^2 - \frac{\left(\int_T^{T+\ell} dx \int_{T_1}^{T_1+\ell} dx' \sigma(t,x)\sigma(t,x') \right)^2}{\int_{T_1}^{T_1+\ell} dx \int_{T_1}^{T_1+\ell} dx' \sigma(t,x)\sigma(t,x')} \right] = 0$$

$$(14.31)$$

The HJM limit is consistent with our intuition that the residual variance should be identically zero for any Libor maturity since all forward interest rates are perfectly correlated. This result is also shown empirically in Section 14.8. Results from hedging with two Libor futures contracts in the HJM model are not presented since a one-factor model cannot be hedged with two instruments. Indeed, M_{ij}^{-1} is singular for the one-factor HJM model.

14.6 Empirical analysis of stochastic hedging

This section illustrates the implementation of stochastic hedging strategies and provides preliminary results on the impact of correlation on the hedge parameters.

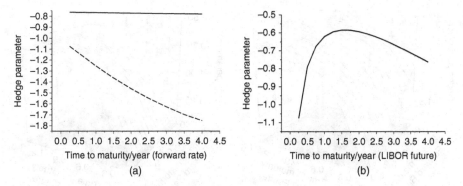

Figure 14.1 Hedge parameter η_1 for stochastic hedging of *caplet*$(t, 1, 4)$ for portfolio $\Pi(t; f_h) = caplet(t, 1, 4) + \eta_1(t)\mathcal{F}(t, T_1)$. (a) η_1 as a function of time to maturity x_h of forward interest rate $f(t, x_h)$ with fixed Libor futures contract maturity T_1; dashed line is for $T_1 = 3$ months and unbroken line is for $T_1 = 4$ years. (b) Hedge parameter η_1 as a function of time to maturity of Libor future T_1, with fixed $f(t, t + \delta)$, where $\delta = 3$ months.

The term structure of the volatility, $\sigma(\theta)$ as well as the parameters η, λ, and μ for the stiff propagator are evaluated using Libor and caplet data [27, 29].

Stochastic hedging mitigates the risk of fluctuations in the pre-specified forward interest rates. The focus of this section is on the stochastic hedge parameters η_i. The optimum portfolio is chosen so that the sum of the hedging parameters, namely $\sum_{i=1}^{N} |\eta_i|$, is minimized. The best hedging strategy entails finding the optimal portfolio for which the portfolio has the smallest possible long and short positions on Libor futures.

14.6.1 Hedging caplet with futures for interest rate

Consider a portfolio with one Libor future and one caplet that is to be hedged against a single forward interest rate $f(t, x_h)$. The portfolio is given by

$$\Pi(t) = caplet(t, t_*, T) + \eta_1(t)\mathcal{F}(t, T_1) \qquad (14.32)$$

The hedging strategy is to stochastic delta hedge the portfolio against forward interest rate $f_h = f(t, x_h)$ using $\partial \Pi(t; f_h)/\partial f_h = 0$.

The η_1 hedge parameter's dependence on x_h and T_1 is shown in Figure 14.1 for *caplet*$(t, 1, 4)$; present time is t (which is set equal to zero), midcurve caplet option maturity time is $t_* = 1$ year, and caplet operational time is $T = 4$ years. The dependence of the hedge parameter η_1 on x_h and T_1 is plotted in Figure 14.1(a) and on Libor maturity time T_1 is plotted in Figure 14.1(b). Figure 14.1(a) shows how the hedge parameter η_1 depends on x_h for two specific values of Libor maturity

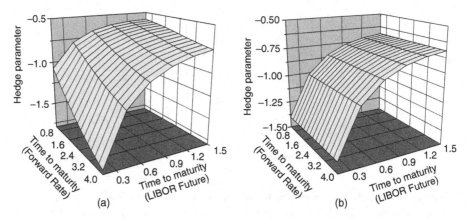

Figure 14.2 Hedge parameter η_1 for stochastic delta hedging of $caplet(t, 1, 4)$ using Libor futures with maturity T_1 and against fluctuations of the forward interest rate $f(t, x_h)$. The portfolio is $\Pi(t) = caplet(t, 1, 4) + \eta_1(t)\mathcal{F}(t, T_1)$. (a) Quantum finance model (b) The HJM limit of $\mathcal{D}(x, x'; t) = 1$; as expected, the hedge parameter η_1 is constant for all x_h since all forward interest rates are perfectly correlated.

time, namely $T_1 = \delta = \frac{1}{4}$ year (three months) and $T_1 = 16 \times \delta = 4$ years are chosen. It is found that $x_h = \delta$ is always the most important forward interest rate to hedge against.

Figure 14.1(b) shows the dependence of the hedge parameter η_1 on different values of T_1, with a fixed value of $x_h = \delta$. The maximun of the hedge parameter η_1 is at $T_1 \simeq 1.5$ years and reflects the maximum of $\sigma(t, x)$ around the same future time. Figure 14.2(a) plots the hedge parameter η_1 against the Libor futures maturity T_1 and the forward interest rate maturity x_h that is being hedged. This figure can be used to select the Libor futures for the optimum portfolio that requires the least number of long and short positions.

The midcurve $caplet(t, t_*, T)$ can be hedged, in general, for different t_* and T values; it is found that, although the value of the parameter changes slightly, the shape of the parametric surfaces are almost identical to the results shown in Figures 14.1 and 14.2(a).

14.6.2 Hedging in quantum finance models compared to HJM

The comparison is carried out for the portfolio given in Eq. (14.32), where one forward interest rate $f(t, x_h)$ is hedged by one Libor future.

To illustrate the contrast between the quantum finance and the single-factor HJM model, the same hedging parameter is plotted in the $\mathcal{D}(x, x'; t) \to 1$ HJM limit. From Figure 14.2(b), the HJM hedge parameter η_1 is independent of the forward interest rate maturity x_h, which is expected since in the HJM model all forward rates $f(t, x_h)$ are perfectly correlated. Therefore, it makes no difference which of the forward interest rates is being hedged against.

14.7 Hedging caplet with two futures for interest rate

The portfolio with two Libor futures is given by

$$\Pi(t) = caplet(t, t_*, T) + \sum_{i=1}^{2} \eta_i(t)\mathcal{F}(t, T_i)$$

Figure 14.3(a) plots the sum $\sum_{i=1}^{2} |\eta_i|$ for hedging the portfolio against one forward interest rate by employing both delta and gamma hedging.

Stochastic delta and gamma hedging are given by

$$\frac{\partial}{\partial f_1}\Pi(t; f_1) = 0 = \frac{\partial^2}{\partial f_1^2}\Pi(t; f_1)$$

A hedge against $f(t, \delta)$ can be constructed such that one obtains an optimal portfolio involving the least amount of short and long positions. The diagonal axis in Figure 14.3(a) has the two Libor futures with the same maturity and hence reduces to delta hedging with one Libor future. The data from which Figure 14.3(a) is plotted

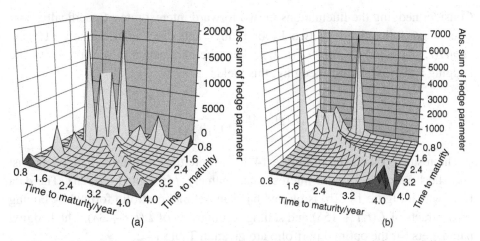

(a)   (b)

Figure 14.3 Summation of absolute hedge parameters $|\eta_1| + |\eta_2|$ for two Libor futures, T_1 and T_2. The portfolio is $\Pi(t) = caplet(t, 1, 4) + \sum_{i=1}^{2} \eta_i(t)\mathcal{F}(t, T_i)$. (a) Stochastic hedge against one forward interest rate with delta and gamma hedging. (b) Stochastic hedging against two forward interest rates, with delta and cross gamma hedging.

Table 14.1 *Parameters for hedging one forward rate with two Libor futures.*

T_1	T_2	x_{h1}	η_1	η_2
1.5 year	0.25 year	0.25 year	-38	71

Table 14.2 *Parameters for hedging two forward rates with two Libor futures and forming an optimum portfolio.*

T_1	T_2	x_{h1}	x_{h2}	η_1	η_2
3.75 year	0.75 year	0.25 year	0.5 year	45	-25

have an optimal portfolio in which one sells 38 contracts of $L(t, t + 6\delta)$ and buys 71 contracts of $L(t, t + \delta)$. The variables in the optimal portfolio are summarized in Table 14.1.

If one chooses the hedged portfolio by minimizing $\sum_{i=1}^{2} \eta_i$ (note no absolute values on the η_is), then the portfolio requires 1500 contracts (long the short maturity and short their long maturity counterparts).

14.7.1 Hedging caplet against two forward interest rates with two Libor futures

Consider hedging the fluctuations in two forward interest rates. Specifically, consider a portfolio given by two Libor futures and one caplet, namely $\Pi(t) = caplet(t, t_*, T) + \sum_{i=1}^{2} \eta_i(t)\mathcal{F}(t, T_i)$ where the parameters η_i are fixed by delta hedging and cross gamma hedging; this yields

$$\frac{\partial}{\partial f_1} \Pi(t; f_1, f_2) = 0 = \frac{\partial^2}{\partial f_1 \partial f_2} \Pi(t; f_1, f_2)$$

The result is displayed in Figure 14.3(b) where the portfolio is hedged against two short maturity forward interest rates, namely $f(t, \delta)$ and $f(t, 2\delta)$. The data from which Figure 14.3(b) is plotted have an optimal portfolio formed by buying 45 contracts of $L(t, t + 15\delta)$ and selling 25 contracts of $L(t, t + 3\delta)$. The hedging parameters for the optimal portfolio are given in Table 14.2.

Figure 14.3 results from minimizing the sum of the absolute values of the hedge parameters, as in Eq. (14.24), which depend on the maturities of the Libor futures T_i. The corresponding empirical results are consistent with the earlier discussion on the importance of forward interest rates in the near future.

14.8 Empirical results on residual variance

The reduction in a portfolio's variance by hedging a caplet with Libor futures is studied empirically for the following portfolio

$$\Pi(t) = caplet(t, T) + \sum_{i=1}^{N} \Delta_i(t)\mathcal{F}(t, T_i)$$

A hedging strategy to minimize the variance $Var[d\Pi(t)/dt]$ of changes in a portfolio's value is considered. The residual variance for hedging a one- and four-year caplet with a Libor future is shown in Figure 14.4(a), along with its HJM counterpart. Observe that the residual variance drops to exactly zero when the same maturity Libor future is used to hedge the caplet. The HJM residual variance is always zero due to the exact correlation of all forward interest rates.

The residual variance for hedging a four-year caplet with two Libor futures is plotted in Figure 14.4(b). It is interesting to note that hedging with two instruments, even with similar maturities, entails a significant decrease in residual variance compared to hedging with one future contract. This is illustrated in Figure 14.4(b), where $\theta = \theta'$ represents hedging with one Libor future. The residual variance in this situation is higher than nearby points, and increases in a discontinuous manner.

One can vary the parameters λ and μ of the stiff propagator $\mathcal{D}(x, x'; t)$ and examine the changes of residual variance. It is found that the neighboring points in

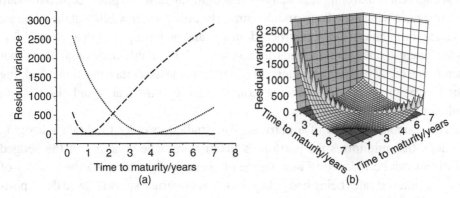

Figure 14.4 (a) Residual variance $Var[d\Pi(t)/dt]$ for $\Pi(t) = caplet(t, T) + \Delta_1(t)\mathcal{F}(t, T_1)$. A one- and four-year caplet is hedged using a Libor future with maturity T_1. The dotted line is the residual variance for a caplet that matures in four years time and the dashed line for a one-year caplet. The unbroken horizontal line is the residual variance in the HJM limit for both caplets. (b) Residual variance for portfolio $\Pi(t) = caplet(t, 4) + \sum_{i=1}^{2} \Delta_i(t)\mathcal{F}(t, T_i)$ in which a four-year caplet is hedged using two Libor futures with maturities T_i.

parameter space have almost the same residual variance and one cannot tell which
parameter offers the better hedge. An explanation of this effect is that forward
interest rates with similar maturities are strongly correlated. The HJM residual
variance for hedging a one-year and a four-year caplet are both identical to zero
and this is consistent with the analytical result obtained in Eq. (14.31).

14.9 Summary

Libor-based interest rate caplets and floorlets are important financial instruments.
The quantum finance model implies that all markets are incomplete since exactly
hedging interest rate instruments is not possible. An infinite number of forward
interest rates need to be hedged that, in principle, need infinitely many securities.
A partial solution is to approximately hedge interest rate instruments. The correlation structure between the forward interest rates was exploited to define an
approximate form of hedging, namely stochastic hedging.

Interest rate instruments are hedged against the movement of a particular forward
interest rate; such hedging is useful in practice and is an important part of managing
interest rate risk. In the path integral formulation of the forward interest rates, the
expectation value of an instrument – conditioned on the value of a particular forward
interest rate – was obtained by introducing an appropriately normalized Dirac-delta
function constraint in the integrand of the functional integral. The quantum finance
formulation, hence, provides a transparent definition of conditional expectation
values of the instruments and yields the results needed for delta and gamma hedging.

An analytical and empirical analysis of hedging against two pre-specified forward
interest rates was carried out to demonstrate the utility and flexibility of the quantum
finance formulation. One can proceed further and hedge a portfolio against N forward interest rates, which leads to a system of N simultaneous equations that
determine the N hedge parameters. In general, stochastic delta and gamma hedging
for increasing N rapidly becomes more and more complicated and closed-form
solutions are difficult to obtain.

The residual variance of a portfolio was studied in order to obtain a hedging
strategy that minimizes the portfolio's fluctuations. The volatility of the hedged
portfolio can be reduced to any degree of accuracy by increasing the number of
forward interest rates being hedged against. The empirical study showed that a portfolio of a caplet hedged with only a few Libor futures is enough to yield a residual
variance with a very small value. One of the main limitations of reducing residual
variance is that this procedure ends up suppressing *all* interest rate fluctuations,
including the 'good' fluctuations that increase the portfolio's value. Instead, the
ideal strategy would consist of hedging against only a subset of forward interest
rate movements that decreases the value of a portfolio.

14.10 Appendix: Residual variance

Consider the variance of a caplet. The delta parameter for the caplet, given in Eq. (14.3), is defined by forward bond price $F = \exp\{-\int_T^{T+\ell} dx f(t, x)\}$ as follows

$$\chi \equiv \frac{\partial \; caplet(t, T)}{\partial F}$$

$$= -V B(t, T) \int_{-\infty}^{+\infty} \frac{dG}{\sqrt{2\pi q^2}} \frac{1}{q^2} \left(G - \int_T^{T+\ell} dx f(t, x) - \frac{q^2}{2} \right)$$

$$\times \left\{ e^{-\frac{1}{2q^2}\left(G - \int_T^{T+\ell} dx f(t,x) - \frac{q^2}{2}\right)^2} (X - e^{-G})_+ \right\}$$

$$= V B(t, T) \left\{ \frac{1}{\sqrt{2\pi q^2}} e^{-d_+^2/2} + \left(\frac{F}{X}\right) \left[N(d_-) - \frac{1}{\sqrt{2\pi q^2}} e^{-d_-^2/2} \right] \right\} \quad (14.33)$$

where $d_\pm = \left(\ln(X/F) \pm q^2/2 \right) / q$. Eq. (14.33) yields the following

$$\frac{1}{\chi F} \frac{d \; caplet(t, T)}{dt} = \frac{1}{\chi F} \frac{\partial \; caplet(t, T)}{\partial F} \frac{\partial F}{\partial t}$$

$$= -\int_T^{T+\ell} \frac{\partial f(t, x)}{\partial t} dx$$

$$= -\int_T^{T+\ell} dx \alpha(t, x) - \int_T^{T+\ell} dx \sigma(t, x) A(t, x)$$

$E[A(t, x)] = 0$ yields that $E[d \; caplet(t, T)/dt] = -\chi F \int_T^{T+\ell} dx \alpha(t, x)$. Therefore

$$\frac{d \; caplet(t, T)}{dt} - E\left[\frac{d \; caplet(t, T)}{dt} \right] = -\chi F \int_T^{T+\ell} dx \sigma(t, x) A(t, x)$$

The variance is given by squaring this expression and taking its expectation value. From Eq. (5.22) $E[A(t, x)A(t, x')] = \delta(0)\mathcal{D}(x, x'; t) = \frac{1}{\epsilon}\mathcal{D}(x, x'; t)$[9] results in the instantaneous caplet price variance being given by

$$Var\left[\frac{d \; caplet(t, T)}{dt} \right] = \frac{1}{\epsilon} \chi^2 F^2 \int_T^{T+\ell} dx \int_T^{T+\ell} dx' \sigma(t, x) \mathcal{D}(x, x'; t) \sigma(t, x')$$

---

[9] On discretizing time $t = n\epsilon$, one has that $dt = \epsilon$ and $\delta(0) = 1/dt = 1/\epsilon$. The quantity ϵ signifies a small step forward in time.

Consider the following Libor portfolio $\hat{\Pi}(t) = \ell V \sum_{i=1}^{N} \Delta_i L(t, T_i)$; its instantaneous variance is

$$\frac{d\hat{\Pi}(t)}{dt} - E\left[\frac{d\hat{\Pi}(t)}{dt}\right] = \sum_{i=1}^{N} \Delta_i \hat{L}(t, T_i) \int_{T_i}^{T_i+\ell} dx \sigma(t, x) A(t, x) \qquad (14.34)$$

where $\hat{L}(t, T_i) = \ell V \exp\{\int_{T_i}^{T_i+\ell} f(t, x) dx\}$ and

$$Var\left[\frac{d\hat{\Pi}(t)}{dt}\right] = \frac{1}{\epsilon} \sum_{i=1}^{N} \sum_{j=1}^{N} \Delta_i \Delta_j \hat{L}(t, T_i) \hat{L}(t, T_j)$$

$$\times \int_{T_i}^{T_i+\ell} dx \int_{T_j}^{T_j+\ell} dx \sigma(t, x) \mathcal{D}(x, x'; t) \sigma(t, x') \quad (14.35)$$

The (residual) variance of the hedged portfolio

$$\Pi(t) = caplet(t, T) + \sum_{i=1}^{N} \Delta_i \mathcal{F}(t, T_i)$$

can be computed in a straightforward manner. Equation (14.35) implies the hedged portfolio's variance equals

$$\chi^2 \int_{T}^{T+\ell} dx \int_{T}^{T+\ell} dx' \sigma(t, x) \sigma(t, x') \mathcal{D}(x, x'; t) \qquad (14.36)$$

$$+ 2\chi \sum_{i=1}^{N} \Delta_i \hat{L}(t, T_i) \int_{T}^{T+\ell} dx \int_{T_i}^{T_i+\ell} dx' \sigma(t, x) \sigma(t, x') \mathcal{D}(x, x'; t)$$

$$+ \sum_{i=1}^{N} \sum_{j=1}^{N} \Delta_i \Delta_j \hat{L}(t, T_i) \hat{L}(t, T_j) \int_{T_i}^{T_i+\ell} dx \int_{T_j}^{T_j+\ell} dx' \sigma(t, x) \sigma(t, x') \mathcal{D}(x, x'; t)$$

14.11 Appendix: Conditional probability for interest rate

Using the results of the bond forward interest rates given in Chapter 5, after a straightforward but tedious calculation, Eqs. (14.14) and (14.17) yield the following results

$$\Psi(G|f_h) = \frac{\int Df \delta(\int_T^{T+l} dx f(t_*, x) - G)\delta(f(t_*, x_h) - f_h)e^S}{\int Df \delta(f(t_*, x_h) - f_h)e^S}$$

$$= \frac{\chi}{\sqrt{2\pi Q^2}} \exp\left[-\frac{1}{2Q^2}(G - G_0)^2\right]$$

$$\Phi(G|f_h; t, T_1) = \frac{\int Df \delta(G - \int_{T_1}^{T_1+\ell} f(t_*, x) dx)\delta(f(t_*, x_h) - f)e^S}{\int Df \delta(f(t_*, x_h) - f_h)e^S}$$

$$= \frac{1}{\sqrt{2\pi Q_1^2}} \exp\left[-\frac{1}{2Q_1^2}(G - G_1)^2\right]$$

The parameters for $\Psi(G|f_h)$ are given below.

$$X = \frac{1}{1 + \ell K}; \quad \tilde{V} = (1 + \ell K)V$$

$$\ln(\chi) = -\int_{th}^{T_n} dx f(t_0, x) - \int_{M_1} \alpha(t, x) + \frac{1}{2}E$$

$$+ \frac{C}{A}\left(f(t_0, x_h) + \int_{t_0}^{th} dt\alpha(t, x_h) - f_h - \frac{C}{2}\right)$$

$$d_+ = (\ln X + G_0)/Q; \qquad d_- = (\ln X + G_0 - Q^2)/Q$$

$$G_0 = \int_{T_n}^{T_n+\ell} dx f(t_0, x) - F - \frac{B}{A}\left(f(t_0, x_h) - C - f_h + \int_{t_0}^{th} dt\alpha(t, x_h)\right) + \frac{q^2}{2}$$

$$Q^2 = q^2 - \frac{B^2}{A}$$

The parameters for $\Phi(G|f_h; t, T_1)$ are given below.

$$G_1 = \int_{T_{n1}}^{T_{n1}+\ell} dx f(t_0, x) + \int_{M_3} \alpha(t, x) - \frac{B_1}{A}\left(f(t_0, x_h) - \int_{t_0}^{th} dt\alpha(t, x_h) - f_h\right)$$

$$Q_1^2 = D - \frac{B_1^2}{A}$$

$$A = \int_{t_0}^{th} dt\sigma(t, x_h)^2 D(t, x_h, x_h; T_{FR})$$

$$B = \int_{M_2} \sigma(t, x_h) D(t, x_h, x; T_{FR})\sigma(t, x)$$

$$B_1 = \int_{\tilde{M}_1} \sigma(t, x_h) D(t, x_h, x; T_{FR})\sigma(t, x)$$

$$C = \int_{M_1} \sigma(t, x_h) D(t, x_h, x; T_{FR}) \sigma(t, x)$$

$$D = \int_{\tilde{\mathcal{Q}}_1} \sigma(t, x) D(t, x, x'; T_{FR}) \sigma(t, x')$$

Caplet volatility is given by

$$q^2 = \int_{\mathcal{Q}_2 + \mathcal{Q}_4} \sigma(t, x) D(t, x, x'; T_{FR}) \sigma(t, x')$$

and

$$E = \int_{\mathcal{Q}_1} \sigma(t, x) D(t, x, x'; T_{FR}) \sigma(t, x')$$

$$F = \int_{t_0}^{t_h} dt \int_{t_h}^{T_n} dx \int_{T_n}^{T_n + \ell} dx' \sigma(t, x) D(t, x, x'; T_{FR}) \sigma(t, x').$$

The domains of integration are given in Figure 14.5. It can be seen that the unconditional probability distribution for the caplet and Libor futures yield volatilities q^2 and D respectively. Hence the conditional expectation reduces the volatility of the caplet by B^2/A, and by B_1^2/A for the Libor futures. This result is expected since the constraint imposed by the requirement of a conditional probability reduces the allowed fluctuations of the instruments.

It could be the case that there is a special maturity time x_h, which causes the largest reduction in conditional variance. The answer is found by minimizing the

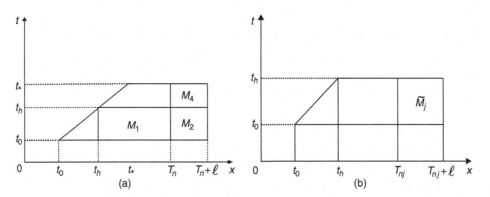

Figure 14.5 (a) Domains of integration M_1, M_2, and for the integration cube $\mathcal{Q}_1$, $\mathcal{Q}_2$, $\mathcal{Q}_4$ where the x' axis has the same limit as its corresponding x axis. (b) Domain of integration $\tilde{M}_j$ and integration cube $\tilde{\mathcal{Q}}_j$ where the x' axis has the same limit as its corresponding x axis.

conditional variance

$$caplet(t_h, t_*, T_n; f_h) = \chi \tilde{V} \left\{ XN(d_+) - e^{-G_0 + \frac{Q^2}{2}} N(d_-) \right\}$$ (14.37)

$$\tilde{L}(t_h, T_{n1}; f_h) = e^{G_1 + \frac{Q_1^2}{2}}$$ (14.38)

Recall the hedging parameter $\eta_1(t)$ is given by Eq. (14.20). Using Eq. (14.38) and setting $t_0 = t$, $t_h = t + \epsilon$, yields an (instantaneous) stochastic delta hedge parameter $\eta_1(t)$ given by

$$\eta_1 = \frac{C}{e^{G_1 + \frac{Q_1^2}{2}} B_1} \left[caplet(t, t_*, T; f_h) - \frac{\eta'}{C} \right]$$ (14.39)

$$\eta' = -B \chi \tilde{V} \left[\frac{XN'(d_+)}{Q} + e^{-G_0 + \frac{Q^2}{2}} N(d_-) - \frac{e^{-G_0 + \frac{Q^2}{2}} N'(d_-)}{Q} \right]$$

14.12 Appendix: Conditional probability – two interest rates

When hedging against two forward interest rates, Eqs. (14.13) and (14.16) require the conditional probability of a caplet given by

$$\Psi(G | f_1, f_2) = \frac{\int Df \delta(\int_T^{T+l} dx f(t_*, x) - G) \prod_{i=1}^2 \delta(f(t_h, x_i) - f_i) e^S}{\int Df \prod_{i=1}^2 \delta(f(t_h, x_i) - f_i) e^S}$$

The conditional probability of Libor, for $j = 1, 2$, is given by

$$\Phi(G | f_1, f_2, T_{nj}) = \frac{\int Df \delta(G - \int_{T_{nj}}^{T_{nj}+\ell} f(t_h, x) dx) \prod_{i=1}^2 \delta(f(t_h, x_i) - f_i) e^S}{\int Df \prod_{i=1}^2 \delta(f(t_h, x_i) - f_i) e^S}$$

A long and tedious calculation yields

$$\Psi(G | f_1, f_2) = \frac{\chi}{\sqrt{2\pi Q^2}} \exp \left[-\frac{1}{2Q^2} (G - G_0)^2 \right]$$

$$\Phi(G | f_1, f_2, T_{nj}) = \frac{1}{\sqrt{2\pi \tilde{Q}_j^2}} \exp \left[-\frac{1}{2\tilde{Q}_j^2} (G - \tilde{G}_j)^2 \right]$$

The distributions for $\Psi(G|f_1, f_2)$ and $\Phi(G|f_1, f_2, T_{nj})$ are given by the following expressions

$$X = \frac{1}{1 + \ell k}; \quad \tilde{V} = (1 + \ell k)V$$

$$\chi = \exp\left\{-\int_{t_h}^{T_n} dx f(t_0, x) - \int_{M_1} \alpha(t, x) + \frac{1}{2}E + \frac{C_{12}}{\tilde{A}_{12}}(R_{12} - \frac{C_{12}}{2})\right\}$$

$$d_+ = (\ln x + G_0)/Q; \qquad d_- = (\ln x + G_0 - Q^2)/Q$$

$$G_0 = \int_{T_n}^{T_n + \ell} dx f(t_0, x) - F - \frac{B_{12}}{\tilde{A}_{12}}(R_{12} - C_{12}) + \frac{q^2}{2}$$

$$Q^2 = q^2 - \frac{B_{12}^2}{\tilde{A}_{12}}$$

$$\tilde{G}_j = \int_{T_{nj}}^{T_{nj}+\ell} dx f(t_0, x) + \int_{\tilde{M}_j} \alpha(t, x) - \frac{\tilde{B}_{12j}}{\tilde{A}_{12}}R_{12} \quad j = 1, 2$$

$$\tilde{Q}_j^2 = D_j - \frac{\tilde{B}_{12j}^2}{\tilde{A}_{12}} \quad j = 1, 2$$

$$R_i = f(t_0, x_i) + \int_{t_0}^{t_h} dt\alpha(t, x_i) - f_i \quad i = 1, 2$$

$$R_{12} = R_1 - \frac{A_{12}}{A_2}R_2$$

$$A_i = \int_{t_0}^{t_h} dt\sigma(t, x_i)^2 D(t, x_i, x_i; T_{FR}) \quad i = 1, 2$$

$$A_{12} = \int_{t_0}^{t_h} dt\sigma(t, x_1)D(t, x_1, x_2; T_{FR})\sigma(t, x_2)$$

$$\tilde{A}_{12} = A_1 - \frac{A_{12}}{A_2}$$

$$B_i = \int_{M_2} \sigma(t, x_i)D(t, x_i, x; T_{FR})\sigma(t, x) \quad i = 1, 2$$

$$B_{12} = B_1 - \frac{A_{12}}{A_2}B_2$$

$$\tilde{B}_{ij} = \int_{\tilde{M}_j} \sigma(t, x_i)D(t, x_i, x; T_{FR})\sigma(t, x) \quad i = 1, 2; \quad j = 1, 2$$

$$\tilde{B}_{12j} = \tilde{B}_{1j} - \frac{A_{12}}{A_2}\tilde{B}_{2j} \quad j = 1, 2, \ldots, 5$$

$$C_i = \int_{M_1} \sigma(t, x_i)D(t, x_i, x; T_{FR})\sigma(t, x) \quad i = 1, 2$$

Furthermore

$$C_{12} = C_1 - \frac{A_{12}}{A_2} C_2$$

$$D_j = \int_{\tilde{Q}_j} \sigma(t,x) D(t,x,x'; T_{FR}) \sigma(t,x') \quad j = 1,2$$

$$q^2 = \int_{Q_2+Q_4} \sigma(t,x) D(t,x,x'; T_{FR}) \sigma(t,x')$$

$$E = \int_{Q_1} \sigma(t,x) D(t,x,x'; T_{FR}) \sigma(t,x')$$

$$F = \int_{t_0}^{t_h} dt \int_{t_h}^{T_n} dx \int_{T_n}^{T_n+\ell} dx' \sigma(t,x) D(t,x,x'; T_{FR}) \sigma(t,x') \tag{14.40}$$

The domains of integration are given in Figure 14.5.

The conditional expectation of a caplet and Libor are given by

$$caplet(t_h, t_*, T_n; f_1, f_2) = \chi \tilde{V} \left(X N(d_+) - e^{-G_0 + \frac{Q^2}{2}} N(d_-) \right) \tag{14.41}$$

$$L(t_h, T_{nj}; f_1, f_2) = e^{\tilde{G}_j + \frac{\tilde{Q}_j^2}{2}} \tag{14.42}$$

14.13 Appendix: HJM limit of hedging functions

The HJM limit of the hedging functions, defined by taking the limit of the propagator $\mathcal{D}(t,x,x') \to 1$ is analyzed for the specific exponential function considered by Jarrow and Turnbull [65]

$$\sigma_{HJM}(t,x) = \sigma_0 e^{\beta(x-t)} \tag{14.43}$$

It can be shown that

$$A = \frac{\sigma_0^2}{2\beta} e^{-2\beta x_h} (e^{2\beta t_h} - e^{2\beta t_0})$$

$$B = \frac{\sigma_0^2}{2\beta^2} e^{-\beta x_h} (e^{-\beta T_n} - e^{-\beta T_n+\ell})(e^{2\beta t_h} - e^{2\beta t_0})$$

$$B_1 = \frac{\sigma_0^2}{2\beta^2} e^{-\beta x_h} (e^{-\beta T_{n1}} - e^{-\beta T_{n1}+\ell})(e^{2\beta t_h} - e^{2\beta t_0})$$

$$C = \frac{\sigma_0^2}{2\beta^2} e^{-\beta x_h} (e^{-\beta t_h} - e^{-\beta T_n})(e^{2\beta t_h} - e^{2\beta t_0})$$

$$D = \frac{\sigma_0^2}{2\beta^3} (e^{-\beta T_{n1}+\ell} - e^{-\beta T_{n1}})^2 (e^{2\beta t_h} - e^{2\beta t_0})$$

$$E = \frac{\sigma_0^2}{2\beta^3}(e^{-\beta T_n} - e^{-\beta t_h})^2(e^{2\beta t_h} - e^{2\beta t_0})$$

$$F = \frac{\sigma_0^2}{2\beta^3}(e^{-\beta T_{n+\ell}} - e^{-\beta T_n})(e^{-\beta T_n} - e^{-\beta t_h})(e^{2\beta t_h} - e^{2\beta t_0})$$

The exponential volatility function given in Eq. (14.43) has the remarkable property, similar to the case found for the hedging of zero coupon bonds in [12], that

$$Q_1^2(HJM) = D_{HJM} - \frac{B_{1HJM}^2}{A_{HJM}} \equiv 0 \qquad (14.44)$$

Hence, the conditional probability for the Libor futures given in Section 4.11 is deterministic. Indeed, once the forward rate f_h is fixed, the following identity is valid

$$L_{HJM}(t_h, T_{n1}; f_h) \equiv L(t_h, T_{n1})$$

In other words, for the volatility function in Eq. (14.43), the Libor futures for the HJM model are exactly determined by only one of the forward interest rates.

However, in general the conditional probability for the caplet is not deterministic since the volatility from t_h to t_*, before the caplet's expiration, is not compensated for by fixing one forward interest rate.

15

Interest rate Hamiltonian and option theory

The Hamiltonian is a differential operator that acts on an underlying state space [70].[1] A Hamiltonian formulation of option theory is discussed and shown to be equivalent to the Black–Scholes approach. In particular, it is shown that the Black–Scholes equation is mathematically identical to the (imaginary time) Schrodinger equation of quantum mechanics [70].

The Hamiltonian formulation of quantum field theory is equivalent to, and independent of, the framework based on the Feynman path integral and the Lagrangian discussed in Chapter 5. A Hamiltonian formulation of interest rates provides another perspective on option theory and interest rates. There are many advantages of having multiple formulations, since for some problems calculations based on the Hamiltonian are more transparent and tractable than using the Lagrangian approach. In particular, the Hamiltonian formulation is useful for exactly solving nonlinear martingale conditions as well as for studying a specific class of debt instruments options, which includes American and barrier options.

15.1 Introduction

The Hamiltonian is introduced by considering option theory for a single equity. Option theory is shown to have a Hamiltonian formulation in which the option price is a function of the matrix elements of the exponential of the Hamiltonian. The Black–Scholes option price is given a Hamiltonian derivation starting from first principles that are reasonable and intuitive.

The interest rate state space and Hamiltonian are derived from the forward interest rates Lagrangian and are a natural generalization of a similar Black–Scholes analysis for equities. However, unlike the case for equity, the interest rate state space and Hamiltonian are time dependent and the Hamiltonian is a differential operator in infinitely many variables.

---

[1] The concepts of state space and operators are briefly reviewed in Appendix A.6.

The martingale condition for interest rates is given a Hamiltonian formulation and the drift is derived for the various choices of numeraires. In the Libor Market Model, the drift is a nonlinear function of the Libors. In the path integral formulation, a nonlinear drift leads to a nonlinear path integral – which, in general, cannot be exactly evaluated due to its computational intractability. In contrast, using the Hamiltonian approach, nonlinear Libor drift is computed *exactly* – from first principles and in a parsimonious manner.

The Hamiltonians of the coupon bond and Libor are derived from the interest rate and bond Hamiltonians. The price of the coupon bond and Libor European and barrier options are computed from the Hamiltonian.

15.2 Hamiltonian and equity option pricing

The central problem in option pricing is the following: given the payoff function $\mathcal{P}$ of some security S at future time T, what is the price of the option at an earlier time $t < T$, namely $C(t, S(t))$? The standard approach for addressing option pricing in mathematical finance is based on stochastic calculus [65] and was expressed in the formalism of quantum mechanics in [12]. A derivation of option pricing, in the framework of quantum mechanics from first principles, is given in [7] and summarized in this section.

A stock S of a company is never negative; the owner of a stock has none of the company's liabilities but a right to dividends and pro rata ownership of a company's assets. Hence $S = e^x \geq 0$; $-\infty \leq x \leq +\infty$. The stock price, at each instant, is considered to have a random value, making it mathematically identical to a quantum particle that at each instant has a random position (when it is not being observed). The real variable x, similar to a quantum system, can consequently be considered to be a degree of freedom describing the behavior of the stock price. The completeness equation for a degree of freedom is discussed in Appendix A.6 and, from Eq. (A.47), is given by

$$\int_{-\infty}^{\infty} dx |x\rangle\langle x| = \mathcal{I}: \text{ completeness equation} \tag{15.1}$$

$|x\rangle$ is a coordinate basis for the state space, denoted by $\mathcal{V}$, which is an infinite-dimensional linear vector space. $\langle x|$ is the basis of the dual state space, denoted by $\mathcal{V}_D$. $\mathcal{I}$ is the identity operator on the tensor product state space $\mathcal{V} \otimes \mathcal{V}_D$.

Option pricing in the framework of quantum mechanics is based on the following assumptions.

- All financial instruments, including the price of the option, are elements of a state space V discussed in Appendix A.6. The stock price $|S\rangle$ has the following representation

$$|S\rangle = \int_{-\infty}^{\infty} dx e^x |x\rangle \langle x|$$
$$S = \langle x|S\rangle = e^x$$

- The final value of an option, at future time T, is given by a payoff function $|P\rangle$ that is an element of the state space V. The option price $|C, t\rangle$ and the payoff function have a coordinate representation

$$|P\rangle = \int_{-\infty}^{\infty} dx P(x)|x\rangle \langle x|; \quad |C, t\rangle = \int_{-\infty}^{\infty} dx C(t, x)|x\rangle \langle x|$$
$$C(t, x) = \langle x|C, t\rangle; \quad P(x) = \langle x|P\rangle$$

Unlike quantum mechanics, the state space V is larger than a normalizable Hilbert space since fundamental financial instruments such as the stock price S are not normalizable.
- The option price is evolved by a Hamiltonian H, which is a linear operator acting on the state space V. Due to the necessity of fulfilling put–call parity, H evolves *all* options on an equity, including both the call and put options.
- The price of the option satisfies the (imaginary time) Schrödinger equation

$$H|C, t\rangle = \frac{\partial}{\partial t}|C, t\rangle \tag{15.2}$$

with the final value fixed by the payoff function as follows

$$|C, T\rangle = |P\rangle; \quad T > t \tag{15.3}$$

It should be emphasized that option pricing is a classical stochastic problem. Unlike the wave function of quantum mechanics, the *option price* $C(t, x)$ is *directly observable*; furthermore, there is no concept of a quantum measurement in option theory. The similarity of option pricing with quantum mechanics, at this stage, is purely mathematical: both can be described by an infinite-dimensional linear vector space V and linear operators like H acting on this vector space.

Integrating Eq. (15.2) yields the following

$$|C, t\rangle = e^{tH}|C, 0\rangle$$

The final value condition given in Eq. (15.3) yields

$$|C, T\rangle = e^{TH}|C, 0\rangle = |P\rangle \Rightarrow |C, 0\rangle = e^{-TH}|P\rangle$$

Hence

$$|C,t\rangle = e^{-(T-t)H}|\mathcal{P}\rangle$$

or, more explicitly, for remaining time $\tau = T - t$

$$C(\tau,x) = \langle x|C,t\rangle = \langle x|e^{-\tau H}|\mathcal{P}\rangle \qquad (15.4)$$

The expression given in Eq. (15.4) shows that the option price is the matrix element of the operator $e^{-\tau H}$, taken between the payoff $|\mathcal{P}\rangle$ and the current price of the stock, namely $\langle x|$.[2]

The option price was obtained earlier in Eq. (3.7) as the expectation value of the payoff function $\mathcal{P}$, discounted by the money market numeraire, and given by

$$C(\tau,x) = E[e^{-r\tau}\mathcal{P}]$$

The expectation value in the equation given above is *conditioned* on the stock price having the value of $S = e^x$ at earlier calendar time t. The conditional expectation value, in the mathematics of quantum mechanics, is represented by the matrix element of the evolution operator $e^{-\tau H}$; hence

$$C(\tau,x) = E[e^{-r\tau}\mathcal{P}] = \langle x|e^{-\tau H}|\mathcal{P}\rangle \qquad (15.5)$$

The Hamiltonian includes the term $e^{-r\tau}$ that arises due to the choice of the money market numeraire.

15.3 Equity Hamiltonian and martingale condition

The fundamental theorem of finance states that for option price to be free from arbitrage opportunities, the Hamiltonian H must yield a *martingale* evolution [59]. The martingale condition is the mathematical expression, in probability theory, of a fair game in which, on the average, a gambler leaves the casino with the money with which she or he enters.

Mathematically, a martingale – discussed in Appendix A.2 – states the following. A (random) stochastic process is a martingale if the expectation value of its future value is equal to its present value. For the money market numeraire, the discounted stock price, namely, $S(t)/e^{rt}$ is a martingale for the risk-free evolution required for pricing options. From Eqs. (3.3) and (15.5), the martingale condition for the discounted stock price $S(t)/e^{rt}$ is given by

$$S = E[e^{-r\tau}S] = \langle x|e^{-\tau H}|S\rangle \qquad (15.6)$$

---

[2] Unlike quantum mechanics, where only the absolute value of the matrix elements are physically observable, for the option price the matrix elements of $e^{-\tau H}$ are directly observable, being the price of options.

Eq. (15.6) can be re-written in state space notation and yields [12]

$$\langle x|S \rangle = \langle x|e^{-\tau H}|S \rangle; \quad \Rightarrow e^{-\tau H}|S \rangle = |S \rangle$$

$$\Rightarrow H|S \rangle = 0 \ : \ \text{martingale condition} \tag{15.7}$$

The requirement of martingale evolution places a condition on the Hamiltonian that it must annihilate the security S. This condition has far-reaching consequences in finance since it holds for more complicated systems such as the forward interest rates.

15.4 Pricing kernel and Hamiltonian

The pricing kernel determines the option price for all European options and clearly displays the central role of the Hamiltonian in the theory of option pricing. The pricing kernel is more complicated for interest rates compared to the case for equity and has been discussed in Section. 5.9. It will be seen, in chapter 16 , that the pricing kernel for infinitesimal time is required for evaluating the price of all path dependent options, including the case for equity and interest rates.

Using the completeness equation given in Eq. (15.1) yields

$$C(t, x) = \int_{-\infty}^{\infty} dx' \langle x|e^{-\tau H}|x' \rangle \langle x'\mathcal{P} \rangle$$

$$= \int_{-\infty}^{\infty} dx' p(x, x'; \tau)\mathcal{P}(x') \tag{15.8}$$

where the pricing kernel is given by

$$p(x, x'; \tau) = \langle x|e^{-\tau H}|x' \rangle \tag{15.9}$$

The pricing kernel $p(x, x'; \tau)$ is the conditional probability, that, given the value $e^{x'}$ at future time $T = t + \tau$, the stock at time t will have the value e^{x}. Eq. (15.9) shows that the pricing kernel is the matrix element of the differential operator $\exp\{-\tau H\}$.

The pricing kernel has been discussed in detail in [12, 15]. The path integral representation for the option price has been discussed in Section 3.11. For action S_{BS} and Lagrangian $\mathcal{L}_{BS}$, the pricing kernel has the following realization

$$p(x, x'; \tau) = \int DX e^{S_{BS}}; \quad S_{BS} = \int_{0}^{\tau} dt \mathcal{L}_{BS} \tag{15.10}$$

Boundary conditions: $x(0) = x'; \ x(\tau) = x$

For an infinitesimal time step ϵ, $S_{BS} \simeq \epsilon \mathcal{L}_{BS}$; the path integral collapses – resulting in no integrations – and yields, for normalization $\mathcal{N}$, the following Dirac–Feynman relation [12]

$$p(x, x'; \epsilon) = \mathcal{N}e^{\epsilon \mathcal{L}_{BS}} = \langle x|e^{-\epsilon H}|x'\rangle \tag{15.11}$$

The action and Lagrangian depend on the system being studied, and, for example, Eq. (3.36) is appropriate for N-correlated equities.

The pricing kernel yields the price of the call and put European options which, for strike price K, are given below

$$Call(\tau, x) = \int_{-\infty}^{\infty} dx' \langle x|e^{-\tau H}|x'\rangle (e^{x'} - K)_+$$

$$Put(\tau, x) = \int_{-\infty}^{\infty} dx' \langle x|e^{-\tau H}|x'\rangle (K - e^{x'})_+ \tag{15.12}$$

15.4.1 Martingale and put–call parity

Using the completeness equation given in Eq. (15.1) one can re-write the martingale condition Eq. (15.7). From Eq. (15.9), the pricing kernel $p(x, x'; \tau) = \langle x|e^{-\tau H}|x'\rangle$ and $e^x = \langle x|S\rangle$ yields

$$\langle x|e^{-\tau H}|S\rangle = \langle x|S\rangle = \langle x|e^{-\tau H} \int_{-\infty}^{\infty} dx' |x'\rangle \langle x'|S\rangle$$

$$\Rightarrow e^x = \int_{-\infty}^{\infty} dx' p(x, x'; \tau) e^{x'} \quad : \text{ martingale condition} \tag{15.13}$$

Put–call parity is the result of very general properties that all option Hamiltonians must satisfy. From Eq. (15.5), if the payoff is a constant equal to say K, then it factorizes out of the expectation value and yields

$$e^{-r\tau} K = E[e^{-r\tau} K] = \langle x|e^{-\tau H}|K\rangle \tag{15.14}$$

From Eq. (4.15)

$$[e^x - K]_+ - [K - e^x]_+ = e^x - K$$

Hence, from Eqs. (15.12), (15.13), and (15.14)

$$Call(\tau, x) - Put(\tau, x) = \int_{-\infty}^{\infty} dx' \langle x|e^{-\tau H}|x'\rangle (e^{x'} - K)$$

$$= e^x - \langle x|e^{-\tau H}|K\rangle$$

$$= S - e^{-r\tau} K$$

which is the result given in Eq. (3.1).

15.5 Hamiltonian for Black–Scholes equation

What should be the form of the Hamiltonian driving the option price of an equity $S = e^x$? Assume that H has the following fairly general form

$$H = -\frac{\sigma^2(x)}{2}\frac{\partial^2}{\partial x^2} + b\frac{\partial}{\partial x} + a \qquad (15.15)$$

where $\sigma^2(x)$ is an arbitrary function of x. The parameter $\sigma^2(x)$ is the volatility of the stock price, and indicates the degree to which the evolution of the stock price is random.

Consider for starters the price of a put option. Suppose the strike price $K \to +\infty$; then the payoff function has the following limit $h(S) = (K - S)_+ \to K$: constant. Hence, from Eqs. (15.4) and (15.15)

$$Put(t,x) = \langle x|e^{-\tau H}|h\rangle \to \langle x|e^{-\tau H}|K\rangle = e^{-a\tau}K$$

For $K \to +\infty$, the put option is certain to be exercised since the holder of the put option, in exchange for the stock, is certain to be paid an amount K at future time T. The present-day value of the put option, from the principle of no-arbitrage, must be the value of K discounted to the present by the risk-free spot interest rate. Hence

$$Put(t,x) \to e^{-r\tau}K \quad \Rightarrow a = r$$

The martingale condition given in Eq. (15.7) yields

$$H|S\rangle = 0 \quad \Rightarrow b = \frac{\sigma^2(x)}{2} - r$$

Collecting the results yields the famous Black–Scholes Hamiltonian [7, 12]

$$H_{BS} = -\frac{\sigma^2(x)}{2}\frac{\partial^2}{\partial x^2} + \left(\frac{\sigma^2(x)}{2} - r\right)\frac{\partial}{\partial x} + r \neq H_{BS}^\dagger \qquad (15.16)$$

The Black–Scholes Hamiltonian H_{BS} is not Hermitian; this is a general feature of all Hamiltonians in finance, the root cause for which arises from the requirement of satisfying the martingale condition.

Note that the Black–Scholes Hamiltonian makes no reference to the market value of the drift of the stock price, which is determined by its rate of return. The reason being the price of the option can only reflect the risk-free rate of return given by r, since otherwise it would be open to arbitrage opportunities [59, 65]. The martingale condition allows the volatility σ^2 to be an arbitrary function of the stock price $S = e^x$, a fact that can be derived by a more standard analysis based on stochastic calculus.

The evolution for the option price, in terms of remaining time $\tau = T - t$, is given, from Eq. (15.2), by the Black–Scholes–Schrodinger equation[3]

$$\frac{\partial C(\tau, x)}{\partial \tau} = -\langle x|H|C \rangle$$

$$= \frac{\sigma^2(x)}{2} \frac{\partial^2 C(\tau, x)}{\partial x^2} - \left(\frac{\sigma^2(x)}{2} - r \right) \frac{\partial C(\tau, x)}{\partial x} - rC(\tau, x)$$

In terms of the variable $S = e^x$ and calendar time t, the Black–Scholes–Schrodinger equation for option pricing is given by

$$\frac{\partial C(t, x)}{\partial t} = -\frac{1}{2} \sigma^2(S) S^2 \frac{\partial^2 C(t, x)}{\partial S^2} - rS \frac{\partial C(t, x)}{\partial S} + rC(t, x)$$

This is the famous Black–Scholes equation derived earlier in Eq. (3.24); it appears in this form since S is taken to be the variable of choice in most of the literature in finance.

The Hamiltonian for N degrees of freedom can be obtained from the Black–Scholes equation for N equities given in Eq. (3.29); in terms of logarithmic degrees of freedom z_i defined by $S_i = e^{z_i}$; $i = 1, 2, \ldots, N$

$$\frac{\partial C}{\partial t} = H_N C$$

$$H_N = -\frac{1}{2} \sum_{ij=1}^{N} \sigma_i \sigma_j \rho_{ij} \frac{\partial^2}{\partial z_i \partial z_j} + \sum_{i=1}^{N} \left(\frac{1}{2} \sigma_i^2 \rho_{ii} - r \right) \frac{\partial}{\partial z_i} + r \quad (15.17)$$

Each equity S_I is a martingale for H_N since

$$H_N S_I = \left[-\frac{1}{2} \sum_{ij=1}^{N} \sigma_i \sigma_j \rho_{ij} \delta_{i-I} \delta_{j-I} + \sum_{i=1}^{N} \left(\frac{1}{2} \sigma_i^2 \rho_{ii} - r \right) \delta_{i-I} + r \right] S_I = 0$$

$$\Rightarrow H_N S_I = 0 \text{: for each } I \quad (15.18)$$

The equity Hamiltonian H_N is non-Hermitian due to the necessity of having a martingale time evolution for the underlying security S_I. H_N acts on the space of

---

[3] In physics, there is an additional factor of i in the Schrodinger equation that can be removed by analytically continuing time to a negative imaginary variable. This analytical continuation is well known in physics and is a precursor for numerically studying the path integral.

functions of $z_1, z_2, \ldots, z_N$, which forms an N-dimensional state (function) space that is time independent – unlike the case for interest rates discussed in Section 15.6 below.

15.6 Interest rate state space $\mathcal{V}_t$

The Hamiltonian and the state space of a system are two independent properties of a quantum theory; the Lagrangian is a result of these two ingredients. The essential features of the interest rates' Hamiltonian and state space are reviewed; a detailed discussion is given in [12].

The state space of a quantum field theory, similar to all quantum systems, is a linear vector space – denoted by $\mathcal{V}_t$ – that consists of all functionals of the field configurations at some fixed time t. The dual space of $\mathcal{V}_t$ – denoted by $\mathcal{V}_{t,\text{Dual}}$ – consists of all linear mappings from $\mathcal{V}_t$ to the complex numbers, and is also a linear vector space. The Hamiltonian $\mathcal{H}_t$ is an operator – the quantum analog of energy – that is an element of the tensor product space $\mathcal{V}_t \otimes \mathcal{V}_{t,\text{Dual}}$ and maps the state space to itself, that is $\mathcal{H}_t : \mathcal{V}_t \to \mathcal{V}_t$.

The Hamiltonians for log Libor $\phi(t, x)$ and bond forward interest rates $f(t, x)$ are far more complicated than the case for equity, discussed in Section 15.5; since $x \in [t, t + T_{FR}]$ the quantum fields $\phi(t, x)$ and $f(t, x)$ exists only for future time, that is for $x > t$. In particular, the interest rates' quantum field has a *distinct* state space $\mathcal{V}_t$ for every instant t. For brevity of notation let $f(t, x)$ denote both log Libor $\phi(t, x)$ as well as bond forward interest rates $f(t, x)$. In all the derivations of this chapter it is assumed that $-\infty \leq f(t, x), \phi(t, x) \leq +\infty$.[4]

For each time slice, the state space is defined for interest rates with $x > t$, as shown in Figure 15.1(a). The state space has a nontrivial structure due to the underlying trapezoidal domain $\mathcal{T}$ of the xt space. On composing the state space for each time slice, the trapezoidal structure for finite time, as shown in Figure 15.1(b), is seen to emerge from the state space defined for each time slice.

The state space at time t is labeled by $\mathcal{V}_t$, and the state vectors in $\mathcal{V}_t$ are denoted by $|f_t>$. For fixed time t, the state space $\mathcal{V}_t$ consists of all possible functions of the interest rates, with future time $x \in [t, t + T_{FR}]$. The elements of the state space of the forward interest rates $\mathcal{V}_t$ includes *all* possible debt instruments that are traded in the market at time t. In continuum notation, the basis states of $\mathcal{V}_t$ are tensor products over the future time x and satisfy the following completeness equation

---

[4] As discussed in Section 6.9, Libor forward interest rates $f_L(t, x) \geq 0$ are nonlinear and positive and will not be discussed in this chapter.

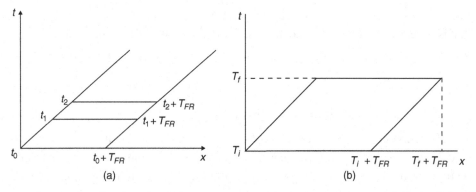

Figure 15.1 The domain of the state space of the interest rates. (a) In the figure the state space $\mathcal{V}_t$ is indicated for two distinct calendar times t_1 and t_2. (b) The trapezoidal domain $\mathcal{T}$ of the forward interest rates required for computing the transition amplitude $< f_{\text{initial}}|\mathcal{T}\left\{\exp - \int_{T_i}^{T_f} \mathcal{H}(t)\,dt\right\}|f_{\text{final}} >$.

$$|f_t > = \prod_{t \leq x \leq t+T_{FR}} |f(t,x) >$$

$$\mathcal{I}_t = \prod_{t \leq x \leq t+T_{FR}} \int_{-\infty}^{+\infty} df(t,x)|f_t\rangle\langle f_t| \equiv \int Df_t|f_t\rangle\langle f_t| \qquad (15.19)$$

Figure 15.1 shows the domain of the state space as a function of time t.

The time-dependent interest rate Hamiltonian $\mathcal{H}(t)$ is the backward Fokker–Planck Hamiltonian and propagates the interest rates *backwards* in time, taking the final state $|f_{\text{final}} >$ given at future calendar time T_f backwards to an initial state $< f_{\text{initial}}|$ at the earlier time T_i.

The transition amplitude Z for a time interval $[T_i, T_f]$ can be constructed from the Hamiltonian and state space by applying the time slicing method. Since the state space and Hamiltonian are both time-dependent one has to use the time-ordering operator $\mathcal{T}$ to keep track of the time dependence: $\mathcal{H}(t)$ for earlier time t is placed to the left of $\mathcal{H}(t)$ that refers to later time. The transition amplitude between a final (coordinate basis) state $|f_{\text{final}} >$ at time T_f to an arbitrary initial (coordinate basis) state $< f_{\text{initial}}|$ at time T_i is given by the following [12]

$$Z = \langle f_{\text{initial}}|\mathcal{T}\left\{\exp - \int_{T_i}^{T_f} \mathcal{H}(t)\,dt\right\}|f_{\text{final}}\rangle \qquad (15.20)$$

Due to the time dependence of the state spaces $\mathcal{V}_t$, the forward interest rates that determine Z form a trapezoidal domain shown in Figure 15.1.

The transition from equity Hamiltonian to interest rates is equivalent to the transition from quantum mechanics to quantum field theory; this is briefly discussed in Appendix A.7.

15.6.1 *Coupon and zero coupon bond state vector*

The coupon and zero coupon bond are important state vectors in the theory of forward interest rates. Consider a risk-free zero coupon bond that matures at time T with a payoff of $1. Recall from Eq. (2.12) that the price of a zero coupon bond at time $t < T$ is given by

$$B(t, T) = e^{-\int_t^T f(t,x)dx}$$

The ket state vector $|B(t, T)>$ is an element of the state space $\mathcal{V}_t$. The zero coupon bond state vector is written as follows

$$B(t, T) \equiv < f_t|B(t, T) >= e^{-\int_t^T f(t,x)dx} \tag{15.21}$$

The coupon bond $|\mathcal{B}>$ is a state vector, with fixed coupons of amount c_i paid at times T_i, and with a final payoff of L at time T. In the state space language, the coupon bond is the following linear superposition of the zero coupon bonds

$$|\mathcal{B}(t) >= \sum_i c_i|B(t, T_i) > +L|B(t, T) >$$

15.7 Interest rate Hamiltonian

Consider the Lagrangian density $\mathcal{L}_\phi(t, x)$ for the log Libor field $\phi(t, x)$ given by Eq. (6.61)

$$\mathcal{L}_\phi(t, x) = -\frac{1}{2}\left[\frac{\partial\phi(t, x)/\partial t - \tilde{\rho}(t, x)}{\gamma(t, x)}\right]D_L^{-1}(t, x, x')\left[\frac{\partial\phi(t, x')/\partial t - \tilde{\rho}(t, x')}{\gamma(t, x')}\right]$$

$$\tilde{\rho} = -\frac{1}{2}\Lambda(t, x) + \rho(t, x); \quad -\infty \le \phi(t, x) \le +\infty$$

The volatility $\gamma(t, x)$ is deterministic and ρ is a stochastic and nonlinear drift term defined in Eq. (6.59). Similarly, a general Lagrangian density for the bond forward interest rates, from Eqs. (5.25) and (5.1), is given by

$$\mathcal{L}_f(t, x) = -\frac{1}{2}\left[\frac{\partial f(t, x)/\partial t - \alpha(t, x)}{\sigma(t, x)}\right]D^{-1}(t, x, x')\left[\frac{\partial f(t, x')/\partial t - \alpha(t, x')}{\sigma(t, x')}\right]$$

$$-\infty \le f(t, x) \le +\infty$$

Neumann boundary conditions, given in Eqs. (5.10) and (6.62), have been incorporated into the expression for (both) the Lagrangians. The derivation for the Hamiltonian is done for an arbitrary propagator $\mathcal{D}^{-1}(t, x, x')$, although for most applications a specific choice, such as the stiff propagator, is made.

Discretizing time into a lattice of spacing ϵ yields $t \rightarrow t_n = n\epsilon$. The Lagrangian $\mathcal{L}(t_n)$ is given by

$$\mathcal{L}(t_n) \equiv \int_{t_n}^{t_n+T_{FR}} dx \mathcal{L}(t_n, x) = -\frac{1}{2\epsilon^2} \int_x \mathcal{A}(t_n, x) \mathcal{D}^{-1}(t, x, x') \mathcal{A}(t_n, x) \quad (15.22)$$

$$\mathcal{A}(t_n, x) = \frac{(f_{t_n+\epsilon} - f_{t_n} - \epsilon\alpha_{t_n})(x)}{\sigma(t_n, x)} \quad \text{or} \quad \frac{(\phi_{t_n+\epsilon} - \phi_{t_n} - \epsilon\rho_{t_n})(x)}{\gamma(t_n, x)} \quad (15.23)$$

$$\int_x \equiv \int_{t_n}^{t_n+T_{FR}} dx \quad (15.24)$$

where $f(t_n, x) \equiv f_{t_n}(x)$; $\phi(t_n, x) \equiv \phi_{t_n}(x)$ has been written to emphasize that time t_n is a parameter for the interest rate Hamiltonian.

The Dirac–Feynman formula relates the Lagrangian $\mathcal{L}(t_n)$ to the Hamiltonian operator by a generalization of Eq. (15.11) and yields

$$< f_{t_n} | e^{-\epsilon\mathcal{H}_f} | f_{t_n+\epsilon} > = \mathcal{N} e^{\epsilon\mathcal{L}(t_n)} \quad (15.25)$$

where $\mathcal{N}$ is a normalization. From the discussion on the pricing kernel in Section 5.9 and, in particular, from Eq. (5.41) a Hamiltonian representation of the interest rates' pricing kernel is provided by Eq. (15.25).

Eq. (15.22) is re-written using Gaussian integration and (ignoring henceforth irrelevant constants), using notation

$$\prod_x \int_{-\infty}^{+\infty} dp(x) \equiv \int Dp$$

yields

$$e^{\epsilon\mathcal{L}(t_n)} = \int Dp e^{-\frac{\epsilon}{2} \int_{x,x'} p(x)\mathcal{D}(t,x,x')p(x') + i \int_x p(x)\mathcal{A}(x)} \quad (15.26)$$

The propagator $\mathcal{D}(t, x, x')$ is the inverse of $\mathcal{D}^{-1}(t, x, x')$.

Consider for concreteness the derivation of the Hamiltonian for log Libor $\phi(t, x)$. Re-scaling the variable $p(x) \rightarrow \gamma(t, x)p(x)$, Eqs. (15.22) and (15.23) yield (up to an irrelevant constant)[5]

$$e^{\epsilon\mathcal{L}(t)} = \int Dp e^{i \int_x p(x)(\phi_{t+\epsilon} - \phi_t - \epsilon\rho_t)(x) - \frac{\epsilon}{2} \int_{x,x'} \gamma(t,x)p(x)\mathcal{D}(x,x';t)\gamma(t,x')p(x')} \quad (15.27)$$

---

[5] Since only two time slices are henceforth considered, the subscript n on t_n is dropped as it is unnecessary.

Hence, the Dirac–Feynman formula given in Eq. (15.25) yields the Hamiltonian as follows

$$\mathcal{N}e^{\epsilon\mathcal{L}(t)} = < \phi_t | e^{-\epsilon\mathcal{H}_\phi} | \phi_{t+\epsilon} > \tag{15.28}$$

$$= e^{-\epsilon\mathcal{H}_t(\delta/\delta\phi_t)} \int Dp e^{i \int_x p(\phi_t - \phi_{t+\epsilon})} \tag{15.29}$$

The Hamiltonian is written in terms of *functional derivatives* in the coordinates of the *dual* state space variables ϕ_t. For each instant of time, there are infinitely many independent interest rates (degrees of freedom) represented by the collection of variables $\phi_t(x), x \in [t, t + T_{FR}]$. Hence, one needs to use functional derivatives, discussed in Appendix A.5, to represent the Hamiltonian as a differential operator.

The degrees of freedom $\phi_t(x)$ refer to time t only through the domain on which the Hamiltonian is defined. Unlike the action $S[f]$ that spans all instants of time from the initial to the final time, the Hamiltonian is an infinitesimal generator in time, and refers to only the instant of time at which it acts on the state space. This is the reason that in the Hamiltonian the time index t can be dropped for the variables $\phi_t(x)$ and replaced by $\phi(x)$ with $t \leq x \leq t + T_{FR}$.

The Hamiltonian for log Libor interest rates, from Eqs. (15.27), (15.28), and (15.29), is given by

$$\mathcal{H}_\phi(t) = -\frac{1}{2} \int_t^{t+T_{FR}} dxdx' M_\gamma(x,x';t) \frac{\delta^2}{\delta\phi(x)\delta\phi(x')}$$

$$- \int_t^{t+T_{FR}} dx \tilde{\rho}(t,x) \frac{\delta}{\delta\phi(x)} \tag{15.30}$$

$$M_\gamma(x,x';t) = \gamma(t,x)\mathcal{D}_L(x,x';t)\gamma(t,x')$$

Similarly, the Hamiltonian for bond forward interest rates is given by

$$\mathcal{H}_f(t) = -\frac{1}{2} \int_t^{t+T_{FR}} dxdx' M_\sigma(x,x';t) \frac{\delta^2}{\delta f(x)\delta f(x')}$$

$$- \int_t^{t+T_{FR}} dx\alpha(t,x) \frac{\delta}{\delta f(x)} \tag{15.31}$$

$$M_\sigma(x,x';t) = \sigma(t,x)\mathcal{D}(x,x';t)\sigma(t,x')$$

The derivation only assumes that the volatilities $\sigma(t,x)$ and $\gamma(t,x)$ are deterministic. The drift terms $\alpha(t,x)$ and $\rho(t,x)$ in the Hamiltonian are completely *general* and can be any (nonlinear) functions of the interest rates.[6]

General considerations related to the existence of a martingale measure rule out any potential terms for the interest rates Hamiltonian [12].[7] The entire dynamics is contained in the kinetic term with the function $M(x,x';t)$ encoding the model chosen for the interest rates; a wide variety of such models have been discussed in [12]. The drift term is completely fixed by the martingale condition and, in particular, by $M(x,x';t)$.

The quantum fields $\phi(t,x)$ and $f(t,x)$ are more fundamental than the velocity quantum field $\mathcal{A}(t,x)$; the Hamiltonian cannot be written in terms of the $\mathcal{A}(t,x)$ degrees of freedom. The reason being that the dynamics of the forward interest rates are contained in the time derivative terms in the Lagrangian, namely terms containing $\partial\phi(t,x)/\partial t$ and $\partial f(t,x)/\partial t$; in going to the Hamiltonian representation, these time derivatives essentially become differential operators $\delta/\delta\phi(t,x)$ and $\delta/\delta f(t,x)$.[8]

15.7.1 Future market time

Future market time is defined by the nonlinear maturity variable $z \equiv z(\theta) = \theta^\nu$; $\theta = x - t$. One can also view θ as a function of z since $\theta(z) = z^{1/\nu}$. The market time Hamiltonian is given by

$$\mathcal{H}_{f,z}(t) = -\frac{1}{2}\int_{z(0)}^{z(T_{FR})} dz\,dz'\,\sigma(t,z)D(z,z')\sigma(t,z')\frac{\delta^2}{\delta f(\theta(z))\delta f(\theta(z'))}$$
$$-\int_{z(0)}^{z(T_{FR})} dz\,\alpha(t,z)\frac{\delta}{\delta f(\theta(z))} \tag{15.32}$$

The $\theta = \theta(z)$ variable is the label of the forward rate functional derivative $\delta/\delta f(\theta)$, but is otherwise replaced everywhere by the nonlinear variable $z(\theta)$. These features of the Hamiltonian are a reflection of the defining equation for market future time discussed in Section 5.6.

---

[6] Drift is fixed by the choice of the numeraire (to be discussed in the next section). The Libor forward interest rate $f_L(t,x)$ that is equivalent to the log Libor rate $\phi(t,x)$ has nonlinear drift as well as stochastic volatility. The Hamiltonian derived in this chapter is for the case of $f(t,x)$ that has deterministic volatility and is valid for bond forward interest rates and for log Libor rate $\phi(t,x)$.

[7] A potential term is a function only of $\phi(t,x)$ or $f(t,x)$; the interest rate Hamiltonian can only depend on the $\delta/\delta\phi(t,x)$ or $\delta/\delta f(t,x)$.

[8] If one wants to use the velocity degrees of freedom $\mathcal{A}(t,x)$ in the state space representation, one needs to use the formalism of phase space quantization [95].

For simplicity of notation all derivations requiring the Hamiltonian are carried out for $\nu = 1$; it is fairly straightforward to re-introduce market time as and when required.

15.7.2 Interest rate and equity Hamiltonians

Interest rate Hamiltonians $\mathcal{H}_\phi(t), \mathcal{H}_f(t)$ are far more complex and qualitatively different from H_N, the Hamiltonian for N equities given in Eq. (15.17).[9] The following is a comparison of the two Hamiltonians.

- Both the Hamiltonians are non-Hermitian, as is typical for the the case of finance.
- Both Hamiltonians have only derivative terms, which is typical of the Fokker–Planck Hamiltonians that arise from 'white noise' [95].
- Both have a kinetic term containing second derivatives and a first derivative 'carrying' the drift term.
- The covariance of the interest rate Hamiltonian, for example $M_\phi(x, x'; t) = \gamma(t, x)\mathcal{D}(x, x'; t)\gamma(t, x')$ is similar to the covariance for equity, namely $\sigma_i \sigma_j \rho_{ij}$. The continuous index x reflects infinitely many independent random variables that drive interest rates and is fundamentally different from the discrete and finite ranged index i for equity. The propagator $\mathcal{D}(x, x'; t)$ is the generalization of the correlation matrix ρ_{ij}.
- A fundamental difference between the interest rate and equity Hamiltonians lies in the time dependence of the interest rate Hamiltonian, which reflects the time dependence of its state space. The interest rate state space $\mathcal{V}_t$ is time dependent and 'moves' in time, whereas the state space of equity $\mathcal{V}$ is fixed and time independent.

15.8 Interest rate Hamiltonian: martingale condition

The existence of a martingale measure is central to the theory of arbitrage-free pricing of financial instruments, and a path integral formulation of this principle has been discussed in Section 5.8. The bond forward interest rate Gaussian Lagrangians discussed in Chapter 5 are quadratic in the fields, and hence the martingale condition for the bond forward interest rates could be solved exactly, as in Section 5.8, by performing a Gaussian path integration.

For the case of nonlinear interest rates, the Lagrangian is nonlinear and hence finding the risk-neutral measure entails the exact evaluation of a nonlinear path integral – in general, an intractable problem. For this reason, the derivation of the risk-neutral measure is reformulated using the Hamiltonian. The Hamiltonian formulation, even for the nonlinear theory of the interest rates with stochastic volatility,

---

[9] For easy reference, recall H_N is given by

$$H_N = -\frac{1}{2}\sum_{ij=1}^{N}\sigma_i\sigma_j\rho_{ij}\frac{\partial^2}{\partial z_i \partial z_j} + \sum_{i=1}^{N}\left(\frac{1}{2}\sigma_i^2\rho_{ii} - r\right)\frac{\partial}{\partial z_i} + r$$

provides an exact solution for the martingale measure [12]. The nonlinear Libor Market Model will be studied in detail using the Hamiltonian approach and the nonlinear Libor drift will be evaluated exactly [3].

Eq. (15.7) shows that the existence of a martingale measure is equivalent to a (risk-free) equity Hamiltonian that annihilates the underlying security S. A similar condition holds for interest rate Hamiltonians, but with a number of complications arising from the nontrivial structure of the time dependent state space $\mathcal{V}_t$ and the fact that the spot rate $r(t) = f(t, t)$ is itself a stochastic quantity.

Consider the money market numeraire $\exp\{\int_{t_0}^{t_*} dt\, r(t)\}$. The martingale condition for the money market numeraire, given in Eq. (3.4), states that the price of the zero coupon bond $B(t_*, T)$ at some future time $T > t_* > t$ is equal to the price of the bond at time t, discounted by the risk-free interest rate $r(t) = f(t, t)$. In other words

$$B(t, T) = E[e^{-\int_t^{t_*} r(t)dt} B(t_*, T)] \tag{15.33}$$

where $E[X]$ denotes the average value of X over all the stochastic variables in the time interval $[t, t_*]$.

In terms of the Feynman path integral, Eq. (15.33) yields

$$B(t, T) = \frac{1}{Z} \int Df e^{-\int_t^{t_*} r(t)dt} e^{S[f]} B(t_*, T) \tag{15.34}$$

In the path integral given in Eq. (15.34), there are two domains; namely the domain for the zero coupon bond that is nested inside the domain of the forward interest rates. These domains are shown in Figure 15.2.

The martingale condition given in Eq. (15.34) is written in an integral form. However, similar to the case of a single security, it is clearly a differential condition

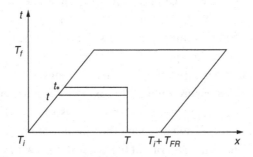

Figure 15.2 Domains for deriving the martingale condition on zero coupon bonds $B(t_*, T)$. The horizontal lines at t_* and t represent $B(t_*, T)$ and $B(t, T)$ respectively. The vertical line at T represents the maturity time of zero coupon bonds. The trapezoid enclosing the zero coupon bonds is the domain for all the forward interest rates.

since it holds for any value of t_*. Hence take $t_* = t + \epsilon$. The reason that one considers only an infinitesimal change for the interest rates is because the Hamiltonian $\mathcal{H}(t)$ is a differential operator. For an infinitesimal evolution in time, the functional integral in Eq. (15.34) collapses to an integration over the final time variables $\tilde{f}_{t+\epsilon}$ on one time slice $t_* = t + \epsilon$. Writing $f(t, x) = f_t(x)$ yields

$$B(t, T) = \mathcal{N} \int D\tilde{f}_{t+\epsilon} e^{-\epsilon f_t(t)} e^{\epsilon \int \mathcal{L}[f, \tilde{f}]} B[\tilde{f}_{t+\epsilon}, T] \qquad (15.35)$$

The above equation is re-written in the language of state vectors. The completeness equation Eq. (15.19) and the Dirac–Feynman relation given in Eq. (15.25) yield the following

$$
\begin{aligned}
&< f_t | B(t, T) > \\
&= \int D\tilde{f}_{t+\epsilon} < f_t | e^{-\epsilon f(t,t)} e^{-\epsilon \mathcal{H}} | \tilde{f}_{t+\epsilon} >< \tilde{f}_{t+\epsilon} | B(t+\epsilon, T) >
\end{aligned} \qquad (15.36)
$$

The completeness equation, from Eq. (15.19), is the following

$$\mathcal{I}_{t+\epsilon} = \int D\tilde{f}_{t+\epsilon} | \tilde{f}_{t+\epsilon} >< \tilde{f}_{t+\epsilon}| \qquad (15.37)$$

Hence from Eq. (15.36)

$$< f_t | B(t, T) >=< f_t | e^{-\epsilon f(t,t)} e^{-\epsilon \mathcal{H}(t)} | B(t+\epsilon, T) > \qquad (15.38)$$

$$\Rightarrow |B(t, T) >= e^{-\epsilon f(t,t)} e^{-\epsilon \mathcal{H}(t)} | B(t+\epsilon, T) > \qquad (15.39)$$

It can be verified, using the explicit representation of the zero coupon bond given in Eq. (15.21), that

$$e^{+\epsilon f(t,t)} | B(t, T) > = |B(t+\epsilon, T) > \qquad (15.40)$$

The discounting factor $e^{-\epsilon f(t,t)}$ plays the remarkable role of matching the zero coupon bonds at two different time slices and yields the following eigenvalue equation

$$|B(t+\epsilon, T) > = e^{-\epsilon \mathcal{H}(t)} | B(t+\epsilon, T) > \qquad (15.41)$$

$$\Rightarrow \mathcal{H}(t) | B(t+\epsilon, T) > = 0 \qquad (15.42)$$

Since there is nothing special about the bond that is being considered, one arrives at the differential formulation of the risk-neutral measure, namely that *all* zero coupon bonds – and consequently all coupon bonds – are eigenfunctions of the Hamiltonian $\mathcal{H}$ that are annihilated by $\mathcal{H}$, that is have zero eigenvalue.[10] That is

$$\mathcal{H}(t)|B(t,T)> = 0 \quad \text{for all } t,\ T \tag{15.43}$$

or more explicitly

$$< f_t|\mathcal{H}(t)|B(t,T) > = \mathcal{H}(t)e^{-\int_t^T dx f_t(x)} = 0 \tag{15.44}$$

The above equation is the field theory generalization of the case of a single security given in Eq. (15.7).

15.9 Numeraire and Hamiltonian

As discussed in Chapter 9, one can choose to discount all financial instruments by any positive valued instrument, and then choose the drift velocity to make the discounted instrument into a martingale. The main results obtained in Chapter 9 are re-derived in the Hamiltonian framework; in particular, it is shown how the various choices of numeraires yield different drift terms.

15.9.1 Money market and forward bond numeraire

The numeraire is $\exp \int_{t_0}^t r(t')dt'$ and one fixes the drift so that the combination $B(t,T)/\exp \int_{t_0}^t r(t')dt'$ is a martingale. Hence the drift α for the bond forward interest rate Hamiltonian $\mathcal{H}_f(t)$, given in Eq. (15.31), is fixed by the martingale condition as in Eq. (15.44), namely[11]

$$\mathcal{H}_f(t)B(t,T) = 0$$

$$\Rightarrow \alpha(t,x) = \int_t^x dx' M(x,x';t)$$

Similarly, for the forward bond numeraire $B(t,t_*)$ with t_* fixed, the combination $B(t,T)/B(t,t_*)$ is a martingale. Hence, from Eq. (15.44)

$$\mathcal{H}_f(t)\left[\frac{B(t,T)}{B(t,t_*)}\right] = 0$$

$$\Rightarrow \alpha(t,x) = \int_{t_*}^x dx' M(x,x';t) \tag{15.45}$$

---

[10] Zero coupon bonds are state vectors of the forward interest rates' state space that are not normalizable.
[11] A detailed derivation is given in [12].

15.9.2 Linear Libor market and forward numeraire

The Libor rate $1 + \ell L(t, T_n + \ell) = \exp\{\int_{T_n}^{T_n+\ell} dx f(t, x)\}$, for the linear Libor market measure, is a martingale for each $T_n = n\ell$, where $\ell = 3$ months. To fix the drift term, the martingale condition is imposed on $L(t, T_n + \ell)$ using the bond forward interest rate Hamiltonian given in Eq. (15.31); hence

$$0 = \mathcal{H}(t) L(t, T_n + \ell)$$

$$\Rightarrow 0 = -\left[\frac{1}{2}\int_t^{t+T_{FR}} dx dx' M(x, x'; t) + \int_t^\infty dx \alpha_L(t, x)\right] e^{\int_{T_n}^{T_n+\ell} dx f(t,x)}$$

$$\Rightarrow \alpha_L(t, x) = -\int_{T_n}^x dx' M(x, x'; t); \quad T_n \leq x < T_n + \ell \tag{15.46}$$

The result stated in Eq. (15.46) has been obtained earlier in Eq. (9.28).

The forward numeraire is chosen to make the forward value of a bond, namely $F(t, T_n) = \exp\{-\int_{T_n}^{T_n+1} dx f(t, x)\}$, into a martingale for each T_n; one can see this measure is very similar to the Libor market measure. Using the martingale condition for the bond forward interest rates Hamiltonian yields the following

$$0 = \mathcal{H}(t) F_L(t, T_n)$$

$$= -\left[\frac{1}{2}\int_t^{t+T_{FR}} dx dx' M(x, x'; t) + \int_t^{t+T_{FR}} dx \alpha_F(t, x)\right] e^{-\int_{T_n}^{T_n+1} dx f(t,x)}$$

$$\Rightarrow \alpha_F(t, x) = \int_{T_n}^x dx' M(x, x'; t) = -\alpha_L(t, x); \quad T_n \leq x < T_n + \ell \tag{15.47}$$

The equation above gives the expected result derived earlier in Eq. (9.14).

15.10 Hamiltonian and Libor Market Model drift

Libor Market Model drift $\rho(t, x)$ derived in Section 6.4 is given an independent derivation in this section based directly on the Libor Market Model Hamiltonian [5]. The derivation for Libor drift $\zeta(t, T_n)$ for the Libor Market Model given in Section 6.7 was quite circuitous; the Libor forward interest rates $f_L(t, x)$ were used as scaffolding and it was not clear why one could not evaluate Libor drift directly using only Libor $L(t, T_n)$. The result is also quite opaque, with the drift having summations and minus signs that do not have a clear explanation. In contrast, the Hamiltonian framework yields a transparent derivation of Libor drift directly using Libor variables $L(t, T_n)$ and the result is intuitively clear.

As in Section 6.4, choose the zero coupon bond $B(t, T_{I+1})$ to be the forward bond numeraire; hence, from Eq. (6.13), for all n

$$X_n(t) \equiv \frac{L(t, T_n) B(t, T_{n+1})}{B(t, T_{I+1})} : \quad \text{martingale}$$

is chosen to be a martingale. Libor time T_n can be less, equal, or greater than T_I and is shown in Figure 6.3.

The drift $\rho(t, x)$ is fixed in the Hamiltonian framework, as in Eq. (15.43), by imposing the martingale condition on $\mathcal{X}_n(t)$, namely that

$$\mathcal{H}_\phi(t)\left[\frac{L(t, T_n)B(t, T_{n+1})}{B(t, T_{I+1})}\right] = 0 \qquad (15.48)$$

Libor maturity is defined at Libor time $T_n = \ell n$, where $\ell = 90$ days; for notational ease, define $L_k = L(t, T_k)$. Write

$$\mathcal{X}_n(t) \equiv \frac{L(t, T_n)B(t, T_{n+1})}{B(t, T_{I+1})} = L_n \frac{B(t, T_{n+1})}{B(t, T_{I+1})}$$

Let t be a Libor time; from the definition of the zero coupon bond given in Eq. (6.5)

$$B(t, T_{n+1}) = \prod_{k=0}^{n}\left(\frac{1}{1 + \ell L_k}\right) \qquad (15.49)$$

The following are the three cases for $\mathcal{X}_n(t)$:

• $n = I$

$$\mathcal{X}_I(t) = L_I = L(t, T_I) \qquad (15.50)$$

• $n > I$

$$\mathcal{X}_n(t) = L_n \prod_{k=I+1}^{n}\left(\frac{1}{1 + \ell L_k}\right) = L_n \exp\left\{-\sum_{k=I+1}^{n} \ln(1 + \ell L_k)\right\} \qquad (15.51)$$

• $n < I$

$$\mathcal{X}_n(t) = L_n \prod_{k=n+1}^{I}(1 + \ell L_k) = L_n \exp\left\{+\sum_{k=n+1}^{I} \ln(1 + \ell L_k)\right\} \qquad (15.52)$$

Eq. (15.48) requires the calculation of $\delta/\delta\phi$ acting on $\mathcal{X}_n(t)$, which in turn needs the following computation

$$\ell L(t, T_n) \equiv \ell L_n = \exp\left\{\int_{T_n}^{T_{n+1}} dx \phi_t(x)\right\} \equiv e^{\phi_n}$$

$$\Rightarrow \frac{\delta}{\delta\phi(x)} L_k = H_k(x)L_k \qquad (15.53)$$

where, as in Eq. (6.58) and shown in Figure 6.8, the characteristic function is

$$
H_k(x) = \begin{cases} 1 & T_k \leq x < T_{k+1} \\ \\ 0 & x \notin [T_k, T_{k+1}) \end{cases}
$$

Recall that the log Libor Hamiltonian is given by Eqs. (15.30) and (6.54); for notational convenience, the drift $\tilde{\rho}(t, x)$ is written such that a 'kinetic' piece of the drift is subtracted out and its dependence on the future Libor time interval $H_n(x)$ is written out as $\rho_n(t, x)$. Hence[12,13]

$$
\mathcal{H}_\phi(t) = -\frac{1}{2} \int_{x,x'} M_\gamma(x, x'; t) \frac{\delta^2}{\delta\phi(x)\delta\phi(x')}
$$
$$
+ \frac{1}{2} \int_x \Lambda(t, x) \frac{\delta}{\delta\phi(x)} - \int_x \rho(t, x) \frac{\delta}{\delta\phi(x)} \tag{15.54}
$$

$$
\rho(t, x) = \sum_{n=0}^{\infty} H_n(x)\rho_n(t, x); \quad \int_x \equiv \int_t^{+\infty} dx \tag{15.55}
$$

$$
\Lambda(t, x) = \sum_{n=0}^{\infty} H_n(x) \int_{T_n}^{T_{n+1}} dx' M_\gamma(x, x'; t) \tag{15.56}
$$

Case (i) $n = I$

As a warm-up for the general derivation, consider a rather detailed derivation for the special case for $n = I$ which yields the martingale condition for $\chi_I = L_I$. From Eqs. (15.48), (15.53), (15.55), and (15.56),

$$
0 = \mathcal{H}_\phi L_I = \Big[-\frac{1}{2} \int_{x,x'} M_\gamma(x, x'; t) H_I(x) H_I(x')
$$
$$
+ \frac{1}{2} \int_x \Lambda(t, x) H_I(x) - \int_x \rho(t, x) H_I(x) \Big] L_I
$$

From Eqs. (15.55) and (15.56)

$$
\int_{x,x'} M_\gamma(x, x'; t) H_I(x) H_I(x') = \int_{T_I}^{T_{I+1}} dx \Lambda_I(t, x)
$$

$$
\int_x \rho(t, x) H_I(x) = \int_{T_I}^{T_{I+1}} dx \rho_I(t, x); \quad \int_x \Lambda(t, x) H_I(x) = \int_{T_I}^{T_{I+1}} dx \Lambda_I(t, x)
$$

---

[12] Henceforth, for notational convenience, the limit of $T_{FR} \to +\infty$ is taken.
[13] The explicit expression for the function $\rho_n(t, x)$ is given in Eq. (6.56).

Hence

$$0 = \mathcal{H}_\phi L_I = \left[-\int_{T_I}^{T_{I+1}} dx \rho_I(t,x) \right] L_I$$

$$\Rightarrow \rho_I(t,x) = 0 \tag{15.57}$$

Case (ii) $n > I$

For the case of $\mathcal{X}_n(t)$, $n > I$, Eqs. (15.51) and (15.53) yield

$$\frac{1}{\mathcal{X}_n(t)} \frac{\delta \mathcal{X}_n(t)}{\delta \phi(x)} = H_n(x) - \sum_{k=I+1}^{n} \frac{e^{\phi_k} H_k(x)}{1 + e^{\phi_k}} \tag{15.58}$$

Note the summation term above is due to the discounting by the forward numeraire $B(t, T_{I+1})$. The second functional derivative yields

$$\frac{1}{\mathcal{X}_n(t)} \frac{\delta^2 \mathcal{X}_n(t)}{\delta\phi(x)\delta\phi(x')} = \left[H_n(x) - \sum_{j=I+1}^{n} \frac{e^{\phi_j} H_j(x)}{1 + e^{\phi_j}} \right] \left[H_n(x') - \sum_{k=I+1}^{n} \frac{e^{\phi_k} H_k(x')}{1 + e^{\phi_k}} \right]$$

$$- \sum_{j=I+1}^{n} \frac{e^{\phi_j} H_j(x) H_j(x')}{1 + e^{\phi_j}} + \sum_{j=I+1}^{n} \left[\frac{e^{\phi_j}}{1 + e^{\phi_j}} \right]^2 H_j(x) H_j(x') \tag{15.59}$$

On applying the log Libor Hamiltonian on $\mathcal{X}_n(t)$, $n > I$, Eqs. (15.56), (15.58), and (15.59) yield, after a few obvious cancellations

$$\frac{1}{\mathcal{X}_n(t)} \mathcal{H}_\phi \mathcal{X}_n(t) = \frac{1}{2} \sum_{j=I+1}^{n} \left[\frac{e^{\phi_j}}{1 + e^{\phi_j}} \right]^2 \Lambda_{jj} - \sum_{j=I+1}^{n} \frac{e^{\phi_j}}{1 + e^{\phi_j}} \Lambda_{jn}$$

$$+ \frac{1}{2} \sum_{j,k=I+1}^{n} \frac{e^{\phi_j + \phi_k}}{[1 + e^{\phi_j}][1 + e^{\phi_k}]} \Lambda_{jk} + \zeta_n - \sum_{j=I+1}^{n} \frac{e^{\phi_j}}{1 + e^{\phi_j}} \zeta_j \tag{15.60}$$

where, from Eq. (6.35)

$$\Lambda_{mn} = \int_{T_m}^{T_{m+1}} dx \int_{T_n}^{T_{n+1}} dx' M_\gamma(x, x'; t)$$

and recall from Eq. (6.55)

$$\zeta_n \equiv \zeta(t, T_n) = \int_{T_n}^{T_{n+1}} dx \rho_n(t,x) \tag{15.61}$$

Inspecting the result in Eq. (15.60) leads to the following ansatz

$$\zeta_n = \sum_{j=I+1}^{n} \frac{e^{\phi_j}}{1+e^{\phi_j}} \Lambda_{jn} \tag{15.62}$$

Hence

$$\sum_{j=I+1}^{n} \frac{e^{\phi_j}}{1+e^{\phi_j}} \zeta_j = \sum_{j=I+1}^{n} \frac{e^{\phi_j}}{1+e^{\phi_j}} \sum_{k=I+1}^{j} \frac{e^{\phi_k}}{1+e^{\phi_k}} \Lambda_{jk} \tag{15.63}$$

The remarkable identity

$$\sum_{j=I+1}^{n} \sum_{k=I+1}^{j} A_{jk} = \frac{1}{2} \sum_{j,k=I+1}^{n} A_{jk} + \frac{1}{2} \sum_{j=I+1}^{n} A_{jj} \tag{15.64}$$

applied to Eq. (15.63) leads to the cancellation of all the terms on the right-hand side of Eq. (15.60) and yields the final result

$$\mathcal{H}_\phi \mathcal{X}_n(t) = 0 \quad : \quad \text{martingale} \tag{15.65}$$

From Eq. (15.61), which states that $\int_{T_n}^{T_{n+1}} dx\, \rho_n(t,x) \equiv \zeta(t,T_n)$, and Eq. (15.62), Libor drift is given by

$$\int_{T_n}^{T_{n+1}} dx\, \rho_n(t,x) = \sum_{j=I+1}^{n} \frac{e^{\phi_j(t)}}{1+e^{\phi_j(t)}} \int_{T_n}^{T_{n+1}} dx \int_{T_j}^{T_{j+1}} dx'\, M(x,x';t)$$

$$\Rightarrow \rho_n(t,x) = \sum_{j=I+1}^{n} \frac{e^{\phi_j(t)}}{1+e^{\phi_j(t)}} \int_{T_j}^{T_{j+1}} dx'\, M(x,x';t) \tag{15.66}$$

$$= \sum_{j=I+1}^{n} \frac{e^{\phi_j(t)}}{1+e^{\phi_j(t)}} \Lambda_j(t,x); \quad T_n \le x < T_{n+1}$$

Case (iii) $n < I$

A derivation similar to Case (ii) yields the result for $\mathcal{X}_n(t)$, $n < I$. One needs to keep track of the relative negative sign in χ_n, given in Eqs. (15.51) and (15.52) arising from the difference in the discounting factor. The following is the final result

$$\rho_n(t,x) = -\sum_{j=n+1}^{I} \frac{e^{\phi_j(t)}}{1+e^{\phi_j(t)}} \int_{T_j}^{T_{j+1}} dx'\, M(x,x';t) \tag{15.67}$$

The exact results given in Eqs. (15.62), (15.66), and (15.67) yield Libor drift, derived earlier in Eq. (6.56), as follows

$$\rho(t,x) = \sum_{n=0}^{\infty} H_n(x)\rho_n(t,x)$$

where

$$\rho_n(t,x) = \begin{cases} \sum_{m=I+1}^{n} \frac{e^{\phi_m(t)}}{1+e^{\phi_m(t)}}\Lambda_m(t,x) & n > I \\[2ex] 0 & n = I \\[2ex] -\sum_{m=n+1}^{I} \frac{e^{\phi_m(t)}}{1+e^{\phi_m(t)}}\Lambda_m(t,x) & n < I \end{cases}$$

15.10.1 *Libor Market Model: Hamiltonian, Lagrangian, and $\mathcal{A}_L(t,x)$*

Libor drift has been obtained in a fairly transparent and direct manner compared to the rather roundabout approach adopted in Section 6.7. The summation that appears in the drift term is due to expressing the ratio $B(t,T_n)/B(t,T_I)$ as a product of Libor variables $L(t,T_k)$. The relative minus sign in the summation term of the drift for $n < I$ and $n > I$ arises from the ratio $B(t,T_n)/B(t,T_I)$ being either the product of $1 + \ell L(t,T_k)$ or of its inverse.

The derivation of Libor drift given in Section 6.7 follows the general spirit of the BGM–Jamshidian derivation. The martingale condition was first expressed in terms of the Libor forward interest rates $f_L(t,x)$ defined in Eq. (6.10); one then did a change of variables and re-expressed the drift in terms of the Libor variables. To carry out this change of variables for the quantum finance case, the Wilson expansion for the velocity quantum field $\mathcal{A}_L(t,x)$ was crucial in capturing the nontrivial correlation terms.

In contrast, in the Hamiltonian derivation of Libor drift, there is no need to employ the $\mathcal{A}_L(t,x)$ field and all the correlation effects are produced by the Hamiltonian. The Libor Hamiltonian is expressed directly in terms of log Libor variables $\phi(t,x)$, making no reference to $f_L(t,x)$. The martingale condition is expressed directly in terms of the Hamiltonian $\mathcal{H}_\phi$ and leads to an exact derivation of Libor drift. The fact that the Jacobian of the transformation from $\mathcal{A}_L(t,x)$ to $\phi(t,x)$ is a constant, as shown in Section 6.14, is essential for obtaining the log Libor Hamiltonian; a nontrivial Jacobian would give rise to new terms.

The Hamiltonian derivation of Libor drift leads to some general conclusions in the context of the Libor Market Model. The martingale condition that the Hamiltonian annihilates the underlying security was first introduced in Eq. (15.7) and then extended to N-securities in Eq. (15.18). The martingale condition subsequently had a nontrivial extension for interest rate instruments due to the need to treat the discounting factor as being stochastic [12] and was given by Eq. (15.65) for the Libor Market Model. The martingale condition was verified in this section by the nontrivial derivation of the nonlinear Libor drift.

In Chapter 6, and Eq. (6.10) in particular, the Libor Market Model was given a differential formulation employing $A_L(t, x)$; the Libor drift was then derived. The next step was a nonlinear change of variables from $A_L(t, x)$ to log Libor $\phi(t, x)$, which yielded the log Libor Lagrangian and path integral; and, finally, the log Libor Hamiltonian was derived from the Lagrangian. The Hamiltonian formulation of the martingale exactly reproduces the earlier result for Libor drift; thus closing the circle, so to speak.

The Hamiltonian derivation of Libor drift provides independent proof of the correctness of the earlier derivation of Libor drift in Section 6.7, which crucially hinged on the Wilson expansion. The Hamiltonian result shows that the Wilson short distance expansion for a Gaussian quantum field is the correct generalization of Ito's calculus and opens the way to applications in theoretical finance.

In summary, the Libor Market Model has been given three different and consistent formulations, namely employing $A_L(t, x)$, $\mathcal{L}[\phi]$, and $\mathcal{H}_\phi(t)$; thus displaying the versatility and flexibility of quantum finance.

15.11 Interest rate Hamiltonian and option pricing

Recall from Eq. (15.5), for the money market numeraire, the option price is given by the following ($\tau = T - t$)

$$C(\tau, x) = E[e^{-r\tau}\mathcal{P}] = \langle x|e^{-\tau H}|\mathcal{P}\rangle \qquad (15.68)$$

For the money market numeraire, the option price follows from the following martingale condition given in Eq. (3.6)

$$\frac{C(t, \cdot)}{\exp(rt)} = E\left[\frac{C(T, \cdot)}{\exp(rT)}\right] = E\left[\frac{\mathcal{P}}{\exp(rT)}\right]$$

The Hamiltonian formulation of option pricing discussed in Section 15.2 needs to be generalized to the case of interest rates. For the coupon bond case, the forward numeraire is given by $B(t, t_*)$; hence, the coupon bond option price, maturing at calendar future time t_*, is given by the following martingale condition

$$\frac{C(t_0, t_*, T, K)}{B(t_0, t_*)} = E\left[\frac{\mathcal{P}_*}{B(t_*, t_*)}\right] = E[\mathcal{P}_*]$$

$$\Rightarrow C(t_0, t_*, T, K) = B(t_0, t_*)E[\mathcal{P}_*] \qquad (15.69)$$

The option price at time t_0 depends on the current value of the interest rates and is given by propagating the payoff function $|\mathcal{P}_*\rangle$ – maturing at time t_* – *backwards* in time, as given in Eq. (15.20), to present time t_0 and discounted by the deterministic zero coupon bond $B(t_0, t_*)$. The initial and final state vectors and payoff function are given as follows

$$|\text{final}\rangle = |\mathcal{P}_*\rangle$$

$$\langle\text{initial}| = \langle\phi^{(0)}| = \prod_{t_0 \leq x \leq \infty} \langle f(x)|$$

$$\mathcal{P}_*[f_*] = \big(\mathcal{B}(t_*, T) - K\big)_+ = \langle f_*|\mathcal{P}_*\rangle$$

The European coupon bond option price, from Eqs. (15.20) and (15.69), is hence given by

$$C_E(t_0, t_*, T, K) = B(t_0, t_*) E[(\mathcal{B}(t_*, T) - K)_+]$$

$$= B(t_0, t_*)\langle f^{(0)}|\mathcal{T}\Big\{\exp - \int_{t_0}^{t_*} dt\mathcal{H}(t)\Big\}|\mathcal{P}_*\rangle \quad (15.70)$$

Note that time is flowing backwards. Using the completeness equation for the state space $\mathcal{V}_*$

$$\mathcal{I} = \int Df_* |f_*\rangle\langle f_*| \quad (15.71)$$

yields

$$\langle f^{(0)}|\mathcal{T}\Big\{\exp - \int_{t_0}^{t_*} dt\mathcal{H}(t)\Big\}|\mathcal{P}_*\rangle$$

$$= \int Df_* \langle f^{(0)}|\mathcal{T}\Big\{\exp - \int_{t_0}^{t_*} dt\mathcal{H}(t)\Big\}|f_*\rangle\langle f_*|\mathcal{P}_*\rangle \quad (15.72)$$

The payoff function state vector $|\mathcal{P}_*\rangle$ is an element of the state space $\mathcal{V}_*$ at future time t_*; in terms of the coordinate basis eigenstate of the dual state space $\mathcal{V}_{*,\text{Dual}}$

$$\langle f_*| \equiv \prod_{t_* \leq x \leq \infty} \langle f(x)| \quad (15.73)$$

the payoff function is given by

$$\langle f_*|\mathcal{P}_*\rangle = \mathcal{P}_*[f_*] = \left(\sum_{i=1}^{N} c_i B(t_*, T_i) - K\right)_+ \quad (15.74)$$

To make the content of the payoff function $\mathcal{P}_*$ more explicit, note that

$$\langle f_*|\mathcal{P}_*\rangle = \begin{cases} \left(\sum_{i=1}^{N} c_i B(t_*, T_i) - K\right)_+, & t_* \leq x \leq T \\[2mm] 0, & x > T \end{cases}$$

$$\equiv \mathcal{P}[f_*]$$

From above, it can be seen that the payoff function $|\mathcal{P}_*\rangle$ has nonzero components in the future direction x only in the interval $t_* \leq x \leq T$.

The domain $\mathcal{R}$ required for computing the matrix element in Eq. (15.72) is given in Figure 5.2(b). The domain $\mathcal{R}$ has the important feature that the state spaces $\mathcal{V}_t$ for all $t \in [t_0, t_*]$ are *fixed* in time and are all identical and equal to $\mathcal{V}_*$, spanned by variables $f(x)$ with $x \in [t_*, T]$. Moreover, on domain $\mathcal{R}$, the Hamiltonian commutes for different times $[\mathcal{H}_*(t), \mathcal{H}_*(t')] = 0$ since only the coefficients $M(x, x'; t)$ and $\alpha(t, x)$ are time dependent.

For these reasons, the forward bond numeraire makes the time ordering in Eq. (15.72) unnecessary and $\mathcal{T}$ can be removed. $\mathcal{H}_*(t)$ is consistently restricted to the domain $\mathcal{R}$ by limiting the range of $x \in [t_*, T]$; at each instant t, $\mathcal{H}_*(t)$ acts on the state space $\mathcal{V}_*$. Hence

$$\mathcal{H}_f(t)|_{\mathcal{R}} = -\frac{1}{2} \int_{t_*}^{T} dx dx' M_\sigma(x, x'; t) \frac{\delta^2}{\delta f(x) \delta f(x')} - \int_{t_*}^{T} dx \alpha(t, x) \frac{\delta}{\delta f(x)}$$

The operator driving the option price is given by a new time integrated operator, namely the evolution operator W_f acting on state space $\mathcal{V}_*$.

$$W_f \equiv \int_{t_0}^{t_*} dt \mathcal{H}_*(t)|_{\mathcal{R}}; \quad W_f : \mathcal{V}_* \to \mathcal{V}_* \tag{15.75}$$

$$\Rightarrow W_f = -\frac{1}{2} \int_{t_*}^{T} dx dx' M_\sigma(x, x') \frac{\delta^2}{\delta f(x) \delta f(x')} - \int_{t_*}^{T} \alpha(x) \frac{\delta}{\delta f(x)}$$

$$M_\sigma(x, x') = \int_{t_0}^{t_*} dt \sigma(t, x) \mathcal{D}(x, x'; t) \sigma(t, x'); \quad \alpha(x) = \int_{t_0}^{t_*} dt \alpha(t, x) \tag{15.76}$$

Hence, the option price is given

$$\frac{C_E(t_0, t_*, T, K)}{B(t_0, t_*)} = \langle f^{(0)}|\mathcal{T}\left\{\exp - \int_{t_0}^{t_*} dt \mathcal{H}(t)\right\}|\mathcal{P}_*\rangle$$

$$= \langle f_*|e^{-W_f}|\mathcal{P}_*\rangle \tag{15.77}$$

The option price is completely determined by the matrix elements of e^{-W_f} taken between two vectors $\langle f_*|$ and $|\mathcal{P}_*\rangle$ and both belong to the same state space $\mathcal{V}_*$. For barrier options, the price is determined by the same matrix element but with the barrier being imposed on the eigenfunctions of W_f.

If, instead of the forward bond numeraire $B(t_0, t_*)$, the money market numeraire $\exp(-\int_{t_0}^{t_*} dt\, r(t))$ is used – where $r(t) = f(t, t)$ – the domain for evaluating the matrix element in Eq. (15.72) is the trapezoidal domain given in Figure 5.2(a). Since the discounting factor $r(t)$ is a random quantity, it is inside the time-ordering, and the discounting factor thus extends the nonzero overlap of the basis state $\langle f_t|$ with $\exp(-\int_{t_0}^{t_*} dt\, r(t))|\mathcal{P}_*\rangle$ to the interval $x \in [t, T]$. This in turn means that the time-ordering symbol T cannot be ignored since the underlying state space and Hamiltonian would now be time dependent; one would need to do a separate (and more complicated) calculation for each $t \in [t_0, t_*]$.

The choice of the appropriate numeraire for a particular problem greatly simplifies all calculations and is analogous to choosing a (coordinate) basis that respects the symmetries of the problem.

15.12 Bond evolution operator

The European coupon bond option price, from Eq. (15.77), is given by

$$C_E(t_0, t_*, T, K) = B(t_0, t_*)\langle f_*|e^{-W_f}|\mathcal{P}_*\rangle$$

The calculation is carried out at calendar time t_0 and all the effects coming from future calendar time from t_0 to t_* are carried by the coefficients $M(x, x')$ and $\alpha(x)$ as in Eq. (15.76). In other words, the option price calculation is carried out in the fixed state space $\mathcal{V}_*$. W_f is a differential operator that contains the correlations of the interest rates in future time direction x.

The state vector $e^{-W_f}|\mathcal{P}_*\rangle$ is the price of the option at time t_0. The operator W_f is the *evolution operator* and e^{-W_f} evolves the payoff state vector, defined at future calendar time t_*, *backwards* in time to its present value at time t_0.

The natural coordinates for the evolution operator in studying coupon bonds and swaptions is the integral of the bond forward interest rates. The *dimensionless bond variable* $g(x)$ is defined by the following

$$B(t_*, x) = e^{-g(x)}; \quad g(x) = \int_{t_*}^{x} dy\, f(y)$$

$$\frac{\delta g(x)}{\delta f(y)} = \int_{t_*}^{x} dy'\, \delta(x - y') = \theta(x - y); \quad x, y \geq t_* \tag{15.78}$$

$$\Rightarrow \frac{\delta}{\delta f(y)} = \int_{t_*}^{T} dx\, \frac{\delta g(x)}{\delta f(y)}\frac{\delta}{\delta g(x)} = \int_{y}^{T} dx\, \frac{\delta}{\delta g(x)} \tag{15.79}$$

where Eq. (A.42) and the chain rule of functional differentiation given in Eq. (A.46) have been used to obtain the above results.

From Eqs. (15.75) and (15.79) and after some simplifications

$$W_g = -\frac{1}{2} \int_{t_*}^{T} dx \int_{t_*}^{T} dx' G(x, x') \frac{\delta^2}{\delta g(x)\delta g(x')} - \int_{t_*}^{T} dx \beta(x) \frac{\delta}{\delta g(x)}$$

$$G(x, x') = \int_{t_*}^{x} dy \int_{t_*}^{x'} dy' \int_{t_0}^{t_*} dt M_\sigma(y, y'; t)$$

$$\beta(x) = \int_{t_*}^{x} dy \int_{t_0}^{t_*} dt \alpha(t, y) = \frac{1}{2} G(x, x) \qquad (15.80)$$

Consider a subspace of the full state space composed of $B(t_*, T_i)$, namely zero coupon bonds that are issued at time t_* and mature at Libor time T_i and their linear span, which includes coupon bonds $\mathcal{B}(t_*)$ issued at t_*. The evolution operator W_g simplifies when it acts on only this subspace. From Eq. (A.42)

$$\frac{\delta g(x)}{\delta g(x')} = \delta(x - x'); \quad x, x \in [t_*, T]$$

and this yields, for an arbitrary function of the bond variables $\mathcal{R}[g_1, \ldots, g_N] = \mathcal{R}[g]$, the following

$$\frac{\delta}{\delta g(x)} \mathcal{R}[g] = \sum_{i=1}^{N} \frac{\delta g(T_i)}{\delta g(x)} \frac{\partial \mathcal{R}[g]}{\partial g_i}$$

$$= \sum_{i=1}^{N} \delta(T_i - x) \frac{\partial \mathcal{R}[g]}{\partial g_i}$$

The delta functions reduce the integrations over $\int dx$ to sums over the bond variables $\sum_i$. In particular, the evolution operator W_g reduces on $\mathcal{R}[g]$ to a partial differential operator with respect to the bond variables g_i. In symbols

$$W_g \mathcal{R}[g] = \left[-\frac{1}{2} \int_{t_*}^{T} dx \int_{t_*}^{T} dx' \sum_{i,j=1}^{N} G_{ij} \delta(T_i - x) \delta(T_j - x') \frac{\partial^2}{\partial g_i \partial g_j} \right.$$

$$\left. - \int_{t_*}^{T} dx \sum_{i=1}^{N} \beta_i \delta(T_i - x) \frac{\partial}{\partial g_i} \right] \mathcal{R}[g] \qquad (15.81)$$

$$\equiv W \mathcal{R}[g]$$

Hence, from Eqs. (15.81) and (15.80)

$$W = -\frac{1}{2} \sum_{i,j=1}^{N} G_{ij} \frac{\partial^2}{\partial g_i \partial g_j} - \sum_{i=1}^{N} \beta_i \frac{\partial}{\partial g_i} \qquad (15.82)$$

where

$$G_{ij} = G(T_i, T_j); \quad \beta_i = \beta(T_i) = \frac{1}{2} G_{ii}$$

Note G_{ij} is the forward bond propagator given in Eq. (11.28) and appears in the price of a coupon bond European option; it is plotted in Figure 11.2.

The coupon bond variables $\mathbf{g} = (g_1, g_2, \ldots, g_N)$, at time t_*, express all the bonds as well as satisfy the completeness equation

$$B(t_*, T_i) = \exp\left\{-\int_{t_*}^{T_i} f(t_*, x) dx\right\} = e^{-g_i}$$

$$\mathcal{B}(t_*, T) = \sum_i c_i B(t_*, T_i) = \sum_i c_i e^{-g_i} = \langle \mathbf{g} | \mathcal{B}(t_*, T)\rangle = \mathcal{B}(t_*, T)[\mathbf{g}]$$

$$g_i \equiv g(T_i) = \int_{t_*}^{T_i} dx f(x); \quad \int_g |\mathbf{g}\rangle\langle\mathbf{g}| = \mathcal{I}; \quad \int_{\mathbf{g}} \equiv \prod_i^{N} \int_{-\infty}^{+\infty} dg_i$$

15.12.1 Martingale and bond evolution operator

The martingale condition has a particularly simple realization for the bond evolution operator. A general coupon bond at time t has the representation

$$\mathcal{B} = \sum_I c_I e^{-g_I}$$

The forward bond numeraire $B(t, t_*)$ requires that

$$F(t, t_*, T_I) = \frac{B(t, T_I)}{B(t, t_*)} = e^{-g_I}$$

be a martingale for all T_I.

The evolution operator W given in Eq. (15.82) yields, as expected

$$We^{-g_I} = c_I \left[-\frac{1}{2} G_{II} + \beta_I\right] e^{-g_I} = 0$$

The time integrated Hamiltonian operator W annihilates $F(t, t_*, T_I)$, the forward price of zero coupon bond $B(t_*, T_I)$, as required for the evolution to be martingale.

15.12.2 Coupon bond Lagrangian

Since the coupon bond evolution operator W is the time integral of the Hamiltonian, calendar time is already contained in W; in particular, the correlation matrix G_{IJ} is the time integral of the coefficients in the Hamiltonian.

The Lagrangian and action are mathematical means for expressing e^{-TW}; the 'time' integrations in S_{CB} is not calendar time; but rather an artifact for constructing a path integral representation of the matrix elements of e^{-TW}. For option pricing, only the special value of $T = 1$ is required.

W is mathematically identical to the Hamiltonian H_N for N-equities given in Eq. (15.17), with $r = 0$ and having the opposite sign for the drift term. Hence, the Lagrangian, action and path integral for coupon bonds, similar to Eq. (3.36), are given by

$$Z_{CB} = \int Dg\, e^{S_{CB}}; \quad S_{CB} = \int_0^T dt \mathcal{L}_{CB}$$

$$\mathcal{L}_{CB} = -\frac{1}{2} \sum_{IJ=1}^{N} G_{IJ}^{-1} \left[\frac{dg_I(t)}{dt} + \frac{1}{2}G_{II} \right] \left[\frac{dg_J(t)}{dt} + \frac{1}{2}G_{JJ} \right]$$

$$\int Dg = \prod_{t=0}^{T} \prod_{I=1}^{N} \int_{-\infty}^{+\infty} dg_I(t)$$

The matrix element $\langle g|e^{-W}|\mathcal{P}_*\rangle$ required for finding the price of an option can be evaluated by the above path integral by putting appropriate boundary conditions of the space of paths that goes into defining $\int Dg$.

15.12.3 Zero coupon bond European option

The option price obtained in Section 11.13 is re-derived to demonstrate the utility of the evolution operator.

The payoff function is given by

$$\mathcal{P}_*[g] = \langle g|\mathcal{P}_*\rangle = \left(e^{-g} - K\right)_+ \tag{15.83}$$

$$g = \int_{t_*}^{T} dx f(t_*, x) \tag{15.84}$$

The zero coupon bond evolution operator simplifies to

$$W = -\frac{1}{2}G\frac{\partial^2}{\partial g^2} - \beta\frac{\partial}{\partial g} \tag{15.85}$$

$$G = \int_{t_*}^{T} dy \int_{t_*}^{T} dy' \int_{t_0}^{t_*} dt\, M(y, y'; t); \quad \beta = \frac{1}{2}G$$

The eigenfunctions and eigenvalues of W are given by

$$e^{ipg} = \langle g|p\rangle$$

$$We^{ipg} = \left(\frac{1}{2}Gp^2 - i\beta p\right)e^{ipg}$$

and completeness, from Eqs. (A.8) and (A.9), by

$$\mathcal{I} = \int_{-\infty}^{+\infty} \frac{dp}{2\pi}|p\rangle\langle p| ; \quad \langle g|p\rangle = e^{ipg}; \quad \langle p|g\rangle = e^{-ipg}$$

The option price, for $f = \int_{t_*}^{T} dx f(t_0, x)$, is given, from Eq. (15.77), by the following

$$C(t_0, t_*, K) = B(t_0, t_*)\langle f|e^{-W}|\mathcal{P}_*\rangle$$

$$= B(t_0, t_*)\int_{-\infty}^{+\infty} \frac{dp}{2\pi}\langle f|e^{-W}|p\rangle \int_{-\infty}^{+\infty} dg\langle p|g\rangle\langle g|\mathcal{P}_*\rangle$$

$$= B(t_0, t_*)\int_{-\infty}^{+\infty} \frac{dp}{2\pi}\int_{-\infty}^{+\infty} dg\, e^{-\frac{1}{2}Gp^2 + i\beta p}e^{i(f-g)}\mathcal{P}_*(g)$$

$$= B(t_0, t_*)\frac{1}{\sqrt{2\pi G}}\int_{-\infty}^{+\infty} dg\, e^{-\frac{1}{2G}(f-g+\beta)^2}\left(e^{-g} - K\right)_+$$

Since $G = q^2 = 2\beta$, the price obtained is equal to the result given in Eq. (11.69).

15.13 Libor evolution operator

Similar to the discussion in Section 15.12, the forward bond numeraire leads to the same simplification for the log Libor Hamiltonian. The operator driving the option price is given by a time integrated operator, namely the Libor evolution operator

U_ϕ acting on Libor state space $\mathcal{V}_*$. From Eq. (15.54)

$$U_\phi \equiv \int_{t_0}^{t_*} dt \mathcal{H}_\phi(t)|_{\mathcal{R}}; \quad U_\phi : \mathcal{V}_* \to \mathcal{V}_*$$

$$\Rightarrow U_\phi = -\frac{1}{2} \int_{t_*}^{T} dx dx' M_\gamma(x, x') \frac{\delta^2}{\delta\phi(x)\delta\phi(x')} - \int_{t_*}^{T} \tilde{\rho}(x) \frac{\delta}{\delta\phi(x)} \quad (15.86)$$

$$M_\gamma(x, x') = \int_{t_0}^{t_*} dt \gamma(t, x) \mathcal{D}(x, x'; t) \gamma(t, x'); \quad \tilde{\rho}(x) = \int_{t_0}^{t_*} dt \left[\rho(t, x) - \frac{1}{2}\Lambda(t, x) \right]$$

Similar to the evolution operator W for the coupon bond sector – derived from the bond forward interest rate Hamiltonian – the log Libor Hamiltonian simplifies when acting on only Libor variables. Consider U_ϕ acting only on functions of Libor, which are of the form $\mathcal{R}[\phi_0(t), \phi_1(t), \dots, \phi_n(t), \dots]$; the log Libor variable $\phi_n(t)$ has been defined in Eq. (6.47). Since the forward bond numeraire yields a fixed state space $\mathcal{V}_*$, in the Hamiltonian framework the variable t in $\phi_n(t)$ can be dropped; hence, from Eq. (A.42), for Libor time starting at T_0

$$\phi_n = \int_{T_n}^{T_{n+1}} dx \phi(x) \quad (15.87)$$

$$\Rightarrow \frac{\delta}{\delta\phi(x)} = \sum_{n=0}^{\infty} \frac{\delta\phi_n}{\delta\phi(x)} \frac{\partial}{\partial\phi_n} = \sum_{n=0}^{\infty} H_n(x) \frac{\partial}{\partial\phi_n} \quad (15.88)$$

where the characteristic function $H_n(x)$ is defined in Eq. (6.57). Substituting Eq. (15.88) into Eq. (15.86) yields, for $U_\phi \mathcal{R}[\phi_n] \equiv U\mathcal{R}[\phi_n]$

$$U = -\frac{1}{2} \sum_{mn} \Delta_{mn} \frac{\partial^2}{\partial\phi_m \partial\phi_n} - \sum_n \left(\rho_n - \frac{1}{2}\Delta_{nn} \right) \frac{\partial}{\partial\phi_n}$$

$$\rho_n = \int_{t_0}^{t_*} dt \int_{T_n}^{T_{n+1}} dx \rho(t, x)$$

$$\Delta_{mn} = \int_{T_m}^{T_{m+1}} \int_{T_n}^{T_{n+1}} dx dx' M_\gamma(x, x')$$

$$= \int_{t_0}^{t_*} dt \int_{T_m}^{T_{m+1}} \int_{T_n}^{T_{n+1}} dx dx' \gamma(t, x) \mathcal{D}(x, x'; t) \gamma(t, x') \quad (15.89)$$

A straightforward derivation yields, from Eqs. (6.56), (15.86), and (15.88)

$$
\rho_n = \begin{cases}
\sum_{m=I+1}^{n} \frac{e^{\phi m}}{1+e^{\phi m}} \Delta_{mn} & T_n > T_I \\[2ex]
0 & T_n = T_I \\[2ex]
-\sum_{m=n+1}^{I} \frac{e^{\phi m}}{1+e^{\phi m}} \Delta_{mn} & T_n < T_I
\end{cases}
\tag{15.90}
$$

15.13.1 Caplet price

A derivation is given of the caplet price using the Libor evolution operator U, which is similar to the derivation of the zero coupon bond option price using the evolution operator W given in Section 15.12.

The payoff function for a caplet maturing at t_*, from Eq. (4.10), is given by

$$
\ell V B(t_*, T_I + \ell)\big[L(t_*, T_I) - K\big]_+
$$

For the forward bond numeraire, from Eq. (8.8) the price of a caplet is given by

$$
\frac{caplet(t_0, t_*, T_I)}{B(t_0, T_{I+1})} = V E\big[\ell L(t_*, T_I) - \ell K\big]_+
\tag{15.91}
$$

Define the effective caplet payoff function by

$$
\mathcal{P}_*[\phi_I] = \langle \phi_I | \mathcal{P}_* \rangle = V\big[e^{\phi_I} - \ell K\big]_+
\tag{15.92}
$$

$$
\ell L(t_*, T_I) = e^{\phi_I}; \quad \phi_I = \int_{T_I}^{T_{I+1}} dx \phi(x)
\tag{15.93}
$$

For the caplet price, the Libor evolution operator U depends only on $\partial/\partial\phi_I$. From Eq. (15.90) $\rho_I = 0$ and hence

$$
U = -\frac{1}{2}q_I^2 \frac{\partial^2}{\partial\phi_I^2} + \frac{1}{2}q_I^2 \frac{\partial}{\partial\phi_I}
$$

$$
q_I^2 = \Delta_{II} = \int_{T_I}^{T_{I+1}} dx \int_{T_I}^{T_{I+1}} dx' \int_{t_0}^{t_*} dt\, M_\phi(x, x'; t)
$$

where q_I^2 has been defined earlier in Eq. (8.10).

The Libor evolution operator U is identical to the Black–Scholes Hamiltonian given in Eq. (15.16) for the case of $r = 0$; in contrast, the bond evolution operator

W given in Eq. (15.85) has the *opposite* sign for the drift term compared to H_{BS} and U. The reason that the Libor and Black–Scholes cases are similar is because in both cases the value of the security is *compounded* as time increases; in contrast, future cash flows of a bond are *discounted* by the bond evolution operator and hence the present-day value of the bond decreases as time increases.

The eigenfunctions and eigenvalues of U are given by

$$e^{ip\phi_I} = \langle \phi_I | p \rangle$$

$$U e^{ip\phi_I} = \frac{1}{2} q_I^2 (p^2 + ip) e^{ip\phi_I}$$

Completeness is given by Eqs. (A.8) and (A.9) as follows

$$\mathcal{I} = \int_{-\infty}^{+\infty} \frac{dp}{2\pi} |p\rangle\langle p|; \quad \langle \phi_I | p \rangle = e^{ip\phi_I}; \quad \langle p | \phi_I \rangle = e^{-ip\phi_I}$$

The option price at time t_0, for $\ell L(t_0, T_I) \equiv e^\phi$, from Eq. (15.77), is given by the following

$$\frac{caplet(t_0, t_*, T_I)}{B(t_0, t_{I+1})} = V \langle \phi | e^{-U} | \mathcal{P}_* \rangle$$

$$= V \int_{-\infty}^{+\infty} \frac{dp}{2\pi} \langle \phi | e^{-U} | p \rangle \int_{-\infty}^{+\infty} dg \langle p | g \rangle \langle g | \mathcal{P}_* \rangle$$

$$= V \int_{-\infty}^{+\infty} \frac{dp}{2\pi} \int_{-\infty}^{+\infty} dg \, e^{-\frac{1}{2} q_I^2 (p^2 + ip)} e^{ip(\phi - g)} \mathcal{P}_*(g)$$

$$= V \frac{1}{\sqrt{2\pi q_I^2}} \int_{-\infty}^{+\infty} dg \, e^{-\frac{1}{2q_I^2}(\phi - g - \frac{1}{2} q_I^2)^2} \left(e^g - \ell K \right)_+$$

$$= V \frac{1}{\sqrt{2\pi q_I^2}} \int_{-\infty}^{+\infty} dg \, e^{-\frac{1}{2q_I^2} g^2} \left(e^{\phi + g - \frac{1}{2} q_I^2} - \ell K \right)_+$$

$$= \ell V \frac{1}{\sqrt{2\pi q_I^2}} \int_{-\infty}^{+\infty} dg \, e^{-\frac{1}{2q_I^2} g^2} \left(L(t_0, T_I) e^{-\frac{1}{2} q_I^2 + g} - K \right)_+ \quad (15.94)$$

The caplet price obtained in Eq. (15.94) is equal to the result given in Eq. (8.11) and yields Black's caplet formula.

15.14 Summary

A complete description of financial instruments is provided by the Hamiltonian that determines the dynamics of the underlying security, together with the security's state space.

The Black–Scholes option pricing theory was expressed completely in terms of the equity Hamiltonian and yields the Black–Scholes equations as a particular realization of the Schrodinger equation of quantum mechanics.

Interest rate state space and Hamiltonian were derived from the forward interest rate Lagrangian and action. The state space is infinite dimensional and the Hamiltonian is a second-order functional differential operator. Both, the bond forward interest rates $f(t, x)$ and the log Libor rates $\phi(t, x)$ have a well-defined Hamiltonian since their volatility is deterministic.

The Hamiltonian realization of the martingale evolution of equities entails that the equity be annihilated by the Hamiltonian. The martingale condition for interest rates leads to a result similar to the case of equities: the interest rate Hamiltonian must annihilate all interest rate instruments that have a martingale evolution. The drifts for various choices of numeraires, fixed by the martingale condition, were evaluated using the Hamiltonian approach. The Hamiltonian is the appropriate framework for imposing the martingale condition for nonlinear interest rates and, in particular, yields the exact expression for the Libor Market Model's nonlinear drift.

A new feature of the interest rates dynamics is that, unlike the case for equity, the interest rates' state space and Hamiltonian are time dependent; this leads to a number of new features for the Hamiltonian and, in particular, that the evolution operator is defined by a time-ordered product. Choosing the forward bond numeraire leads to a major simplification: the interest rates' state space for the forward bond numeraire is equivalent to a fixed state space and one can dispense with time ordering, leading to a time independent evolution operator.

The interest rate evolution operator simplifies for the coupon bond as well as for the Libor sectors, respectively, and in both cases reduces to a second-order partial differential operator, which is equivalent to an N-equity Hamiltonian. The exact price of a zero coupon bond European option and a Libor caplet's price were obtained using the evolution operator.

16

American options for coupon bonds
and interest rates

American options for interest rate caps and coupon bonds are analyzed numerically in the formalism of quantum finance [17, 40]. The main purpose of the analysis is to develop efficient algorithms for analyzing path dependent American options for debt instruments. Managing the proliferation of forward interest rates is the main challenge for algorithms calculating American options. All calculations are carried out using the linear (Gaussian) bond forward interest rates $f(t, x)$, which – due to their simplicity – allows one to focus on the main computational complexities. The algorithms developed for the bond forward interest rates can be extended to the nonlinear Libor Market Model.

Zero coupon bonds, from Eq. (2.12), are given as follows

$$B(t, T) = \exp\left\{-\int_t^T dx f(t, x)\right\}$$

The forward interest rates $f(t, x)$ are allowed to take all real values, namely $-\infty \leq f(t, x) \leq +\infty$; as discussed earlier, this approximation is consistent with all bond prices being strictly positive. The dynamics of $f(t, x)$ are given by Eq. (5.1), namely that

$$\frac{\partial f(t, x)}{\partial t} = \alpha(t, x) + \sigma(t, x)\mathcal{A}(t, x)$$

where $\mathcal{A}(t, x)$ is the velocity Gaussian quantum field. The drift $\alpha(t, x)$ is fixed by the forward numeraire and is given from Eq. (9.14) as follows

$$\alpha(t, x) = \sigma(t, x) \int_{T_n}^x dx' \mathcal{D}(x, x'; t)\sigma(t, x'); \quad T_n \leq x < T_{n+\ell}$$

The interest rate caplet is expressed in terms of the bond forward interest rates $f(t, x)$. Hence, the linear pricing formula given in Eq. (10.1) is used for the testing

365

European caplet prices obtained from the numerical algorithms developed in this chapter.

Calendar and future time are discretized to yield a lattice field theory of bond forward interest rates that provides an efficient numerical algorithm for evaluating the price of American options. The algorithm is shown to hold over a wide range of strike prices and coupon rates. All the theoretical constraints that American options have to obey are shown to hold for the numerical prices of American interest rate caplets and coupon bond options. Nontrivial correlations between the different interest rates are efficiently incorporated into the numerical algorithm. New inequalities are *conjectured* for American coupon bond options, based on the results of numerical studies [17, 40].

16.1 Introduction

American options for debt instruments such as interest rate and coupon bond options are widely traded. An accurate and arbitrage free pricing of American interest rate options has far-reaching applications. American options for the debt instruments are complex since, at any moment in time, there are a large number of future interest rates that exist in the market. All of the interest rates evolve randomly and have strong correlations with the other interest rates.

In the simple case of a European option on equity, the Black–Scholes equation can be explicitly solved to obtain an analytical formula for the price of the option [59]. When one considers financial derivatives that allow anticipated early exercise or depend on the history of the underlying assets, numerical approaches need to be used. Various numerical procedures have been developed in the literature to price exotic financial derivatives on equity with path-dependent features, as discussed in detail in [59]. These procedures involve the use of Monte Carlo simulations, binomial tree (and their improvements), and finite difference methods.

The pricing of European and American options for debt instruments is far more complicated than for equity options. In order to price derivatives of debt instruments, one needs to model the underlying interest rate dynamics. The leading model at present for modeling forward interest rates and their derivatives is the HJM model; for the N-factor model, the interest rates at every instant are driven by N random variables [59, 63]. Numerical techniques for pricing American interest rates options [63] are all based on the generalization of the binomial tree approach [72].

To price American options for equity an efficient computational algorithm, using path integrals, has been developed by Montagna and Nicrosini [76]. The quantum field theory describing the bond forward interest rates is discretized and yields a lattice field theory model; an algorithm that generalizes the path integral approach of [76] to the case of debt instrument options is obtained using the lattice field theory.

16.2 American equity option

The path integral algorithm for pricing American equity options is analyzed as a preparation for the rather complex derivation of the price of American caplets and coupon bond options.

Consider an option on an underlying equity $S(t)$ that matures at some future calendar time t_*, with present time given by t_0; in this chapter $t_0 = 0$. Calendar time is discretized into a lattice, with discrete time $t \to t_n = \epsilon n$, $n = 0, 1, \ldots, M$; the payoff function matures at future calendar time $t_* = \epsilon M$. See Figure 16.1.

Since the payoff is specified at future time t_*, the numerical algorithm is a recursion equation that evolves the payoff function *backwards* in time – with the origin of the calendar time lattice being placed at t_* – so as to produce the option price at earlier time t_0. For this reason, it is convenient to define remaining time $\tau = t_* - t$ that runs *backwards*, decreasing in value as calendar time increases. In terms of remaining time τ, option pricing becomes an initial value problem.

The Black–Scholes Lagrangian, for asset price $S = e^z$ and remaining time $\tau = T - t$, is given by Eq. (3.39) as follows

$$\mathcal{L}_{BS} = -\frac{1}{2\sigma^2} \left[\frac{dz(\tau)}{d\tau} + \alpha \right]^2 \tag{16.1}$$

where $\alpha = r - \sigma^2/2$.

Remaining time is discretized into a lattice $\tau_n = \epsilon(M - n)$, $n = 0, 1, \ldots, M$, with $t_* - t_0 = \epsilon M$. Hence, $\tau_0 = t_*$ is the expiration time of the option and present time is $\tau_M = 0 = t_0$; Figure 16.2 shows the time lattice, with the lattice sites labeled by remaining time τ_n.

Define $z_i \equiv z(\tau_i)$; discretized velocity for remaining time is defined, as in Eq. (3.33), by the finite backward difference $dz/d\tau \simeq (z_i - z_{i-1})/\epsilon$. The lattice Lagrangian is given by

$$\mathcal{L}_{BS}(i - 1) = -\frac{1}{2\sigma^2} \left(\frac{z_i - z_{i-1}}{\epsilon} + \alpha \right)^2 - r \tag{16.2}$$

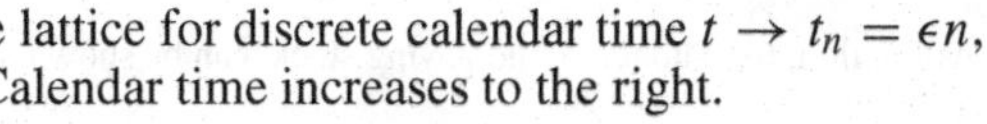

Figure 16.1 The lattice for discrete calendar time $t \to t_n = \epsilon n$, $n = 0, 1, \ldots, M$ with $t_* = \epsilon M$. Calendar time increases to the right.

● ● ● ● ● ● ● ● ●
M M–1 M–2   2   1   0

Figure 16.2 The lattice for remaining discrete time $\tau_n = \epsilon(M-n)$ with $t_* = \epsilon M$ and $t \to t_n = \epsilon n$, $n = 0, 1, \ldots, M$. Calendar time increases to the right and the decreasing numbering of the lattice points are for remaining time.

Let the boundary conditions be given by $z_0 = z$; $z_M = z'$; from Eq. (3.39), the lattice action is given by

$$S_{BS} = \epsilon \sum_{i=0}^{M-1} \mathcal{L}_{BS}(i) = \sum_{i=0}^{M-1} \mathcal{L}(i)$$

$$\mathcal{L}(i) = \epsilon \mathcal{L}_{BS}(i) = -\frac{1}{2s^2}(z_i - z_{i+1} - \tilde{\alpha})^2$$

with dimensionless parameters $s^2 = \epsilon \sigma^2$, $\tilde{\alpha} = \epsilon \alpha$.

The pricing kernel is defined by Eq. (15.9) and from Eqs. (15.10) and (15.11) yields the following [12]

$$p(z', z; M) = \langle z'|e^{-\epsilon MH}|z \rangle = \tilde{\mathcal{N}} \prod_{i=0}^{M-1} \int dz_i e^{S_{BS}} \bigg|_{z(0)=z';z(\tau)=z}$$

$$p(z', z; 1) = \langle z'|e^{-\epsilon H}|z \rangle = \mathcal{N} \exp\{\mathcal{L}\}$$

$$= \sqrt{\frac{1}{2\pi s^2}} \exp\left\{-\frac{1}{2s^2}(z - z' - \tilde{\alpha})^2\right\} \tag{16.3}$$

Consider a European put option P_E with strike price K and payoff function given by $(K - e^z)_+$. The option matures at time $M\epsilon = t_*$ in the future, with present time labeled by $t_0 = 0$.[1] Since remaining time is running backwards, the pricing kernel, from Eq. (3.43), gives the price of the European put option at time $m = M$ by the following equation

$$P_E(z', M) = e^{-M\tilde{r}} \int_{-\infty}^{+\infty} dz\, p(z', z; M)(K - e^z)_+ \tag{16.4}$$

where $\tilde{r} = \epsilon r$.

Consider the case of an American put option with possibility of an early exercise. The payoff of the American option is the same as the European option, with the additional freedom that the holder of the option can exercise the option anytime

---

[1] The price of an American call option, for a nondividend paying stock, can be shown to be equal to the European call option [59].

from the present time t_0 to its maximum maturity date t_*. Time is divided into short intervals of spacing ϵ and early exercise of the option can only take place at the discrete time instants $t_i = i\epsilon$.

To find the price of the American option $P(t)$, one propagates the payoff function backwards in time. At time slice τ_i the American option has a price given by $P(\tau_i)$. To determine the option price at next (earlier) instant τ_{i+1}, one propagates $P(\tau_i)$ (backwards in time) to obtain a trial value of the American option at τ_{i+1}, called $P_I(\tau_{i+1})$. The actual value of the American option at τ_{i+1} is given by the maximum of the (nondiscounted) payoff function and $P_I(\tau_{i+1})$; that is

$$P(\tau_{i+1}) = \text{Max}\{P_I(\tau_{i+1}), (K - e^{z_{i+1}})_+\} \tag{16.5}$$

In the path integral approach the pricing kernel is used for computing the trial option price $P_I(\tau_{i+1})$; Eqs. (16.3) and (16.4) yield

$$P_I(\tau_{i+1}, z') = e^{-\tilde{r}} \int_{-\infty}^{+\infty} dz\, p(z', z; 1) P(\tau_i, z) = e^{-\tilde{r}} N \int_{-\infty}^{+\infty} dz\, e^{\mathcal{L}(z', z)} P(\tau_i, z)$$

$$= e^{-\tilde{r}} \sqrt{\frac{1}{2\pi s^2}} \int_{-\infty}^{+\infty} dz \exp\left\{-\frac{1}{2s^2}\left(z - z' - \tilde{\alpha}\right)^2\right\} P(\tau_i, z) \tag{16.6}$$

Almost all cases of interest have fairly small volatility, that is $s \simeq 0$; for small s, the most efficient procedure for evaluating the integral in Eq. (16.6) is to Taylor expand the function $P(\tau_i, z)$ about the very sharp maximum of the Gaussian part of the integrand located at the point $z' + \tilde{\alpha} \equiv \bar{z}$. Denoting differentiation with respect to z by prime yields the Taylor expansion

$$P(\tau_i, z) = P(\tau_i, \bar{z}) + (z - \bar{z}) P'(\tau_i, \bar{z}) + \frac{1}{2}(z - \bar{z})^2 P''(\tau_i, \bar{z}) + \dots \tag{16.7}$$

Using the fact that

$$\sqrt{\frac{1}{2\pi s^2}} \int_{-\infty}^{+\infty} dz\, e^{-\frac{1}{2s^2}(z-\bar{z})^2} = 1; \quad \sqrt{\frac{1}{2\pi s^2}} \int_{-\infty}^{+\infty} dz\, e^{-\frac{1}{2s^2}(z-\bar{z})^2}(z - \bar{z}) = 0$$

$$\sqrt{\frac{1}{2\pi s^2}} \int_{-\infty}^{+\infty} dz\, e^{-\frac{1}{2s^2}(z-\bar{z})^2}(z - \bar{z})^2 = s^2 \tag{16.8}$$

yields, from Eqs. (16.6), (16.7), and (16.8), the following recursion equation

$$P_I(\tau_{i+1}, z') = e^{-\tilde{r}} \left[P(\tau_i, \bar{z}) + \frac{1}{2}s^2 P''(\tau_i, \bar{z}) \right] + O(s^4) \tag{16.9}$$

Discretizing the values of $\bar{z}$ into a grid of spacing δ of $O(s)$ yields

$$P_I(\tau_{i+1}, z') \simeq e^{-\bar{r}} \left[P(\tau_i, \bar{z}) + \frac{1}{\delta^2}[P(\tau_i, \bar{z} + \delta) - 2P(\tau_i, \bar{z}) + P(\tau_i, \bar{z} - \delta)] \right]$$

(16.10)

To obtain the value of $P_I(\tau_{i+1}, z')$, as in Eq. (16.10), one needs the values of option prices at the earlier time at *three* distinct values of $\bar{z}$, namely $P(\tau_i, \bar{z})$, $P(\tau_i, \bar{z} \pm \delta)$. By induction, it follows that, for each step one recurses (back in time), one loses two points that are on the boundary of the stock price tree; hence, the number of stock values at which the option price can be obtained collapses into a single value. This structure of the recursion equation is shown in Figure 16.3, where the tree reduces to a single point in remaining time. The single point corresponds to a specific value of the stock price, which in Figure 16.3 is indicated by z_M^0.

The purpose of the recursion is to find the option price at some particular value of the stock e^{z_M} and at present calendar time, denoted by remaining time τ_M. Hence, one needs to create a *tree* with specific values z_m of the stock price for each time step; for these specific values of the stock price, the recursion equation evaluates the price of the option; the tree is illustrated in Figure 16.3.

As shown in Figure 16.3, the points on the tree grow linearly with each step in time. The stock values z_i on the tree are taken to have a spacing of $\delta = s$ so that the spread of the z_i values on the tree can span the interval required for obtaining

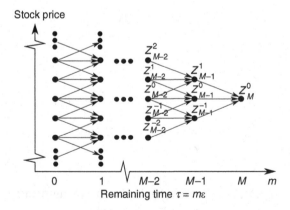

Figure 16.3 Tree of stock values z_m^k at remaining time lattice $\tau = m\varepsilon$, with number of points growing linearly with maturity time. The figure indicates how three points on the tree are required, by the recursion equation, for determining the option price at the next remaining time lattice point.

Table 16.1 *Numerical prices of American and European put options as a function of the possible present time stock prices S. The parameters are* $t_* = 0.5$ *year,* $r = 0.1/year$, $\sigma = 0.4/\sqrt{year}$, $K = 10$, $\epsilon = t_*/100$.

S	American Put	Numerical European Put	Black–Scholes European Put
6.0	4.00	3.558	3.558
8.0	2.095	1.918	1.918
10.0	0.922	0.870	0.870
12.0	0.362	0.348	0.348
14.0	0.132	0.128	0.128

an accurate result from the integration. The tree at time τ_i has the following values for z_i, namely

$$z_m^{(k)} \doteq z_M + \tilde{\alpha} + ks, \qquad k = -(M - m), \ldots, +(M - m) \qquad (16.11)$$

At remaining time τ_m, the tree consists of $2(M - m) + 1$ values of $z_m(k)$, centered on the $S = e^{z_M}$. For $m = M$, this reduces to the value of $z_M^{(0)} = z_M + \tilde{\alpha}$, namely the value of the stock at initial time for which the price of the American option is being computed.

The algorithms expressed in Eqs. (16.5) and (16.10) were numerically tested and yield results, given in Table 16.1, that are fairly accurate as well as consistent with those obtained in [76].

The American and European put option prices, together with the payoff function for the put option, are shown in Figure 16.4(a) and are seen to be consistent with the discussions in [59]; in particular, note that the American put option is always more expensive than the European put option, as indeed it must be since it has more choice; furthermore, the American put option, for small values of the stock price S, has the same slope as the payoff function, hence smoothly joining it.

From [59], the inequalities obeyed by the price of American call and put options C and P respectively, on a stock with stock price S, strike price K, and maturing at future time T, are given by the following

$$S - K \leq C - P \leq S - e^{-rT}K \qquad (16.12)$$

The put–call inequality for American option of a stock is seen in Figure 16.4(b) to hold for the numerical option prices.

All the basic features of the algorithm for pricing an American equity put option appear in the more complex algorithm required for pricing the American coupon bond and interest rate options.

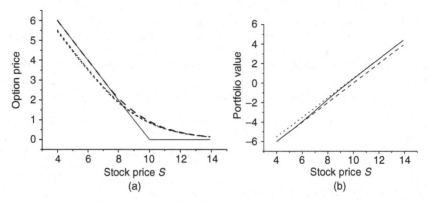

Figure 16.4 (a) Price of the American stock option (dashed line) versus stock price S; the unbroken line is the payoff function and the dotted line is the European option price. (b) Put–call inequality for the American stock option. Unbroken line is $C - P$, dashed line is $S - K$, and dotted line is $S - e^{-rt} K$.

16.3 American caplet and coupon bond options

In the numerical studies of both the caplets and coupon bond options, for simplicity, only options that mature when the instrument becomes operational are studied.[2]

To define the numerical algorithm the time interval $[t_0, t_*]$ is discretized; note $t_0 = 0$ and $t_* = M\epsilon$. Define remaining lattice time by $\tau_m = t_* - (m - 1)\epsilon = (M - m + 1)\epsilon$, $m = 1, 2, \ldots, M + 1$, where $t_1 = t_* = M\epsilon$; present time $t_0 = 0$ yields for lattice time $\tau_{M+1} = 0$. In other words, the option can only be exercised at time $t_1 = M\epsilon$, $t_2 = M\epsilon - \epsilon$, $t_3 = M\epsilon - 2\epsilon$, $\ldots$, $t_M = M\epsilon - (M - 1)\epsilon = \epsilon$. The numerical algorithm recurses 'forward' in remaining time with $\tau_1 \rightarrow \tau_2 \ldots \rightarrow \tau_m, \ldots, \tau_{M+1}$.

Let $C(\tau_m, t_*)$ denote the price of both caplet and coupon bond American option.[3] Choose the *forward measure* with numeraire $B(\tau_m, t_*)$, for which $g(\tau_m) = C(\tau_m, t_*)/B(\tau_m, t_*)$ is a martingale. The trial value of the American option at later remaining time τ_{m+1}, denoted by $\tilde{g}(\tau_{m+1})$, is given from the option price at time τ_m by the martingale property, which from Eq. (3.8), yields the following

$$\tilde{g}(\tau_{m+1}) = E\left[\frac{C(\tau_m, t_*)}{B(\tau_m, t_m)}\right] = E[g(\tau_m)] \qquad (16.13)$$

The tilde in $\tilde{g}(\tau_{m+1})$ denotes the *initial* trial value of the American option at τ_{m+1}. The trial option price is compared with the payoff function (divided by

---

[2] Midcurve options are widely traded in the market. All the numerical procedures developed in this chapter can be generalized in a straightforward manner to the midcurve case.
[3] Since the caplet matures at $T = t_*$, the third argument in $caplet(\tau_i, t_*, T)$ is suppressed.

the appropriate numeraire) and yields the actual value of the option at time τ_{m+1}, which is equal to the maximum of the two [59].

From Eq. (16.13) it can be seen that in all computations, the quantity $g(\tau_m)$ is always equal to the option price $C(\tau_m, t_*)$ divided by the numeraire $B(\tau_m, t_*)$. For the American option, the trial option price needs to be compared with the payoff, with the greater value being retained. Hence, the payoff at intermediate time – between maturity and present time – needs to be defined. At intermediate time τ_m, the scaled payoff is defined by dividing the payoff by the numeraire, to match a similar division in defining $g(\tau_m)$. All zero coupon bond prices in the payoff are replaced by the forward bond prices at time τ_m, as is required for the American option.

16.3.1 Caplet

Consider an interest rate caplet that matures at t_* and caps the interest rate at K for the period T to $T + \ell$.[4] The payoff of the caplet, from Eq. (9.32) is given in terms of $f(t, x)$, by

$$\tilde{V} B(t_*, T) \left[X - F(t_*, T, T + \ell) \right]_+ \tag{16.14}$$

where

$$X = \frac{1}{1 + \ell K}; \quad \tilde{V} = (1 + \ell K) V; \quad F(t_*, T, T + \ell) = \exp\left\{ -\int_T^{T+\ell} dx f(t_*, x) \right\}$$

Figure 16.5 shows the forward interest rates that define the caplet payoff function at different times t_*, t_i, t_0. For the caplet the scaled payoff at remaining time τ_m is given by

$$\tilde{V} \frac{F(\tau_m, t_*, T) [X - F(\tau_m, T, T + \ell)]_+}{B(\tau_m, t_*)}$$

The important point to note is that the form of the payoff does not change with time. The discounting factor at time t_* that appears in the payoff at maturity, namely the bond price $B(t_*, T)$, is changed into the forward bond $F(\tau_m, t_*, T)$ as one moves to an intermediate time τ_m, as shown in Figure 16.5; there is no additional discounting factor. The American option price at time τ_{i+1} is equal to the maximum

---

[4] In the numerical study, only the special case of $T = t_*$, will be considered; for now, the midcurve caplet is analyzed as the formulas are more transparent.

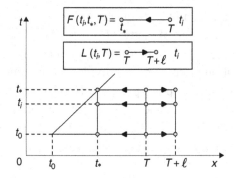

Figure 16.5 $\tilde{V} F[\tau_i, t_*, T](X - F(\tau_i, T, T + \ell))]_+ / B(\tau_i, t_*)$: the scaled payoff function for the caplet at intermediate time $t_i \in [t_0, t_*]$.

of the initial trial option value $g_I(\tau_{m+1})$ and the payoff function at time τ_{m+1}; hence

$$
\frac{C(\tau_{m+1}, t_*)}{B(\tau_{m+1}, t_*)} = g(\tau_{m+1})
$$

$$
= \text{Max}\left[\tilde{g}(\tau_{m+1}), \tilde{V}\frac{F(\tau_{m+1}, t_*, T)\,[X - F(\tau_{m+1}, T, T + \ell)]_+}{B(\tau_{m+1}, t_*)} \right]
$$

$$(16.15)$$

Note from Eq. (16.15) that, in effect, the option price $C(\tau_{i+1}, t_*)$ is being compared with the payoff function $\tilde{V} F(\tau_{i+1}, t_*, T)(X - F(\tau_{i+1}, T, T + \ell))_+$.

16.3.2 Coupon bond

The coupon bond payoff function, from Eq. (4.20), is given by

$$
\left[\sum_{i=1}^{N} c_i B(t_*, T_i) - K \right]_+
$$

The scaled coupon bond payoff, at time τ_i, is given by

$$
\frac{\left[\sum_{j=1}^{N} c_j F(\tau_i, t_*, T_j) - K \right]_+}{B(\tau_i, t_*)}
$$

As is the case for the interest rate caplet, at intermediate time $\tau_m \in [t_0, t_*]$ the bond price $B(t_*, T_j)$ at time t_*, in the payoff function has been replaced, at time τ_m, by the forward bond price $F(\tau_m, t_*, T_j)$. The American option price at time τ_{m+1} is given by

$$\frac{C(\tau_{i+1}, t_*)}{B(\tau_{m+1}, t_*)} = g(\tau_{m+1})$$

$$= Max\left[\tilde{g}(\tau_{m+1}), \frac{\left[\sum_{j=1}^{N} c_j F(\tau_{m+1}, t_*, T_j) - K\right]_+}{B(\tau_{m+1}, t_*)} \right] \quad (16.16)$$

Note the important fact that for both the caplet and coupon bond, the payoff function at each time τ_m is identical to the form of the payoff function at maturity time t_*. In particular, the payoff function is scaled when it is compared with the trial option price $\tilde{g}(\tau_{m+1})$ and this results in the option price being directly compared to the payoff function at intermediate time τ_{m+1}.

16.4 Forward interest rates: lattice theory

The quantum field theory of forward interest rates is defined on the trapezoidal domain in the continuous xt plane, as shown in Figure 5.1. To obtain a numerical algorithm, the xt plane is discretized into a lattice consisting of a finite number of points. The calendar time direction, as mentioned earlier, is discretized into a lattice with spacing ϵ and future time direction x is discretized into a lattice with spacing a.

Recall, from Eqs. (5.4) and (5.6), the stiff action for continuous calendar and future time is given by[5]

$$S = -\frac{1}{2} \int dt \int dx \left[\left(\frac{\partial f/\partial t - \alpha}{\sigma}\right)^2 + \frac{1}{\mu^2}\left\{\frac{\partial}{\partial x}\left(\frac{\partial f/\partial t - \alpha}{\sigma}\right)\right\}^2 \right.$$

$$\left. + \frac{1}{\lambda^4}\left\{\frac{\partial^2}{\partial x^2}\left(\frac{\partial f/\partial t - \alpha}{\sigma}\right)\right\}^2 \right] \quad (16.17)$$

The time lattice is defined as discussed in Section 16.3, with $t_0 = 0$. Future time, similar to calendar time, is labeled running *backwards*, with the origin of future time being placed at the payoff function. In other words, continuous calendar time and future time labels (t, x) are discretized so that lattice remaining time is defined by $t_* - t$ and lattice future remaining time is defined by $T - x$, where T is the maturity time of the underlying instrument. For the caplet, maturity time is when the interest cap becomes operational, whereas for the coupon bond it is the time of the last coupon payment, which is also the time of the principal's payment.

---

[5] Market time is not considered in this chapter since the focus is on developing numerical algorithms and not applications to the financial markets.

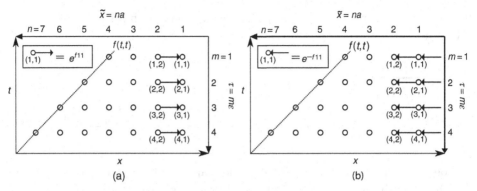

Figure 16.6 The axis labels t and x are calendar and future time, whereas labels τ and $\tilde{x}$ are for remaining calendar and remaining future time. Forward interest rates on the lattice, with each dot representing a forward interest rate $f(t,x) \rightarrow f_{mn}$; an arrow indicates that a forward interest rate $\exp\{f_{mn}\}$ connects lattice (mn) to $(m, n+1)$. The lattice points take values in the range $m = 1, 2, \ldots, M + 1$ and $n = 1, 2, \ldots, N + m$. (a) Payoff function for a caplet. (b) Payoff for the coupon bond.

Given the trapezoidal shape of the forward interest rates domain defined by $x \geq t$ and plotted in Figure 16.6, the range of discretized x_n depends on discretized time τ_m. The calendar and future time lattice have spacing ϵ and a respectively. At maturity time t_*, future time is taken to have $N = (T - t_*)/a$ number of lattice points – corresponding to N forward interest rates that define the payoff function.

Hence, for $t \in [0, t_*]$, the discretized calendar and future time are given by[6]

$$t \rightarrow \tau_m = [M - m + 1]\epsilon; \quad m = 1, 2, \ldots, M + 1$$

$$\Rightarrow 0 \leq \tau_m \leq M\epsilon; \quad \tau_1 = t_* = M\epsilon; \quad \tau_{M+1} = 0$$

$$\tau_m \; : \; x \rightarrow \tilde{x}_n = m\epsilon + [N - n + 1]a; \quad n = 1, 2, \ldots, N + m$$

$$a \leq \tilde{x}_n \leq (N + m)a; \quad T - t_* = Na \tag{16.18}$$

The total number of lattice sites is equal to $N(M + 1) + M(M + 1)/2$; for most numerical calculations one usually takes $\epsilon = a$.

Note the payoff is placed at remaining calendar time $\tau_1 = t_*$, which implies from Eq. (16.18) that $m = 1$; the payoff is always placed at remaining future time $\tilde{x}_{N+m} = a$. Figures 16.6(a) and 16.6(b) show the lattice on which the forward interest rates and the payoff for the caplet and coupon bond are defined.

To define the lattice theory, one needs to rescale the field $f(t, x)$ and all the parameters so that only dimensionless quantities appear in the lattice action. Define

---

[6] A more accurate notation is $x \rightarrow \tilde{x}_n(m)$ since the range of $\tilde{x}_n(m)$ depends on m. However, to avoid cumbersome notation, the index m is assumed for $\tilde{x}_n$.

the following dimensionless lattice quantities

$$f_{mn} = af(\tau_m, \tilde{x}_n) = af\left([M - m + 1]\epsilon, M\epsilon + [N - n + 1]a\right)$$

$$\tilde{\alpha}_{mn} = a\epsilon\alpha(\tau_m, \tilde{x}_n); \quad s_{mn} = \sqrt{\epsilon}a\sigma(\tau_m, \tilde{x}_n)$$

$$\tilde{\mu} = a\mu; \quad \tilde{\lambda} = a\lambda$$

The dimensionless quantum field variables f_{mn} yield the following discretizations

$$\frac{f(t,x)}{\partial t} \simeq \frac{1}{a\epsilon}(f_{m,n} - f_{m+1,n}) \equiv \frac{1}{a\epsilon}\delta_t f_{mn}$$

$$\frac{f(t,x)}{\partial x} \simeq \frac{1}{a^2}(f_{m,n} - f_{m,n+1}) \equiv \frac{1}{a^2}\delta_x f_{mn}$$

Thus, from Eq. (16.17), one obtains the following lattice action S, expressed completely in terms of dimensionless field variables and parameters

$$S_L = -\frac{1}{2}\sum_{m,n}\left[\left(\frac{\delta_t f - \tilde{\alpha}}{s_{mn}}\right)^2\right.$$

$$\left.+\frac{1}{\tilde{\mu}^2}\left(\delta_x\left(\frac{\delta_t f - \tilde{\alpha}}{s_{mn}}\right)\right)^2 + \frac{1}{\tilde{\lambda}^4}\left(\delta_x^2\left(\frac{\delta_t f - \tilde{\alpha}}{s_{mn}}\right)\right)^2\right] \tag{16.19}$$

Doing an integration by parts, the action in Eq. (16.19) yields

$$S_L = -\frac{1}{2}\sum_{m=1}^{M+1}\sum_{n=1}^{N+m}\left(\frac{\delta_t f - \tilde{\alpha}}{s}\right)_{mn}\tilde{D}_{m,nn'}^{-1}\left(\frac{\delta_t f - \tilde{\alpha}}{s}\right)_{mn'} \tag{16.20}$$

where $\tilde{D}_{m,nn'}^{-1}$ is the dimensionless inverse of the propagagtor with dimensionless parameters $\tilde{\mu}, \tilde{\lambda}$. The dimensionless lattice Lagrangian is given from S by the following

$$S_L = \sum_{m=1}^{M+1}\mathcal{L}[\mathbf{f}_{m+1}; \mathbf{f}_m]$$

$$\mathcal{L}[\mathbf{f}_{m+1}; \mathbf{f}_m] = -\frac{1}{2}\sum_{n,n'=1}^{N+m}\left(\frac{\delta_t f - \tilde{\alpha}}{s}\right)_{mn}\tilde{D}_{m,nn'}^{-1}\left(\frac{\delta_t f - \tilde{\alpha}}{s}\right)_{mn'} \tag{16.21}$$

where $\mathbf{f}_m = (f_{m1}, f_{m2}, \ldots, f_{m,N+m})$. Note the number of components of the forward interest rate vector $\mathbf{f}_m$ depends on the time lattice m and has $N + m$-components; the reason being that forward interest rates are defined for all $x \geq t$, which on the lattice implies that the number of forward interest rates depends on τ_m.

The functional integral is discretized and yields the lattice field theory of forward interest rates with the following lattice partition function Z_L

$$Z = \int Df e^S \rightarrow Z_L = \tilde{N} \prod_{m=1}^{M+1} \prod_{n=1}^{N+m} \int_{-\infty}^{+\infty} df_{mn} e^{S_L} \qquad (16.22)$$

$\tilde{N}$ is a normalization constant.

16.5 American option: recursion equation

For the sake of concreteness, the recursion equation for the caplet American option is discussed in the framework of the lattice theory; the analysis carries over without any change to the coupon bond case.

The payoff function of the caplet in Eq. (16.14) is defined for discretized time; at maturity time t_*, the discretized caplet is denoted by $C_1 \equiv C(t_*, t_*, T)$. From Eq. (16.18), the convention being used for remaining future lattice time is that for all τ_m, the last payment of the payoff is given by $T \rightarrow \tilde{x}_1$, that is the minimum value of the future lattice index $\tilde{x}_n$; hence, the zero coupon bond in the payoff function is given by

$$B(t_*, T) \simeq B(\tau_m, \tilde{x}_1) \equiv B(\tau_m, 1) = \exp\left\{-\sum_{n=1}^{N+m} f_{mn}\right\} \qquad (16.23)$$

For simplicity and because Libor data are given only on a future time lattice with spacing ℓ, one takes $a = \ell$. On the lattice, Libor is given by

$$1 + \ell L(t, T) \simeq 1 + \ell L(\tau_m, \tilde{x}_1) = \exp\left\{\frac{\ell}{a} f_{m1}\right\} = \exp\{f_{m1}\}$$

Hence

$$caplet(t_*, t_*, T) = \tilde{V} B(t_*, T)(X - F_*)_+ \equiv C_1$$

$$\Rightarrow C_1 = \tilde{V} \exp\left\{-\sum_{j=1}^{N+1} f_{1j}\right\} (X - e^{-f_{11}})_+ \qquad (16.24)$$

$$= C_1(f_{11}, f_{12}, \ldots, f_{1,N+1}) \equiv C_1(\mathbf{f}_1)$$

The payoff function is evolved backwards in time to obtain the price of the option from the payoff function. To illustrate the general procedure, consider the first step backwards; one starts from the payoff function at time $t_* = \tau_1$ and finds the value of the option at time τ_2 by recursing backwards. Since remaining calendar time runs in the *opposite direction* of calendar time, the index of remaining (lattice) calendar time *increases* as one goes backwards in calendar time. In taking one step backwards in calendar time, the number of independent forward interest rates on the lattice increases by one forward interest rate as follows

$$\mathbf{f}_m \to \mathbf{f}_{m+1}$$

$$(f_{m,1}, f_{m,2}, \ldots, f_{m,N+m}) \to (f_{m+1,1}, f_{m+1,2}, \ldots, f_{m+1,N+m+1})$$

Hence the option price evolves in the following manner

$$C(\mathbf{f}_m) = C_m(f_{m,1}, f_{m,2}, \ldots, f_{m,N+m})$$

$$\to C(\mathbf{f}_{m+1}) = C_{m+1}(f_{m+1,1}, f_{m+1,2}, \ldots, f_{m+1,N+m+1})$$

The expression $\mathcal{N} \exp\{\mathcal{L}[\mathbf{f}_{i+1}, \mathbf{f}_i]\}$ is the *pricing kernel* for the forward interest rates, analogous to the pricing kernel for the (simpler) case of equity given in Eq. (16.3). The pricing kernel yields, similar to Eq. (16.6) for equity options, the option price $C_{i+1}(\mathbf{f}_{i+1})$ at earlier time τ_{i+1} from option price $C_i(\mathbf{f}_i)$ by taking one step backward in time, and generates the initial trial value for the option $\tilde{C}_{i+1}(\mathbf{f}_{i+1})$. From the results derived in Section 15.11 and, in particular, applying Eq. (15.77) to an infinitesimal time step, yields the following

$$\frac{\tilde{C}_{m+1}(\mathbf{f}_{m+1})}{B(\tau_{m+1}, 1)} = \int d\mathbf{f}_m \langle \mathbf{f}_{m+1}|e^{-\epsilon H}|\mathbf{f}_m\rangle \frac{C_m(\mathbf{f}_m)}{B_{m,1}}$$

$$= \mathcal{N} \int d\mathbf{f}_m \, e^{\mathcal{L}[\mathbf{f}_{m+1}, \mathbf{f}_m]} \frac{C_m(\mathbf{f}_m)}{B(\tau_m, 1)} \qquad (16.25)$$

$$= \mathcal{N} \left[\prod_{n=1}^{N+m} \int_{-\infty}^{+\infty} df_{mn} \right] \frac{C_m(\mathbf{f}_m)}{B(\tau_m, 1)}$$

$$\times \exp\left(-\frac{1}{2} \sum_{jk=1}^{N+m} \left(\frac{f_{m,j} - \bar{f}_{m+1,j}}{s_{mj}}\right) \tilde{D}_{m;jk}^{-1} \left(\frac{f_{mk} - \bar{f}_{m+1,k}}{s_{mk}}\right)\right) \qquad (16.26)$$

with

$$\bar{f}_{m+1,n} \equiv f_{m+1,n} + \tilde{\alpha}_{m+1,n}; \quad B(\tau_m, 1) = \exp\left\{-\sum_{n=1}^{N+m} f_{mn}\right\} \qquad (16.27)$$

Note $\tilde{D}_{m;jk}^{-1}$ is the lattice approximation of the continuum propagator and $B(\tau_m, 1)$ has been defined in Eq. (16.23).

The American option price at time τ_{m+1}, from Eq. (16.16) is given by

$$\frac{C(\tau_{i+1}, 1)}{B(\tau_{m+1}, 1)} = \text{Max} \left[\frac{\tilde{C}_{m+1}(\mathbf{f}_{m+1})}{B(\tau_{m+1}, 1)}, \frac{\left[\sum_{j=1}^{N} c_j F(\tau_{m+1}, 1, \tilde{x}_j) - K \right]_+}{B(\tau_{m+1}, 1)} \right]$$

Similar to the case of an American option for an equity discussed in Section 16.2, the interest rate's dimensionless volatility s_{mn} is quite small, that is $s_{mn} \simeq 0$. Hence in the f_{ij} integrations given in Eq. (16.26), the path integral will be dominated by values f_{ij} that are close to $\bar{f}_{i+1,j} = f_{i+1,j} + \tilde{\alpha}_{i+1,j}$. The most accurate way for evaluating the functional integral in Eq. (16.26) is to Taylor expand the function $g_m = C_m/B_{\tau_m,1}$ about $\bar{\mathbf{f}}_{m+1} = (\bar{f}_{m+1,1}, \ldots, \bar{f}_{m+1,N+m+1})$ and yields, for $\bar{g}_m \equiv g_m(\bar{\mathbf{f}}_{m+1})$, the following

$$g_m = \bar{g}_m + \sum_{n=1}^{N+m} (f_{mn} - \bar{f}_{m+1,n}) \frac{\partial \bar{g}_m}{\partial f_{mn}}$$

$$+ \frac{1}{2} \sum_{jk=1}^{N+m} \frac{\partial^2 \bar{g}_m}{\partial f_{mj} \partial f_{mk}} (f_{mj} - \bar{f}_{m+1,j})(f_{mk} - \bar{f}_{m+1,k}) + \ldots$$

The Gaussian integrations over the forward interest rates f_{ij} are carried out using the following results

$$\mathcal{N} \prod_{n=1}^{N+i} \int_{-\infty}^{+\infty} df_{mn} e^{\mathcal{L}[\mathbf{f}_{i+1}, \mathbf{f}_i]} = 1; \quad \prod_{n=1}^{N+i} \int_{-\infty}^{+\infty} df_{mn} \, e^{\mathcal{L}[\mathbf{f}_{i+1}, \mathbf{f}_i]} (f_{ij} - \bar{f}_{i+1,j}) = 0$$

$$\mathcal{N} \prod_{n=1}^{N+i} \int_{-\infty}^{+\infty} df_{mn} e^{\mathcal{L}[\mathbf{f}_{i+1}, \mathbf{f}_i]} (f_{ij} - \bar{f}_{i+1,j})(f_{ik} - \bar{f}_{i+1,k}) = s_{ij} \tilde{D}_{i;jk} s_{ik} \quad (16.28)$$

Hence, Eqs. (16.26) and (16.28) yield the result that

$$\tilde{C}_{m+1} = B_{m+1,1} \left[\bar{g}_m + \frac{1}{2} \sum_{jk=1}^{N+m} \frac{\partial^2 \bar{g}_m}{\partial f_{mj} \partial f_{mk}} s_{mj} \tilde{D}_{m;jk} s_{mk} \right] + \ldots \quad (16.29)$$

Recall that $g_m = g_m[\mathbf{f}_m]$ depends on the vector $\mathbf{f}_m = (f_{m,1}, f_{m,2}, \ldots, f_{m,N+m})$. The value of $\bar{g}_m = \bar{g}_m[\bar{f}_{m+1,n}]$ depends on the entire forward interest rate tree at time τ_{m+1}. Since $g_m[\mathbf{f}_m]$ is being differentiated with respect to only two of the components, namely f_{mj}, f_{mk}, only these (two) components will be explicitly indicated, with the rest of the components in $g_m[\mathbf{f}_m]$ being suppressed.

The second derivative of g_i is numerically estimated using the symmetric second-order difference. The f_is are discretized with spacing δ, which is taken to be $O(s)$, as dictated by the Lagrangian in Eq. (16.26). The symmetric second-order derivative, using

$$\delta^j_+ g(f_m) = \frac{1}{\delta}[g(f_{mj} + \delta) - g(f_m)]$$

$$\delta^j_- g(f_m) = \frac{1}{\delta}[g(f_m) - g(f_{mj} - \delta)]$$

yields the following discretization

$$\frac{\partial^2 \bar{g}_m}{\partial f_{mj} \partial f_{mk}} \equiv \frac{\partial^2 g_m}{\partial f_{mj} \partial f_{mk}}\Big|_{f_{mn}=\bar{f}_{m+1,n}} = \frac{1}{2}(\delta^j_- \delta^k_+ + \delta^k_+ \delta^j_-) g_i\Big|_{f_{mn}=\bar{f}_{m+1,n}}$$

$$= \frac{1}{2\delta^2}\Big[g_m(\bar{f}_{m+1,j} + \delta, \bar{f}_{m+1,k}) - 2g_m(\bar{f}_{m+1,j}, \bar{f}_{m+1,k})$$

$$- g_i(\bar{f}_{m+1,j} + \delta, \bar{f}_{m+1,k} - \delta) + g(\bar{f}_{m+1,j}, \bar{f}_{m+1,k} - \delta)$$

$$+ g(\bar{f}_{m+1,j}, \bar{f}_{m+1,k} + \delta) - g_m(\bar{f}_{m+1,j} - \delta, \bar{f}_{m+1,k} + \delta)$$

$$+ g_i(\bar{f}_{m+1,j} - \delta, \bar{f}_{m+1,k})\Big] \tag{16.30}$$

Figure 16.7(a) is a graphical representation of the seven terms required for the computation of $\partial^2 \bar{g}_m / \partial f_{mj} \partial f_{mk}$. Figure 16.7(b) gives a planar cross-section of the points on the forward interest rate tree.

To evaluate $\partial^2 \bar{g}_m / \partial f_{mj} \partial f_{ik}$, one needs to know the values of g_m at the points $f_{mj} = \bar{f}_{m+1,j}$, for all f_{mj}, $j = 1, 2, \ldots, N + m$. Moreover, as required by Eq. (16.30), the forward interest rate tree at time τ_{m+1} must also contain the following three points, namely $\bar{f}_{m+1,j}, \bar{f}_{m+1,j} \pm \delta$. This feature of the recursion equation

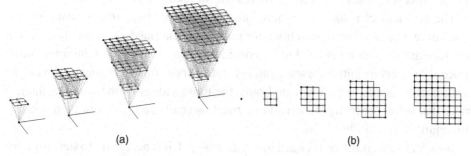

(a)   (b)

Figure 16.7 Calendar time increases upwards and future time increases to the right; the greater the future time, the longer it takes for the tree to terminate. (a) Tree structures for the forward interest rates f_{mn}. Each initial value of the initial forward interest rates evolves as an independent tree, as indicated in the diagram. Each point on the tree is the source of a new set of points. (b) A view of the planar points of the trees of points.

is a reflection of a similar property for the case of the American option for equity as in Eq. (16.10) and shown in Figure 16.3.

The price of the European option for caplets and coupon bonds can be obtained by repeating the backward recursion up to the present time. In Sections 16.8 and 16.9 the numerical price of the European option will be compared with the approximate results obtained from the same algorithm as the American option, so as to assess the accuracy of the algorithm. For the American option, one needs to perform, for *each* step up to the present time, a comparison of the trial option price with the payoff as given in Eqs. (16.15) and (16.16).

16.6 Forward interest rates: tree structure

To minimize the computational complexity and the time of execution, it is mandatory to limit, as far as possible, the number of possible forward interest rates for which the option price is computed.

The definition of the martingale of the forward numeraire yields the following drift

$$\alpha(t,x) = \sigma(t,x) \int_{T_p}^{x} dx' D(x,x';t)\sigma(t,x'); \quad T_p \le x < T_p + \ell$$

For the American option, future time is discretized as $x = na$ and Libor interval $a = \ell$; hence $T_p = p\ell = pa$ lies on the future time lattice with $x_p = T_p$. This in turn, from the above equation, yields as in Eq. (9.15)

$$\alpha(t, x_n) = 0; \quad x_n = T_n \tag{16.31}$$

There is no drift for the lattice on which the American option is being computed. In fact, this simplification is the main reason for taking $a = \ell$.

The option price is fixed by, among other parameters, the initial forward interest rate curve $f(t_0, x)$. The option price for only those intermediate (virtual) values of the forward interest rates need to be considered that contribute to the final option price. For a given initial forward interest rate curve, what this means is that the option price needs to be evaluated only for those values of the forward interest rates that lie on the forward interest rate *tree* [also called a *grid*], similar to the tree for equity given in Figure 16.3.

In order to ascertain the forward interest rates grid, it is necessary to start from the initial forward interest rate curve, which from Eq. (16.18) is given by $f(t_0, x) \rightarrow f_{M+1,n}$, where $n = 1, 2, \ldots, N + M + 1$. Similar to the case for the American equity option discussed in Section 16.2, *each* initial value of the forward interest rates generates an independent tree. Recall that, starting from the initial forward interest rates $f_{M+1,n}$, to reach the forward interest rates at calendar time m, with forward interest rates f_{mn}, one needs to take $M + 1 - m$ steps. Hence the numerical values of the forward interest rates on the forward interest rate tree is given by

$$f_{mn}^k \doteq f_{M+1,n} + k\delta, \quad -(M + 1 - m) \leq k \leq +(M + 1 - m) \quad (16.32)$$

At lattice time m the forward interest rate tree has $2(M + 1 - m) + 1 = 2(M - m) + 3$ number of values for f_{mn}, centered on the initial forward interest rate $f_{M+1,n}$. The spacing of the tree is taken to have a *fixed* value δ, which is of $O(s)$. A fixed value of δ is required to obtain a re-combining tree; conversely, if δ is taken to vary with time, one gets a dense tree with exponentially more points than the recombining tree [65].

The same result as given in Eq. (16.32) is obtained if one recurses backwards from the payoff at calendar time t_* to the initial forward interest rates at t_0. The reason being that for zero drift, that is $\alpha(t, x) = 0$, one has from Eq. (16.30) that the values of the function g_m at the values of the forward interest rates on the grid $f_{mn}, f_{m+1,n}, f_{m+1,n} \pm \delta$ are required to obtain the value of g_{m+1}; if one recurses backwards $M + 1 - m$ times, one hits the initial forward interest rate curve and in effect obtains the result given in Eq. (16.32).

The forward interest rate tree is illustrated in Figure 16.7. Compared with the tree for equity given in Figure 16.3, the forward interest rate at a lattice point (with fixed time t and future time x) has been *expanded* into a tree structure in a direction orthogonal to the xt lattice. The full forward interest rates simultaneously have infinitely many tree structures and all these tree structures are correlated by the action S given in Eq. (16.19).

As one recurses backwards from the payoff function, at any intermediate time the price of the option needs to determined for the forward interest rates only on the grid points. From Eqs. (16.15), (16.16), and (16.30), the initial trial values $\tilde{g}(\tau_m)$ that one needs are the option values from the previous step, with the values of the forward interest rates taken only from the tree structure for lattice time τ_m.

16.7 American option: numerical algorithm

Given the initial forward interest rates, the tree structure from Section 16.6 can be formed and yields the grid of the forward interest rates up to the expiration date of the option. To get the option price today, one starts from expiration time t_* and evolves backwards in calendar time. All computations are carried out using the lattice theory given in Eq. (16.22).[7]

Lattice points are labeled by i, j with i labeling calendar time and j labeling future time. Maturity time t_* is labeled by τ_1, that is $i = 1$, as shown in Figure 16.6.

---

[7] Note that $t_0 = 0$.

The scaled payoff function g_1 is, in general, a function of all forward interest rates f_{1j} with future time taking $N+1$ values, as shown in Figure 16.6, being labeled as $j = 1, 2, \ldots, N+1$. Thus, g_1, which depends on $N+1$ forward interest rates, can be represented as $g_1 = g_1(f_{11}, \ldots, f_{1i}, \ldots, f_{1,N+1})$.

Since the step size ϵ is fixed, the total number of backwards steps in time $M = t_*/\epsilon$ is consequently fixed. At remaining calendar time τ_m the number of tree points for each forward interest rate is given by $2(M - (m-1)) + 1 = 2(M - m) + 3$; there are $N + m$ number of independent forward interest rates. Since each forward interest rate tree has $2(M - m) + 3$ points, this leads to the total number of points of the tree at time m being given by $(2(M - m) + 3)^{N+m}$. The tree is organized into a multidimensional array,[8] with option price $g[2(M - m) + 3]_{N+i}$, being a $N + i$ dimensional array with each index running from $1, 2, \ldots, 2(M - m) + 3$.

The size of the multidimensional array for realistic cases can be very big, requiring a large amount of computer memory and leading to codes for the American option that are inefficient. In order to develop an efficient algorithm the multidimensional array is mapped into a *vector array* with a length of $(2(M - i) + 3)^{N+i}$. The multidimensional matrix representation of the tree has some advantages since the index of each forward interest rate is presented explicitly. Hence in the recursive steps required for evaluating Eqs. (16.15), (16.16), and (16.30), the matrix representation is the most transparent way of keeping track of the grid points from the previous step that are required for deriving the trial option price for the present step.

To go from the matrix representation to the vector array one needs an algorithm for mapping the indices of the matrix to the index of the vector array; in particular, the multidimensional matrix array $g[j_{N+i}] \ldots [j_p] \ldots [j_2][j_1]$ needs to be mapped into a vector array $g[j]$. For time i let the matrix indices $[j_{N+i}] \ldots [j_p] \ldots [j_2][j_1]$ be assigned specific numerical values; the corresponding vector index j is given by the following mapping.

$$g[j_{N+i}] \ldots [j_p] \ldots [j_2][j_1] = g[j]$$

with the mapping of the multi-dimensional indices to the integer j given by

$$[j_{N+i}] \ldots [j_p] \ldots [j_2][j_1] \to j = \sum_{p=1}^{N+i} (j_p - 1)[2(M - (i-1)) + 1]^{p-1}$$

$$(16.33)$$

One should note that the matrix representation is never used in writing the codes for this algorithm. The vector array is used for avoiding the use of the multidimensional

---

[8] $g[M]_N \equiv g \underbrace{[M][M] \ldots [M]}_{N}$.

matrix; the indices of the matrix are only needed as an intermediate step to address grid points in the recursion process.

To find the matrix indices, in particular those that are nearest neighbor and next nearest neighbor as required in evaluating Eq. (16.30), one needs the inverse mapping from the vector array index to matrix indices, which is given by the following. The notation being used is that the vector index j is recursively updated to $j^{(1)}, j^{(2)}, \ldots, j^{(p)}, \ldots, j^{(N+i)}$; recall the notation $j_1, j_2, \ldots, j_p, \ldots, j_{N+i}$ are the indices labeling the multidimensional matrix. The inverse mapping is composed of a two-step algorithm, with the matrix index j_p being determined and the vector index j being updated to $j^{(p)}$.

More precisely, the following is the mapping.

$$g[j] \rightarrow g[j_{N+i}] \ldots [j_p] \ldots [j_2][j_1]$$

Using modular arithmetic for carrying out the mapping from a one-dimensional vector array to the matrix indices yields the following

$$\left\{ \begin{array}{l} j_{N+m} = \text{Integer}[(j-1)/(2(M-m)+3)^{N+m-1}]+1 \\ j^{(1)} = j - \text{Integer}[j/(2(M-m)+3)^{N+m-1}] \end{array} \right.$$

$$\vdots$$

$$\left\{ \begin{array}{l} j_p = \text{Integer}[(j^{(p-1)}-1)/(2(M-m)+3)^{p-1}]+1 \\ j^{(p)} = j^{(p-1)} - \text{Integer}[j^{(p-1)}/(2(M-m)+3)^{p-1}] \end{array} \right.$$

$$\vdots$$

$$\left\{ \begin{array}{l} j_1 = j^{(N+m-1)} \\ j^{(N+m)} = j^{(N+m-1)} - j^{(N+m-1)} = 0 \end{array} \right. \tag{16.34}$$

Note the inverse mapping stops after $N+m$ steps, as indeed it must as this is the total number of forward interest rates at calendar time m. The inverse map returns all the $N+m$ indices j_p of the matrix representation $g[j_{N+m}] \ldots [j_2][j_1]$ from the vector index j of the vector array $g[j]$.

In summary, at step m, the values of American option prices are evaluated and stored in a vector array $g_m[(2(M-m)+3)^{N+m}]$. Evolving one step from τ_m to τ_{m+1}, in order to evaluate the trial value of $g_{m+1}[j]$, one needs to first map the vector index j to matrix indices j_p, $p = 1, 2, \ldots, N+m$ using Eq. (16.34). The matrix indices are needed for tracking those option prices at step m that are required for deriving $g_{m+1}[j]$ on the grid points. Then, one reverts back to the vector array index from these matrix indices by Eq. (16.33), and, furthermore, we obtain the corresponding option values at step m. The recursion process in Eqs. (16.15), (16.16), and (16.30) is then performed to obtain the trial value of $g_{m+1}[j]$.

Completing one recursion step results in trial values $g_I[2(M-m)+3]_{N+m}$ for the $(m+1)$th step. The grid points are *dynamic* in nature since one more forward interest rate, namely $f_{m+1,N+m+1}$, has to be *added* at the $(m+1)$th step, as shown in Figure 16.6. The dimension of the matrix of trial values has to be increased from $N+m$ to $N+m+1$ dimensions, that is $g[j_{N+m}]\ldots[j_1] \to g[j_{N+m+1}][j_{N+m}]\ldots[j_1]$. The new forward interest rate does not directly influence the option values, but enters only through the scaling function $B(\tau_{m+1},1)$.

The expanded matrix has to be assigned numerical values for the new index j_{N+m+1} in the range of 1 to $2(M-((m+1)-1))+1 = 2(M-m)+1$. The way this is done is to make the values of the expanded matrix *independent* of the new forward interest rate; in other words, the following assignment is made for the initial trial option price

$$g[j_{N+m+1}][j_{N+m}]\ldots[j_1] \equiv g[j_{N+m}]\ldots[j_1]$$
$$1 \le j_{N+m+1} \le 2(M-m)+1 \qquad (16.35)$$

For vector array representation, j now takes additional values from $[2(M-m)+1]^{N+m}+1$ to $[2(M-m)+1]^{N+m+1}$. The option values of the vector array for the new values of j, similar to Eq. (16.35), are made independent of the new value of j; hence

$$g[j] \equiv g[j-[2(M-m)+1]^{N+m}]$$
$$\text{where } [2(M-m)+1]^{N+m}+1 \le j \le [2(M-m)+1]^{N+m+1} \quad (16.36)$$

The mapping in Eq. (16.36), in going from the left-hand to the right-hand side of the equation, *shifts*, by a constant, the argument of the new index j and thus brings it back into the old range; as j runs through the (additional) new values, the expanded vector takes values from the old array that are a constant shift from the new values of j. It can be seen by inspection that Eq. (16.36) assigns values to the expanded vector array consistent with the labeling of the new matrix elements given in Eq. (16.35).

The trial option value is compared with the payoff value at the $(m+1)$th step. The final value of $g[j_{N+m+1}][j_{N+m}]\ldots[j_1]$ at the $(m+1)$th step is the higher of the two values, which is the American option at the grid points.

The price of the American option at present time is obtained by repeating the recursion until $m = M+1$. The European option price is given by the same algorithm, but without making the comparison with the payoff value at each step of the recursion.

To start the numerical algorithm, the initial forward interest rates and all the parameters in the lattice Lagrangian have to be specified. The numerical algorithm is given as follows:

- Input initial forward interest rate curve and parameters.
- Generate the payoff for maturity time $i = 1$ and store, for European and American options, in the vector arrays Ge_{old} and Ga_{old}, respectively.
- For $i = 2$ to $M + 1$

  1 Recurse one step back from Ge_{old} and Ga_{old} to get trial values Ge_{new} and Ga_{new} using Eqs. (16.15), (16.16), and (16.30).
  2 Expand Ge_{new} and Ga_{new} from $N + i - 1$ to $N + i$ dimension so as to address the dynamics of the grid using Eq. (16.36).
  3 Compute the payoff value at step i without discounting and store the result in Ga_{old}.
  4 Compare Ga_{old} with Ga_{new} and store the larger one in Ga_{old}. Replace values in Ge_{old} with values in Ge_{new}.
  5 End for.

Option values are assigned to a vector array for each point in the forward interest rate grid. The length of the vector array can be very large, and frequently addressing the elements of the array may cause problems of over the stack or even give wrong values. However, programming languages, in particular C++, have a feature of dynamically addressing the location of the vector array, which helps to avoid these problems.

Note that the first-order term of the recursion contributes significantly to the final value. Furthermore, the drift in forward bond measure is zero at three monthly lattice of points. The tree structure has to be wide enough to include information about the changes in the value of the forward interest rates. One has the freedom to increase the width of the tree by setting the pre-factor of $\delta = O(s)$. Since σ is the volatility for a daily change of the forward interest rate, to obtain the lattice volatility s the real days in each step must be multiplied into σ.

The above numerical algorithm yields, for both the American option and the European option, the price for only one value of the initial forward interest rates. In order to get values for a time series or values depending on different values of the various parameters, the entire algorithm needs to be repeated.[9]

---

[9] The European option price is always evaluated (at the same time as the American option) for carrying out consistency checks on the numerical results.

16.8 American caplet: numerical results

A caplet on Libor and maturing when the caplet becomes operational is analyzed. The initial forward interest rates as well as the volatility function are taken from the Libor market; the propagator $\tilde{D}_{m,jk}$ is assigned numerical values taken from caplet data.[10] For simplicity, take the time lattice $\epsilon = 3$ months. The present time for the caplet is taken to be from 12 September 2003 to 7 May 2004, with maturity at fixed time on 12 December 2004. The American option on the caplet is allowed early exercise at only *five* fixed times.

For M time steps and $N + 1$ forward interest rates in the first step, at step m there are $Q = (2(M - m) + 3)^{N+m}$ option prices that need to be determined. The total number of option prices for the whole algorithm is $\mathcal{O} = \sum_{m=1}^{M+1}(2(M-m)+3)^{N+m}$. Thus for a caplet at 12 September 2003, $N = 0$ and $M = 5$ (since $M = t_*/\epsilon$ and $\epsilon = 3$ months), the number of option prices that need to be determined is $\mathcal{O} = 1,304$. The total number of option prices $\mathcal{O}$ increases rapidly with increasing M, with $\mathcal{O} = 14,758,719$ for $M = 10$.

The caplet tree of the (relevant) forward interest rates is built with $\delta = 2s$; all computations are carried out only for the values of the forward interest rates taking values in the tree. Caplet volatility is taken from the market by moving average on the historical data, and at 12 September 2003 is given in Figure 10.2.

In Chapter 10, the daily prices, from 12 September 2003 to 7 May 2004 of a caplet (option on Eurodollar futures contracts) expiring on 13 December 2004 with a strike price of $98, were computed. The same instrument is studied numerically using the lattice theory of bond forward interest rates [19].

Both European and American caplets are numerically computed. The linear caplet price, given in Eq. (9.36), as well as the numerical algorithm for the American option are derived from the same model of bond forward interest rates. Hence, the analytical expression for the linear European caplet price is used for checking the accuracy of the numerical algorithm.

The American option can be exercised at any time before its expiry day. In principle, one should make the time step ϵ very small and hence M very big; however, doing so would require a huge memory and a very long time to run the program since the possible option values Q for each step would be very large. In Figure 16.8(a) the numerical results of caplet price are shown, and it is seen that the numerical results are quite accurate even for a large value of $\epsilon = 3$ months; for this value of ϵ the program generates 167 daily prices by running for less than two seconds on a desktop. The floorlet numerical price is shown in Figure 16.8(b).

---

[10] Market data are used for pricing caplets to illustrate the empirical application of the numerical algorithm for evaluating the American option.

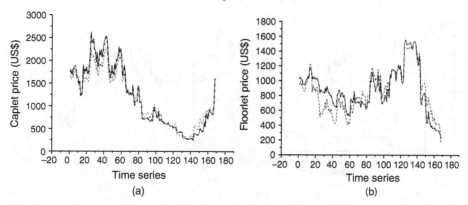

Figure 16.8 (a) American and European caplet prices for fixed maturity at 12 December 2004 versus time t_0 (12 September 2003–7 May 2004). European caplet prices are from the linear caplet formula and the numerical algorithm. The normalized root mean square error for the European caplet price between the numerical value and formula is 7.3%. (b) American and European floorlet prices for fixed maturity at 12 December 2004 versus time t_0 (12 September 2003–7 May 2004) (European floorlet from formula and algorithm). The normalized root mean square error of the European floorlet price from numerical algorithm and formula for floorlet is 8.8%.

Besides accuracy, the numerical results need to be consistent with the general properties of the various options. In particular, the American caplet (put option) must always be more expensive than the European caplet since the American option includes the European option as a special case; however, the American floorlet (call option), in the absence of a dividend, is always equal to the European floorlet [65].

The normalized difference between American and European caplet and floorlet options is shown in Figure 16.9(a). The results are seen to be consistent with the general properties of the American and European options; the normalized difference between the American and European caplet is strictly positive, showing that the American caplet is always more expensive than the European caplet; however, the gap between the American and European floorlet can have negative values, showing that, within the accuracy of the numerical algorithm, their difference is zero.

Although the interval for the time steps ϵ is set equal to three months, one can always decrease this interval to get more accurate results. The American option is more expensive on decreasing the interval ϵ since one needs to pay more to have an option that can be exercised on more occasions before the expiry date.

One can consider the American option being exercised at fixed instants of time as a Bermudan option. A *Bermudan option* can be exercised at a number of pre-fixed times and is equal to a basket of European options, with the difference that once the Bermudan option is exercised, all the remaining European options become invalid.

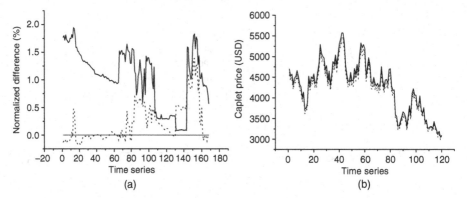

Figure 16.9 (a) The (%) normalized difference between the American and European options for both caplet (unbroken line) and floorlet (dotted line) versus time t_0 (12 September 2003–7 May 2004). Within the numerical accuracy of the computation, the European and American floorlet prices are equal, whereas the caplet prices for the European case are always less than the American case, as expected. (b) Caplet prices versus time t_0 for the European and American options, with four and seven possible exercise times. As required by no-arbitrage, the options with more possible exercise times have higher prices.

A Bermudan option is always cheaper than an American option but more expensive than a European option. Some numerical results for the European, Bermudan, and American caplets are shown in Figure 16.9(b), and are seen to be consistent with the general requirement for these options.

Put–call parity for the European caplet and floorlet are given by Eq. (4.16) as follows

$$caplet(t_0, t_*) - floorlet(t_0, t_*) = \ell V B(t_0, t_* + \ell)[L(t_0, t_*) - K] \quad (16.37)$$

The third argument T, indicating when the caplet becomes operational, is suppressed since the numerical algorithm only studies the price of a caplet and floorlet for $t_* = T$. The result in Figure 16.10(a) verifies that put–call parity is valid for the European option prices generated by the numerical algorithm.

16.9 Numerical results: American coupon bond option

Numerical results for the American coupon bond option are discussed. The main focus of the study of the American coupon bond option is not empirical, but instead is to develop efficient and accurate numerical algorithms. Given the complexity of the instrument, a model is assumed for the volatility of the forward interest rates as well as for the initial value of the forward interest rates. The numerical study analyzes the accuracy of the algorithm.

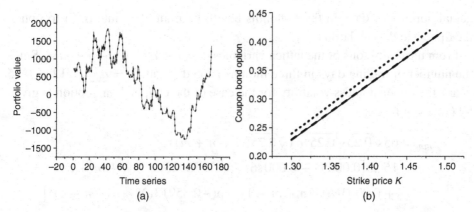

Figure 16.10 (a) Put–call parity for the European caplet and floorlet versus time t_0 (12 September 2003–7 May 2004). The unbroken line is $caplet(t_0, t_*)$ − $floorlet(t_0, t_*)$ and the dashed line is $\ell V B(t_0, t_* + \ell)[L(t_0, t_*) - K]$; the graph shows they are approximately equal, as required by put–call parity. The normalized root mean square error is 3.2%. (b) Prices of a coupon bond American and European put option with payoff function $[K - 0.05B(1, 1.25) - 0.05B(1, 1.5)]_+$ versus strike price K; the dotted line is the American option, dashed line is the European option from the algorithm, and the unbroken line is from the approximate formula. The normalized root mean square error between the numerical value and the formula for the European option is 0.17%.

For the coupon bond American option, the forward interest rate tree is built with the value of $\delta = 6s$; this tree has more elements than the caplet due to the fact that the coupon bond is composed of two zero coupon bonds.

The inital lattice forward interest rate is taken as below

$$f_{mn} = f_0(1 - e^{-\lambda(n-m)}) \tag{16.38}$$

where f_0 is a pre-factor used to make the magnitude the same as the market forward interest rates; let $f_0 = 0.1$ and choose $\lambda = 1$ so that f_{mn} is of the order of 10^{-2}.

Following Bouchaud and Matacz [27], for $\theta = x - t$, the volatility is taken to have the following form (the parameters are fixed by historical forward interest rate data)

$$\sigma(\theta) = 0.00055 - 0.00026 \exp(-0.71826(\theta - \theta_{\min}))$$
$$+ 0.0006(\theta - \theta_{\min}) \exp(-0.71826(\theta - \theta_{\min})) \tag{16.39}$$

where $\theta_{\min} = 3$ months. The volatility of the daily changes in the forward interest rates is given from historical data by $\sigma^2(t, \theta) = \langle \delta f^2(t, \theta) \rangle_c$, $\delta f(t, \theta) = f(t+1, \theta) - f(t, \theta)$. Thus, in building the tree, where each step ϵ is three months, the actual number of days in three months has to be multiplied into $\sigma(\theta)$ to obtain the dimensionless volatility s_{mn}.

From the definitions of the lattice variables $\theta = x - t \to (N - n + m)\epsilon$. Since the number of trading days in three months is 65 days and $\epsilon = a = 3/12 = 0.25$ years, the dimensionless volatility for the case of the coupon bond option is given by $(\theta_{min} = \epsilon)$

$$
\begin{aligned}
s_{mn} &= 65\sqrt{0.25}\sqrt{0.25}\ \sigma(\sqrt{0.25}(N - n + m)) \\
&= 16.25\big[0.00055 - 0.00026\exp(-0.35913(N - n + m - 1), \\
&\quad + 0.0003(N - n + m - 1)\exp(-0.35913(N - n + m - 1))\big]
\end{aligned}
$$

The stiff propagator was used with parameters taken to have the following values $\tilde{\lambda} = 1.790/\text{year}$; $\tilde{\mu} = 0.403/\text{year}$; $\eta = 0.34$.

The numerical study considers $c_1 B(t_*, 1/4) + c_2 B(t_*, 1/2)$ as the coupon bond option that matures in one year's time, that is $t_* - t_0 = t_* = 1$ year and the coupon bond has a duration of six months with three-monthly coupons paid twice; the fixed coupon rate is taken to be equal to c and the principal amount equal to 1. Thus the put option payoff function at time $t_* = 1$ year is given by

$$
\mathcal{P}(t_*) = \left(K - \sum_{i=1}^{2} c_i B(t_*, T_i) \right)_+ \tag{16.40}
$$

where $T_1 = 1.25$ year, $T_2 = 1.5$ year, $c_1 = c$, and $c_2 = c+1$. Taking the $c = 0$ limit converts the coupon bond into a zero coupon bond. For this coupon bond option, $N = 2$ and set $M = 4$; hence, the total number of option prices to be evaluated is equal to $\mathcal{O} = 1,293$. This number increases more rapidly than for the caplet case, and for $M = 10$ it reaches $\mathcal{O} = 118,507,277$.

Comparing the numerical value of the European coupon bond option with the approximate formula given in Eq. (11.39) provides a check on the numerical result. The approximate formula is an expansion in the volatility of the forward interest rates s_{mn}, and as long as this volatility is small, the numerical and approximate results should agree.

For the coupon bond option that is being studied numerically the coefficient C_2 is given by Eq. (11.32); the approximate price for the European coupon bond option is given by Eq. (11.39) as follows

$$
C_2 = \sum_{ij=1}^{2} J_i J_j \left[G_{ij} + \frac{1}{2}G_{ij}^2 \right] + O(G_{ij}^3) \tag{16.41}
$$

Note $J_i = c_i F_i$, with $F_1 = 0.982321$ and $F_2 = 0.963426$.

Table 16.2 *The correlators G_{ij} between different forward bond prices.*

G_{ij}	$i = 1$	$i = 2$
$j = 1$	1.669×10^{-8}	3.624×10^{-8}
$j = 2$	3.624×10^{-8}	7.924×10^{-8}

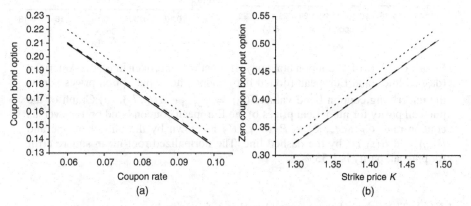

Figure 16.11 Prices of the American coupon bond (dotted line), which are always greater than the European option, from approximate formula (unbroken line) and algorithm (dashed line) are shown. (a) Payoff given by $\left[1.3 - cB(1, 1.25) - (c + 1)B(1, 1.5)\right]_{+}$ versus coupon rate c. The normalized root mean square error between the numerical result and formula for the European option is 0.73%. (b) Prices for a *zero coupon bond* with payoff $\left[K - B(1, 1.5)\right]_{+}$ versus strike price K. The normalized root mean square error between the numerical value and formula for the European option is 0.16%.

The numerical values for G_{ij}, the correlator of the forward bond prices, are computed using Eq. (11.28), with the results given in Table 16.2.

Numerical results for the American coupon bond option prices with changing strike price K and coupon rate c are given in Figures 16.10(b) and 16.11(a). The numerical value of the European coupon bond option is close to the approximate formula given in Eq. (11.39). As required by consistency, the American put option always has a higher price than the European put option. For completeness, the special case of the American option on a zero coupon bond, obtained by setting $c = 0$ in Eq. (16.40), is given in Figure 16.11(b) and shows all the features required by the consistency of the option prices.

The algorithm has been checked for internal consistency by plotting the prices of the American and European coupon bond put options against the coupon rate c

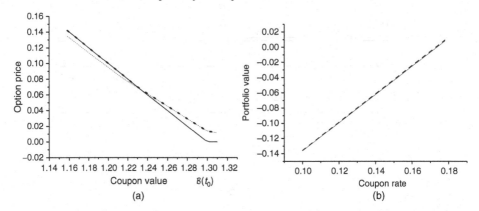

Figure 16.12 (a) The coupon bond payoff function (unbroken line), the American (dashed line), and European (dotted line) coupon bond put option prices versus the underlying coupon bond value $\mathcal{B}(t_0) = \sum_{i=1}^{2} c_i B(t_0, T_i)$. (b) Graph of the put–call parity for numerical prices of the European coupon bond options versus coupon rate. $C_E(t_0, t_*, K) - P_E(t_0, t_*, K)$ is shown by the unbroken line and $\mathcal{B}(t_0) - K B(t_0, t_*)$ by the dashed line. The normalized root mean square error for put–call parity is 1.53%.

and $\mathcal{B}(t_0)$. The price of the coupon bond at t_0 is given by

$$\mathcal{B}(t_0) \equiv \sum_{i=1}^{2} c_i B(t_0, T_i)$$

Figure 16.12(a) shows that the results are consistent with the general properties of these options, with the price of the American option always being higher than the European option; the American option joins the payoff with the same slope as the payoff function for small coupon bond value $\mathcal{B}(t_0)$; for large values of $\mathcal{B}(t_0)$, as expected [59], the American option tends to the European option.

Another check of the algorithm is the put–call parity for European coupon bond options, which obeys the following equation

$$C_E(t_0, t_*, K) - P_E(t_0, t_*, K) = \mathcal{B}(t_0) - K B(t_0, t_*) \tag{16.42}$$

The numerical results in Figure 16.12(b) show that the numerical algorithm satisfies put–call parity with a normalized root mean square error of 1.53%.

16.10 Put–call for American coupon bond option

In analogy with Eq. (16.42) and the inequalities for the case of equity given in Eq. (16.12), one can consider the following inequalities for the case of American coupon bond put P_A and call C_A options

$$\mathcal{B}(t_0) - K \leq C_A(t_0, t_*, K) - P_A(t_0, t_*, K) \leq \mathcal{B}(t_0) - B(t_0, t_*)K : \quad \text{incorrect}$$

(16.43)

On plotting the three expressions in the above equation, as shown in Figure 16.13(a), it is seen that the put–call inequalities are *incorrect*.

Instead of the above incorrect inequality, a *conjecture* is made that the American coupon bond options satisfy the following modified inequalities. The coupon bond value $\mathcal{B}(t_0)$ at time t_0 in Eq. (16.43) is replaced by the forward price of the payoff function, namely $\mathcal{F}(t_0)$ – the payoff function at time t_0 – given by

$$\mathcal{F}(t_0) \equiv \sum_{i=1}^{N} c_i F(t_0, t_*, T_i)$$

The following inequalities have been conjectured in [17]

$$\mathcal{F}(t_0) - K \leq C_A(t_0, t_*, K) - P_A(t_0, t_*, K) \leq \mathcal{F}(t_0) - K B(t_0, t_*) \quad (16.44)$$

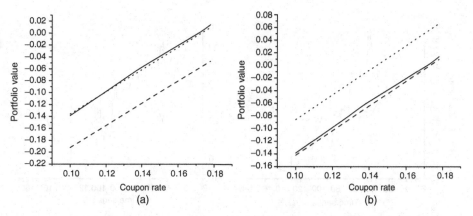

Figure 16.13 (a) Numerical result showing the *incorrectness* of put–call inequalities for American coupon bond options. $C_A(t_0, t_*, K) - P_A(t_0, t_*, K)$ is shown by the unbroken line, $\mathcal{B}(t_0) - K$ is shown by a dashed line, and $\mathcal{B}(t_0) - B(t_0, t_*)K$ is shown by the dotted line. The dotted line crossing the unbroken line violates the inequality given in Eq. (16.43). (b) Numerical result confirming the *conjectured* put–call inequalities of the American coupon bond option. $C_A(t_0, t_*, K) - P_A(t_0, t_*, K)$ is shown by the unbroken line, $\mathcal{F}(t_0) - K$ by the dashed line, and $\mathcal{F}(t_0) - K B(t_0, t_*)$ by the dotted line. The fact that the lines do not cross verifies the conjectured inequalities in Eq. (16.44).

As shown in Figure 16.13(b), the numerical results show the American coupon bond options do, in fact, satisfy the conjectured inequalities. A no-arbitrage argument can be shown to yield inequalities given in Eq. (16.44).

In analogy with the American coupon bond options, the following inequalities were conjectured in for the American caplet and floorlet [17]

$$F(t_0, t_*, t_* + \ell)[L(t_0, t_*) - K] \leq caplet(t_0, t_*) - floorlet(t_0, t_*)$$

$$\leq F(t_0, t_*, t_* + \ell)[L(t_0, t_*) - B(t_0, t_*)K] \quad (16.45)$$

which can also be expressed as follows

$$caplet(t_0, t_*) - floorlet(t_0, t_*) - F(t_0, t_*, t_* + \ell)[L(t_0, t_*) - K] \geq 0$$

$$F(t_0, t_*, t_* + \ell)[L(t_0, t_*) - B(t_0, t_*)K] - caplet(t_0, t_*) - floorlet(t_0, t_*) \geq 0$$

Figure 16.14(a) shows that the conjectured inequalities do, indeed, hold for the numerical prices of the American caplet options.

The conjecture for the American caplet and floorlet is not as significant as the one for the American coupon bond option since the numerical prices also satisfy

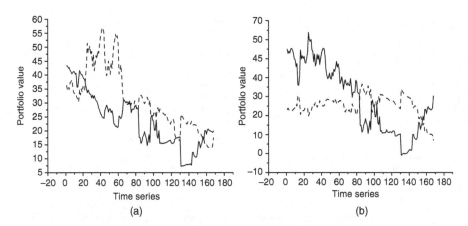

Figure 16.14 From both the diagrams one can see that all the expressions are positive, as is required for the two conjectured put–call inequalities to hold. (a) $caplet(t_0, t_*) - floorlet(t_0, t_*) - F(t_0, t_*, t_* + \ell)[L(t_0, t_*) - K]$ is plotted as the unbroken line and $F(t_0, t_*, t_* + \ell)[L(t_0, t_*) - B(t_0, t_*)K] - caplet(t_0, t_*) - floorlet(t_0, t_*)$ is shown by the dotted line. (b) $caplet(t_0, t_*) - floorlet(t_0, t_*) - B(t_0, t_* + \ell)[L(t_0, t_*) - K]$ is plotted as the unbroken line and $B(t_0, t_* + \ell)[L(t_0, t_*) - B(t_0, t_*)K] - caplet(t_0, t_*) - floorlet(t_0, t_*)$ is shown by the dotted line.

the inequalities that are similar to the equity inequalities in Eq. (16.12), namely

$$B(t_0, t_* + \ell)[L(t_0, t_*) - K] \leq caplet(t_0, t_*) - floorlet(t_0, t_*)$$

$$\leq B(t_0, t_* + \ell)[L(t_0, t_*) - B(t_0, t_*)K] \quad (16.46)$$

The numerical results for the American caplet and floorlet show that they obey the inequalities as shown in Figure 16.14(b).

16.11 Summary

The procedures that are presently widely used by practitioners for calculating American coupon bond options are all based on the HJM model and use variants of the binomial tree to build the tree for the interest rates and coupon bonds; the complexity of the tree in the HJM model is determined by how many factors are driving the interest rates [63].

The approach of quantum finance for evaluating American (coupon bond) options is radically different. One starts with the Hamiltonian and its pricing kernel, for which a model is written from first principles. The payoff function for the American option is propagated (backwards) on a time lattice using the pricing kernel. Propagating the payoff an infinitesimal step backwards in calendar time requires the Hamiltonian. Taking an infinitesimal time step entails numerically performing the path integral, which generates the trial values of the American option on a tree of forward interest rates. At each step, the trial American option value obtained is compared with the payoff function.

The bond forward interest rates defined in Chapter 5 were the basis of all calculations; the simplicity of the Gaussian bond forward interest rates allowed the issues linked to the path dependence of American options to be analyzed without additional complications that arise in studying options in the nonlinear Libor Market Model. The (lattice) forward interest rates $f(t, x) \rightarrow f_{mn}$ are directly involved in the recursion equation. f_{mn} cannot be replaced by a collection of white noise, as is the case for the HJM model, or by the velocity quantum field $\mathcal{A}(t, x)$ – as this would invalidate the entire numerical algorithm. The nontrivial correlations between the changes in the forward interest rates, encoded in the propagator $\tilde{D}_{m,jk}$, are easily incorporated into the recursion equation, as can be seen from Eq. (16.29). The prices of the American option for caplets and coupons were shown to be consistent, obeying all the constraints that follow from the principles of finance.

The comparison of the numerical values of the European caplet and coupon bond options with the analytical linear caplet formula and the approximate coupon bond option formula, respectively, showed that the numerical algorithm is not very accurate for caplets but reaches an accuracy of 99% for the coupon bond options. The numerical results provide an estimate of the accuracy of the volatility perturbation expansion for the coupon bond option price.

The entire computation was carried out on a desktop computer and on a small lattice of about 10 to 20 points; computing the option price for each set of parameters required only a few seconds of computation time. The crude approximations used gave excellent results, showing the possibility of using such algorithms for practical applications.

A conjecture for the put–call inequalities was made for the American coupon bond option and caplet – based on the analysis of the numerical results. The fact that these inequalities seem to hold quite robustly for the numerical results provides evidence for the conjecture.

In conclusion, quantum finance provides a useful and practical framework for developing and implementing efficient and accurate numerical algorithms for pricing American interest rate and coupon bond options.

17

Hamiltonian derivation of coupon bond options

Coupon bond European and barrier options are studied in the Hamiltonian frame-work of quantum finance. In this chapter, the bond forward interest rates' state space and Hamiltonian are used for calculating coupon bond barrier option price. The bond forward interest rates $f(t, x)$ are given by Eq. (5.1)

$$\frac{\partial f(t, x)}{\partial t} = \alpha(t, x) + \sigma(t, x)\mathcal{A}(t, x)$$

All the options studied in this chapter mature at a fixed future time t_*. The forward bond numeraire $B(t, t_*)$, is the most suitable choice; from Eqs. (9.3) and (15.45), the forward bond numeraire yields the following drift

$$\alpha(t, x) = \sigma(t, x) \int_{t_*}^{x} dx' \mathcal{D}(x, x'; t)\sigma(t, x')$$

For the forward bond numeraire, the price of option $C(t_0, t_*, T, K)$ that matures at t_*, from Eq. (3.8), is given by

$$C(t_0, t_*, T, K) = B(t_0, t_*)E\big[\mathcal{P}_*\big] \qquad (17.1)$$

where $\mathcal{P}_*$ is the payoff function.

The coupon bond European and barrier options are calculated using the bond evolution operator, which is the time integrated bond Hamiltonian discussed in Section 15.12. The bond barrier option is encoded in the state space and eigen-functions of the bond evolution operator. A calculation, using an *overcomplete* set of bond eigenfunctions, yields an approximate price for the coupon bond barrier option. Many derivations in this chapter hinge on some remarkable properties of the Dirac-delta function.

17.1 Introduction

The Hamiltonian framework for option pricing has been discussed in Chapter 15 and, in this chapter, this approach is extended to the (more complex) case of coupon bond European and barrier options.

The double barrier option, discussed in Section 3.4, is 'knocked out' (terminated with zero value) if the price of the underlying instrument exceeds or falls below pre-set limits. It is shown that, in general, the constraint function for a coupon bond barrier option can – to a good approximation – be linearized.

The volatility $\sigma^2(t, x)$ of the bond forward interest rates $f(t, x)$ is a small quantity, of the order of 10^{-4}/year. Hence, as discussed in Section 3.14, $\sigma^2(t, x)$ provides a small parameter for generating a volatility expansion for the coupon bond barrier options. The volatility expansion is carried out only to $O(\sigma^2)$ as this is sufficient for developing the techniques that are required for carrying out a computation in the Hamiltonian framework.[1]

The volatility perturbation expansion of the European and barrier option price has the following steps.

- Obtaining the volatility expansion of the payoff function and option price.
- Finding the eigenfunctions of the bond evolution operator. The barrier option is realized by imposing boundary conditions on the eigenfunctions of the bond evolution operator.
- Finding the coefficients for the perturbation expansion of the option price.

17.2 Coupon bond European option price

The European coupon bond option price is derived using the Hamiltonian formulation as a warm-up exercise for the more complex derivation of the barrier option. The price of the coupon bond European option has been obtained to $O(\sigma^4)$ in Chapter 11 using a Feynman perturbation expansion and provides a check for the calculations of this section.

17.2.1 Volatility expansion

The European coupon bond option matures at t_* with strike price K; its price at earlier t_0 is given by $C_E(t_0, t_*, T, K)$. The coupon bond payoff function $\mathcal{P}_*$ is rewritten to isolate the leading term from terms that are higher order in $O(\sigma)$. Similar

---

[1] There is also an empirical basis for computing the option price only to $O(\sigma^2)$. In Chapter 12, an empirical study of the Libor market swaption price, retaining only the second-order $O(\sigma^2)$ term of the expansion, showed that the result has a root mean square error of less than 3%.

to the perturbation expansion in Section 11.2, the zero coupon bond $B(t_*, T_I)$ is expanded about $F(t_0, t_*, T_I)$, its forward price at t_0.

$$P_* = \left(\sum_{I=1}^{N} c_I B(t_*, T_I) - K \right)_+ = (F + V - K)_+ \qquad (17.2)$$

$$V = \sum_{I=1}^{N} J_I [B(t_*, T_I) e^{f_I} - 1] = \sum_{I=1}^{N} J_I [e^{-g_I + f_I} - 1] \qquad (17.3)$$

$$B(t_*, T_I) = e^{-g_I}; \quad g_I = \int_{t_*}^{T_I} dx f(t_*, x)$$

$$F(t_0, t_*, T_I) = e^{-f_I}; \quad f_I = \int_{t_*}^{T_I} dx f(t_0, x)$$

$$J_I = c_I e^{-f_I}; \quad F = \sum_I J_I$$

From Eq. (17.2) the payoff is re-written, as in Eq. (3.64), as follows

$$P_* = (F + V - K)_+$$

$$= \int_{Q, \eta} e^{i\eta(V - Q)} (F + Q - K)_+; \quad \int_{Q, \eta} \equiv \int_{-\infty}^{+\infty} dQ \frac{d\eta}{2\pi}$$

V is of $O(\sigma)$, from Eq. (17.3), and is the only random quantity in the payoff function; consider the following expansion of the payoff in powers of σ

$$P_* \simeq \int_{Q, \eta} e^{-i\eta Q} (F + Q - K)_+ \left\{ 1 + i\eta V - \frac{1}{2} \eta^2 V^2 + O(\sigma^3) \right\} \qquad (17.4)$$

Eqs. (17.1) and (17.4), similar to the analysis in Section 3.14, yield the following expansion for the coupon bond option price

$$\frac{C(t_0, t_*, T, K)}{B(t_0, t_*)} = E[P_*]$$

$$\simeq \int_{Q, \eta} e^{-i\eta Q} (F + Q - K)_+ \left\{ E[1] + i\eta E[V] - \frac{1}{2} \eta^2 E[V^2] + O(\sigma^3) \right\}$$

$$\simeq \int_{Q, \eta} e^{-i\eta Q} (F + Q - K)_+ \left\{ C_0 + i\eta C_1 - \frac{1}{2} \eta^2 C_2 + O(\sigma^3) \right\} \qquad (17.5)$$

Option price given by Eq. (17.5) holds for a wide class of options and, in particular, for both the European and barrier coupon bond options.

The coefficients of the expansion, from Eq. (17.3), are given by

$$C_0 = E[1] \tag{17.6}$$

$$C_1 = E[V] = \sum_{I=1}^{N} J_I E[e^{-g_I + f_I} - 1]$$

$$= \sum_{I=1}^{N} J_I (C_I - C_0); \quad C_I = E[e^{-g_I + f_I}] \tag{17.7}$$

$$C_2 = E[V^2] = E\left[\left\{ \sum_{I=1}^{N} J_I (e^{-g_I + f_I} - 1) \right\}^2 \right]$$

$$= \sum_{I,K=1}^{N} J_I J_K (C_{IK} - 2C_I + C_0); \quad C_{IK} = E[e^{-g_I - g_K + f_I + f_K}] \tag{17.8}$$

17.2.2 Bond evolution operator and eigenfunctions

The European coupon bond option price at time t_0, from Eq. (15.77), is given by

$$C_E(t_0, t_*, T, K, \mathbf{f}) = B(t_0, t_*) E[\mathcal{P}_*]$$

$$= B(t_0, t_*) \langle \mathbf{f} | e^{-W} | \mathcal{P}_* \rangle \tag{17.9}$$

The evolution operator W driving the bond price is expressed in terms of the bond variables g_I; $I = 1, 2, \ldots, N$ and is given, from Eq. (15.82), as follows

$$W = -\frac{1}{2} \sum_{i,j=1}^{N} G_{ij} \frac{\partial^2}{\partial g_i \partial g_j} - \sum_{i=1}^{N} \beta_i \frac{\partial}{\partial g_i} \tag{17.10}$$

$$G_{ij} = G(T_i, T_j); \quad \beta_i = \beta(T_i) = \frac{1}{2} G_{ii} \tag{17.11}$$

The dual vector $\langle \mathbf{f} |$ is constructed from $F(t_0, t_*, T_i)$, the forward bond prices at t_0, as follows

$$f_i \equiv \int_{t_*}^{T_i} dx f(t_0, x); \quad \langle \mathbf{f} | \equiv \langle f_1, f_2, \ldots, f_N | = \prod_{i=1}^{N} \langle f_i | \tag{17.12}$$

The eigenfunctions of W are given by

$$\psi_{\mathbf{p}}[\mathbf{f}] = e^{i \sum_i p_i f_i} \equiv e^{i\mathbf{pf}} = \langle \mathbf{f} | \mathbf{p} \rangle$$

$$W\psi_{\mathbf{p}}[\mathbf{f}] = \langle \mathbf{f} | W | \mathbf{p} \rangle = \left(S_E - i \sum_{i=1}^{N} \beta_i p_i \right) \psi_{\mathbf{p}}[\mathbf{f}]$$

$$S_E \equiv S_E(\mathbf{p}) = -\frac{1}{2} \sum_{ij=1}^{N} p_i G_{ij} p_j \qquad (17.13)$$

The orthogonality and completeness of the eigenfunctions for the 'momentum' $\mathbf{p}$, from Eq. (A.8) and of the bond basis $\mathbf{g}$, from Eq. (A.6) are given by

$$|\mathbf{p}\rangle = |p_1, p_2, \dots, p_N\rangle; \quad \int_{\mathbf{p}} |\mathbf{p}\rangle\langle\mathbf{p}| = \mathcal{I}; \quad \int_{\mathbf{p}} \equiv \prod_{i=1}^{N} \int_{-\infty}^{+\infty} \frac{dp_i}{2\pi} \qquad (17.14)$$

$$|\mathbf{g}\rangle = |g_1, g_2, \dots, g_N\rangle; \quad \int_{\mathbf{g}} |\mathbf{g}\rangle\langle\mathbf{g}| = \mathcal{I}; \quad \int_{\mathbf{g}} \equiv \prod_{i=1}^{N} \int_{-\infty}^{+\infty} dg_i \qquad (17.15)$$

$$\langle\mathbf{p}|\mathbf{p}'\rangle = \prod_{i=1}^{N} [2\pi\delta(p_i - p_i')]; \quad \langle\mathbf{g}|\mathbf{g}'\rangle = \prod_{i=1}^{N} \delta(g_i - g_i')$$

Inserting, in Eq. (17.9), the completeness equation for both the $|\mathbf{p}\rangle$ and $|\mathbf{g}\rangle$ yields

$$C_E(t_0, t_*, T, K) = B(t_0, t_*) \langle \mathbf{f} | e^{-W} | \mathcal{P}_* \rangle$$

$$= B(t_0, t_*) \int_{\mathbf{p}, \mathbf{g}} \langle \mathbf{f} | e^{-W} | \mathbf{p} \rangle \langle \mathbf{p} | \mathbf{g} \rangle \langle \mathbf{g} | \mathcal{P}_* \rangle$$

$$= B(t_0, t_*) \int_{\mathbf{p}, \mathbf{g}} e^{S_E} e^{i(\mathbf{f} + \beta - \mathbf{g})\mathbf{p}} \mathcal{P}_*[\mathbf{g}]$$

The coefficients for the European option are denoted by $C_{E,0}$, $C_{E,1}$, and $C_{E,2}$. The expansion for $E[\mathcal{P}_*]$ given in Eq. (17.5) yields the following expression for the coefficients

$$C_{E,0} = E[1] = \int_{\mathbf{p}, \mathbf{g}} e^{S_E} e^{i(\mathbf{f} + \beta - \mathbf{g})\mathbf{p}} \qquad (17.16)$$

$$C_{E,1} = E[V] = \int_{\mathbf{p},\mathbf{g}} e^{S_E} e^{i(\mathbf{f}+\beta-\mathbf{g})\mathbf{p}} V \qquad (17.17)$$

$$C_{E,2} = E[V^2] = \int_{\mathbf{p},\mathbf{g}} e^{S_E} e^{i(\mathbf{f}+\beta-\mathbf{g})\mathbf{p}} V^2 \qquad (17.18)$$

$$V = \sum_{I=1}^{N} J_I [e^{-g_I + f_I} - 1]$$

17.2.3 Coefficients of perturbation expansion

The expansion coefficients required for evaluating the option price, as given in Eq. (17.5), are computed from the bond evolution eigenfunctions.

The first coefficient is given from Eq. (17.16)

$$C_{E,0} = \int_{\mathbf{p},\mathbf{g}} e^{S_E} e^{i(\mathbf{f}+\beta-\mathbf{g})\mathbf{p}}$$

$$= \int_{\mathbf{p}} e^{S_{E,0}} e^{i(\mathbf{f}+\beta)\mathbf{p}} \prod_{i}[2\pi\delta(p_i)] = e^{\tilde{S}_{E,0}}$$

$$\tilde{S}_{E,0} = S_E(p_i = 0) = 0$$

$$\Rightarrow C_{E,0} = 1$$

The second coefficient, from Eq. (17.17) is given by

$$C_{E,1} = \sum_{I} J_I \int_{\mathbf{p},\mathbf{g}} e^{S_{E,1}} e^{i(\mathbf{f}+\beta-\mathbf{g})\mathbf{p}}[e^{-g_I + f_I} - 1]$$

$$= \sum_{I} J_I \int_{\mathbf{p},\mathbf{g}} e^{S_{E,1}} [e^{-i\mathbf{g}\mathbf{p}} e^{-g_I - \beta_I} - C_{E,0}]$$

$$= \sum_{I} J_I \int_{\mathbf{p}} e^{S_{E,1}} e^{-\beta_I} \left\{ \prod_{i \neq I}[2\pi\delta(p_i)]2\pi\delta(p_I + i) - 1 \right\}$$

$$= e^{\tilde{S}_{E,1} - \beta_I}$$

$$\tilde{S}_{E,1} = S_E(p_I = -i; p_j = 0, j \neq I) = \frac{1}{2}G_{II}$$

$$\Rightarrow C_{E,1} = \sum_{I} J_I[e^{\frac{1}{2}G_{II}} e^{-\beta_I} - 1] = 0$$

since, from Eq. (17.11), $\beta_I = G_{II}/2$. The third coefficient, from Eq. (17.18), is given by

$$
C_{E,2} = \sum_{IK} J_I J_K \int_{\mathbf{p,g}} e^{S_{E,2}} e^{i(\mathbf{f}+\boldsymbol{\beta}-\mathbf{g})\mathbf{p}} [e^{-g_I + f_I} - 1][e^{-g_K + f_K} - 1]
$$

$$
= \sum_{IK} J_I J_K \int_{\mathbf{p,g}} e^{S_{E,2}} e^{-i\mathbf{gp}} [e^{-g_I - g_K - \beta_I - \beta_K} - e^{-g_I - \beta_I} - e^{-g_K - \beta_I} + 1]
$$

Using the results for $C_{E,0}$ and $C_{E,1}$ yields

$$
C_{E,2} = \sum_{IK} J_I J_K \left\{ \int_{\mathbf{p,g}} e^{S_{E,2}} e^{-i\mathbf{gp}} e^{-g_I - g_K - \beta_I - \beta_K} - 1 \right\}
$$

$$
= \sum_{IK} J_I J_K \left\{ \int_{\mathbf{p}} e^{S_{E,2}} e^{-\beta_I - \beta_K} \prod_{i \neq I,K} [2\pi\delta(p_i)] 2\pi\delta(p_I + i) 2\pi\delta(p_K + i) - 1 \right\}
$$

$$
= \sum_{IK} J_I J_K [e^{\tilde{S}_{E,2} - \beta_I - \beta_K} - 1]
$$

The momentum delta functions yield $p_I = -i = p_K$ with all other components $p_i = 0$; $i \neq I, K$. Hence, from Eq. (17.13),

$$
\tilde{S}_{E,2} = S_{E,2}(p_I = -i = p_K)
$$

$$
= -\frac{1}{2} \sum_{ij=1}^{N} p_i G_{ij} p_j = G_{IK} + \frac{1}{2} G_{II} + \frac{1}{2} G_{KK} \tag{17.19}
$$

Combining Eq. (17.19) with $\beta_I = G_{II}/2$ yields, from Eq. (17.19)

$$
C_{E,2} = \sum_{IK} J_I J_K [e^{G_{IK}} - 1] \simeq \sum_{IK} J_I J_K G_{IK} + O(\sigma^3) \tag{17.20}
$$

The results agree, as expected, with the expression given in Eq. (11.29) that was obtained by path integration.

Collecting all the results yields

$$
C_{E,0} = 1; \quad C_{E,1} = 0; \quad C_{E,2} = \sum_{IK} J_I J_K G_{IK} + O(\sigma^3)
$$

Once the coefficients have been computed, the European coupon option price is given by Eq. (3.69).

17.3 Coupon bond barrier eigenfunctions

As discussed in Section 3.4, barrier options are either knock-out or knock-in. The price of only the knock-out options will be evaluated, since the price of the knock-in can be obtained using the relation given in Eq. (3.2).

A double barrier knock-out option has the same payoff $|\mathcal{P}_*\rangle$ as an European option. The option has the *additional conditions* that the option is terminated with zero payoff if, at any instant before the option matures, the price of the underlying coupon bond $\mathcal{B}(t_*, T)$ exceeds a certain maximum value, say U or falls below a minimum value, say L. Figure 17.1 shows the payoff function of the coupon bond barrier option from time t_0 till it matures at time t_*.

The price of the coupon bond barrier option is given by

$$\mathcal{C}_B(t_0, t_*, T, K, \mathbf{f}) = B(t_0, t_*)\langle \mathbf{f}|e^{-W}|\mathcal{P}_*\rangle\big|_{\text{Barrier}} \qquad (17.21)$$

$$L \le \sum_{i=1}^{N} c_i F(t, t_*, T_i) \le U; \quad t_0 \le t \le t_*$$

A coupon bond that is allowed to have its price only in the range of $[L, U]$ is identical to a quantum particle confined in an interval $[a, b]$. In quantum mechanics, the particle's position, denoted by g, is confined by putting the particle inside an infinite potential well such that the potential $U(g)$ is infinite for position g outside the interval $[a, b]$, as shown in Figure 17.2. $U(g)$ should not be confused with upper barrier U. A particle permanently confined inside a potential well is described by eigenfunctions that are zero for all values of the position *outside* the interval $[a, b]$.

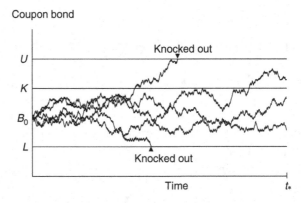

Figure 17.1 The payoff function for a barrier option, with strike price K and maturity time t_*. The initial coupon bond value $\mathcal{B}_0 \in [L, U]$ lies within the barrier, and the option is knocked out if the coupon bond, at any time before maturity at t_*, takes values outside the barrier. Only trajectories that lie within the barrier contribute to the price of the barrier option.

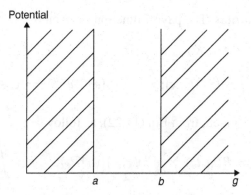

Figure 17.2 The potential well that confines the position of a quantum particle to be inside the interval $[a, b]$.

The barrier condition is incorporated into the barrier option pricing formula – similar to a quantum particle confined to a potential well – by imposing the appropriate boundary conditions on the *eigenfunctions* of the W operator [12, 22]. In particular, for the coupon bond option, let $\langle \mathbf{g} | \psi_n \rangle \equiv \psi_n[\mathbf{g}]$ be a complete set of eigenfunctions of W that satisfy

$$W|\psi_k\rangle = -S_k|\psi_k\rangle; \quad \sum_k |\psi_k\rangle\langle\psi_k| = \mathcal{I}$$

The barrier option is realized by imposing the following two boundary conditions on the eigenfunctions

$$B \equiv \sum_i c_i e^{-g_i}$$

$$\text{B.C.}: \quad \psi_k[\mathbf{g}] = 0 \quad \text{for } B \geq L \text{ and } B \leq U$$

The price of the barrier option is then given by

$$C_B(t_0, t_*, T, K) = B(t_0, t_*)\langle \mathbf{f} | e^{-W} | \mathcal{P}_* \rangle \big|_{\text{Barrier}}$$

$$\Rightarrow C_B(t_0, t_*, T, K) = B(t_0, t_*) \sum_k \langle \mathbf{f} | e^{-W} | \psi_k \rangle \langle \psi_k | \mathcal{P}_* \rangle \qquad (17.22)$$

$$= B(t_0, t_*) \sum_k e^{S_k} \psi_k[\mathbf{f}] \langle \psi_k | \mathcal{P}_* \rangle$$

17.4 Zero coupon bond barrier option price

The zero coupon bond barrier option is the simplest case that illustrates the specific features of the barrier. The option is calculated based on the eigenfunction realization of the barriers. The payoff function is given by

$$\mathcal{P}_* = (e^{-g} - K)_+$$

$$g = \int_{t_*}^T dx f(t_*, x); \quad e^{-g} \in [e^{-b}, e^{-a}] \Rightarrow g \in [a, b]$$

The barrier option is given, from Eq. (17.22), as follows

$$\mathcal{C}_B(t_0, t_*, K) = B(t_0, t_*) \sum_k e^{S_k} \psi_k[f] \langle \psi_k | \mathcal{P}_* \rangle; \quad f = \int_{t_*}^T dy f(t_0, y)$$

$$W \psi_k(g) = -S_k \psi_k(g); \quad \psi_k(a) = 0 = \psi_k(b); \quad a \leq g \leq b$$

The evolution operator, given by Eqs. (17.10) and (15.80), simplifies to

$$W = -\frac{1}{2} G \frac{\partial^2}{\partial g^2} - \beta \frac{\partial}{\partial g}$$

$$G = \int_{t_*}^T dy \int_{t_*}^T dy' \int_{t_0}^{t_*} dt M(y, y'; t); \quad \beta = \int_{t_*}^T dy \int_{t_0}^{t_*} dt \alpha(t, y) = \frac{1}{2} G$$

To incorporate the barrier one solves the Schrodinger eigenfunction equation

$$(W + U(g)) \psi_k(g) = -S_k \psi_k(g)$$

where the potential, as shown in Figure 17.2, is given by

$$U(g) = \begin{cases} 0 & a \leq g \leq b \\ \infty & g \leq a; \, g \geq b \end{cases}$$

Consider the following ansatz for the eigenfunction

$$\langle g | \psi_k \rangle = \psi_k(g) \sim \begin{cases} e^{(ik-\gamma)(g-a)} & a \leq g \leq b \\ 0 & g \leq a; \, g \geq b \end{cases}$$

$$W \psi_k(g) = [\frac{1}{2} G(k^2 - \gamma^2) + ik(G\gamma - \beta) + \beta\gamma] \psi_k(g)$$

γ is chosen to eliminate the term linear in k, which then yields two degenerate solutions $\psi_{\pm k}$. Hence

$$\gamma = \frac{\beta}{G}; \quad \psi_{\pm k}(g) \sim e^{i(\pm k + i\gamma)(g-a)}$$

$$W \psi_{\pm k}(g) = -S_k \psi_{\pm k}(g); \quad -S_k = \frac{1}{2} \left(Gk^2 + \frac{\beta^2}{G} \right) \tag{17.23}$$

To impose the barrier option boundary conditions one superposes the degenerate solutions $\psi_{\pm k}(g)$ to obtain[2]

$$\langle g|\psi_k\rangle = \sqrt{\frac{2}{b-a}}e^{-\gamma g}\sin k(g-a); \quad \langle \psi_k|g\rangle = \sqrt{\frac{2}{b-a}}e^{\gamma g}\sin k(g-a)$$

$$k \equiv k_n = \frac{\pi n}{(b-a)}; \quad n = 1,2,\dots+\infty; \quad \psi_k(a) = 0 = \psi_k(b)$$

Hence, the eigenfunction for all values of g can be written as follows[3]

$$\Psi_k(g) = [\theta(g-a) - \theta(g-b)]\psi_k(g); \quad -\infty \le g \le +\infty \quad (17.24)$$

where

$$\theta(g-a) - \theta(g-b) = \begin{cases} 1 & a < g < b \\ 0 & g \le a; \ g \ge b \end{cases}$$

The eigenfunctions are orthonormal since

$$\langle \psi_{k_n}|\psi_{k_{n'}}\rangle = \frac{2}{b-a}\int_a^b \sin k_n(g-a)\sin k_{n'}(g-a)dg = \delta_{n-n'} \quad (17.25)$$

The Poisson summation formula, given by

$$\sum_{n=-\infty}^{\infty} e^{2\pi inx} = \sum_{n=-\infty}^{\infty} \delta(x-n) \quad (17.26)$$

yields the following

$$\sum_k \langle g|\psi_k\rangle\langle\psi_k|g'\rangle = \frac{2}{b-a}e^{-\gamma(g-g')}\sum_{n=1}^{\infty}\sin k_n(g-a)\sin k_n(g'-a)$$

$$= \frac{1}{2(b-a)}e^{-\gamma(g-g')}\sum_{n=-\infty}^{\infty}\left\{\exp\frac{in\pi}{b-a}(g-g') - \exp\frac{in\pi}{b-a}(g+g'-2a)\right\}$$

$$= \frac{1}{2(b-a)}e^{-\gamma(g-g')}\sum_{n=-\infty}^{\infty}\left[\delta\left(\frac{g-g'}{2(b-a)}-n\right) - \delta\left(\frac{g+g'-2a}{2(b-a)}-n\right)\right]$$

$$= \frac{1}{2(b-a)}e^{-\gamma(g-g')}\delta\left(\frac{g-g'}{2(b-a)}\right) \quad \text{since } a < g,g' < b$$

$$= \delta(g-g') \quad (17.27)$$

---

[2] Note the evolution operator W is not Hermitian; hence, under the duality operation that takes the $|\psi_k\rangle$ to its dual vector $\langle\psi_k|$, the term $e^{-\gamma g}$ switches its sign to $e^{+\gamma g}$.
[3] Since $\partial\theta(g-a)/\partial g = \delta(g-a)$, it follows that $\Psi_k(g)$ is an eigenfunction of W with eigenvalue $-S_k$.

Hence, the eigenfunctions satisfy the completeness equation given by

$$\sum_k |\psi_k\rangle\langle\psi_k| = \mathcal{I} \tag{17.28}$$

Insert both the completeness equation given in Eq. (17.28) as well as the completeness equation for the coordinate eigenstate given by

$$\int_{-\infty}^{+\infty} dg |g\rangle\langle g| = \mathcal{I}$$

into the expression for the barrier option given in Eq. (17.21). Then, for $f = \int_{t_*}^{T} dy f(t_0, y) \in [a, b]$, the following is the exact barrier option price

$$\frac{\mathcal{C}_B(t_0, t_*, K, f)}{B(t_0, t_*)} = \langle f|e^{-W}|\mathcal{P}_*\rangle\big|_{\text{Barrier}}$$

$$= \sum_{k_n} \int_{-\infty}^{+\infty} dg \langle f|e^{-W}|\psi_k\rangle\langle\psi_k|g\rangle\langle g|\mathcal{P}_*\rangle$$

$$= \frac{2}{(b-a)} \int_a^b dg \sum_{k_n} e^{S_k} e^{\gamma(g-f)} \sin k_n (f-a) \sin k_n (g-a) \mathcal{P}_*(g)$$

$$= \frac{e^{-\frac{\beta^2}{2G}}}{2(b-a)} \int_a^b dg \sum_{n=-\infty}^{+\infty} e^{\gamma(g-f)} e^{-\frac{1}{2}Gk_n^2} [e^{ik_n(f-g)} - e^{ik_n(f+g-2a)}]\mathcal{P}_*(g)$$

where recall that $k_n = \pi n/(b-a)$. Hence, since $\gamma = \beta/G$, the exact price of the zero coupon barrier option is given by

$$\mathcal{C}_B(t_0, t_*, K) = B(t_0, t_*) e^{-\frac{\beta^2}{2G}} \int_a^b dg\, e^{\frac{\beta}{G}(g-f)} \mathcal{Q}[g, f; G; a, b]\mathcal{P}_*(g) \tag{17.29}$$

where $\mathcal{Q}[g, f; G; a, b]$ is the barrier function discussed in Section 17.5.

The price of a zero coupon bond barrier option obtained in Eq. (17.29) is similar to the barrier option for equity since both have only one independent degree of freedom [12, 22].

17.5 Barrier function

The barrier function, for $k_n = \pi n/(b-a)$, is defined by the following equation

$$\mathcal{Q}[g, f; G; a, b] = \frac{1}{2(b-a)} \sum_{n=-\infty}^{+\infty} e^{-\frac{1}{2}Gk_n^2} [e^{ik_n(f-g)} - e^{ik_n(f+g-2a)}] \tag{17.30}$$

The barrier function has another representation that is useful for $G \sim 0$. The Poisson summation formula given by Eq. (17.26) yields

$$\frac{1}{2(b-a)} \sum_{n=-\infty}^{+\infty} \exp\left\{ -\frac{1}{2}G\left(\frac{\pi n}{b-a}\right)^2 + i\frac{\pi n\phi}{b-a} \right\}$$

$$= \sqrt{\frac{1}{2\pi G}} \sum_{n=-\infty}^{+\infty} \exp\left\{ -\frac{1}{2G}[\phi - 2(b-a)n]^2 \right\}$$

The representation of the barrier function, which is rapidly convergent for an expansion around $G = 0$, is given by

$$\mathcal{Q}[g, f; G; a, b] = \frac{1}{\sqrt{2\pi G}} \sum_{n=-\infty}^{+\infty} \left[\exp -\frac{1}{2G}[f - g - 2(b-a)n]^2 \right.$$

$$\left. - \exp -\frac{1}{2G}[f + g - 2a - 2(b-a)n]^2 \right]$$

The barrier function is not a smooth function of its arguments, especially for the case of $G \simeq 0$. To illustrate the key feautures of $\mathcal{Q}[g, f; G; a, b]$ the following cases are plotted.

- In Figure 17.3(a), $\mathcal{Q}[g, f; G; a, b]$ is plotted for $G = 0.01$; it can be seen that the barrier function is an irregular function of the arguments f and g.
- In Figure 17.3(b), $\mathcal{Q}[g, f; G; a, b]$ is plotted for $G = 0.6$ and the function varies very smoothly as a function of f and g.

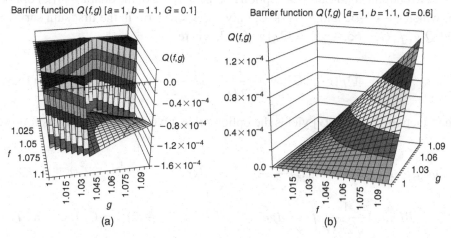

Figure 17.3 Barrier function $\mathcal{Q}[g, f; G; a, b]$ for $a = 1$; $b = 1.1$ with (a) $G = 0.01$ and (b) $G = 0.6$.

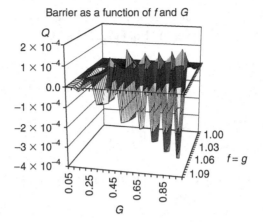

Figure 17.4 Barrier function for $a = 1$; $b = 1.1$. The values of $f = g$ are plotted for different values of G.

- Figure 17.4 plots $\mathcal{Q}[g, f; G; a, b]$ as a function of G and for the diagonal case of $f = g$; it is seen that the barrier function is a very irregular function of G going through sharp changes.

The barrier function figures show that any hedging as a function of volatility G needs to be considered very carefully due to the irregular dependence of $\mathcal{Q}[g, f; G; a, b]$ on G.

17.5.1 Limiting cases of $\mathcal{Q}[g, f; G; a, b]$

The zero coupon bond European option is a special case of the double barrier option, for which $a \to -\infty$ and $b \to +\infty$; only the $n = 0$ term in the first sum survives and $\mathcal{Q}[g, f; G; -\infty, +\infty] = \mathcal{Q}_E[g, f; G]$, where

$$\mathcal{Q}_E[g, f; G] = \sqrt{\frac{1}{2\pi G}} \exp\left\{ -\frac{1}{2G}(f - g)^2 \right\} \qquad (17.31)$$

The function $\mathcal{Q}_E[g, f; G]$ yields the option price, given in Eq. (11.49), as follows

$$\mathcal{C}_B(t_0, t_*, K, f) \to B(t_0, t_*) \frac{e^{-\frac{\beta^2}{2G}}}{\sqrt{2\pi G}} \int_{-\infty}^{+\infty} dg\, e^{\frac{\beta}{G}(g-f)} e^{-\frac{1}{2G}(f-g)^2} (e^{-g} - K)_+$$

$$= B(t_0, t_*) \frac{1}{\sqrt{2\pi G}} \int_{-\infty}^{+\infty} dg\, e^{-\frac{1}{2G}(f+\beta-g)^2} (e^{-g} - K)_+ = C_E(t_0, t_*, K, f)$$

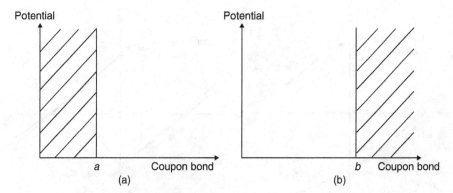

Figure 17.5 (a) Barrier option for a knock-out option for values of payoff below a minimum value of a. (b) Barrier option for a knock-out option for values of payoff greater than a maximum value of b.

As expected, on removing the barriers, the barrier option reduces to the zero coupon bond European option.

A single barrier on the left at the position a, as shown in Figure 17.5(a), is given by taking $b \to +\infty$; only the $n = 0$ for the first term and $n = 0$ for the second term survive in $Q[g, f; G; a, +\infty] = Q_a[g, f; G]$ where

$$Q_a[g, f; G] = \sqrt{\frac{1}{2\pi G}} \left[\exp\left\{ -\frac{1}{2G}(f - g)^2 \right\} - \exp\left\{ -\frac{1}{2G}(f - g - 2a)^2 \right\} \right]$$

Similarly, a single barrier on the right at b, shown in Figure 17.5(b), is obtained by taking $a \to -\infty$; only the terms with $n = 0$ for the first term and $n = 1$ for the second term survive in $Q[g, f; G; -\infty, b] = Q_b[g, f; G]$ and yield

$$Q_b[g, f; G] = \sqrt{\frac{1}{2\pi G}} \left[\exp\left\{ -\frac{1}{2G}(f - g)^2 \right\} - \exp\left\{ -\frac{1}{2G}(f - g - 2b)^2 \right\} \right]$$

17.6 Barrier linearization

The exact price of the zero coupon bond barrier option could be obtained because the barrier is a *linear* function of the forward interest rates. In contrast, the case of the coupon bond is far more complicated due to the nonlinear nature of the barrier.

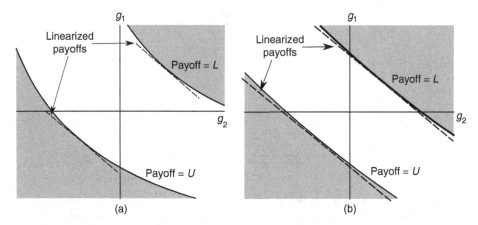

Figure 17.6 (a) The nonlinear barrier condition. The coupon bond trajectories, from time t_0 to t_*, take values only in the (unshaded) permitted domain – lying in-between U and L; all other trajectories do not contribute to the option price. (b) Linearized barrier constraint.

Recall, from Eqs. (15.82) and (17.22), the evolution operator and eigenfunctions for the barrier option are given by

$$W = -\frac{1}{2} \sum_{i,j=1}^{N} G_{ij} \frac{\partial^2}{\partial g_i \partial g_j} - \sum_{i=1}^{N} \beta_i \frac{\partial}{\partial g_i} \qquad (17.32)$$

$$W\psi_k[\mathbf{g}] = -S_k \psi_n[\mathbf{g}]$$

$$\text{B.C.:} \quad \psi_k[\mathbf{g}] = 0 \text{ for } \sum_{I=1}^{N} c_I e^{-g_I} \geq L \text{ and } \sum_{I=1}^{N} c_I e^{-g_I} \leq U$$

The evolution operator W is similar to the Laplacian operator in N Euclidean dimensions, and the barriers define two $N-1$-dimensional subspaces via the nonlinear constraint equations $\sum_I c_I e^{-g_I} = L$ and $\sum_I c_I e^{-g_I} = U$. The nonlinear barriers for the coupon bond are shown in Figure 17.6(a).

The boundary conditions that the eigenfunctions of the coupon bond are zero on these two $N-1$-dimensional subspaces require that the eigenfunctions of the W operator are zero on the nontrivial and nonlinear subspaces – in general, an intractable problem. For these reasons, analytically finding the exact price of a coupon bond barrier option is, in practice, almost impossible.

Due to the specific form of the coupon bond price an approximate solution for the bond barrier option can be found that is leading order in $\sigma(t, x)$, the forward interest rate volatility.

A small value of forward interest rate volatility σ implies that the fluctuations of the coupon bond about its initial value are of $O(\sigma)$; this was the reason that the initial value of the coupon bond was subtracted from the payoff function in the volatility expansion given in Eq. (3.62). Hence, a similar subtraction of the initial coupon bond price should yield $O(\sigma)$ fluctuations for the payoff function.

The barrier constraint is linearized about the leading term in the barrier; consider the following combination

$$g_I - f_I - \beta_I = \int_{t_*}^{T_I} dx [f(t_*, x) - f(t_0, x)] - \beta_I \qquad (17.33)$$

The representation of the bond forward interest rates given in Eq. (5.2) yields

$$f(t_*, x) = f(t_0, x) + \int_{t_0}^{t_*} dt\, \alpha(t, x) + \int_{t_0}^{t_*} dt\, \sigma(t, x) A(t, x)$$

$$\text{with} \quad E\big[A(t, x)A(t', x')\big] = \delta(t - t') \mathcal{D}(x, x'; t)$$

Hence

$$g_I = f_I + \beta_I + \int_{t_*}^{T_I} dx \int_{t_0}^{t_*} dt\, \sigma(t, x) A(t, x)$$

The variance is given by

$$E\Big[(g_I - f_I - \beta_I)^2\Big] = \int_{t_0}^{t_*} dt \int_{t_*}^{T_I} dx \int_{t_*}^{T_I} dx'\, \sigma(t, x) \mathcal{D}(x, x'; t) \sigma(t, x')$$

$$= G_{II} \sim O(\sigma^2)$$

The calculation above shows that all the fluctuations of the random quantity $g_I - f_I - \beta_I$ are of $O(\sigma)$; the β_I term needs to be subtracted to account for the drift of g_I from time t_0 to maturity time t_*. One has the following *linearization* of the barrier condition

$$B(t_*, T) = \sum_I c_I e^{-g_I} = \sum_I c_I e^{-f_I - \beta_I} e^{-(g_I - f_I - \beta_I)}$$

$$= \sum_I d_I e^{-(g_I - f_I - \beta_I)} \simeq \sum_I d_I (1 + f_I + \beta_I) - \sum_I d_I g_I + O(\sigma^2)$$

$$\text{where} \quad d_I \equiv c_I e^{-f_I - \beta_I}$$

Define the new barrier limits

$$a = \sum_I d_I (1 + f_I + \beta_I) - U; \quad b = \sum_I d_I (1 + f_I + \beta_I) - L \qquad (17.34)$$

Figure 17.6(b) shows that for small values of the bond variables the linearization of the barrier function yields a good approximation; the coefficients d_I are chosen to ensure that the linearization takes into account the leading value of the coupon bond. Only fluctuations (configurations) that are near the leading value are small.

The new linearized barrier conditions are now defined by $a \leq gd \leq b$; more precisely

$$\text{B.C.} : \ \psi_k[\mathbf{g}] = 0 \text{ for } gd \leq a \text{ and } gd \geq b; \quad gd \equiv \sum_I g_I d_I \quad (17.35)$$

The barrier for the coupon bond is the same as given in Figure 17.2, except now the linearized value of the coupon bond gd is constrained to takes values in the interval $[a, b]$.

Note that the linearization of the barrier cannot be systematically improved by, say, expanding the barrier to quadratic or higher terms in the bond variables g_i; the reason being that there are no systematic techniques that can generate the eigenfunctions on nontrivial domains that result from including higher-order nonlinear terms of the barrier condition.

17.7 Overcomplete barrier eigenfunctions

The linearized barrier constraints can be implemented via eigenfunctions of W in a manner similar to the one used for the zero coupon bond barrier option. There is, however, an additional feature of coupon bonds that is not present for the zero coupon case. For the coupon bond, the linear sum of *all* the bond variables, namely $gd = \sum_I d_I g_I$, needs to be constrained. A symmetric combination of all the coordinates implies that a change of variables from the g_i; $i = 1, 2, \ldots, N$ to another set of N variables will, in general, not place all the g_is on an equal footing. Hence, a change of variables will not simplify the constraint equation.

One way out of this conundrum is to *increase* the space of complete eigenfunctions by including *another* crucial eigenfunction of gd, namely the eigenfunction that carries the barrier condition. This leads to more eigenfunctions than are required for providing a complete basis for the state space and is compensated by adding a *constraint* in the completeness equation – analogous to the construction of coherent states in quantum mechanics [70].

Similar to the zero coupon bond case given in Eq. (17.24), consider the following ansatz for the coupon bond eigenfunctions

$$\langle \mathbf{g} | \Psi_{\mathbf{p},k} \rangle = \Psi_{\mathbf{p},k}(\mathbf{g}) = e^{i \mathbf{p} \mathbf{g}} \psi_k(gd); \quad -\infty \leq g_i \leq +\infty \quad (17.36)$$

$$\psi_k(gd) \sim [\theta(gd - a) - \theta(gd - b)] e^{i(k+i\gamma)(gd-a)}; \quad -\infty \leq gd \leq +\infty$$

The eigenfunctions $e^{i\mathbf{pg}}$ form a complete basis, as given in Eq. (17.14), and including the eigenfunction $\psi_k(gd)$ makes the eigenfunctions $\Psi_{\mathbf{p},k}(\mathbf{g})$ *overcomplete*.

The bond evolution operator W acting on the eigenfunctions, using a notation for later convenience, yields the following

$$W[e^{i\mathbf{pg}}\psi_k(gd)] = -(S + i\beta p)[e^{i\mathbf{pg}}\psi_k(gd)] \tag{17.37}$$

$$-S = \frac{1}{2}pGp + \frac{v^2}{2}(k + i\gamma)^2 + (k + i\gamma)pGd - ik\beta d + \gamma\beta d$$

$$pGp \equiv \sum_{I,J=1}^{N} p_I G_{IJ} p_J; \quad pGd \equiv \sum_{I,J=1}^{N} p_I G_{IJ} d_J$$

$$\beta d = \sum_{I=1}^{N} \beta_I d_I; \quad \beta p = \sum_{I=1}^{N} \beta_I p_I; \quad v^2 \equiv dGd = \sum_{I,J=1}^{N} d_I G_{IJ} d_J$$

As in the zero coupon bond case, γ is chosen to eliminate the terms in S that are *linear* in k so that one can obtain two degenerate solutions $\psi_{\pm k}(gd)$. This yields

$$\gamma = \frac{1}{v^2}(\beta d + ipGd) \tag{17.38}$$

and which, in turn, gives the following result

$$-S = \frac{1}{2}pGp + \frac{v^2}{2}k^2 + \frac{1}{v^2}[(\beta d)^2 - (pGc)^2 + i(\beta d)pGd]; \quad v^2 \equiv dGd \tag{17.39}$$

To impose the barrier option boundary conditions, following the case of the zero coupon bond, one superposes the degenerate solutions $\psi_{\pm k}(gd)$ to obtain, for $a \leq gd \leq b$, the following[4]

$$\langle \mathbf{g}|\psi_k \rangle \equiv \psi_k(gd) = [\theta(gd - a) - \theta(gd - b)]\sqrt{\frac{2}{b-a}}e^{-\gamma gd}\sin k(gd - a)$$

$$\langle \psi_k|\mathbf{g} \rangle = [\theta(gd - a) - \theta(gd - b)]\sqrt{\frac{2}{b-a}}e^{\gamma gd}\sin k(gd - a) \tag{17.40}$$

$$k \equiv k_n = \frac{\pi n}{(b-a)}; \quad n = 1, 2, \ldots + \infty$$

$$\text{Boundary conditions}: \ \psi_k(gd)\Big|_{gd=a} = 0 = \psi_k(gd)\Big|_{gd=b}$$

---

[4] As in the zero coupon bond case, the evolution operator W is not Hermitian. The duality operation takes $|\psi_k\rangle$ to its dual vector $\langle\psi_k|$; the term $e^{-\gamma gd}$ switches its sign and goes to $e^{+\gamma gd}$.

Unlike the case of the zero coupon bond barrier option, the constraint equation for the coupon bond involves N variables and is realized by the following representation

$$\theta(gd - a) - \theta(gd - b) = \int_a^b dh\delta(h - gd)$$

$$= \int_a^b dh \int_{-\infty}^{+\infty} \frac{d\xi}{2\pi} e^{i\xi(h-gd)} \equiv \int_{h,\xi} e^{i\xi(h-gd)} \qquad (17.41)$$

The completeness equation is given by[5]

$$\mathcal{I} = \left(\sum_{i=1}^N d_i\right) \int_{\mathbf{P}} 2\pi\delta \left(\sum_{i=1}^N p_i\right) \sum_{k=1}^\infty |\Psi_{\mathbf{p},k}\rangle\langle\Psi_{\mathbf{p},k}| \qquad (17.42)$$

$$= \left(\sum_{i=1}^N d_i\right) \int_{\mathbf{P}} 2\pi\delta \left(\sum_{i=1}^N p_i\right) \sum_{k=1}^\infty |\mathbf{p}\rangle|\psi_k\rangle\langle\psi_k|\langle\mathbf{p}|$$

There is an extra $\sum_k$ sum over the additional eigenfunctions $|\psi_k\rangle$ that gives rise to overcompleteness. Unlike Eq. (17.14), the *constraint* $(\sum_{i=1}^N d_i)\delta(\sum_{i=1}^N p_i)$ has been introduced in the completeness equation given in Eq. (17.42) to compensate for the overcomplete set of eigenfunctions.

To prove the completeness equation, consider the following

$$\langle\mathbf{f}|\mathcal{I}|\mathbf{g}\rangle = \left(\sum_{i=1}^N d_i\right) \int_{\mathbf{P}} \sum_{k=1}^\infty 2\pi\delta \left(\sum_{i=1}^N p_i\right) \langle\mathbf{f}|\mathbf{p}\rangle\langle\mathbf{f}|\psi_k\rangle\langle\psi_k|\mathbf{g}\rangle\langle\mathbf{p}|\mathbf{g}\rangle$$

$$= \left(\sum_{i=1}^N d_i\right) \int_{\mathbf{P}} 2\pi\delta \left(\sum_{i=1}^N p_i\right) e^{i\mathbf{p}(\mathbf{f}-\mathbf{g})} \mathcal{F} \qquad (17.43)$$

$$\mathcal{F} = \sum_{k=1}^\infty \langle\mathbf{f}|\psi_k\rangle\langle\psi_k|\mathbf{g}\rangle$$

$$= \frac{2}{b-a} \sum_{k=1}^\infty \int_{h,\xi,h',\xi'} e^{i\xi(h-fd)} e^{i\xi'(h'-gd)} e^{-\gamma(fd-gd)}$$

$$\times \sin(k(fd - a)) \sin(k(gd - a))$$

---

[5] Since $k \equiv k_n = \frac{\pi n}{(b-a)}$, $\sum_{k=1}^\infty$ means a sum over n, that is $\sum_{n=1}^\infty$.

The bond state $|\mathbf{g}\rangle$ is taken to be unrestricted. The initial state $\langle\mathbf{f}|$ of the coupon bond is within the barrier; that is, $a \le fd \le b$ as shown in Figure 17.1, and which implies that $\int_{h,\xi} e^{i\xi(h-fd)} = 1$; from Eq. (17.27) and above

$$
\mathcal{F} = \frac{2e^{-\gamma(fd-gd)}}{b-a} \sum_{k=1}^{\infty} \sin(k(fd-a)) \sin(k(gd-a)) \int_{h,\xi} e^{i\xi(h-gd)}
$$

$$
= \delta(fd-gd) \int_{h,\xi} e^{i\xi(h-gd)} \tag{17.44}
$$

Using the representation $2\pi\delta(\sum_{i=1}^{N} p_i) = \int_{-\infty}^{+\infty} d\zeta \exp(i\zeta \sum_{i=1}^{N} p_i)$, incorporating the result of Eq. (17.44) into Eq. (17.43) and doing the $\int_{\mathbf{p}}$ integrations yields

$$
\langle\mathbf{f}|\mathcal{I}|\mathbf{g}\rangle = \left(\sum_{i=1}^{N} d_i\right) \int_{-\infty}^{+\infty} d\zeta \prod_{i=1}^{N} \delta(f_i - g_i + \zeta) \int_{h,\xi} e^{i\xi(h-gd)} \delta(fd-gd)
$$

$$
= \left(\sum_{i=1}^{N} d_i\right) \int_{-\infty}^{+\infty} d\zeta \prod_{i=1}^{N} \delta(f_i - g_i + \zeta) \int_{h,\xi} e^{i\xi(h-gd)} \delta\left(\zeta \sum_{i=1}^{N} d_i\right)
$$

$$
= \int_{h,\xi} e^{i\xi(h-gd)} \prod_{i=1}^{N} \delta(f_i - g_i)
$$

$$
= \prod_{i=1}^{N} \delta(f_i - g_i): \quad \text{completeness equation} \tag{17.45}
$$

where the last equation above is a consequence of $a \le fd \le b$.

Hence, Eq. (17.45) confirms that the overcomplete set of eigenfunctions in Eq. (17.42) yield the correct completeness equation. For the zero coupon bond barrier options, Eq. (17.42) reduces to the one given in Eq. (17.28) since the constraint $\delta(p)$ removes the extra eigenfunctions e^{ipg}.

The completeness equation requires only that the bond vector $\langle\mathbf{f}|$ must satisfy the condition that $a \le fd \le b$, leaving the bond vector $|\mathbf{g}\rangle$ completely free; this more general result is required when the completeness equation Eq. (17.42) is used for evaluating the coupon bond barrier option price.

17.8 Coupon bond barrier option price

The coupon bond barrier option is given by Eqs. (17.22) and (17.42) as follows[6]

$$\frac{\mathcal{C}_B(t_0, t_*, T, K)}{B(t_0, t_*)} = \langle \mathbf{f} | e^{-W} | \mathcal{P}_* \rangle \big|_{\text{Barrier}}$$

$$= \left(\sum_{i=1}^{N} d_i \right) \int_{\mathbf{p}} 2\pi \delta \left(\sum_{i=1}^{N} p_i \right) \sum_{k=1}^{\infty} \langle \mathbf{f} | e^{-W} | \Psi_{\mathbf{p},k} \rangle \langle \Psi_{\mathbf{p},k} | \mathcal{P}_* \rangle$$

$$= \left(\sum_{i=1}^{N} d_i \right) \int_{\mathbf{p}.\mathbf{g}} 2\pi \delta \left(\sum_{i=1}^{N} p_i \right) \sum_{k=1}^{\infty} \langle \mathbf{f} | e^{-W} | \Psi_{\mathbf{p},k} \rangle \langle \Psi_{\mathbf{p},k} | \mathbf{g} \rangle \langle \mathbf{g} | \mathcal{P}_* \rangle \quad (17.46)$$

$$\Rightarrow \frac{\mathcal{C}_B(t_0, t_*, T, K)}{B(t_0, t_*)} \equiv \int_{\mathbf{g}} \mathcal{K}[\mathbf{f}, \mathbf{g}] \mathcal{P}_*[\mathbf{g}] \quad (17.47)$$

where the last equation defines the *pricing kernel* $\mathcal{K}$ for the barrier option. The linearized boundary conditions for the barrier option eigenfunctions, from Eq. (17.35), are given by

boundary conditions: $\Psi_{\mathbf{p},k}[\mathbf{g}] = 0$ for $gd \geq b$ and $gd \leq a$

Consider an initial value of the coupon bond, as shown in Figure 17.1, that lies within the barrier; the linearized approximation implies that $a \leq fd \leq b$; Eqs. (17.36), (17.37), (17.39), (17.40), and (17.41) yield the following

$$\langle \mathbf{f} | e^{-W} | \Psi_{\mathbf{p},k} \rangle = e^{S + i\beta \mathbf{p}} \langle \mathbf{f} | \Psi_{\mathbf{p},k} \rangle$$

$$= \sqrt{\frac{2}{b-a}} e^{S + i(\beta + \mathbf{f})\mathbf{p} - \gamma f d} \sin k(fd - a)$$

$$S = -\frac{1}{2} pGp - \frac{v^2}{2} k^2 - \frac{1}{v^2} \left[(\beta d)^2 - (pGc)^2 + i(\beta d) pGd \right]$$

$$\langle \Psi_{\mathbf{p},k} | \mathbf{g} \rangle = \sqrt{\frac{2}{b-a}} e^{-i g \mathbf{p} + \gamma g d} \sin k(gd - a) \int_{\xi, h} e^{i\xi(h - gd)}$$

$$\mathcal{P}_*[\mathbf{g}] = \left[\sum_I c_I e^{-g_I} - K \right]_+ ; \quad v^2 = dGd$$

---

[6] Eq. (17.46) is obtained by using $\int_{\mathbf{g}} |\mathbf{g}\rangle \langle \mathbf{g}| = \mathcal{I}$.

Eqs. (17.47) and (17.4) yield the following expansion for the barrier option

$$\frac{\mathcal{C}_B(t_0, t_*, T, K)}{B(t_0, t_*)} = \int_{\mathbf{g}} \mathcal{K}[\mathbf{f}, \mathbf{g}] \mathcal{P}_*[\mathbf{g}]$$

$$= \int_{Q, \eta} e^{-i\eta Q}(F + Q - K)_+ \left[C_{B,0} + i\eta C_{B,1} - \frac{1}{2}\eta^2 C_{B,2} + O(\sigma^3) \right] \qquad (17.48)$$

The coefficients of the expansion for the barrier option, from Eqs. (17.17), (17.7), and (17.8), are given by

$$C_{B,0} = E[1]$$

$$C_{B,1} = E[V] = \sum_{I=1}^{N} J_I(C_{B,I} - C_{B,0})$$

$$C_{B,2} = E[V^2] = \sum_{I,K=1}^{N} J_I J_K (C_{B,IK} - 2C_{B,I} + C_{B,0}) \qquad (17.49)$$

A detailed derivation of the coefficients is given in Section 17.11 and yields the following results

$$C_{B,0} = E[1] = \int_{\mathbf{g}} \mathcal{K}[\mathbf{f}, \mathbf{g}]$$

$$= e^{S_{B,0}} \int_a^b dh \exp\left\{ \frac{1}{v^2} \beta d(h - fd) \right\} \mathcal{Q}[h, fd; v^2; a, b] \qquad (17.50)$$

$$S_{B,0} = -\frac{1}{2v^2}(\beta d)^2$$

Unlike the case of the European option, the coefficient $C_{B,0}$ is not equal to 1; the reason being that even in the absence of any payoff function, the barrier constrains the allowed paths and reduces $C_{B,0}$ from the unconstrained value of 1.

The second coefficient is given by

$$C_{B,I} = E[e^{-g_I + f_I}] = \int_{\mathbf{g}} \mathcal{K}[\mathbf{f}, \mathbf{g}] e^{-g_I + f_I}$$

$$= e^{S_{B,I}} \int_a^b dh \exp\left\{ \frac{1}{v^2} \left(\beta d - \sum_J G_{IJ} d_J \right)(h - fd) \right\} \mathcal{Q}[h, fd; v^2; a, b] \qquad (17.51)$$

$$S_{B,I} = -\frac{1}{2v^2} \left(\beta d - \sum_J G_{IJ} d_J \right)^2$$

The barrier function $\mathcal{Q}[h, fd; v^2; a, b]$ is given in Eq. (17.31).

Unlike the case of the European option, the coefficient $C_{B,1}$ is not equal to 1; the reason being that the martingale condition that led to $C_{B,1}$ is only valid for an unconstrained path; the barrier constrains the expectation and reduces it from the value of 1.

The third coefficient is given by

$$C_{B,IK} = E[e^{-g_I - g_K + f_I + f_K}] = \int_{\mathbf{g}} \mathcal{K}[\mathbf{f}, \mathbf{g}] e^{-g_I - g_K + f_I + f_K}$$

$$= e^{S_{B,IK}} \int_a^b dh\, \mathcal{Q}[h, fd; v^2; a, b] \qquad (17.52)$$

$$\times \exp\left\{ \frac{1}{v^2} \left(\beta d - \sum_J G_{IJ} d_J - \sum_J G_{KJ} d_J \right) (h - fd) \right\}$$

$$S_{B,IK} = G_{IK} - \frac{1}{2v^2} \left(\beta d - \sum_J G_{IJ} d_J - \sum_J G_{KJ} d_J \right)^2$$

To extract the perturbative expansion of the option price to $O(\sigma)$ from the coefficients $C_{B,0}, C_{B,1}$, and $C_{B,2}$, the leading order term for the barrier option has to be isolated. On inspecting the coefficients, it is clear that in fact $C_{B,0}$ is a term of $O(1)$, with $C_{B,1}$ and $C_{B,2}$ being of order $O(\sigma)$ and $O(\sigma^2)$ respectively.

Collecting the results for the coefficients given in Eqs. (17.48), (17.49), (17.50), (17.51), and (17.52) yields the following

$$\frac{C_B(t_0, t_*, T, K)}{B(t_0, t_*)} = e^{-\frac{1}{2v^2}(\beta d)^2} \int_a^b dh \exp\left\{ \frac{1}{v^2} \beta d(h - fd) \right\} \mathcal{Q}[h, fd; v^2; a, b]$$

$$\times \int_{Q,\xi} e^{-iQ\eta} \left[1 + i\eta D_1 - \frac{1}{2}\eta^2 D_2 + O(\sigma^3) \right] (Q + F - K)_+ \qquad (17.53)$$

where coefficients D_1 and D_2 are given in Eqs. (17.54) and (17.55), respectively.

Eq. (17.53) is one of the most important results of this chapter, namely the barrier option price is the integral of the following two factors:

- the function $\exp\{-\frac{1}{2v^2}(\beta d)^2\} \exp\{\frac{1}{v^2}\beta d(h - fd)\} \mathcal{Q}[h, fd; v^2; a, b]$ that encodes the properties of the barrier and
- the factor $\int_{Q,\xi} e^{-iQ\eta}[1 + i\eta D_1 - \frac{1}{2}\eta^2 D_2 + O(\sigma^3)](Q + F - K)_+$ that encodes the properties of the payoff function.

Each factor has been evaluated approximately and one can improve the price of the barrier option by improving the approximation for each of these factors.

The coefficient D_1, to $O(\sigma^3)$, is given by Eqs. (17.50), (17.51), and (17.53)

$$D_1 = \sum_I J_I \left[e^{S_{B,I} - S_{B,0}} e^{-\frac{1}{v^2}(\sum_J G_{IJ} d_J)(h - fd)} - 1 \right] \tag{17.54}$$

$$\simeq -\frac{1}{v^2} \left[\sum_{I,J} J_I G_{IJ} d_J \right] (h - fd) + \Gamma$$

$$\Gamma = \frac{\beta d}{v^2} \sum_{I,J} J_I G_{IJ} d_J + \frac{1}{2v^4} \sum_I J_I \left(\sum_J G_{IJ} d_J \right)^2 [(h - fd)^2 - v^2]$$

The coefficient D_2, to $O(\sigma^3)$, is given by Eqs. (17.48) to (17.52)

$$D_2 = \sum_{I,K} J_I J_K \left[e^{S_{B,IK} - S_{B,0}} e^{-\frac{1}{v^2}(\sum_J G_{IJ} d_J + \sum_J G_{KJ} d_J)(h - fd)} \right. \tag{17.55}$$

$$\left. - e^{S_{B,I} - S_{B,0}} e^{-\frac{1}{v^2}(\sum_J G_{IJ} d_J)(h - fd)} - e^{S_{B,K} - S_{B,0}} e^{-\frac{1}{v^2}(\sum_J G_{KJ} d_J)(h - fd)} + 1 \right]$$

$$\simeq \sum_{I,K} J_I J_K G_{IK} - \frac{1}{v^2} \left[\sum_{IJ} J_I G_{IJ} d_J \right]^2 + \frac{1}{v^4} \left[\sum_{IJ} J_I G_{IJ} d_J \right]^2 (h - fd)^2$$

The results for D_1 and D_2 yield

$$D_2 - D_1^2 \simeq \sum_{I,K} J_I J_K G_{IK} - \frac{1}{v^2} \left[\sum_{IJ} J_I G_{IJ} d_J \right]^2$$

$$+ \Gamma \left[\frac{2}{v^2} \sum_{I,J} J_I G_{IJ} d_J (h - fd) - \Gamma \right] + O(\sigma^3) \tag{17.56}$$

Collecting all the terms yields, from Eq. (3.69), the following *main result* for the approximate price of the barrier option

$$\frac{C_B(t_0, t_*, T, K)}{B(t_0, t_*)} = \frac{e^{-\frac{1}{2v^2}(\beta d)^2}}{\sqrt{2\pi}} \int_a^b dh \exp \left\{ \frac{1}{v^2} \beta d(h - fd) \right\} Q[h, fd; v^2; a, b]$$

$$\times I(X) \sqrt{D_2 - D_1^2} + O(\sigma^3)$$

$$X = (K - F - D_1) / \sqrt{D_2 - D_1^2}$$

From Eq. (3.71) one has $I(X) = 1 + O(X)$ and this yields, to $O(X)$, the leading order price of the barrier option

$$
\frac{C_B(t_0, t_*, K)}{B(t_0, t_*)} \simeq \frac{e^{-\frac{1}{2v^2}(\beta d)^2}}{\sqrt{2\pi}}
$$

$$
\times \int_a^b dh \exp\left\{\frac{1}{v^2}\beta d(h - fd)\right\} \mathcal{Q}[h, fd; v^2; a, b]\sqrt{D_2 - D_1^2} \qquad (17.57)
$$

17.9 Barrier option: limiting cases

The Hamiltonian formulation yields a derivation, as in Sections 17.2 and 17.4, of the European option and zero coupon barrier option prices. The result of the coupon bond barrier option is examined for the following limiting cases.

- The HJM limit of exactly correlated bond forward interest rates.
- The zero coupon bond barrier option.
- Coupon bond European option.

17.9.1 The one-factor HJM model

The HJM model [56, 65] of the forward interest rates is widely used in finance and is a special case of the quantum finance model. In the HJM approach all the forward interest rates, in the language of quantum finance, are *exactly* correlated and this implies that $\mathcal{D}(x, x'; t) \to 1$ and hence $M(x, x', t) = \sigma(t, x)\sigma(t, x')$. Furthermore, in the one-factor HJM model the volatility function is taken to have an exponential form given by $\sigma(t, x) = \sigma_0 e^{-\lambda(x-t)}$.

Taking the HJM limit of the bond correlator yields

$$
G_{ij} = \int_{t_0}^{t_*} dt \int_{t_*}^{T_i} dx \int_{t_*}^{T_j} dx' M(x, x', t)
$$

$$
\to G_{ij}^{HJM} = \sigma_0^2 \int_{t_0}^{t_*} dt \int_{t_*}^{T_i} dx\, e^{-\lambda(x-t)} \int_{t_*}^{T_j} dx'\, e^{-\lambda(x'-t)} = \sigma_R^2 Y_i Y_j
$$

$$
Y_i \equiv Y(t_*, T_i) = \frac{1}{\lambda}[1 - e^{-\lambda(T_i - t_*)}]; \qquad \sigma_R^2 = \frac{\sigma_0^2}{2\lambda}[1 - e^{-2\lambda(t_* - t_0)}]
$$

The first two terms in Eq. (17.56) cancel in the HJM limit and yield the following result

$$
[D_2 - D_1^2]_{HJM} = \Gamma_{HJM}\left[2\frac{JY}{Yd}(h - fd) - \Gamma_{HJM}\right]
$$

$$\Gamma_{HJM} = \frac{\beta d}{Yd} JY + \frac{1}{2} \frac{JY^2}{(Yd)^2} [(h - fd)^2 - \sigma_R^2 (Yd)^2]$$

$$Yd = \sum_i Y_i d_i; \quad JY = \sum_i J_i Y_i; \quad JY^2 = \sum_i J_i Y_i^2$$

17.9.2 Zero coupon bond barrier option

The coupon bond barrier option reduces to the zero coupon case when only one coupon, say the final coupon, is nonzero; hence $c_N = 1$ and $c_i = 0$, $i \neq N$. For the barrier functions, the coefficients have the following limits $d_N = 1$ and $d_i = 0$, $i \neq N$ and yield

$$\beta d \to \beta; \quad v^2 = \sum_{IJ} d_I G_{IJ} d_J \to G_{NN} = G$$

$$\sum_{IJ} G_{IJ} d_J \to G_{NN} = G; \quad G_{IK} \to G_{NN} = G$$

The martingale condition yields $\beta = G/2$ and hence

$$S_{B,0} \to -\frac{\beta^2}{2G}; \quad S_{B,I} \to -\frac{(\beta - G)^2}{2G} = -\frac{\beta^2}{2G}$$

$$S_{B,IK} \to G - \frac{(\beta - 2G)^2}{2G} = -\frac{\beta^2}{2G}$$

$$\mathcal{Q}[h, fd; v^2; a, b] \to \mathcal{Q}[h, f; G; a, b]$$

$$\mathcal{P}_* \to [J(e^{-(g-f)} - 1) + F - K]_+; \quad J = e^{-f} = F$$

Collecting the results above yields, from Eq. (17.53), the following zero coupon limit of the coupon bond barrier option

$$C_B(t_0, t_*, T, K) \to B(t_0, t_*) \int_{Q,\eta} e^{-i\eta Q} (F + Q - K)_+ \mathcal{Z}$$

$$\mathcal{Z} = C_{B,0} + i\eta C_{B,1} - \frac{1}{2} \eta^2 C_{B,2} + O(\sigma^3)$$

$$= e^{-\frac{\beta^2}{2G}} \int_a^b dg \, \mathcal{Q}[g, f; G; a, b] e^{\frac{\beta}{G}(g-f)} \left(1 + i\eta[Je^{-(g-f)} - 1]\right.$$

$$\left. - \frac{1}{2}\eta^2 [J^2 e^{-2(g-f)} - 2Je^{-(g-f)} + 1]\right)$$

The results given above yield that

$$\frac{\mathcal{C}_B(t_0, t_*, T, K)}{B(t_0, t_*)} \rightarrow e^{-\frac{\beta^2}{2G}} \int_a^b dg \, e^{\frac{\beta}{G}(g-f)} \mathcal{Q}[g, f; G; a, b] \mathcal{P}_*(g) + O(\sigma^3)$$

This is the expected approximation of the exact result, which is given in Eq. (17.29).

17.9.3 Coupon bond European option limit

Consider the limit of $a \rightarrow -\infty$ and $b \rightarrow \infty$; the function $\mathcal{Q}$ reduces to a single term, namely

$$\mathcal{Q}[h, fd; v^2; a, b] \rightarrow \frac{1}{\sqrt{2\pi v^2}} \exp\left\{-\frac{1}{2v^2}[h - fd]^2\right\}$$

$$\text{and} \quad \int_a^b dh \rightarrow \int_{-\infty}^{+\infty} dh \tag{17.58}$$

To show the perturbative barrier option – in the limit of $a \rightarrow -\infty$ and $b \rightarrow +\infty$ – to be equal to the European option to $O(\sigma^3)$ one re-writes Eq. (17.53) by exchanging the order of integration in the following manner

$$\frac{\mathcal{C}_B(t_0, t_*, T, K)}{B(t_0, t_*)} = \frac{e^{-\frac{1}{2v^2}(\beta d)^2}}{\sqrt{2\pi v^2}} \int_{\mathcal{Q}, \xi} e^{-i\mathcal{Q}\eta} \int_{-\infty}^{+\infty} dh \exp\left\{\frac{1}{v^2}\beta d(h - fd)\right\}$$

$$\times \exp\left\{-\frac{1}{2v^2}[h - fd]^2\right\}\left[1 + i\eta D_1 - \frac{1}{2}\eta^2 D_2 + O(\sigma^3)\right](Q + F - K)_+$$

Performing the Gaussian integrations over h yields

$$\mathcal{C}_{B,0} \rightarrow \frac{e^{-\frac{1}{2v^2}(\beta d)^2}}{\sqrt{2\pi v^2}} \int_{-\infty}^{+\infty} dh \exp\left\{\frac{1}{v^2}\beta d(h - fd)\right\} \exp\left\{-\frac{1}{2v^2}[h - fd]^2\right\} = 1$$

$$\mathcal{C}_{B,1} \rightarrow \frac{e^{-\frac{1}{2v^2}(\beta d)^2}}{\sqrt{2\pi v^2}} \int_{-\infty}^{+\infty} dh \exp\left\{\frac{1}{v^2}\beta d(h - fd)\right\} \exp\left\{-\frac{1}{2v^2}[h - fd]^2\right\} D_1$$

$$\simeq \frac{1}{4}\sum_I J_I(\beta d)^2 = 0 + O(\sigma^4)$$

$$C_{B,2} \to \frac{e^{-\frac{1}{2v^2}(\beta d)^2}}{\sqrt{2\pi v^2}} \int_{-\infty}^{+\infty} dh \exp\left\{\frac{1}{v^2}\beta d(h - fd)\right\} \exp\left\{-\frac{1}{2v^2}[h - fd]^2\right\} D_2$$

$$\simeq \sum_{IK} J_I J_K G_{IK} + \frac{1}{2v^4} \sum_{IK} J_I J_K (\beta d)^2 \sum_J G_{IJ} d_J \sum_J G_{KJ} d_J$$

$$= \sum_{IK} J_I J_K G_{IK} + O(\sigma^4)$$

Hence, to $O(\sigma^4)$

$$C_{B,0} \to 1 = C_{E,0}$$
$$C_{B,1} \to 0 = C_{E,1}$$
$$C_{B,2} \to \sum_{I,J} J_I J_K G_{IK} = C_{E,2}$$

and the European option is the limit of the barrier option; namely

$$C_B(t_0, t_*, T, K) \to C_E(t_0, t_*, T, K)$$

17.10 Summary

The Hamiltonian formulation of the quantum field theory of bond forward interest rates provides an efficient computational tool for analyzing the coupon bond European and barrier options. The earlier result for the European coupon bond option obtained in Chapter 11 is seen to emerge in a straightforward manner in the Hamiltonian approach.

The zero coupon barrier option price was obtained exactly by imposing the constraint of the barrier on the eigenfunctions of the Hamiltonian, or more accurately, of the bond evolution operator. The bond evolution operator was expressed in terms of the 'bond variables' that are more natural for studying coupon bonds.

The computation of the coupon bond barrier option turned out to be fairly complicated for two reasons: *firstly*, because the linear superposition of different zero coupon bonds constitutes a coupon bond, thus making the constraint a function of many variables and; *secondly*, because the barrier on the coupon bond imposes a nonlinear constraint on the bond forward interest rates. It was shown that, under very general conditions, the linearized payoff function yields the leading contribution to the coupon bond barrier option price. An overcomplete set of eigenfunctions of the evolution operator was used for imposing the linearized barrier on the evolution of the bond variables.

Eq. (17.57) shows that the entire calculation for the barrier option factorizes into two connected and nontrivial components, one factor reflecting the properties of the barrier and the other that of the payoff function. This feature of the barrier option seems to be very general and could prove useful in analyzing options with more complex barriers and payoffs.

The coupon bond barrier option price is an explicit function of the initial values of the zero coupon bonds, the strike price, the duration of option, and the barrier; hence all the hedging parameters can be (approximately) evaluated analytically from the results obtained.

The approximate price of the coupon bond barrier option can be tested numerically as well as be used for empirical studies. The various limiting cases for the coupon bond barrier option provide useful formulas for testing numerical algorithms that can then be used to explore the option's nonperturbative and nonlinear regimes.

The framework of quantum finance is a flexible and fruitful approach for computing coupon bond barrier and European option prices. The nontrivial correlations of the bond variables, encoded in the imperfect correlator G_{IJ}, determine the price of the barrier option. Taking the bond variables to be exactly correlated, as is the case with the HJM model, yields systematic errors for all pricing and hedging parameters.

17.11 Appendix: Barrier option coefficients

The coefficients for the coupon bond barrier option are defined in Section 17.8 as follows

$$C_{B,0} = E[1] = \int_{\mathbf{g}} \mathcal{K}[\mathbf{f}, \mathbf{g}]; \quad C_{B,I} = E[e^{-g_I + f_I}] = \int_{\mathbf{g}} \mathcal{K}[\mathbf{f}, \mathbf{g}] e^{-g_I + f_I}$$

$$C_{B,IK} = E[e^{-g_I - g_K + f_I + f_K}] = \int_{\mathbf{g}} \mathcal{K}[\mathbf{f}, \mathbf{g}] e^{-g_I - g_K + f_I + f_K}$$

The initial conditions imply that $a \leq fd \leq b$ and Eqs. (17.36), (17.37), (17.39), (17.40), and (17.41) yield the following

$$\mathcal{K}[\mathbf{f}, \mathbf{g}] = \frac{2 \sum_{i=1}^{N} d_i}{b - a} \sum_{k} \int_{\xi, h} \int_{\mathbf{p}, \mathbf{g}} 2\pi \delta \left(\sum_{i=1}^{N} p_i \right) e^{i\xi(h - gd)} e^{S + i(\beta + \mathbf{f})\mathbf{p} - i\mathbf{g}\mathbf{p}} e^{\gamma(gd - fd)}$$

$$\times \sin k(fd - a) \sin k(gd - a)$$

$$
= \frac{\sum_{i=1}^{N} d_i}{2(b-a)} \sum_{n=-\infty}^{+\infty} \int_{\xi,h} \int_{\mathbf{p,g}} 2\pi \delta \left(\sum_{i=1}^{N} p_i \right) e^{i\xi(h-gd)} e^{S+i(\beta+\mathbf{f})\mathbf{p}-ig\mathbf{p}} e^{\gamma(gd-fd)}
$$

$$
\times \left[e^{-ik_n(gd-fd)} - e^{ik_n(gd+fd-2a)} \right]
$$

where $k \equiv k_n = \pi n/(b-a)$. Furthermore

$$
S = -\frac{v^2}{2} k_n^2 + S_p ; \quad S_p \equiv -\frac{1}{2} pGp - \frac{1}{v^2} \left[(\beta d)^2 - (pGc)^2 + i(\beta d) pGd \right]
$$

$$
\gamma = \frac{1}{v^2} (\beta d + ipGd); \quad v^2 = dGd
$$

Some details are given of the derivation of the coefficients $C_{B,IK}$; the other coefficients need a similar, but simpler, calculations. Shifting the variable $g_i \to g_i + f_i + \beta_i$ in the expression for $C_{B,IK}$ and then performing the $\int_{\mathbf{g}}$ integrations yields the following

$$
C_{B,IK} = \int_{\mathbf{g}} \mathcal{K}[\mathbf{f,g}] e^{-g_I - g_K}
$$

$$
= \frac{\sum_{i=1}^{N} d_i}{2(b-a)} \sum_{n=-\infty}^{+\infty} e^{-\frac{v^2}{2} k_n^2} \int_{\xi,h} e^{i\xi(h-gd-fc-\beta d)} \int_{\mathbf{p,g}} 2\pi \delta \left(\sum_{i=1}^{N} p_i \right) e^{S_p} e^{-ig\mathbf{p}}
$$

$$
\times e^{\gamma(gd+\beta d)} \left[e^{-ik_n(gd+\beta d)} - e^{ik_n(gd+\beta d+2fd-2a)} \right] e^{-g_I - g_K}
$$

$$
= \frac{1}{2(b-a)} \sum_{n=-\infty}^{+\infty} e^{-\frac{v^2}{2} k_n^2} \int_{\xi,h} e^{i\xi(h-fd-\beta d)} \left[A e^{-ik_n \beta d} - B e^{ik_n(\beta d+2fd-2a)} \right]
$$

$$
\tag{17.59}
$$

where

$$
A = \left(\sum_{i=1}^{N} d_i \right) \int_{\mathbf{p}} e^{S_p + \gamma \beta d} 2\pi \delta \left(\sum_{i=1}^{N} p_i \right) \delta(p_K + \Lambda_A d_K - i) \delta(p_I + \Lambda_A d_I - i)
$$

$$
\times \prod_{i \neq I, K} \delta(p_i + \Lambda_A d_i)
$$

$$
\Lambda_A = i\gamma + \xi + k
$$

$$
\tag{17.60}
$$

and

$$
B = \left(\sum_{i=1}^{N} d_i\right) \int_{\mathbf{p}} e^{S_p + \gamma \beta d} 2\pi \delta\left(\sum_{i=1}^{N} p_i\right) \delta(p_K + \Lambda_B d_K - i)\delta(p_I + \Lambda_B d_I - i)
$$

$$
\times \prod_{i \neq I, K} \delta(p_i + \Lambda_B d_i)
$$

$$
\Lambda_B = i\gamma + \xi - k
$$

17.11.1 Dirac-delta functions

One needs to explicitly solve the delta functions and fix the values of p_i, which are required for determining the action S_0 and the drift γ; furthermore, the explicit values of all the p_is are required for carrying out the N-integrations, namely $\int_{\mathbf{p}}$ so as to evaluate A and B. The reason the delta functions apparently look intractable is because the function γ appears in the delta functions and γ is itself a function of all the p_is: the delta functions, in effect, yield a set of (apparently intractable) simultaneous equations for the p_is.

However, by recursively solving the delta functions we have the rather remarkable result that the delta functions explicitly fix, in the following manner, the value of all the p_is. Consider

$$
\left(\sum_{i=1}^{N} d_i\right) 2\pi \delta\left(\sum_{i=1}^{N} p_i\right) \delta(p_I + \Lambda_A d_I - i)\delta(p_K + \Lambda_A d_K - i)
$$

$$
\times \prod_{i \neq I, K} \delta(p_i + \Lambda_A d_i)
$$

$$
= \left(\sum_{i=1}^{N} d_i\right) 2\pi \delta\left(\Lambda_A \sum_{i=1}^{N} d_i - 2i\right) \delta(p_I + \Lambda_A d_I - i)\delta(p_K + \Lambda_A d_K - i)
$$

$$
\times \prod_{i \neq I, K} \delta(p_i + \Lambda_A d_i)
$$

$$
= 2\pi \delta\left(\Lambda_A - \frac{2i}{\sum_i d_i}\right) \delta\left(p_I + \frac{2i d_I}{\sum_i d_i} - i\right) \delta\left(p_K + \frac{2i d_K}{\sum_i d_i} - i\right)
$$

$$
\times \prod_{i \neq I, K} \delta\left(p_i + \frac{2i d_i}{\sum_i d_i}\right) \tag{17.61}
$$

Eq. (17.61) uniquely fixes, within the delta functions, *all* the p_is and, hence, one can perform the $\int_{\mathbf{p}}$ integration to explicitly obtain A. A similar result is obtained

Table 17.1 *The drift and cross-terms of p_is with d_is for the different expansion coefficients.*

	$ipGd = i\sum_{IJ} p_I G_{IJ} d_J$	$\gamma = \frac{1}{v^2}(\beta d + ipGd)$
$C_{B,0}$:	0	$\frac{1}{v^2}\beta d$
$C_{B,I}$:	$\frac{v^2}{\sum_I d_I} - \sum_J G_{IJ} d_J$	$\frac{1}{v^2}(\beta d + \frac{v^2}{\sum_I d_I} - \sum_J G_{IJ} d_J)$
$C_{B,IK}$:	$2\frac{v^2}{\sum_I d_I} - \sum_J G_{IJ} d_J - \sum_J G_{KJ} d_J$	$\frac{1}{v^2}(\beta d + 2\frac{v^2}{\sum_I d_I} - \sum_J G_{IJ} d_J - \sum_J G_{KJ} d_J)$

for B with Λ_B replacing Λ_A; note the constraints on the p_is do not depend on Λ_A and hence the values of S_p and γ are the same for coefficients A and B.

There are $N+1$ delta functions in Eq. (17.61) and performing the $\int_p$ integrations leaves over one delta function, which for the A term is given by $\delta(\Lambda_A - 2i/\sum_i d_i) = \delta(i\gamma + \xi + k - 2i/\sum_i d_i)$, where γ has been fixed by Eq. (17.61) and given in Table 17.1. Using this delta function, and a similar one for the B term, to perform the ξ integration in Eq. (17.59) yields the following

$$C_B^{IK} = \frac{e^{S_{IK}+\gamma\beta d}}{2(b-a)} \sum_k \int_h e^{-\frac{v^2}{2}k^2} [e^{-i(k+i\gamma)(h-fd-\beta d)} e^{-ik\beta d}$$

$$- e^{-i(k-i\gamma)(h-fd-\beta d)} e^{ik(\beta d + 2fd - 2a)}]$$

$$= \frac{e^{S_{IK}}}{2(b-a)} \int_h e^{\frac{1}{v^2}(\beta d - \sum_J G_{IJ} d_J - \sum_J G_{IK} d_J)(h-fd)}$$

$$\sum_{n=-\infty}^{+\infty} e^{-\frac{v^2}{2}k_n^2} [e^{-ik_n(h-fd)} - e^{ik_n(h+fd-2a)}]$$

$$= e^{S_{IK}} \int_a^b dh\, e^{\frac{1}{v^2}(\beta d - \sum_J G_{IJ} d_J - \sum_J G_{IK} d_J)(h-fd)} \mathcal{Q}[h, fd, v^2; a, b]$$

where recall $k = k_n = n\pi/(b-a)$ and $\mathcal{Q}[h, fd, v^2; a, b]$ is the barrier function given by Eq. (17.30).

After a long and tedious calculation one obtains

$$S_{IK} = G_{IK} - \frac{1}{v^2}\left(\beta d - \sum_J G_{IJ} d_J - \sum_J G_{IK} d_J\right)^2 \qquad (17.62)$$

The results for the different coefficients are obtained in the same manner as for $C_{B,IK}$; a remarkable result is that for all the coefficients the barrier function $\mathcal{Q}[h, fd, v^2; a, b]$ completely factorizes, leading to perturbative coefficients D_1 and D_2 that are evaluated in Section 17.8.

Table 17.2 *Value of quadratic function of the p_is for the different expansion coefficients.*

$$pGp = \sum_{IJ} p_I G_{IJ} p_J$$

$C_{B,0}$:	0
$C_{B,I}$:	$-\frac{v^2}{\sum_I d_I} + \frac{2}{\sum_I d_I} \sum_J G_{IJ} d_J - G_{II}$
$C_{B,IK}$:	$-2G_{IK} - G_{II} - G_{KK} - 4(\frac{v^2}{\sum_i d_i})^2 + \frac{4}{\sum_i d_i}(\sum_i G_{iI} d_i + \sum_i G_{iK} d_i)$

Table 17.3 *The value of the 'action' for the different expansion coefficients.*

$$S_p = -\tfrac{1}{2} pGp - \tfrac{1}{v^2}\left[(\beta d)^2 - (pGd)^2 + i(\beta d)pGd\right]$$

$S_{B,0}=$	$-\frac{1}{2v^2}(\beta d)^2$
$S_{B,I}=$	$-\frac{1}{2v^2}(\beta d - \sum_J G_{IJ} d_J)^2$
$S_{B,IK}=$	$G_{IK} - \frac{1}{2v^2}(\beta d - \sum_J G_{IJ} d_J - \sum_J G_{IK} d_J)^2$

The main results required for evaluating $C_{B,0}$, $C_{B,1}$, and $C_{B,2}$ are given in Tables 17.1 to 17.3; Tables 17.1 and 17.2 yield the values for $ipGd$, γ, and pGp. S_p, for the different coefficients, is given in Table 17.3.

Epilogue

At present, mainstream theoretical finance is almost completely dominated by stochastic calculus. *Quantum finance* [12], in contrast, presents a formulation of finance that is completely independent of stochastic calculus. *Quantum finance* addressed only a few problems of finance. The onus thus fell on me to demonstrate that a wide class of problems of finance, which at present are understood in terms of stochastic calculus, have a natural quantum finance formulation and generalization. It needed to be established that the quantum formulation of finance is equivalent to the one based on stochastic calculus and – if it is to be useful for the practitioners – is conceptually transparent and computationally tractable. Furthermore, it needed to be shown that quantum finance is not merely equivalent to stochastic calculus but goes *beyond* it; namely, that models based on quantum finance are more general than those based on stochastic calculus.

Interest rates and coupon bonds and their derivatives form a major component of the debt market. They are, by far, the most complex and intricate types of financial instruments that also have a rich mathematical structure. Debt instruments provided an ideal testing ground for *quantum finance*. I gravitated to the study of interest rates and coupon bonds, since these financial instruments provide an excellent arena for exhibiting the main features of quantum finance – as well as for illustrating its point of departure from stochastic calculus.

No attempt has been made to survey the subject of debt instruments, for which there exists a vast corpus of literature. Instead, this book has focused on only *new* results that have been obtained by applying quantum finance to debt instruments. The single most important new result, from my point of view, is the quantum formulation of the logarithmic Libor quantum field $\phi(t, x)$. In particular, the introduction of the nonlinear Hamiltonian, Lagrangian, and Feynman path integral for $\phi(t, x)$ opens up a potentially new field of study by providing a comprehensive platform for analyzing and pricing all interest rate instruments.

As has been demonstrated by the multifarious calculations carried out in the preceding chapters, the quantum formulation of finance allows for derivations that are straightforward and yields analytical results that are readily calibrated. Predictions made by quantum finance models are well suited to empirical studies and, in general, are seen to be more accurate than similar models based on stochastic calculus.

There are three quantum fields at the root of all calculations, namely the velocity field $\mathcal{A}(t, x)$, the bond forward interest rates $f(t, x)$ and log Libor $\phi(t, x)$.

The bond forward interest rates $f(t, x)$ drive the coupon bond market, whereas the Libor forward interest rates $f_L(t, x)$ and their associated logarithmic Libor rates $\phi(t, x)$ determine interest rates for cash time deposits. Although both the forward interest rates are based on the concept of discounting cash flows, these two are fundamentally different; $f(t, x)$ is modeled as a Gaussian quantum field, whereas $f_L(t, x)$ is a nonlinear quantum field, having stochastic volatility and stochastic drift.

The Libor forward interest rate $f_L(t, x)$ is related to log Libor $\phi(t, x)$ by Eq. (10.7)

$$\exp\left\{\int_{T_n}^{T_n+\ell} dx f_L(t, x)\right\} = 1 + \exp\left\{\int_{T_n}^{T_n+\ell} dx \phi(t, x)\right\}$$

The minimum tenor of $\ell \neq 0$ for cash deposits gives rise to a nonlinear mapping from the Libor forward interest rates $f_L(t, x)$ to $\phi(t, x)$. The nonlinear connection between the Libor forward interest rates $f_L(t, x)$ and $\phi(t, x)$ is the reason that modeling the two distinct sectors, namely bonds and interest rates, gives rise to many interesting and challenging problems.

The velocity field $\mathcal{A}(t, x)$ is a two-dimensional quantum field that is a natural generalization of white noise. Both $f(t, x)$ and $\phi(t, x)$, from Eqs. (5.1) and (6.58), can be expressed in terms of $\mathcal{A}(t, x)$ as follows

$$\frac{\partial f(t, x)}{\partial t} = \alpha(t, x) + \sigma(t, x)\mathcal{A}(t, x)$$

$$\frac{\partial \phi(t, x)}{\partial t} = \rho(t, x) + \gamma(t, x)\mathcal{A}_L(t, x)$$

The dynamics of the velocity field $\mathcal{A}(t, x)$ are specified by the stiff Lagrangian $\mathcal{L}[\mathcal{A}]$ that incorporates market time. From Eq. (5.25)

$$\mathcal{L}[\mathcal{A}] = -\frac{1}{2}\left\{\mathcal{A}^2(t, z) + \frac{1}{\mu^2}\left(\frac{\partial \mathcal{A}(t, z)}{\partial z}\right)^2 + \frac{1}{\lambda^4}\left(\frac{\partial^2 \mathcal{A}(t, z)}{\partial^2 z}\right)^2\right\}$$

$$z = (x - t)^\eta$$

The Libor velocity quantum field $\mathcal{A}_L(t,x)$ has the same stiff action as $\mathcal{A}(t,x)$, except the parameters μ_L, λ_L, and η_L have different empirical values.

The interest rate quantum fields $f(t,x)$ and $\phi(t,x)$ are more fundamental than the velocity field $\mathcal{A}(t,x)$. All calculations based on the path integral can be carried out using either $\mathcal{A}(t,x)$ or the interest rate quantum fields, since they are related by a change of variables. However, the interest rate state space and Hamiltonian operator can be written only in terms of interest rates quantum fields $f(t,x)$ or $\phi(t,x)$ and not in terms of $\mathcal{A}(t,x)$. For path dependent options, such as the American and barrier options, the numerical algorithms can only be defined in terms of $f(t,x)$ or $\phi(t,x)$. Since $\mathcal{A}(t,x)$ is a Gaussian quantum field, its entire content is encoded in its propagator, which appears in the interest rate Hamiltonian. Hence, $\mathcal{A}(t,x)$ is effectively incorporated into the Hamiltonian dynamics.

Theoretical finance is a rather unique subject, being the fusion and synthesis of quantitative laws – the hallmark of the natural sciences – with the political economy and psychology of human behavior. Quantum finance provides a meeting ground of mathematical regularity, embodied in the modeling of financial instruments by Lagrangians, Hamiltonians, and Feynman path integrals, with the subjective dimension of human beings, represented by the parameters of the theory.

In conclusion, the principles formulated in *quantum finance* are deeply grounded in the fundamentals of finance and firmly rooted in quantum mathematics; these principles enrich the field of quantitative finance and provide powerful theoretical and mathematical tools for analyzing financial instruments. This book, in particular, has demonstrated that quantum finance yields many fruitful results in the study and analysis of debt instruments.

Appendix A

Mathematical background

A few essential mathematical topics are discussed so as to have a complete and self-contained presentation of all the material required for following the derivations in the book.

A.1 Dirac-delta function

The Dirac-delta function is useful in the study of continuum spaces, and some of its essential properties are reviewed. Dirac-delta functions are not ordinary Lebesgue measureable functions since they have support on a measure zero set; rather they are generalized functions also called distributions. In essence, the Dirac-delta function is the continuum generalization of the discrete Kronecker-delta function.

Consider a continuous line labeled by coordinate x such that $-\infty \le x \le +\infty$, and let $f(x)$ be an infinitely differentiable function. The Dirac-delta function, denoted by $\delta(x - a)$, is defined by the following

$$\delta(x - a) = \delta(a - x) \ : \ \text{even function}$$

$$\delta(c(x - a)) = \frac{1}{|c|}\delta(x - a)$$

$$\int_{-\infty}^{+\infty} dx f(x)\delta(x - a) = f(a) \tag{A.1}$$

$$\int_{-\infty}^{+\infty} dx f(x)\frac{d^n}{dx^n}\delta(x - a) = (-1)^n \frac{d^n}{dx^n} f(x)|_{x=a} \tag{A.2}$$

The Heaviside step function $\Theta(t)$ is defined by

$$\Theta(t) = \begin{cases} 1 & t > 0 \\ \frac{1}{2} & t = 0 \\ 0 & t < 0 \end{cases} \tag{A.3}$$

From its definition $\Theta(t) + \Theta(-t) = 1$. The following is a representation of the δ-function.

$$\int_{-\infty}^{b} \delta(x - a) = \Theta(b - a) \tag{A.4}$$

$$\Rightarrow \int_{-\infty}^{a} \delta(x - a) = \Theta(0) = \frac{1}{2} \tag{A.5}$$

where the last equation is due to the Dirac-delta function being an even function. From Eq. (A.4)

$$\frac{d}{db}\Theta(b - a) = \delta(b - a)$$

A representation of the δ-function based on the Gaussian distribution is

$$\delta(x - a) = \lim_{\sigma \to 0} \frac{1}{\sqrt{2\pi\sigma^2}} \exp\left\{-\frac{1}{2\sigma^2}(x - a)^2\right\} \tag{A.6}$$

Moreover

$$\delta(x - a) = \lim_{\mu \to \infty} \frac{1}{2}\mu \exp\left\{-\mu|x - a|\right\}$$

The definition of Fourier transform yields a representation of the Dirac-delta function that is widely used in various chapters for representing the payoff of financial instruments. It can be shown that

$$\delta(x - a) = \int_{-\infty}^{+\infty} \frac{dp}{2\pi} e^{ip(x-a)} \tag{A.7}$$

A proof of Eq. (A.7) is found in many books on quantum mechanics [70]. One can perform the following consistency check of Eq. (A.7). Integrate both sides of

Eq. (A.7) over x as follows

$$\text{L.H.S.} = \int_{-\infty}^{+\infty} dx e^{-ikx} \delta(x-a) = e^{-ika}$$

$$\text{R.H.S.} = \int_{-\infty}^{+\infty} dx e^{-ikx} \int_{-\infty}^{+\infty} \frac{dp}{2\pi} e^{ip(x-a)}$$

$$= \int_{-\infty}^{+\infty} \frac{dp}{2\pi} e^{-ipa} 2\pi \delta(p-k) = e^{-ika}$$

where Eq. (A.7) was used in performing the x integration for the right-hand side. Hence, one can see that Eq. (A.7) is self-consistent.

The scalar product of the basis state $|x'\rangle$ and its dual $\langle x|$ is a Dirac-delta function; the Fourier representation yields

$$\langle x|x'\rangle = \delta(x-x') = \int_{-\infty}^{+\infty} \frac{dp}{2\pi} e^{ip(x-x')}$$

$$= \int_{-\infty}^{+\infty} \frac{dp}{2\pi} \langle x|p\rangle \langle p|x'\rangle$$

and hence, for momentum space basis $|p\rangle$, the completeness equation is given by

$$\int_{-\infty}^{\infty} \frac{dp}{2\pi} |p\rangle \langle p| = \mathcal{I} \tag{A.8}$$

The scalar product of the dual basis state $\langle x|$ with the momentum basis state $|p\rangle$ is given by

$$\langle x|p\rangle = e^{ipx}; \quad \langle p|x\rangle = e^{-ipx} \tag{A.9}$$

To see the relation of the Dirac-delta function with the discrete Kronecker delta, recall for n, m integers

$$\delta_{n-m} = \begin{cases} 0 & n \neq m \\ 1 & n = m \end{cases}$$

Discretize continuous variable x into a lattice of discrete points $x = n\epsilon$, and let $a = m\epsilon$; then $f(x) \to f_n$. Discretizing Eq. (A.1) gives

$$\int_{-\infty}^{+\infty} dx f(x)\delta(x-a) \to \epsilon \sum_{n=-\infty}^{+\infty} f_n \delta(x_n - a_m)$$

$$= f_m = \sum_{n=-\infty}^{+\infty} \delta_{n-m} f_m$$

$$\Rightarrow \delta(x-a) \to \frac{1}{\epsilon} \delta_{n-m} \tag{A.10}$$

Taking the limit of $\epsilon \to 0$ in the equation above yields

$$\delta(x-a) = \lim_{\epsilon \to 0} \frac{1}{\epsilon} \delta_{n-m} = \begin{cases} 0 & x \neq a \\ \infty & x = a \end{cases}$$

A.2 Martingale

A martingale refers to a special category of stochastic processes. An arbitrary discrete stochastic process is a collection of random variables X_i, $i = 1, 2, \ldots, N$ that is described by a joint probability distribution function $p(x_1, x_2, \ldots, x_N)$. The stochastic process is a martingale if it satisfies

$$E[X_{n+1}|x_1, x_2, \ldots, x_n] = x_n : \quad \text{martingale} \tag{A.11}$$

In other words, the expected probability of the random variable X_{n+1} – conditioned on the occurrence of x_i for random variable X_i, $i = 1, 2, \ldots, n$ – is simply x_n itself.

One can think of the martingale as describing a gambling game; the given condition x_n is the amount of money that the gambler has on the conclusion of the nth game, and the random variable X_{n+1} represents the various possible outcomes of the $(n + 1)$th game. The martingale condition states that the *expected value* of the gambler's money at the end of the $(n + 1)$th game is equal to the money that he enters the $(n + 1)$th game with, namely x_n. Using Eq. (A.11), one can prove the following result.

$$E[X_{n+1}] = \int dx_1 dx_2 \ldots dx_n dx_{n+1} E[X_{n+1}|x_1, x_2, \ldots, x_n] p(x_1, x_2, \ldots, x_{n+1})$$

$$= \int dx_1 dx_2 \ldots dx_n dx_{n+1} \, x_n \, p(x_1, x_2, \ldots, x_{n+1}) = E[X_n]$$

$$\Rightarrow E[X_{n+1}] = E[X_n] = E[X_{n-1}] = \cdots = E[X_1]$$

$$\Rightarrow E[X_n] = E[X_1] \tag{A.12}$$

Suppose there is a boundary condition specified for the stochastic process such that $X_1 = x_1$ is fixed; hence $E[X_1] = x_1$ and

$$E[X_n] = E[X_1] = x_1 \qquad\qquad (A.13)$$

In a fair game obeying the martingale condition, the gambler, on the average, neither loses nor wins, and leaves the casino with the money that she or he came in with, namely x_1.

For the case of a security $S(t)$, the random variable $X(T) = e^{-\int_0^T dt\, r(t)} S(T)$ follows a martingale process. From Eq. (A.13)

$$E[X(T)] = X(0) = S(0)$$
$$\Rightarrow S(0) = E[e^{-\int_0^T dt\, r(t)} S(T)|S(0)] \qquad (A.14)$$

The price of the security at $t = 0$, that is $S(0)$, is usually taken as the initial condition for the stochastic process. Eq. (A.14) plays a central role in determining the risk-neutral martingale measure in finance.

Consider a general stochastic differential equation for a two-dimensional function $\chi(t, x)$

$$\frac{\partial \chi(t, x)}{\partial t} = d(t, x) + \int dx'\, G(x, x'; t) v(t, x') \mathcal{A}(t, x') \qquad (A.15)$$

where $\mathcal{A}(t, x)$ is the Gaussian two-dimensional quantum field defined by the action given in Eq. (5.4). $G(x, x'; t)$ is a deterministic function and the quantities $d(t, x)$ and $v(t, x)$ can, in general, depend on $\chi(t, x)$. Eq. (A.15) is a stochastic differential equation that is encountered in Chapter 6 in the study of Libor.

An initial (or final) condition needs to be specified to obtain a solution for Eq. (A.15); for applications in finance, the initial condition is specified as follows

$$\text{Boundary condition:} \quad \chi(t_0, x) = \text{ fixed}$$

Discretizing time in Eq. (A.15) yields, for infinitesimal ϵ

$$\chi(t + \epsilon, x) = \chi(t, x) + \epsilon d(t, x) + \epsilon \int dx'\, G(x, x'; t) v(t, x') \mathcal{A}(t, x') \quad (A.16)$$

The martingale condition given in Eq. (A.11) requires that the expectation value of $\chi(t + \epsilon, x)$ is taken conditioned on $\chi(t, x)$ having a *fixed* value. In taking the expectation value of Eq. (A.16), the functions $d(t, x)$ and $v(t, x)$ are deterministic, since they depend only on $\chi(t, x)$ and hence can be taken outside the expectation value. Since $E[\mathcal{A}(t, x')] = 0$, taking the conditional expectation value of both sides of Eq. (A.16) yields the following

$$E[\chi(t + \epsilon, x)|\chi(t, x)] = \chi(t, x) + \epsilon d(t, x)$$

$$+ \epsilon \int dx' G(x, x'; t)v(t, x')E[\mathcal{A}(t, x')]$$

$$= \chi(t, x) + \epsilon d(t, x)$$

The martingale condition given in Eq. (A.11) requires $E[\chi(t + \epsilon, x)|\chi(t, x)] = \chi(t, x)$; hence, from Eqs. (A.11) and (A.12)

$$d(t, x) = 0: \quad \text{martingale condition}$$

$$\Rightarrow E[\chi(t + \epsilon, x)] = E[\chi(t, x)]$$

$$\Rightarrow E\left[\frac{\partial \chi(t, x)}{\partial t}\right] = 0: \quad \text{martingale} \tag{A.17}$$

In summary, for $\chi(t, x)$ to be a martingale, it is sufficient that the drift is zero, namely

$$d(t, x) = 0: \quad \text{martingale condition} \tag{A.18}$$

$$\Rightarrow \frac{\partial \chi(t, x)}{\partial t} = \int dx' G(x, x'; t)v(t, x')\mathcal{A}(t, x') \tag{A.19}$$

Option theory hinges on the martingale property of option pricing. Suppose $\mathcal{P}$ is the payoff of an option $C(t)$ that matures at T. For numeraire $M(t)$, the discounted option $C(t)/M(t)$ must follow a martingale evolution for its price to be free from arbitrage opportunities. The martingale property given in Eq. (A.13) then requires that the price at present time t_0 is given by

$$\frac{C(t_0)}{M(t_0)} = E\left[\frac{C(T)}{M(T)}\right] = E\left[\frac{\mathcal{P}}{M(T)}\right]$$

$$\Rightarrow C(t_0) = M(t_0)E\left[\frac{\mathcal{P}}{M(T)}\right] \tag{A.20}$$

The prices of all options are obtained from the application of Eq. (A.20) to various instruments and payoff functions.

A.3 Gaussian integration

Gaussian integration permeates all of theoretical finance, as well as forming one of the foundations of quantum theory. One-dimensional and multi-dimensional Gaussian integration are briefly reviewed. Gaussian integrals have the remarkable property that they can be generalized to infinite dimensions, which is briefly discussed.

A.3.1 One-dimensional Gaussian integral

Consider the one-dimensional definite Gaussian integral

$$Z[j] = \mathcal{N} \int_{-\infty}^{+\infty} e^{-\frac{1}{2}\lambda x^2 + jx} dx$$

All the moments of x can be obtained by

$$E[(x^n)] = \frac{d^n Z[j]}{dj^n}\bigg|_{j=0}$$

and hence $Z[j]$ is called the generating function for the Gaussian distribution.

The normalization constant $\mathcal{N}$ is chosen so that $Z(0) = 1$. Squaring $Z[0]$, and converting it to polar coordinates, gives

$$Z^2[0] = \mathcal{N}^2 \int_{-\infty}^{+\infty} \int_{-\infty}^{+\infty} e^{-\frac{1}{2}\lambda(x^2+y^2)} dx dy = \mathcal{N}^2 \int_0^\infty \int_0^{2\pi} r e^{-\frac{1}{2}\lambda r^2} dr d\theta$$

$$1 = \mathcal{N}^2 2\pi \int_0^\infty d\xi \, e^{-\lambda\xi} = \mathcal{N}^2 \frac{2\pi}{\lambda} \Rightarrow \mathcal{N} = \sqrt{\frac{\lambda}{2\pi}}$$

Shifting $x \to x - \dfrac{j}{\lambda}$ leaves the integration measure invariant and yields the final result

$$Z[j] = e^{\frac{1}{2\lambda}j^2} \mathcal{N} \int_{-\infty}^{+\infty} e^{-\frac{1}{2}\lambda x^2} dx$$

$$= e^{\frac{1}{2\lambda}j^2} \tag{A.21}$$

A.3.2 Higher-dimensional Gaussian integrals

Consider the general n-dimensional Gaussian integral, with variables $x_1, x_2, \ldots, x_n$. The Gaussian integral can be written as

$$Z[J] = \mathcal{N} \int_{-\infty}^{+\infty} e^S dx_1 dx_1 dx_2 \ldots dx_n$$

$$S = -\frac{1}{2} \sum_{i,j=1}^n x_i A_{ij} x_j + \sum_{i=1}^n J_i x_i$$

Choose the normalization constant so that $Z(0) = 1$. In quantum theory S is called the action.

A is an $n \times n$ positive symmetric matrix can be diagonalized by an orthogonal matrix M and yields

$$A = M^T \mathrm{diag}(\lambda_1, \lambda_2, \ldots, \lambda_n) M$$

$$MM^T = \mathcal{I}_{n \times n}$$

where $\mathcal{I}_{n \times n}$ is an $n \times n$ unit matrix, and M^T is the transpose of M.

Only matrices A with positive eigenvalues $\lambda_i \geq 0$ are considered. A change of variables

$$x_i = \sum_{j=1}^{n} M_{ij} z_j$$

$$\prod_{i=1}^{n} dx_i = \det(M) \prod_{i=1}^{n} dz_i = \prod_{i=1}^{n} dz_i$$

yields for the n-dimensional Gaussian integral

$$Z[J] = \mathcal{N} \prod_{i=1}^{n} \left[\int_{-\infty}^{+\infty} dz_i \, e^{-\frac{1}{2}\lambda_i z_i^2 + \tilde{J}_i z_i} \right]$$

$$\tilde{J}_i \equiv \sum_{j=1}^{n} J_j M_{ji}^T$$

The n-dimensional Gaussian integral has completely factorized into a product of one-dimensional Gaussian integrals, all of which can be evaluated by the result given in Eq. (A.21). Hence

$$Z[J] = \mathcal{N} \prod_{i=1}^{n} \left[\sqrt{\frac{2\pi}{\lambda_i}} e^{\frac{1}{2\lambda_i} \tilde{J}_i^2} \right] \tag{A.22}$$

In matrix notation

$$\mathcal{N} \prod_{i=1}^{n} \sqrt{\frac{2\pi}{\lambda_i}} = \mathcal{N}(2\pi)^{n/2} \frac{1}{\sqrt{\det A}} = 1$$

$$\sum_{i=1}^{n} \frac{1}{\lambda_i} \tilde{J}_i^2 = J \frac{1}{A} J \equiv J A^{-1} J$$

Hence, the final result can be written as

$$Z[J] = \exp\left(\frac{1}{2} J A^{-1} J \right) \tag{A.23}$$

A.3.3 Normal (Gaussian) random variable

The normal, or Gaussian, random variable – denoted by $N(\mu, \sigma)$ – is a variable x that has a probability distribution given by

$$P(x) = \frac{1}{\sqrt{2\pi\sigma^2}} \exp\left\{-\frac{1}{2\sigma^2}(x-\mu)^2\right\} \qquad (A.24)$$

From Eq (A.21)

$$E[x] \equiv \int_{-\infty}^{+\infty} x P(x) = \mu : \text{mean}$$

$$E[(x-\mu)^2] \equiv \int_{-\infty}^{+\infty} (x-\mu)^2 P(x) = \sigma^2 : \text{variance}$$

Any normal random variable is equivalent to the $N(0,1)$ random variable via the following linear transformation

$$X = N(\mu, \sigma);\ Z = N(0,1)$$

$$\Rightarrow X = \mu + \sigma Z$$

All the moments of the random variable $Z = N(0,1)$ can be determined by the generating function given in Eq. (A.21); namely

$$E[z^n] = \frac{d^n}{dJ^n} Z[J]|_{J=0}$$

The cumulative distribution for the normal random variable $N(x)$ is defined by

$$\text{Prob}(-\infty \le z \le x) = N(x) = \frac{1}{\sqrt{2\pi}} \int_{-\infty}^{x} e^{-\frac{1}{2}z^2} dz \qquad (A.25)$$

A sum of normal random variables is also another normal random variable

$$Z_1 = N(\mu_1, \sigma_1);\ Z_2 = N(\mu_2, \sigma_2);\ \ldots;\ Z_n = N(\mu_n, \sigma_n)$$

$$\Rightarrow Z = \sum_{i=1}^{n} Z_i = N(\mu, \sigma) \ \Rightarrow\ \mu = \sum_{i=1}^{n} \mu_i;\ \ \sigma^2 = \sum_{i=1}^{n} \sigma_i^2$$

The result above can be proved using the generating function given in Eq. (A.21).

A.3.4 Infinite-dimensional Gaussian integrations

Consider a continuum number of integration variables $x(t)$, with $-\infty \le t \le +\infty$, and with the "action" given by

$$S = -\frac{1}{2} \int_{-\infty}^{+\infty} dt\,dt'\, x(t) D^{-1}(t,t') x(t') + \int_{-\infty}^{+\infty} dt\, J(t) x(t) \qquad \text{(A.26)}$$

By discretizing the variable t, following the steps taken in the derivation of the $n \times n$ case, and then taking the limit of $n \to \infty$ yields

$$Z[J] = \mathcal{N} \prod_{t=-\infty}^{+\infty} \int_{-\infty}^{+\infty} dx(t) e^{S} = \exp\left\{ \frac{1}{2} \int_{-\infty}^{+\infty} dt\,dt'\, J(t) D(t,t') J(t') \right\} \qquad \text{(A.27)}$$

$$\int_{-\infty}^{+\infty} ds\, D^{-1}(t,s) D(s,t') = \delta(t - t')$$

The normalization $\mathcal{N}$ is now a divergent quantity, that ensures the usual normalization $Z(0) = 1$. In discussions on quantum theory, Eq. (A.27) plays a central role.

Consider the special case of N continuous $x_i(t)$, with $i = 1, 2, \ldots, N$ and action given by

$$S = -\frac{1}{2} \sum_{ij}^{N} \int_{-\infty}^{+\infty} dt\, D_{ij}^{-1} x_i(t) x_j(t) + \sum_{i}^{N} \int_{-\infty}^{+\infty} dt\, J_i(t) x_i(t) \qquad \text{(A.28)}$$

From Eq. (A.27), the generating functional is given by

$$Z[J] = \exp\left\{ \frac{1}{2} \sum_{ij}^{N} \int_{-\infty}^{+\infty} dt\,dt'\, J_i(t) D_{ij}(t,t') J_j(t') \right\} \qquad \text{(A.29)}$$

$$E[x_i(t) x_j(t')] = \frac{1}{Z} \int DX e^{S} x_i(t) x_j(t') \Big|_{J_i=0} = D_{ij}(t,t') \qquad \text{(A.30)}$$

The fundamental reason why Gaussian integration generalizes to infinite dimensions is because the measure is invariant under translations, that is under $x(t) \to x(t) + \xi(t)$; one can easily verify that this symmetry of the measure yields the result obtained in Eq. (A.27).

Consider the action of the 'harmonic oscillator' given by

$$S = -\frac{m}{2} \int_{-\infty}^{+\infty} dt \left[\left(\frac{dx(t)}{dt} \right)^2 + \omega^2 x^2(t) \right] \tag{A.31}$$

$$= -\frac{m}{2} \int_{-\infty}^{+\infty} dt\, x(t) \left(-\frac{d^2}{dt^2} + \omega^2 \right) x(t)$$

$$\Rightarrow D^{-1}(t,t') = m \left(-\frac{d^2}{dt^2} + \omega^2 \right) \delta(t - t')$$

where an integration by parts was done, discarding boundary terms at $\pm\infty$, to obtain the second equation above. The propagator $D(t,t')$ is given by

$$D(t,t') = \frac{1}{2\pi m} \int_{-\infty}^{+\infty} dp \frac{e^{ip(t-t')}}{p^2 + \omega^2}$$

$$= \frac{1}{2m|\omega|} e^{-|\omega||t-t'|}$$

The result above can be verified by using Eq. (A.7).

A.4 White noise

The salient properties of Gaussian white noise are reviewed. The defining equations for white noise are the following

$$E[R(t)] = 0; \quad E[R(t)R(t')] = \delta(t - t') \tag{A.32}$$

The random variables $R(t)$ are shown in Figure A.1; each point t on the vertical line represents one independent random variable $R(t)$.

To write the probability measure for $R(t)$, let time $t \in [t_1, t_2]$ take discrete values in the finite interval, which depends on the problem of interest; discretize $t \to m\epsilon$, with $m = 1, 2, \ldots, M$ where $M = [(t_2 - t_1)/\epsilon]$, and with $R(t) \to R_m$. From Eqs. (A.32) and (A.10), the expectation value for white noise is given by

$$E[Rn] = 0; \quad E[R_m R_n] = \frac{1}{\epsilon} \delta_{m-n} \Rightarrow E[R_n^2] = \frac{1}{\epsilon} \tag{A.33}$$

The discussion in Section 5.10 can be reduced to the case of white noise and explains in what sense a random quantity like $R^2(t)$ can be considered to be deterministic.

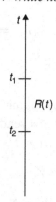

Figure A.1 One independent (integration) random variable $R(t)$ corresponds to each point of the t-axis. The collection of independent variables constitutes white noise.

White noise is singular for equal time, given by $t = t'$, since $\delta(0) = \infty$, and gives rise to the results of Ito calculus [12]. From Eq. (A.33), the equal time product of white noise, to leading order in ϵ is *deterministic* and yields

$$R_n^2 = \frac{1}{\epsilon} + \text{ random terms of } O(1) \tag{A.34}$$

From Eq. (A.33) it follows that for each n, R_n is an independent Gaussian random variable $N(0, 1/\sqrt{\epsilon})$. The joint probability distribution for M independent Gaussian random variables from Eq. (A.24) is, hence, given by

$$P[R] = \left[\sqrt{\frac{\epsilon}{2\pi}}\right]^M \exp\left\{-\frac{\epsilon}{2}\sum_{m=1}^{M} R_m^2\right\}$$

$$\int dR = \prod_{m=1}^{M}\int_{-\infty}^{+\infty} dR_m$$

$$\Rightarrow E[R_n R_m] = \int dR\, P[R] R_n R_m = \frac{1}{\epsilon}\delta_{n-m}$$

The normalization of $P[R]$ factorizes out of all calculations. In the path integral formulation, the normalization is taken into account by defining an action S_0 and dividing out by the 'partition function' $\tilde{Z}$ as follows

$$\tilde{S}_0 = -\frac{\epsilon}{2}\sum_{m=1}^{M} R_m^2; \quad \int dR = \prod_{m=1}^{M}\int_{-\infty}^{+\infty} dR_m \tag{A.35}$$

$$\Rightarrow E[R_n R_m] = \frac{1}{\tilde{Z}} \int dR e^{\tilde{S}_0} R_n R_m = \frac{1}{\epsilon}\delta_{n-m} \qquad (A.36)$$

$$\tilde{Z} = \int dR e^{\tilde{S}_0}$$

The continuum limit is obtained by taking $\epsilon \to 0$. For purposes of rigor, the continuum limit needs to be taken by first rendering the path integral into discrete multiple integrals as given above. For $t_1 < t < t_2, \epsilon \to 0$ yields

$$\tilde{S}_0 \to S_0; \ \tilde{Z} \to Z \qquad (A.37)$$

$$S_0 = -\frac{1}{2}\int_{t_1}^{t_2} dt R^2(t) \qquad (A.38)$$

$$\int dR \to \int DR = \prod_{t=t_1}^{t_2}\int_{-\infty}^{+\infty} dR(t) \qquad (A.39)$$

$$E[R_n R_m] \to E[R(t)R(t')] = \frac{1}{Z}\int DR e^{S_0} R(t)R(t')$$

$$= \delta(t - t')$$

The expression above for $E[R(t)R(t')]$ is one of the simplest examples of a path integral correlator.

The action functional S_0 is ultra-local with all the variables being decoupled; generically, $\int DR$ stands for the (path) integration over all the random variables $R(t)$ which appear in the problem.

The path integral can be extended to N correlated Gaussian white noises $R_i(t)$, which are defined by

$$E[R_i(t)] = 0; \ \ E[R_i(t)R_j(t')] = \rho_{ij}\delta(t - t'); \ \ 0 \le t, t' \le T$$

The path integral that yields the white noise correlators, from Eqs. (A.29) and (A.30), given by

$$E[R_i(t)R_j(t')] = \frac{1}{Z}\int DR R_i(t)R_j(t') e^S = \rho_{ij}\delta(t - t')$$

$$S = \int_0^T dt\mathcal{L}; \ \mathcal{L} = -\frac{1}{2}\sum_{ij=1}^N \rho_{ij}^{-1} R_i(t)R_j(t): \text{ Lagrangian} \quad (A.40)$$

$$\int DR = \prod_{t=0}^{T}\prod_{i=1}^{N}\int_{-\infty}^{+\infty} dR_i(t); \ Z = \int DR e^S$$

All expectation values of financial instruments are evaluated using the path integral. For an instrument $\mathcal{O}$, its average is given as follows

$$E[\mathcal{O}] = \frac{1}{Z} \int DR\mathcal{O}e^{S_0}$$

A useful formula for the generating functional for R, obtained as a special case of Eq. (A.27), is the following

$$Z[j] = \frac{1}{Z} \int DR e^{\int_{t_1}^{t_2} dt j(t) R(t)} e^{S_0[R, t_1, t_2]}$$

$$= \exp\left\{\frac{1}{2} \int_{t_1}^{t_2} dt j^2(t)\right\} \tag{A.41}$$

A.5 Functional differentiation

Consider variables f_n, $n = 0, \pm 1, \pm 2, \dots, \pm N$ that satisfy

$$\frac{\partial f_n}{\partial f_m} = \delta_{n-m}$$

Let $t = n\epsilon$, with $N \to \infty$. The limit $\epsilon \to 0$ yields

$$\frac{\partial f_n}{\partial f_m} \to \frac{\delta f(t)}{\delta f(t)'} \equiv \lim_{\epsilon \to 0} \frac{1}{\epsilon} \frac{\partial f_n}{\partial f_m}$$

$$\Rightarrow \frac{\delta f(t)}{\delta f(t')} = \lim_{\epsilon \to 0} \frac{1}{\epsilon} \delta_{n-m} \to \delta(t - t') \tag{A.42}$$

In general, the *functional derivative* of $\Omega[f]$ – an arbitrary functional of $f(t)$ – is denoted by $\delta/\delta f(t)$ and is defined by

$$\frac{\delta \Omega[f]}{\delta f(t)} = \lim_{\epsilon \to 0} \frac{\Omega[f(t') + \epsilon \delta(t - t')] - \Omega[f]}{\epsilon} \tag{A.43}$$

Note that ϵ has the dimensions of $[f] \times [t]$. In the notation of state space one has

$$\left\langle f \middle| \frac{\delta}{\delta f(t)} \middle| \Omega \right\rangle = \frac{\delta}{\delta f(t)} \langle f | \Omega \rangle = \frac{\delta \Omega[f]}{\delta f(t)}$$

- Consider the simplest function $\Omega[f] = f(t_0)$; then, from Eq. (A.43)

$$\frac{\delta\Omega[f]}{\delta f(t)} = \frac{\delta f(t_0)}{\delta f(t)} = \lim_{\epsilon \to 0} \frac{f(t_0) + \epsilon\delta(t - t_0) - f(t_0)}{\epsilon} = \delta(t - t_0)$$

- Let $\Omega[f] = \int d\tau\, f^n(\tau)$; from above

$$\frac{\delta\Omega[f]}{\delta f(t)} = \int d\tau n f^{n-1}(\tau)\frac{\delta f(\tau)}{\delta f(t)} \tag{A.44}$$

$$= \int d\tau n f^{n-1}(\tau)\delta(t - \tau) = n f^{n-1}(t) \tag{A.45}$$

A.5.1 Chain rule

The chain rule for calculus of many variables has a generalization to functional calculus. Consider a change of variables from f_n to g_n; the chain rule of calculus yields

$$\frac{\partial}{\partial f_n} = \sum_{m=1}^{N} \frac{\partial g_m}{\partial f_n}\frac{\partial}{\partial g_m}$$

As before, let $t = n\epsilon$, $t' = m\epsilon$; re-write the above expression as follows

$$\frac{1}{\epsilon}\frac{\partial}{\partial f_n} = \epsilon \sum_{m=1}^{N} \left[\frac{1}{\epsilon}\frac{\partial g_m}{\partial f_n}\right]\left[\frac{1}{\epsilon}\frac{\partial}{\partial g_m}\right]$$

Taking the limit of $N \to \infty$ and $\epsilon \to 0$ yields

$$\lim_{\epsilon \to 0} \frac{1}{\epsilon}\frac{\partial}{\partial f_n} \to \frac{\delta}{\delta f(t)} = \int dt' \frac{\delta g(t')}{\delta f(t)}\frac{\delta}{\delta g(t')} \; : \; \text{Chain rule} \tag{A.46}$$

A.6 State space $\mathcal{V}$

The option price $C(t, x)$ on some underlying security S is an element of a linear vector space – called a *state space* and denoted by $\mathcal{V}$. The state space consists of all possible functions of the security S. For the case when the security is a stock price, $S = e^x$, with $x \in \mathcal{R}$, and the state space $\mathcal{V}$ consists of all possible functions $f(x)$, with $x \in \mathcal{R}$.

The coordinate x is called a *degree of freedom*. The dual space of $\mathcal{V}$ – denoted by $\mathcal{V}_{\text{dual}}$ – consists of all linear mappings from $\mathcal{V}$ to the complex numbers, and is also a linear vector space.

In Dirac's bracket notation for state vectors, an element g of V is denoted by the vector $|g\rangle$ and an element of V_{dual} by the dual vector $\langle p|$. The scalar product is defined for any two vectors from the state space and its dual, and is given by a complex number equal to $\langle p|g\rangle = \langle g|p\rangle^*$, where $*$ stands for complex conjugation. Both V and its dual V_{dual} are referred to as the state space of the system.

The state space in option pricing is of central importance since all option prices belong to a state space defined by the underlying security. The simplest description of the state space, similar to the description of a finite-dimensional vector space, is to enumerate a complete set of basis vectors so that any arbitrary vector can then be represented in terms of these basis states. The completeness equation is a statement that one has a complete set of linearly independent basis vectors.

Consider the possible values of a stock price $S = e^x$; suppose for now that the stock can have only a discrete set of values. Let $x = na$ with lattice spacing a; since $-\infty \leq x \leq +\infty$, n can be any integer. The basis states are labeled by $|n\rangle$ and the dual basis states by $\langle n|$. The discrete values of the stock price are represented by an infinite column vector with the only nonzero entry being unity in the nth position. Hence

$$|n\rangle \ : \ n = 0, \pm 1, \pm 2, \ldots, \pm\infty$$

$$|n\rangle = \begin{bmatrix} \cdots \\ 0 \\ 1 \\ 0 \\ \cdots \end{bmatrix} \ : \ n\text{th position}; \quad \langle m|n\rangle = \delta_{n-m} \equiv \begin{cases} 1 & n = m \\ 0 & n \neq m \end{cases}$$

$$\sum_{n=-\infty}^{+\infty} |n\rangle\langle n| = \mathcal{I} = \text{diagonal}(1, 1, \ldots) \ : \ \text{completeness equation}$$

where $\mathcal{I}$ above is the infinite-dimensional unit matrix. The completeness is also referred as the *resolution of the identity* since only a complete set of basis states, taken together, can construct the identity operator on state space.

The allowed values of the stock price S correspond to x taking any real value, that is $x \in \mathcal{R}$, and hence the limit of $a \to 0$ needs to be taken. The state vector for the particle is given by the 'ket vector' $|x\rangle$, with its dual given by the 'bra vector' $\langle x|$. In terms of the underlying lattice ($x = na$)

$$|x\rangle = \lim_{a \to 0} \frac{1}{\sqrt{a}} |n\rangle; \quad -\infty \leq x \leq \infty$$

The scalar product, for $x = na$ and $x' = ma$, in the limit of $a \to 0$, is given, from Eq. (A.10), by the Dirac-delta function

$$\langle x | x' \rangle = \frac{1}{a} \delta_{m-n} \to \delta(x - x') \equiv \begin{cases} \infty & x = x' \\ 0 & x \neq x' \end{cases}$$

The completeness equation is given by

$$\sum_{n=-\infty}^{+\infty} |n\rangle\langle n| \to a \sum_{n=-\infty}^{+\infty} |x\rangle\langle x|$$

$$\Rightarrow \int_{-\infty}^{+\infty} dx |x\rangle\langle x| = \mathcal{I} \ : \ \text{completeness equation}$$

where $\mathcal{I}$ is the identity operator on (function) state space.

A more direct derivation of the completeness equation is to consider the scalar product of two functions, namely

$$\langle f | g \rangle \equiv \int dx f^*(x) g(x)$$

$$= \langle f | \left\{ \int_{-\infty}^{+\infty} dx |x\rangle\langle x| \right\} | g \rangle$$

and this yields the completeness equation

$$\mathcal{I} = \int_{-\infty}^{+\infty} dx |x\rangle\langle x| \qquad\qquad (A.47)$$

The completeness equation given by Eq. (A.47) is a key equation that is central to the analysis of the state space. For the case of two equities $S = e^x$ and $Q = e^y$, the values x, y obey the completeness equation given by

$$\mathcal{I} = \int_{-\infty}^{+\infty} dx dy |x, y\rangle\langle x, y| \qquad\qquad (A.48)$$

where $|x, y\rangle \equiv |x\rangle \otimes |y\rangle$. The generalization to N equities is straightforward.

The vector $|f\rangle$ and its dual $\langle f|$ have the important property that they define the 'length' $\langle f|f\rangle$ of the vector. The completeness equation Eq. (A.47) yields the following[1]

$$\langle f|f\rangle = \left\langle f \int_{-\infty}^{+\infty} |x\rangle\langle x|f\right\rangle$$

$$= \int_{-\infty}^{+\infty} f(x)^* f(x) \geq 0$$

The bra and ket vectors $\langle x|$ and $|x\rangle$ are the basis vectors of the V_{dual} and V respectively. An element of the state space V is the ket vector $|f\rangle$, and can be thought of as an infinite-dimensional vector with components given by $f(x) = \langle x|f\rangle$. The vector $|f\rangle$ can be mapped to a unique *dual* vector denoted by $\langle f| \in V_{\text{dual}}$. In components $f^*(x) = \langle f|x\rangle$. The vector $|f\rangle$ has the following representation in the $|x\rangle$ basis

$$|f\rangle = \int_{-\infty}^{+\infty} dx\langle x|f\rangle|x\rangle$$

$$= \int_{-\infty}^{+\infty} dx f(x)|x\rangle \qquad (A.49)$$

A.6.1 Operators: Hamiltonian

An operator is defined as a linear mapping of the state space V onto itself, and is an element of the tensor product space $V \otimes V_{\text{dual}}$. For a two-state system, the state space is a two-dimensional Euclidean space and operators are 2×2 matrices. Consider a state space that consists of all functions of single (real) variable x, namely $V \equiv \{f(x)|x \in \Re\}$, where $\langle x|f\rangle = f(x)$; operators on this state space are infinite-dimensional generalizations of $N \times N$ matrices, with $N \to \infty$.

One of the most important operators is the coordinate operator $\hat{x}$ that simply multiplies $f(x) \in V$ by x, that is $\hat{x} f(x) \equiv x f(x)$. Another important operator is the differential operator $\partial/\partial x$ that maps $f(x) \in V$ to its derivative $\partial f(x)/\partial x$. All the operators that will be studied are functions of the operators $\hat{x}$ and $\partial/\partial x$.

Similar to a $N \times N$ matrix M that is fully specified by its matrix elements $M_{ij}, i, j = 1, \ldots, N$, an operator is also specified by its matrix elements. For the

---

[1] In quantum mechanics, only the subspace of V consisting of state vectors that have unit norm, defined by $\langle f|f\rangle = 1$, correspond to physical systems and is called a Hilbert space. In finance the state space is larger than the Hilbert space since many financial instruments are represented by state vectors, such as the price of a stock given by e^x, that do not have a finite norm.

operators $\hat{x}$ and $\partial/\partial x$, in the notation of Dirac

$$\hat{x} f(x) = x f(x)$$

$$\Rightarrow \langle x|\hat{x}|f\rangle = x\langle x|f\rangle = x f(x)$$

$$\langle x|\frac{\partial}{\partial x}|f\rangle = \frac{\partial f(x)}{\partial x}$$

In other words, the matrix element $\langle x|\hat{x}|f\rangle$ of the operator $\hat{x}$ is given by $x f(x)$. Choose the function $|f\rangle = |x'\rangle$ that yields

$$\langle x|\hat{x}|x'\rangle = x\langle x|x'\rangle = x\delta(x - x')$$

Pursuing the analogy with matrices further, it is known that a matrix M has a Hermitian conjugate defined by $M_{ij}^{\dagger} \equiv M_{ji}^{*}$. Similar to a matrix, the Hermitian conjugate of an arbitrary operator $\mathcal{O}$ is defined by[2]

$$\langle f|\mathcal{O}^{\dagger}|g\rangle \equiv \langle g|\mathcal{O}|f\rangle^{*} \tag{A.50}$$

Furthermore, similar to matrices, the Hermitian adjoint of a sum of operators is given by $(A + B + \ldots)^{\dagger} = A^{\dagger} + B^{\dagger} \ldots$, and of a product of operators is given by $(AB \ldots)^{\dagger} = \ldots B^{\dagger} A^{\dagger}$.

The state space, completeness equation, and operators are a generalization of finite-dimensional linear algebra. Finite-dimensional vectors are generalized to infinite-dimensional vectors with continuous labels such as $|x\rangle$; functions, such as $f(x)$, are interpreted as infinite-dimensional vectors with $f(x) = \langle x|f\rangle$. Matrices that act on finite-dimensional linear vector space are generalized into infinite-dimensional differential operators, the most important of these being the coordinate operator $\hat{x}$ and the differential operator $\partial/\partial x$; operators such as the Hamiltonian are built out of these coordinate and differential operators.

In quantum finance, the Hamiltonians describing stochastic financial instruments are not Hermitian. Furthermore, the state space of financial instruments on which the Hamiltonian acts is not a positive normed Hilbert space, but, instead, is much larger with many financial instruments having a divergent, infinite norm.

A.7 Quantum field

Interest rates, in the framework of quantum finance, are modeled as a two-dimensional quantum field. The concept of a quantum field is discussed so that its essential mathematical features can be addressed in some generality

---

[2] The reason for studying Hermitian conjugation is because one needs to know the space that an operator acts on, namely whether it acts on $\mathcal{V}$ or on its dual $\mathcal{V}_{\text{dual}}$. For non-Hermitian operators, and these are the ones that occur in finance, the difference is important.

A classical field is a deterministic function of its arguments; in other words, it has only one configuration. A typical example of a classical field is the density of air, which has a unique value at every point in a room. A quantum field is the collection of all possible configurations of the classical field; the quantum field corresponding to air's density is a collection of all possible densities at every point in the room.

A quantum field consists of *all possible configurations* of a classical field; the natural question arises as to how does one describes such an object? A quantum field is taken to be a *random function*, with its different configurations having different likelihoods of occurrence. How does one assign probabilities to the occurrence of different configurations? The Lagrangian and Hamiltonian of a system are two equivalent ways of assigning such probabilities.

A.7.1 Lagrangian formulation

Consider for example a nonrelativistic (one-dimensional) string, and let its displacement from equilibrium at time t and at position x be denoted by $\phi(t, x)$, as shown for a particular instant t_0 in Figure A.2.

Let the string's initial position at time t_1 be given by $\phi_1(x) = \phi(t_1, x)$, and its final position at time t_2 be given by $\phi_2(x) = \phi(t_2, x)$. Suppose the string has mass per unit length given by ρ, and string tension (energy per unit length) given by T. A general expression for the action of the string, namely S_{string} is given by [95]

$$S[\phi] = \int_{t_1}^{t_2} dt \int_{-\infty}^{+\infty} dx \mathcal{L}(t, x) \qquad (A.51)$$

$$\equiv S_{\text{kinetic}} + S_{\text{potential}}$$

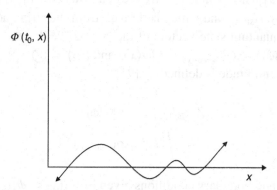

Figure A.2 A typical string configuration.

$$\mathcal{L}(t,x) = -\frac{1}{2}\left[\rho\left(\frac{\partial\phi}{\partial t}\right)^2 + T\left(\frac{\partial\phi}{\partial x}\right)^2 + V(\phi)\right] \qquad (A.52)$$

$\mathcal{L}(t,x)$ is the Lagrangian density for ϕ and $V(\phi)$ is the potential (function) of the field ϕ.

A classical string has a unique shape. The string quantum field, at each instant of its evolution, can take *all possible shapes*. Hence, for example, to compute the average values of functions of the quantum string one needs to integrate over all possible values for the string's position at each point x and for each instant t. The quantum field theory of the string field $\phi(t,x)$ is defined by the Feynman path integral, which is the *functional integration* over all possible configurations of the quantum field. In particular, the correlation of the quantum string is heuristically given by

$$E[\phi(t,x)\phi(t',x')] \equiv \; <\phi(t,x)\phi(t',x')>$$

$$= \frac{1}{Z}\int D\phi e^{S[\phi]}\phi(t,x)\phi(t',x')$$

$$Z = \int D\phi e^{S[\phi]}; \int D\phi = \prod_{t=t_1}^{t_2}\prod_{x=-\infty}^{+\infty}\int_{-\infty}^{+\infty} d\phi(t,x)\bigg|_{\phi(t_1,x)=\phi_1(x)}^{\phi(t_2,x)=\phi_2(x)}$$

The probability for the occurrence of different configurations for the quantum field $\phi(t,x)$ is heuristically given by $e^{S[\phi]}/Z$; $\{\phi(t,x)\}$ is called a quantum field, since – unlike a classical string which has a determinate and fixed value for every x and t – the quantum field takes all possible values for each x and t [47].

A.7.2 Hamiltonian formulation

The dynamics of a quantum field can also be determined by the Hamiltonian of the string, denoted by $\hat{H}_{\text{string}}$ and which is derived from the string action S_{string}. The initial and final quantum state vectors of the (string) field are given by its initial and final shapes $|\phi_1\rangle = \bigotimes_{-\infty<x<+\infty}|\phi(x)\rangle$ and $|\phi_2\rangle = \bigotimes_{-\infty<x<+\infty}|\phi'(x)\rangle$.

The transition amplitude is defined by [95]

$$Z \equiv \langle\phi_2|e^{-\tau\hat{H}_{\text{string}}}|\phi_1\rangle \qquad (A.53)$$

$$= \int D\phi \exp(S_{\text{string}}) \qquad (A.54)$$

with $\tau = t_2 - t_1$ and boundary conditions given by $\phi_1(x) = \phi(t_1,x)$ and $\phi_2(x) = \phi(t_2,x)$.

For a single particle the state space of states for quantum mechanics depends on one variable, given by $|x\rangle$, whereas for a single field ϕ, the quantum field's state space depends on infinitely many independent variables given by the infinite tensor product $|\phi\rangle = \bigotimes_{-\infty<x<+\infty} |\phi(x)\rangle$.

Quantum mechanics is a system that, at a given instant in time, has only a finite number of random variables; a quantum field, in contrast, is a system that, at a given instant in time, has infinitely many independent random variables. This, in essence, is the difference between quantum mechanics and quantum field theory.

From a more mathematical point of view there is no measure theoretic interpretation of the expression $\prod_{t_1<t<t_2} \prod_{-\infty<x<+\infty} \int_{-\infty}^{+\infty} d\phi(t,x)$. The only rigorous definition of Eq. (A.54) is to limit the volume of spacetime to be finite, and then discretize spacetime so that the infinite-dimensional integration given in Eq. (A.54) is reduced to an ordinary finite-dimensional multiple integral.

If the action S is only a quadratic function of the quantum field ϕ, the theory is said to be a free (Gaussian) quantum field, and one can take the continuum limit without having to address the problem of renormalization. The quantum field for the bond forward interest rates $f(t,x)$ is Gaussian. The quantum field for the Libor forward interest rates $f_L(t,x)$ – and its representation by the log Libor quantum field $\phi(t,x)$ in the Libor Market Model – is *highly nonlinear* and non-Gaussian. The computational simplicity of the bond forward rates is the reason it has been used extensively for illustrating various theoretical and empirical properties of the forward interest rates. However, for all applications to the interest rate Libor and Euribor markets, the nonlinear log Libor quantum field $\phi(t,x)$ has to be analyzed.

A.8 Quantum mathematics

Quantum mathematics originates in quantum physics, which treats all physical phenomena as being *intrinsically* uncertain and random. In contrast, a classical system is completely deterministic. Quantum mathematics is a theoretical and mathematical framework for describing the inherent uncertainty of nature. There are many subtleties of quantum uncertainty that do not appear in any classical system; but the quantum world has one common link with classical phenomena: if a classical system has sufficient complexity so that it behaves like a random system, then this random system can be described in a comprehensive and effective manner by the mathematics of quantum physics.

Risk is fundamental to finance and financial instruments, since the future is always uncertain. One may think of the uncertainty in finance as arising from a classical phenomenon – namely, the social, economic, and financial system – that has become so complex that its future outcome is effectively random. It is precisely this uncertainty, this randomness in what the future holds, that makes quantum mathematics a natural and powerful formalism for analyzing finance.

One of the essential features of quantum mathematics is the *synthesis* of *calculus* with *linear algebra*. To illustrate this synthesis, note that in linear algebra the representation of a vector in N-dimensional Euclidean space $\mathfrak{R}^N$ is given in terms of N linearly independent basis vectors $\mathbf{e}_i$. An arbitrary vector is expressed by its components in the following manner

$$\mathbf{v} = \sum_{i=1}^{N} v_i \mathbf{e}_i; \quad \mathbf{e}_i \cdot \mathbf{e}_j = \delta_{i-j}$$

In quantum mathematics, an infinite-dimensional generalization of linear algebra is made to functional analysis by generalizing Euclidean space $\mathfrak{R}^N$ to state space $\mathcal{V}$, which is a space of functions. There are now a continuous infinity of independent basis vectors $|x\rangle$, where x is governed by rules of calculus. The 'vector' $|f\rangle$ belonging to $\mathcal{V}$, from Eq. (A.49), has the following representation in the $|x\rangle$ basis

$$|f\rangle = \int_{-\infty}^{+\infty} dx f(x)|x\rangle; \quad \langle x|x'\rangle = \delta(x - x')$$

The Dirac notation provides a transparent representation of the infinite-dimensional generalization of linear algebra, which naturally combines it with calculus. Function $f(x)$ of a continuous variable x, the mainstay of calculus, in quantum mathematics is endowed with a *linear* structure that is inherited from state (function) space $\mathcal{V}$. The Dirac-delta function plays a crucial role in creating this synthesis. Section A.5 on functional differentiation shows how the concepts of differentiation are generalized in quantum mathematics; the concept of state space $\mathcal{V}$ leads to the infinite-dimensional generalization of matrices to operators acting on the state space. The Feynman path integral is an infinite-dimensional generalization of integration – consisting of summing over all paths between two points and is equivalent to integrating over infinitely many independent integration variables.

Quantum field theory lifts quantum mathematics to an entirely new and higher plane by introducing a state space that consists of the infinite tensor product of the single particle state space $\mathcal{V}$ and entails studying functions of infinitely many independent variables. The Hamiltonian is a functional differential operator, with functional differentiation being the infinite-dimensional generalization of differentiation. The path integral also generalizes to functional integration that consists of summing over all possible functions defined on higher-dimensional underlying manifolds. Many new features are present in quantum field theory, such as phase transitions, that are absent in quantum mechanics. For the case of finance, forward interest rates are modeled as a quantum field that consists of all possible functions on a two-dimensional trapezoidal manifold; functional integration consists of summing over this collection of functions.

A parallel to the power of quantum mathematics is the case of calculus. Newton discovered calculus in formulating his equations of motion; in subsequent centuries, calculus came to permeate all fields of quantitative science. Quantum mathematics is, similarly, finding applications in many fields – with finance being just one of these [82].

The utility of quantum mathematics for solving problems beyond its original domain has been increasing with every passing day. A precedent in using quantum mathematics outside the quantum domain is the case of classical phase transitions, which are the result of the fluctuations of classical random fields. Planck's constant $\hbar$, which is present in all quantum phenomena, does not appear in classical phase transitions. Nevertheless, the mathematical structure of renormalizable quantum field theories accurately describes the fluctuations of the classical random field and provides a microscopic explanation of phase transitions. In particular, field theory calculations provide quantitative results that have been experimentally verified. Conversely, the theory of classical phase transitions throws new light on the concept of renormalization – a procedure essential for making sense of nonlinear relativistic quantum field theories [95]. All renormalizable quantum field theories are the continuum limit of a lattice system undergoing a classical second-order phase transition.

Appendix B

US debt markets

The US financial system has been going through major transformations during the last 30 years and is thought to be a precursor of the changes that all mature financial systems will go through. The US financial system has been an innovator of new and novel financial instruments and the impact of these instruments on the economy can be studied by examining their effect on the US economy.

The primary focus of this book is the debt market. The US debt markets are analyzed as an exemplar of the characteristics of a global leader of the debt markets.

Another reason for studying the US debt market is that extensive data on financial instruments and on the derivatives market are available in the public domain. All the data for the graphs and diagrams in this appendix are taken from publicly listed sources.

B.1 Growth of US debt market

One of major changes in the structure of the US economy during the last 30 years has been the increasing importance of the financial sector in generating corporate earnings. As shown in Table B.1, the fraction of corporate earnings from the financial sector has grown almost 400% over the last 60 years and 50% over the last 20 years.

The total debt of the US, both internal and external, has undergone a dramatic increase. As shown in Figure B.1(a) the total US public and private debt over the last 30 years has grown from US$3 trillion in 1975 to US$42 trillion in 2005, close to the entire world's 2005 Gross Domestic Product (GDP) of about US$44 trillion. In January 2007, the public debt of the United States in the form of US Treasury Bonds stood at $5 trillion, of which 44% was owned by foreign countries. The GDP of the US in 2005 was about US$12.5 trillion with the total debt being about 340% of GDP. The total US GDP in January 2008 was US$14 trillion, with US federal debt standing at US$9.5 trillion, composed of US$5.3 trillion in Treasury Bonds and US$4.2 trillion in intragovernment debt such as Social Security IOUs.

Table B.1 *Changes in the total earnings of US corporations from the domestic manufacturing and financial sectors.*

	US corporate earnings			
	1950	1965	1990	2007
Financial sector	8%	12.5%	20%	35%
Domestic manufacturing	–	50%	30%	12%

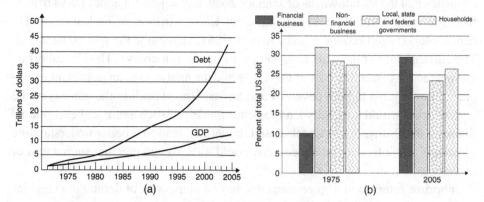

Figure B.1 (a) The growth of the total public and private debt in the United States compared to the GDP. (b) The composition of private and public debt in the United States.

Figure B.1(b) shows that from 1975 to 2005 the US financial sector has increased its share of the total private and public debt from 10% to about 30%, an almost 300% rise. The debt issued by US financial companies in 2005 was about US$13 trillion with the US local, state and federal government issuing debt of about US$10 trillion (the US federal government's debt in 2006 was US$3 trillion). The debt of financial companies and of the US government is mostly in the form of bonds and credit derivatives.

The developments in the US show the increasing importance of the financial sector in generating profit for the private sector as well as being the repository of the liquidity of the economy. The 2008 financial crisis, however, has been a major setback to the 'financialization' of the US economy. The expansion and profitability of the financial sector is explained by some experts as a reflection of an enormous financial bubble that has yet to run its full course [78]. It is doubtful if the two decades long expansion of the US financial sector will be repeated. Rather, the US financial sector is faced with a crisis of liquidity and credibility that may take a long time to be reversed. Furthermore, as discussed in Section 2.3, in 2007 New York lost its pre-eminence to London as the world's leading center of finance. The US capital markets, greatly weakened by the 2008 economic crisis, have to now face other competing, and new, centers of finance, as well as other factors that may further erode the competitiveness and importance of the US capital markets.

B.2 2008 Financial meltdown: US subprime loans

It is well known that the September–October 2008 meltdown of the US capital markets and financial institutions started in the subprime mortgage sector. IMF estimates that the meltdown, as of January 2009, had wiped out about US$4 trillion from the world's stock markets, with about US$2.8 trillion being lost in the US capital markets and economy and a loss of US$1.2 trillion in the rest of the world. The current global financial crisis has yet to run its full course. The US subprime mortgage loans are at the root of this crisis and are analyzed as an important exemplar of how such a bubble develops and bursts. Understanding the genesis of the current crisis would hopefully lead to better models of such a rare and exceptional financial instability. Such models could, hopefully, lead to a better understanding of how, if possible, the global economy could forecast and preempt such a turn of events.

Subprime refers to mortgage loans that have a higher risk of default than regular residential mortgages. Subprime loans are loans given to borrowers who have a patchy and opaque credit history, do not have regular employment, do not have any significant financial assets, and would normally not qualify for a standard housing mortgage. Subprime loans are thus lower in quality to prime housing loans. The US subprime mortgage bubble began in 2000 and started to come apart in July 2007. The subprime mortgage loan crisis resulted from the coincidence of a global savings glut with the explosion in financial innovations – made possible by ever-more sophisticated mathematical models and by information technology. There was a misplaced view of some investors that 'risk management' could handle all forms of risk and this misconception fueled the bubble to new heights.

The 1997 East Asian financial crisis had taught a hard lesson to the affected countries, namely that one needs to have substantial national savings if one is not to be held hostage to Western 'bailouts'. Most East Asian countries amassed huge foreign currency reserves that were subsequently invested in the international capital markets. The surplus capital that flooded the capital markets came, largely, from East Asian economies and from the cash-flush oil and gas producers.

In 2007, the US residential mortgage market was worth about US$10 trillion, with about 75% of it being repackaged securities – issued mostly by two US-government sponsored mortgage giants Freddie Mac and Fannie Mae. During the 2006–2008 period, subprime US housing loans were thought to be worth about US$650 billion, with some estimates putting it at US$1 trillion. Risky mortgages as a whole accounted for about US$1.7 trillion.

B.2.1 Five stages of the subprime loans bubble

The subprime crisis shows all the typical stages and features of a financial bubble that has been postulated by Kindelberger and Aliber [68] – based on their theory of financial instability. The following is a brief summary of the five distinct phases of a financial bubble, with the subprime bubble being used as an exemplar [69, 89].

1   A *new financial instrument* is introduced. In the case of subprime loans, most of the mortgages were 'securitized' through a new instrument, namely collateralized debt obligation (CDO) and sold off as coupon bearing bonds. The CDOs were used for rating the different tranches of subprime bonds, with the best possible rating of AAA being assigned to the least risky component and with the most risky 'toxic' component being retained by the bank. Rating agencies gave a AAA rating to about 80% of these bonds [90]. The banks sold off all the securitized mortgages to investors, retaining only the toxic component. These same AAA bonds later were found, starting in 2007, to be as risky as the 'toxic' component, showing the complicity of rating agencies in the subprime crisis. The investment grade AAA rating allowed for a massive inflow of investments from insurance companies and sovereign, pension, and hedge funds, and others. A large fraction of the AAA mortgage tranche was sold off to European banks and pension funds.

Although some critics blame securitization for the subprime crisis, this is not completely accurate. The rating of essentially junk bond category instruments, which by right should have all been rated as BBB and below, were given AAA rating by the incorrect view that separating 'junk bonds' into tranches could somehow reduce their measure of risk. When the crisis struck, it was found that all the subprime mortgage tranches, from AAA down to the 'toxic' component were all strongly correlated, with defaults happening for all tranches with equal likelihood. In short, securitization was carried out incorrectly, with corporate greed and quick profits creating instruments that were doomed to fail.

2   The second stage was a *rapid expansion* of credit. The housing bubble expanded due to extremely low interest rates – starting from about 6% in 2000 and reaching, by July 2003, a post-Second World War record low of 1% – and staying there for a full year. Securitization led to large-scale leveraging, with banks paying only 10% of an assets value, the rest being funded by low interest credit. Structured investment vehicles (SIVs) were created by the banks to provide off-balance-sheet funds to home buyers for making the initial down payment of about 15%; in effect, a subprime mortgage could be obtained with no cash down payment and, due to low interest rates, serviced by what would otherwise be monthly rentals. Numerous borrowers, who normally would not be entitled to mortgage loans, entered the real

estate market constantly inflating the housing bubble. Mortgage brokers and lenders made a commission on completing a loan, but with no responsibility for the full recovery of the loan, thus further fueling the bubble.

3   The bubble reached its peak during a *speculative frenzy*. The steady appreciation of US real estate assets – 50% from 2000 to mid-2005 – attracted a diverse range of investors, who were all betting on a continuing rise of housing prices. Many households refinanced their loans and withdrew cash to spend on consumption and to speculate on the housing market. It is estimated that by 2005, 40% of homes were purchased for speculating on the real estate market – either by investors or as second homes. By December 2005, at the height of the bubble, new mortgage borrowings increased to US$1.11 trillion, almost 7% of the US GDP.

4   The bubble started its *deflation* in 2006, when US housing prices started to fall, triggered by a rise in US interest rates, which had risen to 5.25% by June 2006. As risk premiums had fallen and spreads between borrowing and lending narrowed, central bankers and regulators warned of dangers to the mortgage market. What could not be foreseen was that the collapse of US subprime mortgage lending would be a catalyst for a sudden bust.

Investors began to have doubts about the sustainability of the housing 'boom'. Seeking to hedge their risks, investors sought protection in credit default swaps (CDS), which is an instrument that swaps mortgage stream payments with payments guaranteed by the banks. By 2007, a large number of CDS had been issued to cover a notional value of real estate assets to the tune of US$42.5 trillion. When the bubble burst, it was found that most of the CDS were worthless, leaving the investors with no protection whatsoever.

5   Meltdown of the bubble, leading to *fear and flight*. In February 2007 HSBC acknowledged a loss of US$10.8 billion in its US real estate portfolio. In July 2007, two of Bears Stearn's hedge funds defaulted on about US$10 billion of financial obligations, wiping out US$1.5 billion of investors, money. By August 2007 there was large-scale panic in the financial markets resulting in flight from the US real estate market. Banks were no longer willing to provide liquidity to other banks since it was not clear which financial institution was holding what quantity of 'toxic' subprime mortgages. British bank Northern Rock went bankrupt and was nationalized in September 2007. Millions of homeowners in the US had defaulted on their mortgage payments and were in danger of losing their property. The financial panic spread throughout the world, having the worst impact on US and European financial institutions.

B.2.2 The 2008 financial crisis

By April 2008, US interest rates had been lowered to 2% and by November 2008 to 1%, but the crisis still showed no signs of abating. From April 2007 to July 2008, US housing prices fell another 4.9%. Global financial losses due to the residential mortgage meltdown of US subprime loans was estimated by the IMF, in June 2008, to be about US$950 billion – as well as costing half a percentage of the 2008 world economic growth. In July 2008, IndyMac Bancorp mortgage lender was taken over by the US government, the second largest bankruptcy in US history, which was soon eclipsed, in September 2008, by the much larger bankruptcy of Lehman Brothers, one of the leading US investment banks.

The US government sponsored mortgage giants Freddie Mac and Fannie Mae have issued about US$5.2 trillion of mortgage backed securities, almost half of the total US home mortgage outstanding debt, with about US$1 trillion being held by East Asian countries. In the year preceding July 2008, their share prices had fallen by 87% and 79% respectively, with no end in sight. The US government's decision, taken in September 2008, to underwrite these mortgage giants, in effect adds a liability of US$5 trillion to the US taxpayers and could even trigger a chain reaction of financial bankruptcies [39].

In September 2008, the US government committed US$ 700 billion toward rescuing the leading financial institutions and injecting liquidity into the US economy. Similar steps were taken by the UK and other European Union countries, thus creating an enormous liability for the public. Furthermore, the US rescue package provided no support to either domestic or foreign investors who had suffered massive losses due to the US financial meltdown.

Some market watchers predict that the worst will not be over until the end of 2009. Of the 7 million subprime loans outstanding, if there is no state intervention, it is estimated that by 2010, about 40% will default. The main holders of subprime loans are the poorer sections of US society, with 54% of African American homeowners holding subprime loans compared to 47% for Hispanic homeowners and 18% for Caucasians [90]. The subprime mortgage loans have caused the value of the US Dollar to decline thus creating an inflation worldwide in food and commodity prices – since these are denominated in the US currency. Needless to say, the social problems that have been created by the subprime loan crisis are enormous.

The large-scale infusion of liquidity into the capital markets need not, necessarily, have resulted in a financial meltdown. It is thought by many experts that the subprime crisis is a reflection of deep structural problems with the current global financial system [90]. Be that as it may, what is clear is that corruption, manipulative policies, and corporate greed of US financial institutions are the immediate causes for the US financial meltdown and, in particular, of the mortgage loan crisis.

Making credit available at historically low interest rates, allowing credit standards to collapse, permitting mortgages to people with low or no credit ratings, creating AAA 'investment grade' vehicles from what were essentially C grade junk bond type of instruments – and all of this with the active connivance of bankers, mortgage brokers, lawyers, and rating agencies – are just a few examples of the policies that have all contributed to, if not caused, the current crisis.

Are derivative instruments responsible for the subprime crisis? Was the crisis caused by CDOs, SIVs, and CDSs? The answer is clearly no. An instrument is only as good as the assumptions that go into its formulation. If intrinsically high risk loans are equated to AAA bonds by rating agencies, all the results that follow from the model will be prone to failure. Investors holding CDSs had no direct knowledge of the loan portfolios involved in the swap – they knew only what the banks told them. With hindsight, the models were made mostly based on false premises. A more serious criticism is that financial instruments designed for the US housing market have not been rigorous, not following a proper financial engineering approach that is *ex-ante* (before the fact). Instead, most real estate models are constructed *post-facto*, based on default rates of mortgages and hence are quite worthless for either predicting or preventing a crisis such as the subprime one [46, 64, 74]. Investors had no realistic idea of the liabilities the CDSs entailed and, once the crisis struck, were quite unprepared for meeting their onerous financial obligations.

The view that 'risk management', using sufficiently complicated hedging instruments, can protect the investor from all forms of risk is incorrect. Effective risk management has to be based on an accurate representation of the real economy, and models for estimating and hedging risks must respect this requirement. The lesson from the subprime crisis for quantitative financial modeling is that one must first get the basics correct before one embarks on any form of modeling, which needs to be *ex-ante* and not *post-facto*.

Investors were hit particularly hard by the economic crisis. Since the US government's bail out plans did not offer any protection or compensation to the investors, the financial crisis greatly eroded international and institutional investors' confidence in the US financial system and capital markets. From a long-term perspective, the US financial meltdown has shown many fundamental flaws in the Anglo-American model of capitalism. In particular, the rather one-sided and uncritical view that unfettered and unregulated financial markets are the best way of achieving an optimum allocation of capital has been shown to be incorrect. Left unregulated, capital markets seem to invariably lead to the formation of financial bubbles – leading to an equally inevitable 'correction' that results in the large-scale destruction of capital and to a long-running economic downturn.

It is worth recalling that the real estate bubble in Japan burst in the early 1990s and, to date (2009), the Japanese economy has not yet fully recovered. Similar to Japan's case, a collapse of the US housing market could cause a decade-long US economic downturn – as is convincingly argued by Morris [78]. There are, of course, differences between the US and Japan, a primary one being that the US Dollar is an international reserve currency. However, an erosion of the US Dollar's value would lead to a flight of global savings to the Euro and other assessts, leading to a further intensification of the current US economic crisis. Furthermore, if China goes into a recession due to the economic crisis in the US and Europe, then this could possibly result in a prolonged global recession.

B.2.3 Open questions

The subprime crisis invariably points to the question of how should the financial markets be regulated? How can the global capital markets serve the needs of the world economy without spreading financial contagions that are outside international regulations? How can society reap benefits from financial instruments without these same instruments being used to obscure and hide the facts? How can the global financial system be protected from corporate greed, from the manipulations of unscrupulous lawyers, brokers, and bankers? The US and many European taxpayers are faced with a fairly obvious question: why should the public's money be spent on paying for the losses made by unscrupulous and incompetent banks when the profit these banks make is privately divided amongst their shareholders? This question, at present, remains largely unanswered.

Glossary of physics terms

Action The time integral of the Lagrangian. The exponential of the action is proportional to the probability of different random configurations.

Bra and ket vectors The notation with the 'bra' vector $\langle b|$ representing an element of the dual state space and the 'ket' vector $|k\rangle$ representing a vector from the state space, and with the inner product ('bracket') $\langle b|k\rangle$ being a complex number.

Classical field A deterministic function of two or more variables, denoted by say $\alpha(t, x)$; the classical field, in general, depends on time t as well as another variable x.

Completeness equation An equation that results from the set of basis vectors having a linear span that covers the entire state space.

Dual state space A space associated with a vector space, consisting of all mappings of the state space into the complex numbers.

Eigenfunctions Special state vectors that are associated with an operator such that under the action of this operator, they are only changed up to a multiplicative constant, called the *eigenvalue*.

Functional A quantity that depends on a complete function, for example the integral of a function is a functional.

Gaussian distribution Generic term for probability distributions that are given by an exponential of a quadratic function of the random variables. The normal distribution is the simplest example.

Generating functional A functional from which all the moments of random variables can be produced by differentiation.

Hamiltonian A differential operator that acts on the state space; in particular, it evolves the system in time. In finance the system is sometimes evolved backwards in time, since this is required for obtaining the price of an option.

Hermitian conjugation The transposition and complex conjugation of operators.

Lagrangian A functional related to the probabilities for various occurrences of a random path or a random field configuration.

Lattice quantum field theory A quantum field theory that is defined on an underlying lattice of (discrete) points.

Linear quantum field theory Theories with a Lagrangian that is a quadratic function of the quantum fields. Linear theories are also called free, or Gaussian, quantum field theories.

Nonlinear quantum field theory Theories with a Lagrangian that has terms that are cubic or higher in the quantum field. Also called interacting, or non-Gaussian, quantum field theories.

Operators The generalization of matrices. Operators act on the elements of a state space. In quantum mechanics, empirically observable quantities are represented by operators.

Partition function The summation of the exponential of the action over all possible configurations of a quantum field.

Path The trajectory followed by a quantity evolving in time, denoted by $x(t)$, where t is usually time.

Path integral A summation over all the possible random paths. For a field that takes random values on say a plane, the path integral is an integration over all possible functions on the plane. Path integration is also called functional integration.

Potential A potential is a term that conditions the movement of a quantum particle or, more generally, the outcome of random variables – making some configurations more favorable than others.

Potential barrier Potential barriers can be realized by imposing boundary conditions of the eigenfunctions of the Hamiltonian.

Quantum field The collection of all possible configurations (functions) of a classical field. A quantum field is a random function that takes value on the plane or higher-dimensional spaces.

Quantum field theory The theory of quantum fields; observable quantities are obtained by averaging over all possible configurations of the quantum fields.

Quantum mechanics The theory of the atomic realm based on the concepts of probability, with physical quantities being represented by state vectors and operators.

Random path A collection of all possible deterministic paths.

State space A linear vector space, the generalization of a finite-dimensional vector space, that describes the state of a quantum (random) system.

Wilson expansion The expansion of products of quantum fields that are at nearby points.

Glossary of finance terms

American option An option that can be exercised at any time before the pre-set expiry date of the contract.

Arbitrage Gaining a risk-free profit, above the capital market's risk-free return, by simultaneously entering into two or more financial transactions.

Asian option An option that has a payoff function that depends on the average value of the security for the duration of the option.

Barrier option An option that has a fixed maturity and is terminated with zero value before maturity time if the security breaches pre-set limits to its value.

Bermudan option An option that is allowed early exercise before maturity time only at pre-set times.

Bond forward interest rates A model of the forward interest rates driven by Gaussian quantum fields without strictly positive rates.

Capital Economic value of real assets in society.

Capital market Market for trading in all forms of financial instruments; in particular, for trading in equity, debt, and derivative instruments.

Coupon bond A promissory note for a pre-determined series of cash flows.

Derivative securities Financial assets that are derived from other financial assets.

Discounting A factor relating the future value of money to its present value.

Efficient market hypothesis For a financial market in equilibrium, changes in the prices of all securities are random.

Equity A share in the ownership of a real asset, like a company.

European option An option that can be exercised only at a pre-set expiry date of the contract.

Exotic option Options that are more complex than vanilla options, such as the American, Asian, barrier options, and so on, are called exotic options.

Financial assets Paper that entitles its holder to a claim on a fraction of real assets, and to the income (if any) that is generated by the underlying real assets.

Financial instrument A specific form of a financial asset – be it a stock or a bond.

Financial market Market for buying and selling financial assets and instruments.

Fixed income securities Instruments of debt issued by corporations and governments that promise either a single fixed payment or a stream of fixed payments. Also called bonds.

Forward bond price The forward bond price is the price – contracted and fixed today – of a (zero coupon) bond that is to be issued some time in the future.

Forward contract An obligatory contract between a buyer and a seller, in which the seller agrees to provide the commodity or financial instrument at some future time for a price fixed at present time, with only a single cash flow when the contract matures.

Forward interest rates The *strictly positive* forward interest rate $f(t,x)$ is the future interest rate for an overnight loan at a *future time x*, with the contract being fixed at an earlier time $t < x$.

Futures contract A contract similar to a forward contract. A major difference is that a futures contract, for the duration of the contract, is marked to the market with a series of cash flows.

Hedging A general term for the procedure of *reducing* and, if possible, eliminating random fluctuations in the price of a financial instrument by including it in a portfolio together with other related (correlated) instruments.

Ito calculus Another term for stochastic calculus.

Libor (Euribor) London (Euro) interbank offered rate.

Libor forward interest rates A model of strictly positive forward interest rates $f_L(t,x)$ such that all Libor rates and coupon bonds are strictly positive.

Log Libor (Euribor) The logarithm of Libor (Euribor) yields logarithmic interest rates $\phi(t,x)$ that can take any real value.

Market equilibrium For a market in equilibrium, all information has been assimilated leading to all securities having their fair price. Theoretically, it is expected that all trading ceases for a market in equilibrium.

Martingale process A martingale is a stochastic process such that the expectation value at present, conditioned on the occurrence of a value of the random variable in the previous step, is equal to the previous value. Martingale is the mathematical formulation of a fair game.

Money market Market for trading in money market instruments, such as short-term debt, cash, foreign currency transactions, and so on.

Numeraire The discounting factor used in computing the present value of money from its future value.

Option A contract, with a fixed maturity, in which the buyer has the right to – but is not obliged to – either buy or sell a security to the seller of the option at some pre-determined (but not necessarily fixed) strike price. Options can be written on underlying financial instruments such as stocks and bonds as well as on other options.

Pricing kernel The conditional probability for the final value of a financial instrument, given its present value.

Principle of no arbitrage No risk-free financial instrument can yield a rate of return above that of the market's risk-free rate.

Random variable A variable that has no fixed value, but instead takes a whole range of values. The likelihood of its various outcomes is given by a probability distribution.

Real assets Capital goods, skilled management, raw material, land, labor force, and so on, that are necessary for producing goods and services.

Return The profit obtained from an investment.

Risk The uncertainty in obtaining return on investment.

Security A financial asset.

Stochastic calculus Calculus of functions that depend on stochastic variables.

Stochastic process A (time-ordered) collection of random variables with outcomes governed by a joint probability distribution. The collection may have a discrete or a continuous labeling.

Stochastic variable Another term for a random variable.

Stochastic volatility Volatility of a random variable that is itself a stochastic variable.

Stocks and shares Financial instruments representing equity.

Tenor Duration of a cash time deposit.

Treasury Bond A zero coupon bond with no risk of default, that is a risk-free bond. Treasury Coupon Bonds are similarly risk-free coupon bonds.

Vanilla option A European option is also called a vanilla option.

Volatility The standard deviation (square root of variance) of any random variable (including financial instruments).

White noise A set of random variables, indexed by time, that at every instant have a probability distribution given by the normal distribution.

Zero coupon bond A financial instrument that is a promissory note for a predetermined single fixed payoff of say €1 at some future time T.

Zero coupon yield curve Provides simple interest earnings on a fixed deposit that are annually or semi-annually compounded.

Symbols

Only new symbols introduced in a chapter are listed. A consistent system of notation has been used as far as possible.

Chapter 2 Interest rates and coupon bonds

$B(t,T)$	zero coupon bond
$\mathcal{B}(t)$	coupon bond
$f(t,T_1,T_2)$	forward interest rate, at calendar time t, for a deposit from future time T_1 to T_2
$f(t,x)$	forward interest rate, at calendar time t, for an instantaneous deposit at future time x
$F(t,T_1,T_2)$	forward price, at time $t < T_1$, of a zero coupon bond $B(T_1,T_2)$.
$Z(t,T)$	zero coupon yield curve (ZCYC)
y	coupon bond yield to maturity
$L(t;T_1,T_2)$	Libor, at time t, for deposit from future time T_1 to T_2
ℓ	Libor tenor ℓ, usually taken to be 90 days
$L(t;T)$	Libor, at time t, for deposit from future time T to $T + \ell$
$s(t)$	average value of Libor at time t
$w(t)$	white noise for Libor data at time t

Chapter 3 Options and option theory

$S(t)$	stock price at time t
$C(t,S(t))$	call option price at time t
$P(t,S(t))$	put option price at time t
$\mathcal{P}$	option payoff function
$R(t)$	white noise
$\alpha(t)$	drift of stock price $S(t)$
$\Pi(t,S(t))$	portfolio

473

Δ	delta parameter
$S_i(t)$	ith stock price $S_i(t)$
$R_i(t)$	ith white noise
ρ_{ij}	white noise correlation
$\Pi_H(t, S(t))$	hedged portfolio
τ	remaining time
$z_n(\tau)$	logarithm of nth stock price
z_{ni}	Fourier coefficient of $z_n(\tau)$
$\int DZ$	path integral over all z_n
Z_{BS}, Z_{MG}	Black–Scholes and Merton–Garman partition function
$\mathcal{L}_{BS}, \mathcal{L}_{MG}$	Black–Scholes and Merton–Garman Lagrangian
S_{BS}, S_{MG}	Black–Scholes and Merton–Garman action
$\alpha(t, x)$	drift of forward interest rates $f(t, x)$
$\sigma(t, x)$	volatility of forward interest rates $f(t, x)$
$I(X)$	function derived from normal cumulative distribution $N(x)$

Chapter 4 Interest rate and coupon bond options

R_S	fixed rate for interest rate swap
R_P	par rate for interest rate swap
$\Theta(x)$	Heaviside step function
$\mathcal{F}(t_0)$	forward price of coupon bond $\mathcal{B}(t)$
$C_L(t_0, T_0, R_S)$	floating receiver swap option
$C_R(t_0, T_0, R_S)$	fixed receiver swap option
$C_{HJM}(t_0, t_*, K)$	HJM coupon bond call option
$N(d)$	normal cumulative distribution
$Y(t_*, T_i)$	HJM function
$\sigma_E(t, x)$	HJM volatility function

Chapter 5 Quantum field theory of bond forward interest rates

T_{FR}	maximum future time
$\mathcal{A}(t, x)$	forward interest rate velocity quantum field
$S[f], \mathcal{L}[f]$	action and Lagrangian for forward interest rates
$S[\mathcal{A}], \mathcal{L}[\mathcal{A}]$	action and Lagrangian for $\mathcal{A}(t, x)$
$\mathcal{N}(t, x, x')$	general propagator for $S[f]$
$\mathcal{D}(t, x, x')$	forward interest rate propagator
$Z[h]$	generating function for $\mathcal{A}(t, x)$
$\theta = x - t$	remaining future time
$z(\theta) = \theta^\eta$	market future time
η	index for future time

$G(t, x, x')$	forward interest rate stiff propagator
$\theta_{\pm} = \theta \pm \theta'$	combination of future time
$M(t, x, x')$	covariance of forward interest rates
$\alpha_F(t, x)$	drift for forward numeraire
$\mathcal{K}[f(t_i, \cdot), f(t_f, \cdot)]$	pricing kernel of forward interest rates
$L(t, T_k)$	Libor from T_k to T_{k+1}
$\zeta_k(t)$	Libor drift
$\gamma_k(t)$	BGM–Jamshidian volatility
$D(t_*, T)$	discounted zero coupon bond $B(t_*, T)$

Chapter 6 Libor Market Model of interest rates

$f_L(t, x)$	Libor forward interest rates
$\mu(t, x)$	Libor forward interest rates' stochastic drift
$v(t, x)$	Libor forward interest rates' stochastic volatility
$\mathcal{A}_L(t, x)$	velocity quantum field for Libor Market Model
$\chi_n(t)$	martingale instrument
$B(t, T_{I+1})$	zero coupon bond forward numeraire
$\mu_I(t, x)$	drift for forward numeraire $B(t, T_{I+1})$
$M_v(t, x, x')$	covariance of time deposit forward interest rates
$\zeta(t, x)$	Libor Market Model drift
$\Lambda_{mn}(t, x)$	Libor Market Model correlator
q_n^2	Libor Market Model kinematical drift
$\phi(t, x)$	logarithm of Libor
$\rho(t, x)$	drift of logarithm of Libor
$H_n(x)$	characteristic function for $[T_n, T_{n+1}]$
$S[\phi], \mathcal{L}[\phi]$	action and Lagrangian for $\phi(t, x)$
J	Jacobian of transformation $\mathcal{A}_L(t, x) \rightarrow \phi(t, x)$
$\mu_0(t, x)$	drift for $\ell f(t, x) >> 1$
t_s, t_f	singular and crossover time of Libor forward interest rates

Chapter 7 Empirical analysis of forward interest rates

$< \mathcal{O} >$	expectation of $\mathcal{O}$
$< \delta f >$	daily changes of forward interest rates $f(t, x)$
$\sigma_E(\theta)$	empirical volatility of forward interest rates $f(t, x)$
$\zeta(\theta)$	scaling factor
$\tilde{\sigma}(\theta)$	rescaled volatility
$C(\theta, \theta')$	normalized propagator
$\delta L(t, \theta)$	daily changes in $L(t, \theta)$
$\delta \ln L(t, \theta)$	daily changes in the logarithm of $L(t, \theta)$

$G(\theta, \theta')$	normalized stiff propagator
$R(\theta_+)$	curvature of $G(\theta, \theta')$
$C_z(z(\theta), z(\theta'))$	normalized propagator for market time $z(\theta)$
$z_\pm = z(\theta) \pm z(\theta')$	combination of future market time
$\Delta \mathcal{D}(\theta, \theta')$	difference in two normalized propagators
$\chi(\theta)$	expectation value of $\ell L(t,\theta)/[1 + \ell L(t,\theta)]$

Chapter 8 Libor Market Model of interest rate options

σ_B^2	Black's caplet volatility
β_n	integrated Libor drift
$\beta_n^{(0)}$	Libor drift β_n to $O(\gamma^2)$
A_n	stochastic term in the expansion of zero coupon bond $B(t,T)$
V	stochastic term in the expansion of an interest rate swap
C_1, C_2	expectation value of V and V^2
$\mathcal{P}_{Asn}$	Asian swaption payoff
V_{Asn}	stochastic term in the expansion of an Asian swap
X_{Asn}	effective Asian strike price
$\tilde{K}_{Asn}$	modified Asian strike price
$\zeta_n(t)$	BGM–Jamshidian Libor drift
Γ_n	BGM–Jamshidian Libor correlation
$Z(\eta)$	BGM–Jamshidian Libor partition function

Chapter 9 Numeraires for bond forward interest rates

$M(t, t_*)$	money market numeraire
$\alpha_M(t, x)$	money market interest rate drift
$\alpha_I(t, x)$	forward numeraire interest rate drift
S_M	money market action
S_I	forward numeraire action
$\mathcal{M}$	domain for money market numeraire
$\alpha_M(t, x)$	money market measure drift
$\alpha_L(t, x)$	Libor market measure drift
$\alpha_F(t, x)$	forward bond numeraire drift
$\Psi_F(G)$	pricing kernel for forward bond numeraire
$\Psi_L(G)$	pricing kernel for Libor measure

Chapter 10 Empirical analysis of interest rate caps

σ_H	caplet historical volatility
σ_I	caplet implied volatility

$\Pi(t_0)$	caplet and floorlet portfolio
$Y(t, T_n)$	logarithm of $1 + \ell L(t, T_n)$

Chapter 11 Coupon bond European and Asian options

$\mathcal{P}_*$	payoff for coupon bond option
$\alpha_*(t, x)$	drift for forward numeraire $B(t, t_*)$
F_i	forward bond price of $B(t_*, T_i)$
R_i	domain of forward interest rates
Q_i	integral over domain R_i of $\sigma \mathcal{A}(t, x)$
β_i	integral over domain R_i of $\alpha_*(t, x)$
V	stochastic terms in coupon bond payoff
$Z(\eta)$	partition function for coupon bond option
$C_0, C_1 \ldots C_4$	coefficients in the expansion of $\ln Z(\eta)$
$M(x, x'; t)$	covariance of forward interest rates
$\mathcal{F}(t)$	forward coupon bond price
$\mathcal{P}_{Asn}$	payoff for coupon bond Asian option
C_{BGM}	BGM–Jamshidian limit of coupon bond option

Chapter 12 Empirical analysis of interest rate swaptions

C_I, C_{II}	swaptions on forward interest rates
$\delta C_I, \delta C_{II}$	daily changes in C_I, C_{II}
$C_{I,2}, C_{II,2}$	second moments of C_I, C_{II}
$\mathcal{I}$	integral of forward interest rate covariance
m_1, m_2, m_3	limits of integration for $\mathcal{I}$
d_1, d_2	limits of integration for $\mathcal{I}$
$Y(t_0, t_*, T)$	logarithm of forward bond price $F(t_0, t_*, T)$
$W(t_*, T_i)$	HJM option function
$z(t, x)$	analog of ZCYC for trial forward interest rates

Chapter 13 Correlation of coupon bond options

$\mathcal{P}_1, \mathcal{P}_2$	payoff functions for two coupon bonds
$\mathcal{M}_{12}$	connected discounted correlator
$\mathcal{M}_1, \mathcal{M}_2$	discounted expectation value of $\mathcal{P}_1, \mathcal{P}_2$
$Z(\eta_1, \eta_2)$	partition function for two coupon bond options
ρ	correlation coefficient
a_1, a_2	drift of coupon bonds
$\mathcal{F}_1, \mathcal{F}_2$	forward price of coupon bonds
X_1, X_2	rescaled and shifted strike price

$m_0, \ldots, m_3$	coefficients in the expansion of $\mathcal{M}_{12}$
M_1, M_2	normalized discounting functions
V_1, V_2	stochastic terms in the payoff function
A_{ij}	correlation of discounted V_1, V_2
$\mathcal{T}_{12}$	domain for discounting two coupon bonds
Ω_{12}	average of discounting over domain $\mathcal{T}_{12}$
$\mathcal{J}_1, \mathcal{J}_2$	rescaled values of J_1, J_2
$\sigma^2(\mathcal{P}_1)$	coupon bond auto-correlation
$\tilde{P}_1$	discounted coupon bond payoff function $\mathcal{P}_1$

Chapter 14 Hedging interest rate options

$\alpha(t, x)$	drift for Libor market measure
$\mathcal{F}_F(t, T_i)$	Libor futures
$\Psi(G, T, T + \ell)$	pricing kernel for caplet
$\Pi(t)$	portfolio for caplet hedging
$\delta\Pi(t)$	daily changes in portfolio
$\Psi(G\|f_h)$	expectation value of G conditioned on the occurrence of $f(t_*, x_h) = f_h$
$\tilde{L}(t, T_1, f_h)$	Libor future conditioned on $f(t, x_h) = f_h$
$\Phi(G\|f_h, t, T_1)$	expectation value of G conditioned on the occurrence of $\tilde{L}(t, T_1, f_h)$
η_1	stochastic delta hedging parameter
K_i, M_{ij}	coefficients of variance of $d\Pi/dt$
V_R	residual variance
χ	coefficient of caplet delta parameter

Chapter 15 Interest rate Hamiltonian and option theory

$\langle x\|$	bra vector
$\|x\rangle$	ket vector
$\mathcal{V}$	state space
$\mathcal{V}_D$	dual state space
$\mathcal{I}$	identity operator on state space
$\|C, t\rangle$	option price state vector
$\langle x\|C, t\rangle$	option price state vector in x basis
$\|\mathcal{P}\rangle$	option payoff state vector
$\langle x\|\mathcal{P}\rangle$	option payoff state vector in x basis
H	Hamiltonian operator
$H^\dagger$	Hermitian adjoint Hamiltonian

$\lvert S\rangle$	stock price state vector
H_{BS}	Black–Scholes Hamiltonian
H_N	Hamiltonian for N equities
$\lvert f_t\rangle$	interest rate state vector
$\mathcal{V}_t$	interest rate state space at calendar time t
$\mathcal{I}_t$	identity operator on state space $\mathcal{V}_t$ at calendar time t
Z	partition function for interest rates
$\lvert B(t,T)\rangle$	zero coupon bond vector at time t
$\lvert \mathcal{B}(t)\rangle$	coupon bond vector at time t
$\mathcal{L}_\phi(t,x)$	Lagrangian density for log Libor $\phi(t,x)$
$\mathcal{L}_f(t,x)$	Lagrangian density for forward interest rates $f(t,x)$
$\mathcal{L}(t_n,x)$	Lagrangian density
$\mathcal{H}_\phi(t)$	Hamiltonian for log Libor $\phi(t,x)$
$\mathcal{H}_f(t)$	Hamiltonian for forward interest rates $f(t,x)$
$\mathcal{H}_*(t)$	Hamiltonian for forward numeraire $B(t,t_*)$
ϕ_j	integral of $\phi(t,x)$ over interval $[T_i,T_{i+1}]$
W_f	evolution operator for forward interest rates $f(t,x)$
W	evolution operator for coupon bonds
$g(x)$	bond variable
g_i	bond variable $B(t_*,T_i)=e^{-g_i}$
$G(x,x')$	bond correlator
Z_{cb}	coupon bond partition function
$\mathcal{L}_{cb}$	coupon bond Lagrangian

Chapter 16 American options for coupon bonds and interest rates

$p(z,z';1)$	pricing kernel for equity for time step ϵ
$P(\tau_i)$	American put option price
$g(\tau_i)$	American caplet or coupon bond price
$m,\,n$	remaining calendar and future lattice time
$B_{m+1,1}$	lattice forward numeraire
f_{mn}	lattice forward interest rates
$\tilde{\alpha}_{mn}$	lattice forward interest rate drift
s_{mn}	lattice forward interest rate volatility
S_L	lattice forward interest rate action
$\mathcal{L}[\mathbf{f}_{m+1},\mathbf{f}_m]$	lattice forward interest rate Lagrangian density
$C_m(f_{mn})$	option price at remaining calendar time $m\epsilon$
$\tilde{C}_m(f_{mn})$	trial option price at remaining calendar time $m\epsilon$
f_{mn}^k	values of the lattice forward interest rate tree
$g[j]$	American option price on the interest rate tree

Chapter 17 Hamiltonian derivation of coupon bond options

f_I	log of the forward zero coupon bond $F(t_0, t_*, T_I) = e^{-f_I}$
C_{IK}	coefficients for option price
$\langle \mathbf{f}|$	state vector basis states for initial coupon bond
W	evolution operator
G_{ij}	correlation term in the evolution operator W
β_i	drift term in the evolution operator W
$\Psi_{\mathbf{p}}[\mathbf{f}]$	eigenfunction of the evolution operator W
$S_E[\mathbf{p}]$	real part of the eigenvalue of the evolution operator W
C_E	European coupon bond option price
$C_{E,0}, C_{E,1}, C_{E,2}$	coefficients for C_E
C_B	coupon bond barrier option price
$\psi_k(g)$	barrier eigenfunction for a zero coupon bond
$\psi_k(gd)$	barrier eigenfunction for a coupon bond
$\mathcal{Q}$	barrier function
$\mathcal{Q}_E$	European limit of barrier function
$\mathcal{Q}_a, \mathcal{Q}_b$	single barrier limit of barrier function
a, b	linearized barrier limits
$\Psi_{\mathbf{p},k}$	overcomplete barrier eigenfunctions
$\mathcal{K}[\mathbf{f}, \mathbf{g}]$	barrier option pricing kernel
$C_{B,0}, C_{B,1}, C_{B,2}$	expansion coefficients for barrier option C_B
$S_{B,0}, S_{B,1}, S_{B,2}$	real part of the eigenvalue of the overcomplete barrier eigenfunctions
D_1, D_2	coefficients for barrier option price

References

[1] L. Anderson and J. Andresean, 'Volatility Skews and Extensions of the Libor Market Model', *Applied Mathematical Finance* **7**(1) 1–32 (2000).

[2] B. E. Baaquie and Tang Pan, 'Monte Carlo Simulation of interest rates in Quantum Finance' (submitted for publication) (2009).

[3] B. E. Baaquie, 'Libor Market Model in Quantum Finance: 1: The Wilson Expansion' (to be published in *Physical Review E*) (2009).

[4] B. E. Baaquie, 'Libor Market Model in Quantum Finance: 2: The Hamiltonian' (to be published in *Physical Review E*) (2009).

[5] B. E. Baaquie, 'Libor Market Model in Quantum Finance: 3: Swaptions and Bond Options' (to be published in *Physica A*) (2009).

[6] B. E. Baaquie and Cao Yang, 'Empirical Analysis of Quantum Finance Interest Rate Models', *Physica A* **388** (2009) 2666–2681.

[7] B. E. Baaquie, *Quantum Mechanics and Option Pricing: Proceedings of the Second Quantum Interaction Symposium*, College Publications (2008) 49–53.

[8] B. E. Baaquie, 'Quantum Finance Hamiltonian for Coupon Bond European and Barrier Options', *Physical Review E* **77** (2008).

[9] B. E. Baaquie, 'Feynman Perturbation Expansion for the Price of Coupon Bond Options and Swaptions in Quantum Finance. I. Theory', *Physical Review E* **75** (2007).

[10] B. E. Baaquie, 'Price of Coupon Bond Options in a Quantum Field Theory of Forward Interest Rates', *Physica A* **370**(1) (2006) 98–103.

[11] B. E. Baaquie, 'A Common Market Measure for Libor and Pricing Caps, Floors and Swaps in a Field Theory of Forward Interest Rates', *International Journal of Theoretical and Applied Finance* **8**(8) (2005) 999–1018.

[12] B. E. Baaquie, *Quantum Finance*, Cambridge University Press (2004).

[13] B. E. Baaquie, 'Quantum Field Theory of Forward Rates with Stochastic Volatility', *Physical Review E* **65** (2002).

[14] B. E. Baaquie, 'Quantum Field Theory of Treasury Bonds', *Physical Review E* **64** (2001).

[15] B. E. Baaquie, 'A Path Integral Approach to Option Pricing with Stochastic Volatility: Some Exact Results', *Journal de Physique* I **7**(12) (1977) 1733–1753 (France).

[16] B. E. Baaquie and J. P. Bouchaud, 'Stiff Interest Rate Model and Psychological Future Time', *Wilmott Magazine* (April 2004) 2–6.

[17] B. E. Baaquie and Cui Liang, 'American Option pricing for Interest Rate Caps and Coupon Bonds in Quantum Finance', *Physica A* **38** (2007) 285–316.

[18] B. E. Baaquie and Cui Liang, 'Feynman Perturbation Expansion for the Price of Coupon Bond Options and Swaptions in Quantum Finance. II. Empirical', *Physical Review E* **75** (2007).

[19] B. E. Baaquie and Cui Liang, 'Empirical Investigation of a Field Theory Formula and Black's Formula for the Price of an Interest Rate Caplet', *Physica A* **374**(1) (2007) 331–348.

[20] B. E. Baaquie, Cui Liang and M. Warachka, 'Hedging Libor Derivatives in a Field Theory Model of Forward Interest Rates', *Physica A* **374**(2) (2007) 730–748.

[21] B. E. Baaquie and Cui Liang, 'Correlated Coupon Bond Options', Unpublished (2006).

[22] B. E. Baaquie, C. Coriano and M. Srikant, 'Hamiltonian and Potentials in Derivative Pricing Models: Exact Results and Lattice Simulations', *Physica A* **334**(3–4) (2004) 531–557.

[23] B. E. Baaquie and M. Srikant, 'Comparison of Field Theory Models of Interest Rates with Market Data', *Physical Review E* **69** (2004).

[24] B. E. Baaquie, M. Srikant, and M. Warachka, 'A Quantum Field Theory Term Structure Model Applied to Hedging', *International Journal of Theoretical and Applied Finance* **6**(5) (2003) 443–467.

[25] Bank of International Settlements, 'OTC derivatives market activity in second half of 2007', Monetary and Economic Department Report (May 2008).

[26] F. Black and M. Scholes, 'The Pricing of Options and Corporate Liabilities', *Journal of Political Economy* **81** (1973) 637–654.

[27] J. P. Bouchaud and A. Matacz, 'Explaining the Forward Interest Rate Term Structure', *International Journal of Theoretical and Applied Finance* **3** (2000) 381.

[28] J. P. Bouchaud and M. Potters, *Theory of Financial Risks*, Cambridge University Press (2000).

[29] J. P. Bouchaud, N. Sagna, R. Cont, N. El-Karoui, and M. Potters, 'Phenomenology of the Interest Rate Curve', *Applied Financial Mathematics* **6** (1999) 209–232.

[30] J. P. Bouchaud and A. Matacz, 'An Empirical Investigation of the Forward Interest Rate Term Structure', *International Journal of Theoretical and Applied Finance* **3** (2000) 703–729.

[31] Z. Bodie and R. C. Merton, *Finance*, Prentice Hall (2000).

[32] A. Brace, D. Gatarek, and M. Musiela, 'The Market Model of Interest Rate Dynamics', *Mathematical Finance* **7** (1996) 127–154.

[33] D. Brigo and F. Mercurio, *Interest Rate Models – Theory and Practice*, Springer (2007).

[34] A. J. G. Cairns, *Interest Rate Models*, Princeton University Press (2004).

[35] J. Y. Campbell, A. S. Low, and A. C. Mackinlay, *The Econometrics of Financial Markets*, Princeton University Press (1997).

[36] M. Choudhury, *Fixed Income Markets*, Wiley Finance (2004).

[37] N. Chriss, *Black–Scholes and Beyond: Option Pricing Models*, Irwin Professional Publishing (1997).

[38] P. Collin-Dufresne and Robert S. Goldstein, 'Pricing Swaptions within an Affine Framework', The Journal of Derivatives **10**(1) (2002).

[39] C. Crook, 'Regulation Needs More Tuning', *Financial Times* (7 April 2008) 13.

[40] Cui Liang, '*Investigation of Interest Rate Derivatives by Quantum* Finance', Ph.D. thesis, National University of Singapore (2007).

[41] A. Duran and G. Caginalp, 'Overreaction Diamonds: Precursors and AfterShocks for Significant Price Changes', *Journal Quantitative Finance*, **7**(3) (2007) 321–342.

[42] F. De Jong, J. Driessen, and A. Pelsser, 'Libor Market Models Versus Swap Market Models for the Pricing of Interest Rate Derivatives: An Empirical Analysis', *European Finance Review* **5**(3) (2001) 201–237.

[43] D. Donoho, 'Nonlinear Wavelet Methods for Recovery of Signals from Noisy Data', *Proceedings of Symposia in Applied Mathematics*, **47** (1993).

[44] J. Driessen, P. Klaassen, and B. Melenberg, 'The Performance of Multi-Factor Term Structure Models for Pricing and Hedging Caps and Swaptions', *Journal of Financial and Quantitative Analysis*, **38**(3) (2003) 635–672.

[45] D. Duffie, *Dynamic Asset Pricing Theory*, Princeton University Press (1994).

[46] M. S. Ebrahim and I. Mathur, 'Pricing Home Mortgages and Bank Collateral', *Journal of Economic Dynamics and Control* **31**(4) (2007) 1217–1244.

[47] R. P. Feynman and A. R. Hibbs, *Quantum Mechanics and Path Integrals*, McGraw-Hill (1965).

[48] B. Flesker, 'Testing of the Heath–Jarrow–Morton/Ho-Lee Model of Interest Rate Contingent Claims Pricing', *Journal of Financial and Quantiative Analysis* **38** (1993) 483–495.

[49] H. Geman and M. Yor, *Mathematical Finance* **3** (1999) 55.

[50] H. Geman, N. E. Khouri, and J.-C. Rochet, 'Changes of Numeraire, Changes of Probability Measure and Option Pricing', *Journal of Applied Probability* **32** (1995) 443.

[51] P. Glasserman and N. Merener, 'Cap and Swaption Approximation in Libor Market Models with Jumps', *Journal of Computational Finance* (2003).

[52] P. Goldstein, 'The Term Structure of Interest Rates as a Random Field', *Journal of Financial Studies* **13**(2) (2000) 365.

[53] I. Grubsic, 'Interest Rate Theory: The BGM Model', MSc Thesis, Leiden University (2002).

[54] P. S. Hagan, D. Kumar, A. S. Lesniewski, and D. E. Woodward, 'Managing Smile Risk', *Wilmott Magazine* (September 2002) 84–108.

[55] P. Hunt and J. Kennedy, *Financial Derivatives in Theory and Practice*, Wiley Finance (2000).

[56] D. Heath, R. Jarrow, and A. Morton, 'Bond Pricing and the Term Structure of Interest Rates: A New Methodology for Contingent Claim Valuation', *Econometrica* **60** (1992) 77–105.

[57] M. Henard, 'Explicit Bond Formula in Heath–Jarrow–Morton One Factor Model', *International Journal of Theoretical and Applied Finance* **6**(1) (2003) 57–72.

[58] Henry-Labordere, 'Combining the SABR and LMM Models', *Risk* (October 2007) 102–107.

[59] J. C. Hull, *Options, Futures, and Other Derivatives*, Fourth Edition, Prentice Hall (2000).

[60] F. J. Fabozzi, *The Handbook of Fixed Income Securities*, McGraw-Hill (2005).

[61] J. James and N. Weber, *Interest Rate Modeling*, John Wiley (2001).

[62] F. Jamshidian, 'Libor and Swap Market Models and Measures', *Finance and Stochastics* **1**(14) (1997) 293–330.

[63] R. A. Jarrow, *Modelling Fixed Income Securities and Interest Rate Options*, McGraw-Hill (1995).

[64] D. Jaffee and J. Stiglitz, 'Credit Rationing', in B. M. Friedman and F. H. Hahn, (eds.), *Handbook of Monetary Economics Volume II*, Elsevier Science B.V. (1990) pp. 838–888.

[65] R. Jarrow and S. Turnbull, *Derivative Securities*, Second Edition, South-Western College Publishing (2000).

[66] M. Joshi and R. Rebonata, 'A Displaced-Diffusion Stochastic Volatility Libor Market Model: Motivation, Definition and Implementation', *Quantitative Finance* **3** (2003) 458–469.

[67] D. P. Kennedy, 'Characterizing Gaussian Models of the Term Structure of Interest Rates', *Mathematical Finance* **7** (1997) 107–118.

[68] C. P. Kindelberger and R. Aliber, *Manias, Panics and Crashes*, John Wiley (2005).

[69] J. Landa, 'Subprime Collapse Part of Economic Cycle', *San Antonio Business Journal* (26 October 2007).

[70] L. D. Landau and L. M. Lifshitz, *Quantum Mechanics: Non-Relativistic Theory*, Volume 3, Third Edition, Elsevier Science (2003).

[71] J.-P. Laurent and O. Scaillet, 'Variance Optimal Cap Pricing Models', working paper (2003).

[72] A. Li, P. Ritchken, and L. Sankarasubramanian, 'Lattice Models for Pricing American Interest Rate Claims', *The Journal of Finance* **50**(2) (1995).

[73] M. Livingstone, *Bonds and Bond Derivatives*, Blackwell Publishing (2005).

[74] S. Malpezzi and S. M. Wachter, 'The Role of Speculation in Real Estate Cycles', Working Paper, Wharton School of Management, University of Pennsylvania, Philadelphia (2002).

[75] R. Merton, 'Option Pricing when Underlying Stock Returns are Discontinuous', *Journal of Financial Economics* **3** (1976) 125–144.

[76] G. Montagna and O. Nicrosini, 'A Path Integral Way to Option Pricing', *Physica A* **310** (2002) 450–466.

[77] J. Montier, *Behavioural Finance*, Wiley Finance (2002).

[78] C. R. Morris, 'The Trillion Dollar Meltdown: Easy Money, High Rollers, and the Great Credit Crash', *Public Affairs* (2008).

[79] M. Overhaus, A. Bermudez, H. Buehler, A. Ferraris, C. Jordinson, and A. Lamnouar, *Equity Hybrid Derivatives*, Wiley Finance (2007).

[80] A. F. Perold and J. N. Musher, *The World Market Portfolio* (2002).

[81] A. M. Polyakov, 'Fine Structure of Strings', *Nuclear Physics* **B268** (1986) 406.

[82] *Proceedings of the Second Quantum Interaction Symposium* (QI-2008) College Publications (2008).

[83] R. Rebonata, *Modern Pricing of Interest-Rate Derivatives*, Princeton University Press (2002).

[84] R. Rebonato, *Interest-Rate Option Models*, Wiley (1996).

[85] P. Santa-Clara and D. Sornette, 'The Dynamics of the Forward Interest Rate Curve with Stochastic String Shocks', *Journal of Financial Studies* **14**(1) (2001) 149.

[86] D. F. Schrager and A. A. J. Pelsser, 'Pricing Swap', *Mathematical Finance*, forthcoming.

[87] A. Shleifer, *Inefficient Markets: An Introduction to Behavioral Finance*, Oxford University Press (2000).

[88] S. E. Shreve, *Stochastic Calculus for Finance II: Continuous-Time Models*, Springer Finance (2008).

[89] R. J. Shiller, *The New Financial Order: Risk in the 21st Century*, Princeton University Press (2003).

[90] G. Soros, 'The New Paradigm for Financial Markets', *Public Affairs* (2008).

[91] H. D. Soto, *The Mystery of Capital*, Black Swan (2001).

[92] M. Srikant, 'Stochastic Processes in Finance: A Physics Perspective', M.Sc. thesis, National University of Singapore (2003).

[93] K. G. Wilson, 'Non-lagrangian models in current algebra', *Physical Review* **179** (1969) 1499–1512.

[94] K. G. Wilson and W. Zimmermann, 'Operator Product Expansions and Composite Field Operators in the General Framework of Quantum Field Theory', *Communications in Mathematical Physics* **24**(2) (1972) 87–106.

[95] J. Zinn-Justin, *Quantum Field Theory and Critical Phenomenon*, Oxford University Press (1993).

Index

Printed in the United States
by Baker & Taylor Publisher Services